CRC HANDBOOK OF
Medicinal Spices

James A. Duke
with
Mary Jo Bogenschutz-Godwin
Judi duCellier
Peggy-Ann K. Duke – *"Illustrator"*

CRC PRESS

Boca Raton   London   New York   Washington, D.C.

Peggy-Ann K. Duke has the copyright to all black and white line illustrations.

Library of Congress Cataloging-in-Publication Data

CRC handbook of medicinal spices / James A. Duke ... [et al.].
      p. cm.
    Includes bibliographical references and index.
    ISBN 0-8493-1279-5 (alk. paper)
     1. Materia medica, Vegetable--Handbooks, manuals, etc. 2. Spices--Therapeutic
use--Handbooks, manuals, etc. 3. Herbs--Therapeutic use--Handbooks, manuals, etc. I.
Duke, James A., 1929-

RS164 .C826 2002
615′.321--dc21   2002067412

Visit the CRC Press Web site at www.crcpress.com

© 2003 by CRC Press LLC

No claim to original U.S. Government works
International Standard Book Number 0-8493-1279-5
Library of Congress Card Number 2002067412
Printed in the United States of America 1 2 3 4 5 6 7 8 9 0
Printed on acid-free paper

Acknowledgments

Perhaps it is unusual to acknowledge one's coauthors in a new book, but I sure wish to acknowledge mine for their patience and perseverance with this new book. To Mary Jo Bogenschutz-Godwin for tidying up my most untidy first drafts, and for querying our database, after updating the database at the United States Department of Agriculture (USDA), working with my former USDA colleagues Jimmie Mowder, Ed Bird (deceased), and Quinn Sinnott. I am of course indebted to the USDA for maintaining the database these many years and to Dr. Alan Stoner for facilitating this. Readers of the book will realize the importance of the USDA database in shaping some of our new concepts and even for suggesting new indications for old spices. To Judi duCellier who, for more than 25 years, has quietly, and without complaining, struggled not only with my illegible handwritten notes, complete incompetence at the computer, quick reprioritizations, and now terminal dyslexia (double meaning intended), and produced useful documents that I mold into first drafts. To Peggy-Ann K. Duke, botanist and co-compiler, for closer to five decades, to whom both the world and I are indebted for her talented art, learned as a botanist while sharing with me the wonders of botany at the University of North Carolina, under out great teachers, alphabetically, Drs. J. E. Adams, C. R. Bell, J. N. Couch (who swore I could not be both a botanist and a musician; my music proves him right), Victor Greulach, A. E. Radford, and H. R. Totten, who kept us interested in botany. That interest is still today reflected, in the seven plus decades of Peggy's and my lives, in my Green Farmacy Garden in Maryland and the ReNuPeRu Garden in Peru which I started nearly a decade ago. It now functions fine without me, thanks to Pamela Bucur de Arevalo and the wonderful workers at the Explorama Lodges of Amazon Peru, where Peggy and I shared the turning of the New Millennium. Coincidentally, we may be leading a course there at the ReNuPeRu garden next year, teaching Latin Americans how to better grow and process some herbs, medicinal plants and spices covered in this book. As I have struggled with this book, I have had the marvelous luck to have acquired a new director for my Green Farmacy Garden, phytopathologist Holly Shull Vogel. In a sense, she keeps the Green Farmacy Garden alive through unfailing labors, too often all her own. She shares my vision of teaching America about the best and safest medicines, like some of the spices in this book. Illustrations for the onion, frankencense, cassia, cinnamon and cassia, garlic, bayleaf, and myrtle are used with the permission of Duke, J., *Medicinal Plants of the Bible*, Trado-Medic Books, Buffalo, New York, 1983. All other illustrations in this book are courtesy of Peggy-Ann K. Duke. Our thanks to those patient people at CRC who tolerated our frequent changes of direction, especially Barbara Norwitz, Sara Kreisman, and Joette Lynch. And to you, the reader, and your health, may the spices of life prolong and enhance the quality of your lives, saving you from what is believed to be America's biggest killer, Adverse Drug Reactions (ADR's) according to *The Journal of the American Medical Association*, May 1, 2002.

James A. Duke

The Author

James A. "Jim" Duke, Ph.D., is a Phi Beta Kappa graduate of the University of North Carolina, where he received his Ph.D. in Botany. He then moved on to postdoctoral activities at Washington University and the Missouri Botanical Gardens in St. Louis, Missouri, where he assumed professor and curator duties, respectively. He retired from the United States Department of Agriculture (USDA) in 1995 after a 35-year career there and elsewhere as an economic botanist. After retiring, he was appointed Senior Scientific Consultant to *Nature's Herbs* (A Twin Labs subsidiary), and to an online company, ALLHERB.COM. He currently teaches a master's degree course in botanical healing at the Tai Sophia Institute in Columbia, Maryland.

Dr. Duke spends time exploring the ecology and culture of the Amazonian Rain Forest and sits on the board of directors and advisory councils of numerous organizations involved in plant medicine and the rainforest. He is updating several of his published books and refining his online database, http://www.ars-grin.gov/duke/, still maintained at the USDA. He is also expanding his private educational Green Farmacy Garden at his residence in Fulton, Maryland.

Contents

Introduction

You remember the days a decade ago when I celebrated the 500th anniversary of Columbus' "Discovery of America" and the Native Americans who had colonized it some 25,000 years earlier and been visited by Scandinavians a bit earlier. I believe Leif Ericsson also encountered Native Americans when he landed nearly 500 years earlier, up around Vinland, north of the United States. In reflecting the anniversary of Columbus' voyage, I often make the comment that Columbus set sail seeking black pepper and black Indians and instead found red Indians and red pepper, changing the cuisine and the medicine of the world and reshaping everyone's food basket and medicine chest significantly.

The travels of Columbus opened up one of the world's greatest exchanges of flora and fauna, and yes even germs, including some lethal smallpox germs, as well as higher plants (many never having been seen before outside America) and animals. This has laxly been termed the "Columbian exchange," the rapid movement, to and fro, of useful plants and animals, some for the first time, from continent to continent.

Frequently, the major producers of spices are not regions to which the species originally belonged, but areas of introduction as a result of the Columbian exchange of plants and animals around the world. I got very excited at what I learned in preparing my talk, *Spice rack/medicine chest—Five hundred years after Columbus*, presented under the auspices of Oldways in Spain the following year (Duke, 1991, 1992). Spices are important medicines that have withstood the empirical tests of millennia. New books come out every year embracing the time proven medicinal efficacy of one spice or another. Chile, garlic, ginger, onion, pepper, and turmeric are almost as popular, and deservedly sso, as medicines as they are as spices.

I'll freely dispense sage advice:
Sage is an herb, not a spice!
Herbs are tasty temperate shoots!
Spices, barks, buds, seeds, roots, and fruits!
That's why spices are much higher priced!

I'll not labor with the technical and varying definitions of spices as opposed to culinary herbs, but I summarized much of it in the verse above. Overgeneralizing, culinary herbs are temperate leafy shoots used culinarily to flavor other dishes. And I know of no culinary herb that lacks medicinal activities. (*Mentha requienii* is so small that it seems not to have evolved any serious medicinal folklore; its the only popular herb for which I found no published medicinal folklore.) And overgeneralizing, spices are more often tropical and involve other plant parts, not just the leaves and shoots. But there is no fine line between spice and herb, and furthermore no fine line between, herb, spice, food, and medicine. Chile, garlic, ginger, onion, pepper, and turmeric are all herbaceous in the botanical sense of the word, i.e., not producing any wood; they are all often included in the spice charts and statistics of the world; they are all foods; they are all medicines.

I have intentionally omitted from this book many of the better-known temperate culinary herbs. *Anethum graveolens* (dill), *Brassica* sp. (mustard), *Coriandrum sativum* (coriander), *Cuminum cyminum* (cumin), *Foeniculum vulgare* (fennel), *Mentha* spp. (peppermint, spearmint, etc), *Origanum vulgare* (oregano), *Ocimum basilicum* (basil), *Papaver somniferum* (poppy), *Petroselinum crispum* (parsley), *Pimpinella anisum* (aniseed), *Salvia officinalis* (sage), and *Thymus vulgaris* (thyme). These are clearly culinary and medicinal herbs, and all are carried in the USDA spice statistics. Most are also covered in detail and illustrated in Ed. 2 of our CRC *Handbook of Medicinal*

Herbs (Duke et al., 2002). Many of them are also covered in my books on *Medicinal Plants of the Bible* (Duke, 1983, 1999), *Culinary Herbs* (Duke, 1985), and *Living Liqueurs* (Duke, 1987).

Under indications, I list most published iindications that crossed my desk, alphabetically, with each indication followed by the 'f' or numerical score for efficacy, followed by the citation for the source. It was with some trepidation that I converted more specific terms such as arthritis to arthrosis, and bronchitis to bronchosis; but I think that was a more economical (space wise) was of presenting the data. Classically the suffix 'itis' means inflammation and 'osis' means ailment of. Thus arthritis is inflammation of the joint, and arthrosis is broader, meaning an ailment in the joint. Where some author just said "for joint problems," that became 'arthrosis,' but where they were more specific and said inflammation of the joint it means the more specific 'arthritis.' Towards the end I aggregated both under 'arthrosis.' Many people will dislike the fact I converted all the more specific -itis etntries to -osis, rather than somewhat redundantly include both.

In the indications paragraph, you see parenthetical numbers followed by three-letter abbre-viations (abbreviation of source) or an alphanumeric X-1111111 to identify PubMed citations. A parenthetical efficacy score of (1) under an activity or indication means that a chemical in the plant or an extract of the plant has shown the activity or proven out experimentally (animal, not clinical) for the indication. This could be *in vitro* animal or assay experiments. A hint; not real human proof! Nothing clinical yet! I score (2) here if the aqueous extract, ethanolic extract, or decoction or tea derived from the plant has been shown to have the activity or to support the indication in clinical trials. Commission E (KOM) and Tramil Commission (TRA) approvals were automatically scored (2) also, as they represented consensus opinions of distinguished panels. The rare (3) scoring for efficacy means that there are clinical trials showing that the plant itself (not just an extract or phytochemical derived therefrom) has the indication or activities. The solitary (f) in many of the citations means that it is unsupported folk medicine, or I have not seen the science to back it up. The three-letter abbreviations are useful short citations of the references consulted in arriving at these numbers. I have by no means cited every source here. But unlike KOM and hopefully better than PDR for Herbal Medicines, ed. 1 and 2 (PHR and PH2), I indicate at least one source for every indication and activity I report. Commission E (Blumenthal et al., 1998) did not list sources.

And after much soul searching, I have decided to spare our readers the long list of all the phytochemicals reported from each of these spices. These are available for your purview on the USDA phytochemical database (http://www.ars-grin.gov/duke/). Many of these are detailed para-graphically in Duke and duCellier (1993). Instead, I have pulled forward for you some of the major compounds that may underlie many of the reported activities of these species. These data, too, are available on our USDA website, where I also list the source of each data entry. Another new feature is the addition of our Multiple Activity queries, not yet available on the USDA database. With the able assistance of Sue Mustalish, R.N., and Leigh Broadhurst, Ph.D. and certified nutritionist, I have accumulated many of the activities that might contribute to the alleviation, correction and/or prevention of an ailment. The computer then searches for phytochemicals reported from that spice that have the desired activities. As you will see, this shows that the spice is a menu of biologically active compounds that might help the malady. I suspect the body is skillful at sifting through those phytochemicals with which your genes have co-evolved for so many millions of years. This does not prove that the spice will help; it just proves that the spice contains phytochemicals, often by the dozens, that have been shown to have useful activities.

In this book, I focus on the medicinal application of spices. If you need to know more about quality specifications and the like, I suggest you consult Purseglove et al. (1981) or Tainter and Grenis (1993). I feel a bit stronger about the medicinal potential of spices than did Purseglove, who said, "Spices are no longer as important medicines as they were in the past, but some have minor uses in modern pharmacopoeias, of which probably the most important is *Capsicum*."

(*Capsicum* shows up in proprietary preparations from A to Z, Axsain to Zostrix, JAD.) As a matter of fact, I agree with Purseglove that the spices did suffer a decline in both medicinal importance and relative value. But I predict that such spices as capsicum, cinnamon, garlic, ginger, onion, and turmeric will assume relatively more medicinal importance again, as the economic costs and knowledge of the side-effects of prescription pharmaceuticals increase. You see, each spice contains thousands of useful phytochemicals. Pharmaceuticals usually contain only one or two.

I actually believe that many educated Americans, after reading this book, may sometimes head to the spice chest for minor ailments instead of the medicine chest. When one considers that 80% of the world cannot afford our pharmaceuticals, I'll speculate that already more humans use spices as medicines than use prescription pharmaceuticals. I'll even put the spices up against the pharmaceuticals, the garlic against the statins for high cholesterol, the ginger against antacids for ulcer and even for morning sickness (they don't have an approved pharmaceutical), capsaicin vs. Acyclovir for shingles, and turmeric vs. Vioxx for arthritis and vs. Cognex for Alzheimer's.

I could start my spice story ~500 years ago when Columbus discovered America, or 50,000–60,000 years ago, when humans were learning that wrapping their food in leaves kept the ashes off, retained the juices, and sometimes improved the flavor, or even tenderized tough meat; or ca. 5000 years ago, when garlic and onion were contributing to Egypt's pyramids, ginger joining early Chinese medicine chests, pepper penetrating Ayurvedic medicine chests, and sesame spicing Assyrian wines. Babylonians, ca. 2700 B.P. (before present), were familiar with cardamom, coriander, garlic, saffron, thyme, and turmeric. Assyrians, ca. 2650 B.P., were familiar with anise, cardamom, coriander, cumin, dill, garlic, myrrh, poppy, saffron, sesame, thyme, and turmeric. Around 2400 B.P., the father of medicine, Hippocrates, said "let food be your medicine, medicine your food." Already, he was familiar with cinnamon, coriander, marjoram, mint, saffron, and thyme. In those days, spices were as important for medicine, embalming, preserving food, and masking bad odors, as they were for more mundane culinary matters. Now, in the new millennium, I may be reverting to the Hippocratean corollary: let food be your medicine. Many Americans are a bit alarmed by *Journal of the American Medical Association* statistics (*JAMA* p. 2891, 1987) that the "prevalence of fatal drug reactions has been estimated at 0.01% for surgical in-patients and 0.1% for medical in-patients." That indicates that at least 1 in 1000 patients in a hospital will die of iatrogenic causes. With medicine getting more and more expensive and impersonal, and high iatrogenic death rates as quoted from JAMA, people are actually afraid of their doctors and/or health plans. I've been with the same health plan for nearly two decades. On visits to my neurosurgeon, my charts and magnetic resonance imagery (MRI) were lost, so that I wasted an afternoon. On one visit to the GP, I saw the erroneous comment that I was on the contraceptive pill. When one GP prescribed a sulfa drug for sinusitis, he had to ask me if I was allergic to sulfa. This should already be in his computer, as should my blood type. He could produce neither. Having been with a plan that long, I find it disgraceful that they can't tell me instead of asking me. That's why I'm inclined to listen to news that garlic might cure sinusitis, rhinitis, even meningitis. And that's why all this renewed interest in spices and foods as pharmaceuticals, an aversive reversion to Hippocrates. So spices are working their way back into the medicine chest, with many good reasons, economic, gustatory, and salutory.

FORMAT

After some deliberation I have limited the entries on individual spice species to five major headings — Medicinal Uses, Indications, Other Uses, Cultivation, and Chemistry.

MEDICINAL USES

Since the medicinal uses are the most important to me, they are first. I cover some of the major historical and / or new facts from current findings, chemical or clinical abstracts. The folk and real medicinal potential of the species are noted.

INDICATIONS

The indications are listed in a concise format. The indications are followed by a parenthetical score and abbreviated reference citation(s). The scores are: f = folkloric only; 1 = with in vitro, animal or chemical but no clinical rationale; 2 = with positive clinical trials (or Commission E approval) of extracts of the spice; or 3 = with positive clinical trials (or Commission E approval) for the spice itself, as in whole garlic. The score is followed by an abbreviation of one or more references that helped me arrive at the score. The biological activities have been omitted, many of which are covered in my CRC Handbook of Medicinal Herbs, 2nd ed., published in July 2002. About 75% of these spices were covered, including sections on activities; indications; dosage; contraindications, interactions and side effects; and extracts. For the better known spices, I include a few of the dozens of possible Multiple Activity queries of the USDA database, delineating the compounds in a plant that might contribute to several biological activities, each of which might contribute to the resolution of an " indication".

OTHER USES

I discuss the historical and/or conventional culinary uses and other non-medicinal uses of the spice. Here you will also find occasional references to pesticidal activities.

CULTIVATION

I have compiled information on cultivation of the spice, sometimes tempered by my hands-on experiences in the Green Farmacy Garden, in Howard County Maryland, where I have grown more than half of the spices covered in this book.

CHEMISTRY

All of the chemicals are not listed. Remember that each plant species contains hundreds of chemicals in the part per million levels, and thousands in the parts per trillion levels. Here, I select a few of those I deem more important, or about which there is current breaking news. Rarely, if ever, is any phytochemical working alone; more likely, phytochemicals are acting synergistically with other compounds in a species. Often such phytochemicals protect the spice from its natural enemies, synergetically. Then, we humans borrow them, to protect us from our enemies, synergetically. For extended chemical information, readers are referred to earlier CRC compendia:

Duke, J.A., *CRC Handbook of Phytochemical Constituents in GRAS Herbs and Other Economic Plants*, CRC Press, Boca Raton, FL, 1992

Duke, J.A., *CRC Handbook of Biologically Active Phytochemicals and Their Activities*, CRC Press, Boca Raton, FL, 1992

Beckstrom-Sternberg, S. and Duke, J. A., *Handbook of Mints (Aromathematics): Phytochemicals and Biological Activities*, CRC Press, Boca Raton, Fl, 1996

USDA database (http://www.arsgrin.gov/duke)

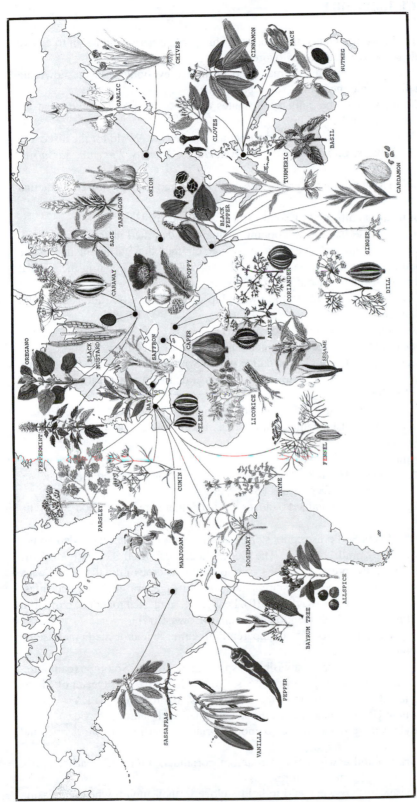

CENTERS OF ORIGIN FOR SOME HERBS AND SPICES

DATELINE — SPICE TIMETABLE

- ~60,000,000 B.P. Flowering plants, the subject of this book, emerge and begin to evolve phytochemicals defensive against phytovores.
- ~6,000,000 B.P. Primates evolve into man who begins co-evolving with the flowering plants, some edible, some medicinal, some poisonous.
- ~50,000 B.P. Man learns of the culinary attributes of leaves wrapped around meat for cooking (McCormick, 1981).
- ~18,000 B.P. Man crosses the Bering Bridge, opening up the New World for the discovery of the New World spices, *Capsicum, Cunila, Osmorrhiza, Peumus, Pimenta, Sassafras, Vanilla.*
- ~12,000 B.P. Boldo Man in Monte Verde Chile, with boldo and a couple dozen medicinal plants.
- ~7000 B.P. Hot peppers cultivated in South America (Wood, 1993).
- ~6000 B.P. Sumerians use licorice and opium; Fenugreek identified in Iraq.
- ~5000 B.P. Charak, the father of Ayurvedic medicine, claimed that garlic "maintains the fluidity of blood and strengthens the heart" (Rahman, 2001). Not known in the wild, it was cultivated in the Middle East at least 5000 years ago.
- ~5000 B.P. Ancient historians equate the ownership of ginger or its trade routes with prosperity (Schulick, 1996).
- ~4700 B.P. Cassia recorded in China (Bown, 2001).
- ~4500 B.P. Andean Indians already using coca, the source of cocaine.
- ~4000 B.P. Shen Nong first Ben Cao or native herbal with 365 drugs, Cassia, ephedra, ginseng, rhubarb. Garlic already in use in China (consumed with raw meat), introduced into Japan (Rivlin, 2001).
- ~3900 B.P. Sesame oil expressed in Urartu (now Armenia) (TAD).
- ~3730 B.P. Joseph sold to Ishmaelites with camels taking spices, balm, and myrrh to Egypt (Genesis 37) (PEA).
- ~3600 B.P. Cassia recorded in Egypt.
- ~3500 B.P. Art at Queen Hatshepsut's temple at Luxor showed potted frankincense, possibly used in rejuvenating face masks (Bown, 2001).
- ~3500 B.P. King Tut's Tomb contained 6 cloves garlic (Fulder dates it ~ 4000 B.C.; Rivlin 1500).
- 3500 B.P. Papyrus Ebers 800 prescriptions; 700 mostly plant drugs compounded sometimes with beer, honey, milk, or wine. The Codex Ebers, ca. 1500 B.C., is one of the earliest sources indicating prescription of garlic to treat cancerous growths. The Codex also suggests garlic, as I do today, for circulatory ailments, general malaise, and infestations with insects and parasites (Rivlin, 2110). Fenugreek suggested to induce childbirth (Bown, 2001). Probably the first mention of sesame (Bown, 2001).
- ~3400 B.P. Poppy, its seeds or its opium, apparently in culinary and/or medicinal use by Cretans, Egyptians, and Sumerians.
- ~3000 B.P. Solomon immortalizes many Biblical spices in his song of Solomon (camphire, cinnamon, frankincense, myrrh, pomegranate, saffron, spikenard). The Queen of Sheba visits, bringing as gifts of state "camels that bear spices" (Chronicle 9) (PEA).
- ~2775 B.P. Olympic Games founded, champions crowned with laurel (=bayleaf).
- ~2700 B.P. Babylonians familiar with cardamom, coriander, garlic, saffron, thyme, and turmeric.
- ~2650 B.P. Assyrians familiar with anise, cardamom, coriander, cumin, dill, garlic, myrrh, poppy, saffron, sesame, thyme, and turmeric.
- ~2500 B.P. Chinese courtiers were said to hold a clove in their mouth when addressing an emperor.

- ~2500 B.P. Sasruta writings in India mention cardamom, cinnamon, pepper, and turmeric.
- ~2400 B.P. Hippocrates—"First, do no harm. Second, let food be your farmacy." Used garlic for pulmonary complaints, abdominal and uterine growths (read cancer), and as a cleansing agent and purgative. Also familiar with cinnamon, coriander, marjoram, mint, saffron, and thyme.
- ~2350 B.P. Aristotle catalogued medicinal properties of many herbs and spices.
- ~2335 B.P. Alexander the Great's army plunders Gaza, sending its frankincense to Greece.
- ~2300 B.P. Theophrastus, "Father of Botany," described medicinal attributes of many spices, including black pepper and long pepper. He noted that hot sunny regions produced the most aromatic spices. Licorice suggested for asthma, bronchoses, cough, pulmonoses (FAY).
- ~2165 B.P. Death of Chinese Princess Tai, buried with cinnamon, galangal, ginger, and pepper.
- ~2050 B.P. Mithridates, "The Royal Toxicologist," rhizomatists offering ginger.
- ~2050 B.P. First mustard seed brought to England by the Romans (McCormick, 1981).
- ~2050 B.P. Caius Plinus Secundus (A.D. 23–79), "Pliny the Elder," *Natural History*—Medical Bot, listed 23 uses for garlic against infections and toxins (anticipating p450–2E1 activities on toxins (Rivlin, 2001) and maybe even antianthrax (JAD). Pliny considered licorice native to Sicily).
- Birth of Jesus. Wise men bring Frankincense and Myrrh.
- Crucifixion. Christ given "gall" on the cross, perhaps opium.
- Many herbs and spices mentioned in the Bible (almond, anise, bay, black cumin, caper, carob, cassia, cinnamon, coriander, cumin, dill, fenugreek, frankincense, galbanum, garlic, juniper, leek, marjoram or Biblical hyssop, mint, mustard, myrrh, myrtle, onion, poppy, rue, saffron, sage, spikenard, storax, possibly turmeric and wormwood).
- Cloves reach China, India, Rome as spice (Bown, 2001).
- ~47 A.D. Pedanios Dioscorides (first century A.D.) "Materia Medica," followed the Roman armies. Recommended garlic to clean the arteries, for GI disorders, for joint disease and seizures (Rivlin, 2001). Boiled garlic and oregano for bedbugs and lice (Fulder, 1997). Ginger suggested as aphrodisiac (Schulick, 1996). Sesame seed sprinkled on bread (Bown, 2001).
- ~130–200 A.D. Galen medicinal extracts of galenicals containing dozens of ingredients (opium).
- ~200 A.D. Ginger taxed in Rome, first listed as medicinal in China (Bown, 2001).
- ~300 A.D. Bower Manuscript. Garlic for debility, dyspepsia, fatigue, and infections.
- ~410 A.D. Alaris, the Visigoth, subjugates Rome and demands 3,000 lb peppercorns as tribute (McCormick, 1981).
- ~570 A.D. Birth of Mohammad.
- ~595 A.D. Mohammad marries Khadija, and they run a Meccan shop trading in oriental spices, frankincense and myrrh. Muslims consolidate monopoly on spice trade, which lasts for centuries. The Qu'ran hints that ginger (not identified in the Bible) is a heavenly and spiritual beverage (Schulick, 1996).
- ~600 A.D. Clove and nutmeg listed as medicinal herb in China (Bown, 2001).
- ~720 A.D. Cardamon mentioned in China as medicine (Bown, 2001).
- Late Eighth Century. First drug stores in Bagdad; Muslims and Arabs rescued many of the books from Christian/Roman wars; the first apothecary camphor, cloves, cubebs, nutmeg, tamarind.
- ~980–1037 A.D. Avicennia ("Ibn Sina"), Arabic scholar, Persian physician, poet, writer Unani or Greek Canon Of Medicine. 200 Publications.
- ~1098–1179 A.D. Hildegard of Bingen, Benedictine nun in Rhineland. She believed that raw garlic was more effective than cooked. I still believe this. She predicts use of celery in "gicht." She also covered cinnamon, cloves, cubeb (a type of pepper), fenugreek, galangal, ginger, licorice, nutmeg, pepper, zedoary, etc., plus dozens of culinary herbs.
- ~1280 A.D. Marco Polo observed ginger cultivated in China and India.

- ~1305 A.D. Edward I levies a tax on licorice to help pay for the London Bridge (FAY).
- ~1350 A.D. Black Death kills ~25 million Europeans (PEA). Spices widely tried but unsuccessfully, with the possible exception of garlic.
- ~1368–1654 A.D. Bastard cardamom first mentioned as Chinese medicine. Ming Dynasty (Bown, 2001).
- ~1447 A.D. English outlaw adulteration of spices (McCormick, 1981).
- ~1475 A.D. Bjornnson's Icelandic manuscript, before the invention of gin, prescribes juniper-spiced wine for cold and headache (CEB).
- ~1492 A.D. Columbus seeks a short route to the orient and black Indians and black pepper but instead finds red Indians and red pepper. Folklore says he was guided to Terra Firma by the aroma of sassafras.
- ~1500 (1493–1541) A.D. Paracelsus, the first reductionist (chemicals are responsible for medicinal activities of herb) disliked the vogue fascination with exotic imported herbs.
- ~1502 A.D. Ferdinand and Isabella tell Columbus re his fourth voyages, "All…spices and other products must be delivered to Francisco de Porras" (PEA).
- ~1512 A.D. Portuguese take Moluccas, consolidating monopoly on nutmeg (Bown, 2001).
- ~1513 A.D. Oviedo reaches Darien Panama, first to document capsicum peppers in Tierra Firme (Andrews, 1995).
- ~1536 A.D. Portuguese invade Ceylon to cement monopoly on cinnamon.
- ~1560 A.D. Spaniards employ sassafras for venereal disease (Bown, 2001).
- ~1567 A.D. Nutmeg poisoning reported in British pregnant lady who ingested 10–12 fruits and became deliriously inebriated (Bown, 2001).
- ~1569 A.D. Hungarians change name of red pepper to paprika (Andrews, 1995).
- ~1575 A.D. Monardes Seville says of sassafras, "It is almost incredible, for with the naughtie meates and drinkying of the rawe waters, and slepying in the dewes, the most parts of them came to fall into continual agues" (Erichsen-Brown, 1989).
- ~1597 A.D. John Gerard (1546–1607) writes his herbal; London ~1300 medicinal species; ginger noted to "provoke venerie" (Schulick, 1996; Griggs, 1998); "horseradish…commonly used among the Germans for sauce to eate fish" (Bown, 2001).
- ~1600 A.D. "And had I but one penny in the world, thou shouldst have it to buy gingerbread." Shakespeare; Love's Labors Lost (Schulick, 1996).
- ~1600 A.D. Henry IV of France, who regularly chewed garlic, has "breath that would fell an ox at twenty paces" (Fulder, 1997).
- ~1600 A.D. George II of England bans melegueta pepper as injurious to the health (Bown, 2001).
- ~1630 A.D. Jesuits recognize the febrifugal capacity of cinchona, long before the discovery of gin and tonic (Bown, 2001).
- ~1653 A.D. Nicholas Culpepper's *The English Physician* seemed to speak more of herbs than spices, calling Capsicum bird pepper, cayenne pepper, and guinea pepper.
- ~1787 A.D. Annatto introduced for cultivation in India (Bown, 2001).
- ~1800–1850 A.D. Shaker era in America. Physics Garden at New Lebanon with horehound, marjoram, poppy, sage, and savory.
- ~1803 A.D. Serturner isolates and identifies morphine from poppy.
- ~1820 A.D. Wintergreen leaves listed in U.S. Pharmocopoeia (until 1894) (Bown, 2001).
- ~1820 A.D. Caventou and Pelletier separated quinine and cinchonine from Peru Bark.
- ~1835 A.D. Texans develop chili powder combining various ground peppers (McCormick, 1981). Salicylic acid synthesized for the first time. (Chile contains salicylic acid.)
- ~1869 A.D. Boldo "first investigated by a French physician" (Bown, 2001).
- ~1884 A.D. Freud and then Koller discover anesthetic activity of cocaine.
- ~1915 A.D. 25,000 physicians, the Eclectics, embrace ginger and other natural medicines (Schulick, 1996).

I could have started my spice story ca. 2300 B.P., when Theophrastus, Father of Botany, was describing the medicinal attributes of many of our spices. He noted that hot, sunny regions produced the most aromatic of spices. Even today, I often find that the same species, grown in a dry tough environment, has more of the aromatic phytochemicals than the same species grown in a shadier, more humid environment. A pampered herb or spice may have more primary metabolites and proportionately fewer secondary metabolites. Translation: the pampered organic herb or spice may be the better food plant, but the tough wiry unpampered herb or spice may be the better medicinal. I could start my intro with Dioscorides, leading Greek physician of the first century A.D., whose works on botany were to be the standards until the days of Columbus. Some would suggest I should start with the birth of Jesus, others with the birth of Mohammad (A.D. 570). Bethlehem and Mecca are both suffering from proximity to the ravishes of war today. And both have experienced millennia of spice caravans and spice trading.

At the birth of Christ, wise men brought oriental spices, frankincense and myrrh. Most of the Old World spices had already traveled far and wide as spices, if not as plants. The temperate spices were widely distributed as plants, but the tropical tree spices still were dear and shrouded in mystery. Among spices mentioned in my *Medicinal Plants of the Bible* (Duke, 1983; Duke, 1999) are almond, anise, bay, black cumin, caper, carob, cassia, cinnamon, coriander, cumin, dill, fenugreek, frankincense, galbanum, garlic, juniper, leek, marjoram or Biblical hyssop, mint, mustard, myrrh, myrtle, onion, poppy, rue, saffron, sage, spikenard, storax, and wormwood. If your ancestors came from that part of the world, your genes have probably known the phytochemical contents of many of these Biblical herbs and spices for more than two millennia. Your genes have not yet known synthetic medicines for two centuries.

The fall of the Roman Empire was clear when Alexandria was occupied by the Arabs in A.D. 641. But one very important man, born A.D. 570, probably affected early spice trade even more than Columbus, certainly long before Columbus. As a young man, this important figure worked with Meccan merchants involved with spices in Arabia and Syria. Then he "graduated" to the role of camel driver and caravan leader for the widow Khadija, 15 years his senior. After their marriage in 595, he served as a partner in a Meccan shop trading in oriental spices, frankincense, and myrrh.

So, oriental spices were important to Mohammad, founder of one great religion, as they were to Jesus, namesake of another great religion. Yet I don't find the spices, not even Arabian frankincense and myrrh, listed in the very useful website hosting the Hadith. I do find some mention of black cumin, hyssop, leek, mustard, and onion, but no cinnamon, cassia, caraway, and coriander. Nothing seems to receive higher praise than the black cumin, *Nigella sativa* (from the database): "I heard Allah's Apostle saying, 'There is healing in black cumin for all diseases except death'."

I do not have a searchable Qu'ran on my computer like my searchable Bible. Unlike the Bible, the Qu'ran was compiled over a very short period of time and is entirely orientated toward revelation and the word of Allah. The Qu'ran does mention the benefits of consuming certain foods such as honey and the abstinence of alcohol, but it contains very little specific information on health and disease. Prophetic medicine was mostly prayer. The Hadith, however, details guidelines on diet and the treatment of simple ailments. One can search MSA-USC Hadith Database: http://www.usc.edu/dept/MSA/reference/searchhadith.html

By 750, the Mohammedan religion stretched 7,000 miles from the borders of China to Spain. And Arab traders had a monopoly on oriental spices and gold alike, just as they nearly attained monopolies, collaborating with other OPEC nations, on petroleum in more recent decades. The Port City of Basra, about which I heard so much in the Persian Gulf Wars of the 1980s and 1990s, was founded in the year 635 where the Euphrates and Tigris meet. The great physician Rhazes (850–925) became chief of the great hospital in Bagdad, and his works, like those of Avicennia (ca. 1000), influenced European medicine heavily. By 1096, the first of the Crusades began nibbling at the Muslim empire.

The USDA once defined spices as parts of plants (dried seeds, buds, fruit or flower parts, bark or roots) usually of tropical origin. They contrasted herbs as leafy parts of temperate species. King Charlemagne once defined an herb and/or spice as "the friend of the physician and the pride of cooks."

That was back in the days when herbs and spices were major medicinals and friend of the physician. Today, herbs and spices may be viewed with less than disdain by the physician and the pharmaceutical industry, as they are more and more proving to prevent disease as well as cure it. From *Allium* for preventing cancer, high blood pressure, high blood sugar, and maybe even the common cold and the more-and-more-common yeast; through *Glycyrrhiza* for preventing caries, diabetes, maybe even ulcers; to *Zingiber* for preventing seasickness and ulcers and alleviating morning sickness, there is increased interest in preventive medicine and designer foods to prevent and/or alleviate curable and incurable ailments. All spices have folk medicinal reputations, and extracts of most species have exciting biological activities. All spices contain important curative phytochemicals.

Could a spice rack prove to be a medicine chest? I discourage that. No one should self-diagnose and self-medicate. They should, however, seek out for their physicians those holistic physicians who are intelligent enough to consider dietary and lifestyle modifications to prevent and/or treat disease, before they sucker the patient into the synthetic-pill-a-day-for-life syndrome so pleasing to the pharmaceutical firms and pharmacophilic allopaths.

ANTISEPSIS

Eating, with or without spices, is one of the most dangerous things I do every day. Germs and their toxins are everywhere. I often read of deaths and illnesses associated with contaminated foods, be it hamburger or apple juice. All this despite modern preservatives, refrigeration, and hygienic food preparation. "Phytochemicals are legacies of multiple co-evolutionary races between plants and their enemies—parasites, pathogens, and herbivores. These chemical cocktails are the plants' recipes for survival" (Sherman and Flaxman, 2001).

After examining 43 spices in more than 4500 meat-based recipes from traditional cookbooks of 36 countries, Sherman and Billing (1999) concluded that spices are used because of their antimicrobial properties. Many spices have antimicrobial (especially antibacterial) properties. Spice use is greater in hot climates, where meats spoil relatively quickly, than in cool climates. Recipes from hot climates use more of the most highly inhibitory spices. Spices are often used in quantities sufficient to kill microbes and in ways that preserve their antimicrobial properties. Cookbooks provide records of our co-evolutionary race with foodborne pathogens (Chasan, R., 1999).

Still, foodborne bacteria (especially species of *Clostridium, Escherichia, Listeria, Salmonella, Staphylococcus,* and *Streptococcus*), or their toxins, debilitate millions of people annually and kill thousands. During 1971–1990, food poisoning, primarily bacterial, affected 29.2 out of every 100,000 Japanese but only 3.0 out of every 100,000 Koreans. The Korean meat-based recipes are spicier than the Japanese (Sherman and Billing, 1999). Even here at home in the U.S., foodborne illnesses afflict an estimated 80 million people per year, and 1 in 10 Americans experiences bacteria-related food poisoning annually (Hui et al., 1994). Ten thousand or more Americans will die of food poisoning, said CSPI one year. Moreover, new foodborne pathogens continually are evolving, along with resistance among existing pathogens to monochemical bactericides. Still, bacteria are more frequently implicated in foodborne disease outbreaks than yeasts or fungi. All 30 spices tested were found to kill or inhibit at least 25% of the bacterial species on which they had been tested, and 15 inhibited at least 75% of bacterial species (Sherman and Billing, 1999). Garlic, onion, allspice, and oregano were most potent. They inhibited or killed every bacterium tested. One study tested eugenol, menthol, and anethole on three pathogenic bacteria, *Salmonella typhimurium, Staphylococcus aureus,* and *Vibrio parahaemolyticus*. Each spice component inhibited the bacteria differently (and this points to synergies with spice mixtures). Eugenol was more active than thymol, which was more active than anethole. Eugenol and isoeugenol are sporostatic to *Bacillus subtilis* at the 0.05–0.06% level. Gingerol and zingerone also have sporostatic activity, but at 0.8–-0.9%. Inhibition effectiveness was related inversely to the molecular weight of the phenolic (Tainter and Grenis, 1993). The longer the side chain on the phenolic ring structure, the less the antimicrobial activity; 0.12% ground glove and 0.02% eugenol decreased the rate and extent of germination of *Bacillus subtilis* spores.

Carnosol and ursolic acid were tested on six strains of foodborne bacteria and yeast. Their antimicrobial activity was compared to BHA and BHT, also known to have antioxidant and antimicrobial activity. Carnosol was more effective than BHA or BHT. Ursolic acid was more effective than BHT (Tainter and Grenis, 1993).

SYNERGY

Pepper and citric acid play special roles as synergists. "Citric acid potentiates the antibacterial effects of other spices, because low pH disrupts bacterial cell membranes" (Sherman and Billing, 1999). "Black pepper comes from *Piper nigrum*, an exclusively tropical plant which has several useful properties. For example, the compound piperine inhibits the ubiquitous, deadly bacterium *Clostridium botulinum*. Black pepper is also a 'bioavailability enhancer,' meaning that it acts synergistically to increase the rate at which cells, including microorganisms, absorb phytotoxins" (Sherman and Billing, 1999). Khan and Balick (2001) note that tamarind increased bioavailabilty of other drugs, including, I presume, herbal.

Many spices are more potent when mixed. French "quatre epices" (pepper, cloves, ginger, and nutmeg) is often used to make sausages and may in fact make the sausage last longer. Curry powder (which contains 22 different spices), pickling spice (15 spices), and chili powder (10 spices) are broad-spectrum "antimicrobial melanges" (Sherman and Billing, 1999). Andrews (1995) elaborates on this spice called curry. Originating in India, curry is a combination of freshly ground spices, principally chili pepper, with as few as 5 or as many as 50 spice ingredients. Slightly roasted ground chillis are powdered and mixed in with ground turmeric (for color) and adding coriander, along with other spices, alphabetically, allspice, anise, bay, caraway, cardamom, celeryseed, cinnamon, cloves, cubeb, curry leaf, dill, fennel, fenugreek (both leaves and seeds), garlic, ginger, juniper, mace, mint, mustard, nutmeg, pepper (both black and white), poppyseed, saffron, sumac, zedoary, not to mention salt. Andrews lists a simpler chile powder, a blend of several peppers, with garlic powder, oregano, cumin, cayenne and paprika, garlic (Andrews, 1995).

AVOIDANCE

Sherman and Flaxman (2001) stress that, even in those countries using spices heavily and regularly, pre-adolescent children and women in their first trimester typically avoid highly spiced foods. Morning sickness may reduce maternal intake of foods containing teratogens during the early embryogenesis, when delicate fetal tissues are most susceptible to chemical disruption. Women who experience morning sickness are less likely to miscarry than women who do not. A possible negative corollary to the Sherman and Flaxman hypothesis might tend to discourage the use of ginger to avoid hyperemesis gravidarum during the first trimester. See ginger for its role in avoiding morning sickness.

SPICE STATISTICS

Today, spice use is ubiquitous, but spices are far more important in some cuisines than others. "Japanese dishes are often described as delicate, Indonesian and Szechwan as 'hot,' and middle European and Scandinavian dishes as 'bland'" (Sherman and Billing, 1999).

"Cookbooks generally distinguish between seasonings (spices used in food preparation) and condiments (spices added after food is served), but not between herbs and spices" (Sherman and Billing, 1999). Herbs "are defined botanically (as plants that do not develop woody, persistent tissue), usually are called for in their fresh state, whereas spices generally are dried." 93% of recipes call for at least one spice. "On average, recipes called for nearly four, but some lacked spices, especially in temperate countries or in vegetarian dishes. Others had up to 12 spices. In 10 countries, Ethiopia, Kenya, Greece, India, Indonesia, Iran, Malaysia, Morocco, Nigeria, and Thailand, every meat-based recipe required at least one spice. In Scandinavia, one-third of the recipes had no spices. Vegetable dishes are almost always less spicy than meat dishes, a clue that leads Sherman and Hash (2001) to argue that the spices evolved as antimicrobial agents. I agree.

Black pepper and onion were used more frequently (63 and 65%) than garlic, 35%, chilis, 24%, lemon and lime juice, 23%, parsley, 22%, ginger, 16%, and bay leaf, 13%. Then came coriander, cinnamon, cloves, nutmeg, thyme, paprika, sweet pepper, cumin, celery, turmeric, allspice, mustard, cardamom, saffron, mint, dill, oregano, basil, lemongrass, sesame, tamarind, sage, rosemary, anise, marjoram, caraway, capers, tarragon, juniper, fenugreek, horseradish, fennel, and savory (Sherman and Hash, 2001). Those that I fail to include in this book, I have covered earlier in my *Culinary Herbs* (Duke, 1985) and/or *Living Liqueurs* (Duke, 1987). I anticipate a CRC *Handbook of Medicinal Culinary Herbs* as a sequel to this spice book, lamenting that there is no clear-cut line between the definitions of spice and culinary herb. But all are medicinal.

Here I use statistics more appropriate for the 500th anniversary of Columbus' arrival in America. For more recent statistics, see the USDA web site: *http://www.fas.usda.gov/ustrade*

Spices ranked according to the most valuable imports to the U.S. are:

1.	Black Pepper	$115 million	Tropical Fruit
2.	Vanilla	62	Tropical Fruit
3.	Sesame	26	Subtropical Seed
4.	Cinnamon (and cassia)	22	Tropical Bark
5.	Capsicum	22	Tropical Fruit
6.	Nutmeg and mace	12	Tropical Seed
7.	Mustard	9	Temperate Seed
8.	Oregano	7	Temperate Shoot
9.	Caper	7	Subtropical Bud
10.	Sage	6	Temperate Shoot
11.	Ginger	5	Tropical Root
12.	Cumin	4	Subtropical Fruit
13.	Clove	4	Tropical Bud
14.	Poppy	4	Temperate Seed

Some valuable spices produced by the U.S. are:

1. Sesame
2. Capsicum
3. Mustard
4. Oregano
5. Sage
6. Poppy
7. Ginger — Hawaii produced 3697 MT in 1989 from 67 harvested acres

When first approached about a lecture on Columbus' effects on the spice trade, I decided to restrict my talk to seeds. But then I went through the then recent version of U.S. Spice Trade (FTEA 1–90). When you classify the part of the plant that gets into the spice trade, there are few real seed, mustard, nutmeg, poppy, and sesame as noteworthy exceptions. And then there are those "seeds" in the carrot family, aniseed, caraway, coriander, cumin, dill, fennel, and the like, that are really one-seeded fruits. Just a technicality; those are the same "seed" I plant if I want more anise, caraway, etc.

And post-Columbian activities have resulted in Brazil closing in on Indonesia, Malaysia, and India as a leading black pepper producer.

Cloves, though native to Indo-Malaysia, are mostly produced for export by Brazil, Madagascar, and Tanzania. Indonesia produces a lot but is also the world's largest consumer. Cinnamon and cassia are still mostly produced from the Asian subcontinent, with some cinnamon in the Seychelles.

Nutmegs and mace, both from the same tree, come to us mostly from their native Indian subcontinent, but Grenada and Trinidad are making small contributions to the U.S. market.

Asian turmeric is still largely provided us by India, but Latin America is largely independent for its "azafran." America has supplanted Asia as the source of cardamoms to the U.S.; Guatemala supplied more than any other producer to the export market, 3 to 1. India still produces a lot, but most is for local consumption.

In the pages that follow, I tabulate the four major producers of each of our spice imports, italicizing those countries where the plant is not native. Figures have been very generously rounded. A summary table of the nativity of these four main producers of each of our main "spice" imports follows. These figures were applicable ca. 1992, 500 years after Columbus discovered America. Where available, I have added year 2000 import statistics from FAS, 2002.

Spice	Nativity
Allspice	Native
Anise	Mostly native
Basil	Alien
Caper	Native
Capsicum	Mostly alien
Caraway	Mostly native
Cardamom	Mostly alien
Cassia/cinnamon	Mostly native
Celery seed	Mostly alien
Cloves	Mostly alien
Coriander	Mostly native
Cumin	Mostly native
Dill	Mostly native
Fennel	Mostly native
Garlic	Mostly native
Ginger	Mostly alien
Laurel (bayleaf)	Native
Licorice	Mostly alien (specifically, not generic)
Mace	Mostly native
Marjoram	Mostly native
Mint	Mostly native
Mustard	Mostly alien
Nutmeg	Mostly native
Onion	Alien
Oregano	Alien
Parsley	Mostly alien
Pepper	Mostly alien
Poppyseed	Mostly alien
Rosemary	Native
Saffron	Mostly alien (and incredible)
Sage	Native
Savory	Native
Sesame	Mostly alien
Tarragon	Alien
Thyme	Mostly native
Turmeric	Mostly native
Vanilla	Mostly alien

U.S. SPICE IMPORTS, CERCA

500 years of the Columbian Exchange, updated from my Columbian Exchange lectures (Duke, 1991, 1992).

ALLSPICE

Native American, allspice is still mostly produced in America, Grenada being the major producer. Allspice, of which I imported more than 1000 tons in 1989, worth nearly $2 million, is essentially the dried unripe fruit of tropical *Pimenta racemosa,* assigned to Zeven and Zhukovsky's Middle American Center of Diversity.

Jamaica	310 tons worth ca.	$725,000
Honduras	260 tons worth ca.	$325,000
Mexico	240 tons worth ca.	$325,000
Guatemala	185 tons worth ca.	$225,000
Total	**1100 tons worth ca.**	**$1,800,000**

Appropriately first in my discussion, this is the only spice exclusively cultivated in the western hemisphere, and one of only three native to America. As with other commodity groups, Latin America has contributed more to the World Food Basket (allspice, capsicum, vanilla) than North America, which contributed briefly, only sassafras, also called cinnamonwood, and spicebush, namesake of the nutmeg state. In the *Journal of Columbus' First Voyage* (4 Nov. 1492), I read that Columbus showed Kuna Indians of San Blas peppercorns and they, by sign language, apparently indicated that there was a lot of it around. Early Spanish explorers found this tree in the West Indies in the 1500s. Apparently, introductions to Asia failed to flower and consequently fruit, so introductions were all but abandoned. Long before Columbus, Maya Indians used allspice in embalming. Still under the influence of the *Piper*mania, Francisco Hernandez called it *Piper Tabasci*, having found it in Tabasco Mexico between 1571 and 1577. That's also why it was called pimienta, later corrupted to pimento. And like pepper (*Piper*), the allspice fruits were used to preserve meats on long voyages. These preservative activities are due to some of the aromatic and antiseptic compounds which abound in allspice (anethole, caryophyllene, eugenol, linalool, pinene, and terpinene).

ANISE

This seed (1106 tons worth $1.777 million) is in reality the fruit of temperate *Pimpinella anisum,* assigned to the Near Eastern Center of Diversity.

Turkey	700 tons	$1,000,000
Spain	170 tons	$400,000
China	150 tons	$275,000
Hong Kong	35 tons	$64,000
Total	**1100 tons**	**$1,800,000**

First century Romans ingested aniseed cakes after feasts to prevent indigestion. Anise is said to have helped repair London Bridge way before Columbus. In 1305, King Edward I put a toll on anise. By the time of Edward IV, anise was used to perfume his personal linens. Oil from aniseed gives most of the flavor to licorice, at least in the U.S.

In 2000, the U.S. imported ~1500 MT of aniseed worth more than 3 million dollars (FAS, 2002).

BASIL

Basil (1992 tons worth $2.47 million) is the dried (or fresh) leaves of temperate *Ocimum basilicum,* assigned to Zeven and Zhukovsky's Indochina-Indonesia Center of Diversity, although I still think of it as a Mediterranean herb.

Egypt	1800 tons	$1,900,000
Albania	45 tons	$125,000
Yugoslavia	40 tons	$150,000
Mexico	20 tons	$90,000
Total	**2000 tons**	**$2,500,000**

Basil, with marjoram, mint, sage, savory, and thyme have long been used since ancient time to flavor foods. Dioscorides even added that a little basil wine was good for the eyes. Basil rivals oregano as a pizza herb and is, of course, indispensable to pesto. But it contains estragole, closely related to safrole.

In 2000, the US imported ~3300 MT of basil worth ca. 5.5 million dollars (FAS, 2002).

CAPERS

Capers (1246 tons worth ca. $7.857) are pickled flower buds of Mediterranean *Capparis spinosa,* assigned appropriately to the Mediterranean Center of Diversity:

Spain	500 tons	$3,500,000
Morocco	300 tons	$1,800,000
Italy	10 tons	$60,000
Denmark	10 tons	$70,000
Total	**1250 tons**	**$7,900,000**

Known as "Desire" in Ecclesiastes 12, and still today produced mostly in the Mediterranean, this spice is one of several that were important in the Bible. Other Biblical spices include black cumin, cassia, cinnamon, coriander, cumin, dill, fenugreek, frankincense, galbanum, garlic, hyssop (debatable), laurel, mint, mustard, myrrh, myrtle, onion, oregano (sensu lato), poppy, and saffron.

In 2000, the U.S. imported ~ 450 MT of capers worth more than 1.5 million dollars (FAS, 2002).

CAPSICUMS

Capsicums (22,868 tons worth $42,132.7 million) are fruits of subtropical capsicum species, native to Latin America. California alone produced 10,261 tons of capsicum worth $11 million. Categorization skews the import data.

Mexico was main source of "Anaheim" and "Anco" imports, in 1990 providing 1250 tons worth $2,350,000.

Paprika:

Spain	2700 tons	$5,000,000
Hungary	950 tons	$1,650,000
Total	**4200 tons**	**$7,500,000**

Other ground capsicum:

Mexico	2700 tons	$1,200,000
Pakistan	285 tons	$525,000
India	280 tons	$460,000
China	280 tons	$625,000
Total	**4200 tons**	**$4,000,000**

Other unground capsicums:

China	3300 tons	$6,500,000
India	3215 tons	$5,575,000
Mexico	2925 tons	$5,000,000
Pakistan	1125 tons	$1,800,000
Total	**13,200 tons**	**$28,000,000**

In 1493, Peter Martyr reported back that Columbus had brought back "peppers more pungent than that from Caucasus." By 1650, capsicum cultivation had spread to Africa, Asia, and Europe. Specialization led to paprika in Hungary and sweet peppers in Spain, moving us back into the realm of vegetables.

In 2000, the U.S. imported ~21,000 MT of capsicum pepper worth ~28 million dollars. And in 2000, the U.S. imported nearly 9000 MT of paprika worth nearly 18 million dollars and ~1500 MT of pimento worth ca. 5.5 million dollars (FAS, 2002).

CARAWAYS

Caraways (3446 tons worth $2.507 million) are ripe fruits of temperate *Carum carvi*, supposedly native to the Eurosiberian Center of Diversity.

Netherlands	2750 tons	$2,000,000
Hungary	255 tons	$200,000
Egypt	250 tons	$165,000
Poland	60 tons	$40,000
Total	**3500 tons**	**$2,500,000**

In 2000, the U.S. imported ~3300 MT of caraway worth nearly 3 million dollars (FAS, 2002).

CARDAMOMS

Cardamoms (164 tons worth $545 thousand) are the dry whole fruits or decorticated seed of tropical *Elettaria cardamomum,* assigned to the Indochina-Indonesia and Hindustani Centers of Diversity. Here's a spice that moved to America with a vengeance.

Guatemala	125 tons	$400,000
Costa Rica	20 tons	$50,000
India	20 tons	$90,000
Morocco	2 tons	$10,000
Total	**165 tons**	**$550,000**

Early reports of Ayurvedic medicine mention cardamoms for dysuria and obesity. It was already in Greek commerce. In the first century A.D., Rome imported cardamom from India. In Alexandria, taxes were levied on Indian cardamoms in A.D. 176. In Rosenthal's day, cardamoms were the third most expensive spice (then $6.00 per kilo, country of origin, now [1989] $3.23 per kilo f.o.b. NY) topped by saffron (then $225/kilo, now $100/kg) and vanilla (then $9.80/kg country of origin, now $41.65 f.o.b. NY).

And cardamom coffee, known as "gahwa," was a symbol of hospitality, served and received with ritual. You are supposed to drink at least three cups, audibly slurping, before any business transpires. Bedouins roast green coffee beans and crush them in a brass mortar and pestle. Then green cardamom pods are broken so that the seeds can be dropped into a pot of hot water, with a dash of saffron or cloves, sugar, and the ground coffee. Boil 2 to 3 minutes, strain, and serve. Poor Saudi's are said to prefer to be without rice than to be without cardamoms, perhaps because it is believed (1) to cool the body during extreme heat, (2) to help digestion, and, (3) to be aphrodisiac. Non-Arab Scandinavians are said to chew cardamoms after excessive consumption of alcohol, hoping to deceive the noses of their spouses.

In 2000, the U.S. imported ~325 MT of cardamoms worth nearly 3.5 million dollars (FAS, 2002).

CASSIA AND CINNAMON

Cassia and cinnamon (14,796 tons worth plus $37.289 million) are dried bark of tropical *Cinnamomum cassia*, assigned to the Indochina-Indonesia Center, and *C. verum,* assigned to the China-Japan Center. But most American cinnamon is cassia, and around 1992, they became aggregated with it in the FTEA statistics.

Indonesia	10,900 tons whole cassia	$29,000,000
China	1450 tons	$3,050,000
Sri Lanka	550 tons	$1,900,000
Madagascar	700 tons	$600,000
Indonesia	165 tons ground cassia	$425,000
Madagascar	55 tons	$55,000
Sri Lanka	50 tons	$90,000
Total	**15,000 tons**	**$37,000,000**

In Exodus 30:23–5, the Lord told Moses to use cinnamon and cassia et al., to anoint the tabernacle of the children of Israel. In 1990, the DOC and USDA aggregated cinnamon and cassia in their statistics, since most American cinnamon was in fact the related species cassia. Most of the "cinnamon" purchased in the U.S. is said to be "cassia," so perhaps I should talk about "cassia" buns, "cassia" toast, and "cassia" teas. The cinnamon toast my wife takes for upset distress and the cinnamon tea some people take for hangovers is more probably cassia. Both cassia and cinnamon contain carminative compounds.

In 1264 London, cassia (fit for commoners, usually cheaper than cinnamon, once fit for lords) sold for 10 shillings a pound, cf 12 shillings for sugar, 18 shillings for ginger, and only 2 shillings for more temperate cumin. In 1971, Rosenthal said the contemptuous "commoner" evaluation for cassia was no longer valid, but in the last year of separate record in FTEA documents (1988), cassia bark was ca. $2.00 a kilo f.o.b. NY, while cinnamon still commanded closer to $3.00.

Sadam Hussain may have burned a year's supply of (U.S.) oil during the Gulf War. In his grief over the loss of his wife, fiddling Nero is said to have burned a year's supply of cinnamon. France was receiving cinnamon as early as 761, to be assigned to various monasteries. Ninth century Swiss chefs used cinnamon cloves and pepper to season fish. Cinnamon played a big bad role in Sri Lanka's history. As the most sought after spice in fifteenth and sixteenth century explorations, it, with the black pepper, played a role in the colonization of Ceylon and the discovery of America. Portuguese colonialists forced Ceylonese to pay tribute with cinnamon bark in 1505 when they seized it.

In 2000, the U.S. imported ~15,000 MT of cinnamon/cassia worth ~16 million dollars (FAS, 2001).

CELERY SEED

Celery seed (2901 tons worth $2.211 million) is the fruit of temperate *Apium graveolens*, assigned to the Mediterranean Center of Diversity.

India	2800 tons	$2,100,000
China	65 tons	$40,000
Egypt	14 tons	$7,000
France	11 tons	$22,000
Total	**2900 tons**	**$2,200,000**

Wild celery was woven into some garlands found in Egyptian tombs of the twentieth dynasty. Romans and Greeks grew it more for food than medicine. It does have carminative, hypotensive, and sedative activities.

CLOVES

Cloves (1134 tons worth $2.328 million) are dried flower buds of tropical *Syzygium aromaticum*, assigned to the Indochina-Indonesia Center of Diversity.

Madagascar	380 tons	$750,000
Brazil	300 tons	$525,000
Indonesia	175 tons	$250,000
Comores	60 tons	$120,000
Total	**1150 tons**	**$2,500,000**

Native to the Spice Islands, cloves are mentioned in oriental literature of the Chinese Han period. Chinese courtiers, ca.. 2500 B.P., were said to hold a clove in their mouths when addressing an emperor. The clove's Chinese name meant chicken-tongue, while its French, Portuguese, and Spanish names implied nails. By A.D. 176, cloves were imported to Alexandria, and they were well known in Europe by the fourth century. The Portuguese controlled the Spice Islands from 1514 until 1605, when the Dutch expelled them. By 1651, the Dutch adopted strict measures to control their clove and nutmeg monopolies. Any person illegally growing or trading cloves was killed.

When a child was born in the Molucas, a clove tree was planted to keep a rough record of its age. Death of the tree was a bad omen. You can imagine what the new Dutch law requiring destruction of unauthorized clove trees did for the Moluccans. The French broke the Dutch monopoly by smuggling seeds and/or plants to some of the French colonies of Bourbon and Mauritius.

Indonesians invented their kreteks in the late nineteenth century, mixing two parts tobacco to one part ground cloves. Although Indonesia is still a producer today, most of its cloves are imported for the kreteks, unfortunately for the health of other nations. In Rosenthal's day, ca. half the worlds cloves went into Indonesia kreteks.

Containing the dental analgesic eugenol, cloves have quite a medicinal reputation. USDA's Richard Anderson (*Am. Health,* Nov. 1989, p. 96) reports that bayleaf, cinnamon, cloves, and turmeric all can treble insulin activity, hinting that as little as 500 mg might be enough to have some effect. A tea of 500 mg each of these spices, with coriander and cumin, should be enough to treble insulin activity, possibly helping in late-onset diabetes.

In 2000, the U.S. imported ~1000 MT of cloves worth more than 4 million dollars (FAS, 2002).

CORIANDER

Coriander (2418 tons worth $1.230 million) is the fruit of temperate *Coriandrum sativum*, assigned to the Mediterranean Center of Diversity.

Morocco	1175 tons	$630,000
Romania	400 tons	$175,000
Canada	250 tons	$150,000
Argentina	250 tons	$75,000
Total	**2500 tons**	**$1,250,000**

Both coriander and cumin have hypoglycemic activity in experimental animals.

In 2000, the U.S. imported ~4000 MT of coriander worth more than 2 million dollars (FAS, 2002).

CUMIN

Cumin (4707 tons worth $4.539 million) is the fruit of temperate or subtropical *Cuminum cyminum*, assigned to the Mediterranean and adjacent Centers of Diversity.

Turkey	3150 tons	$2,650,000
Pakistan	700 tons	$900,000
India	550 tons	$725,000
China	150 tons	$160,000
Total	**4700 tons**	**$4,500,000**

In 2000, the U.S. imported ~8000 MT of cumin worth ~14.5 million dollars (FAS, 2001).

DILL

Dill (615 tons worth $525,000), although referred to as dillweed (the herb) and dillseed (the fruit), is more appropriately called dillfruit and dillweed, from temperate *Anethum graveolens*, assigned to the Mediterranean and Hindustani Centers of Diversity.

India	475 tons	$350,000
Egypt	45 tons	$50,000
Pakistan	30 tons	$15,000
Sweden	25 tons	$60,000
Total	**625 tons**	**$550,000**

In 2000, the U.S. imported ~700 MT of dill worth more than 1 million dollars (FAS, 2002).

FENNEL

Fennel (2810 tons worth $2.964 million) is really the fruit of temperate *Foeniculum vulgare*, assigned to the Mediterranean Center of Diversity.

Egypt	1900 tons	$1,900,000
India	700 tons	$700,000
Turkey	70 tons	$60,000
Taiwan	50 tons	$150,000
Total	**2800 tons**	**$3,000,000**

In 2000, the U.S. imported ~3300 MT of fennel worth ca. 3.7 million dollars (FAS, 2002).

GARLIC

Garlic (3196 tons worth $3.917 million) is, in this case, the dehydrated bulb of temperate and subtropical *Allium sativum*, assigned to the CJ, CE, NE Centers of Diversity, but clearly in the Mediterranean in Biblical times.

For dehydrated garlic:

China	2750 tons	$3,350,000
Mexico	200 tons	$300,000
Guatemala	90 tons	$80,000
Hong Kong	80 tons	$100,000
Total	**3250 tons**	**$4,000,000**

GINGER

Ginger (5865 tons worth $6.643 million) is the root or rhizome of tropical *Zingiber officinale*, assigned to the Hindustani Center of Diversity. Latin Americans grow much of their own ginger, Indonesia, Thailand and Taiwan, being the major producers. Hawaii produced nearly 4000 tons in 1989, worth nearly 6 million dollars. Fiji was the major source of U.S. imports, followed by China and Brazil. For whole ginger, major sources are:

Fiji	1500 tons	$1,500,000
China	1100 tons	$1,100,000
Brazil	950 tons	$1,050,000
India	500 tons	$500,000
India also provided 280 tons ground ginger worth $100,000.		
Total Ginger	**6000 tons**	**$6,750,000**

Confucius, ca. 2500 years B.C., mentioned the chiang, still important in Chinese cookery and medicine. By the second century A.D., ginger was imported to Alexandria. It was mentioned in Anglo-Saxon leech books. Since live rhizomes were imported from the East Indies, it was logical that ginger would be one of the first post-Columbian introductions to the West Indies.

In 2000, the U.S. imported ~19,000 MT of ground ginger worth ~15 million dollars (FAS, 2001).

LAUREL

Laurel (1701 tons worth $3.061 million) is the bayleaf of Mediterranean *Laurus nobilis,* appropriately assigned to the Mediterranean Center of Diversity.

Spain	750 tons	$1,750,000
Turkey	650 tons	$875,000
Morocco	125 tons	$170,000
France	50 tons	$170,000
Total	**1750 tons**	**$3,000,000**

Mythology generated the Greek name for the plant Daphne. Apollo was said to have relentlessly pursued the unwilling nymph Daphne until merciful Gods turned her into a laurel tree. At the Olympic Games, founded ca. 2775 B.P., champions were crowned with laurels. Our *Baccalaureate* means nothing more than *laurel berries* (more appropriately *drupes*), rather suggestive of American sassafras drupes. Roman legionaires, in atonement, wiped blood from their swords with laurel leaves.

In 2000, the U.S. imported ~110 MT of bayleaf worth ca. 230 thousand dollars (FAS, 2002).

LICORICE

Licorice is the root of temperate species of the genus *Glycyrrhiza*, most often *Glycyrrhiza glabra*, assigned to the Mediterranean and Eurosiberian Centers of Diversity. Ninety percent of our imports go into flavoring tobacco, hence, I treat it as spice.

China	5000 tons	$3,000,000
Afghanistan	2650 tons	$1,150,000
Turkey	1120 tons	$900,000
Pakistan	425 tons	$250,000
Total (1988)	**9500 tons**	**$5,500,000**

Note how much bigger Afghanistan is today than Pakistan. In 1985–1986, their roles were reversed.

MACE

Mace (294 tons worth $1.863 million) is the aril of the seed of tropical *Myristica fragrans*, assigned to the Indochina-Indonesia Center of Diversity. Nutmeg, *q.v.*, is a different part of the same fruit.

Indonesia	175 tons	$1,500,000
Singapore	90 tons	$250,000
France	15 tons	$75,000
Netherlands	10 tons	$25,000
Total	**300 tons**	**$2,000,000**

Native to the Spice Islands, nutmeg reached Constantinople commerce ca. A.D. 540. Once again the Arabs interjected themselves and showed the origins in myth. Note the Dutch and French connection above. As with cloves, the Dutch attained the monopoly on nutmeg and mace shortly after driving off the Portuguese. Then, by subterfuge, French explorers arrived back in France in 1770 with a lot of seeds and seedlings of clove and nutmegs from the Spice Islands. Plants found their way to Bourbon, Cayenne, and Syechelles. Nutmegs were introduced to Zanzibar in 1818 from Mauritius or Reunion.

MARJORAM

Marjoram (1988 imports of 380 tons worth $481,500) is foliage of temperate *Origanum majorana*, assigned to the Mediterranean Center of Diversity. Data reporting was discontinued. 1988 imports were:

Egypt	350 tons	$420,000
France	20 tons	$30,000
Canada	12 tons	$12,000
Indonesia	4 tons	$3000
Total	**400 tons**	**$500,000**

Mint (Leaves)

Mint leaves (1988 imports of 171.8 tons worth ca. $580,000) are leaves of various temperate and subtropical *Mentha* hybrids, especially *Mentha x piperita*, assigned to the Mediterranean and Eurosiberian Centers of Diversity. (But, in 1989, the U.S. produced 3017 tons of peppermint oil worth $86 million and 1510 tons spearmint oil worth $25.6 million.) Data reporting has been discontinued on imported mint leaf following 1988, data for which follow:

Germany	75 tons	$375,000
Egypt	65 tons	$65,000
Turkey	20 tons	$25,000
Taiwan	6 tons	$100,000
Total	**175 tons**	**$600,000**

In 2000, the U.S. imported nearly 400 MT of mints worth more than 1 million dollars (FAS, 2002).

Mustard (Seed)

Mustard seed (53,479 tons worth $21.858 million in 1989) is the first true seed on the list, from temperate species of *Brassica* and *Sinapis*. The new "Canola" variety puts this Old World spice in the hands of Canada, as far as unprocessed seed are concerned.

Canada	46,500 tons	$12,500,000

But when it comes to prepared and/or ground mustard:

Canada	3425 tons	$4,800,000
France	2500 tons	$3,500,000
Germany	300 tons	$450,000
Thailand	130 tons	$150,000
Japan	50 tons	$275,000
Total	**53,000 tons**	**$21,500,000**

By 2000, whole mustard seed was the top-volume spice import to the U.S., at ca. 51,000 MT. Prepared mustard was imported at ~7000 MT worth nearly 9 million dollars (FAS, 2002).

Nutmeg

Nutmeg (1915 tons worth $11.073 million) is the seed of tropical *Myristica fragrans*, assigned to the Indochina-Indonesia Center of Diversity.

Indonesia	1600 tons	$9,500,000
L & W Islands	100 tons	$600,000
India	70 tons	$300,000
France	60 tons	$350,000
Total	**1900 tons**	**$11,000,000**

In 2000, the U.S. imported ~1.900 MT of nutmeg worth more than 12 million dollars, and nearly 200 tons of mace worth ca. 1.8 million dollars (FAS, 2002).

ONIONS (DEHYDRATED)

Onions (1371 tons worth $1.115 million) are the bulbs of temperate and subtropical *Allium cepa,* assigned to the Mediterranean and Central Asian Centers of Diversity.

Mexico	1200 tons	$850,000
Hungary	70 tons	$150,000
Germany	60 tons	$30,000
Yugoslavia	25 tons	$50,000
Total	**1375 tons**	**$1,115,000**

OREGANO

Oregano (4405 tons worth $6.545 million) is dried leaf of several mostly temperate species, not all in the same families. Most people associate oregano with *Origanum vulgare*, a temperate species assigned to the Eurosiberian Center of Diversity but clearly at home in a Mediterranean climate.

Turkey	2100 tons	$3,200,000
Mexico	1550 tons	$1,900,000
Greece	310 tons	$525,000
Israel	200 tons	$550,000
Total	**4400 tons**	**$6,500,000**

In 2000, the U.S. imported ~360 MT of oregano worth more than 900,000 dollars (FAS, 2002).

PARSLEY (ADVANCED)

Parsley (268 tons worth $561,000 in 1989) is the dried leaf of temperate *Petroselinum sativum,* assigned to the Mediterranean Center of Diversity. Separate statistics were abandoned for whole unground parsley following 1988 when imports were:

Mexico	750 tons	$550,000
Israel	75 tons	$200,000
Canada	25 tons	$20,000
Turkey	10 tons	$25,000
Total (1988)	**975 tons**	**$1,100,000**

Advanced parsley such as parsley flakes were in 1989.

Israel	140 tons	$340,000
Mexico	80 tons	$100,000
Germany	30 tons	$80,000
Brazil	10 tons	$15,000

In 2000, the U.S. imported ~1700 MT of parsley worth nearly 6 million dollars (FAS, 2002).

PEPPER (BLACK AND WHITE)

Pepper (37,753 tons worth $95.211 million) is the dried fruit of tropical *Piper nigrum*, assigned to the Hindustani Center of Diversity.

Black Pepper:

Indonesia	11,000 tons	$28,000,000
Brazil	11,000 tons	$21,500,000
Malaysia	6750 tons	$19,000,000
India	1275 tons	$3,700,000
Total	**32,000 tons**	**$77,000,000**

White Pepper:

Indonesia	5275 tons	$16,000,000
Singapore	90 tons	$325,000
Malaysia	60 tons	$200,000
China	40 tons	$100,000
Total	**5500 tons**	**$17,000,000**

Purseglove et al. (1981) note that pepper was one of the first oriental spices introduced to Europe. Theophrastus (ca. 2300 B.P.) alluded to black pepper and long pepper. At the time of Christ, pepper probably traveled from India through the Persian Gulf to Charax, or up through the Red Sea to Egypt, thence overland to Alexandria and the Mediterranean. By A.D., customs was levied on long pepper and white pepper but not black pepper. Hindu colonists took pepper to Java. In his 1298 memoirs, Marco Polo describes pepper cultivation in Java and mentions Chinese sailing vessels trading in pepper. By the Middle Ages, pepper was big in Europe, to preserve and season meats, and, with other spices, "to overcome the odours of bad food and unwashed humanity." Toward the end of the tenth century, England required Easterlings, early German spice traders in England, to pay tribute including 10 pounds of pepper for the privilege of trading with the Brits. Under Henry II, 1180, a pepperer's guild was founded in London. This gave way to the spicer's guild and finally, in 1429, the present Grocer' Company. The pepperers and spicers were the forerunners of the apothecaries, emphasizing the "vital role that spices formerly played in occidental medicine." There's a return to the spicerack for medicine, especially with capsicum, cloves, garlic, ginger, licorice, onions and turmeric.

In 2000, the U.S. imported 43,500 MT of black pepper worth ~205 million dollars and 7300 MT of white pepper worth ~37 million dollars (FAS, 2001).

POPPYSEED

Poppyseed (4160 tons worth $3.718 million) is in reality a seed, the same as the temperate opium poppy, *Papaver somniferum*, early moved about but assigned to the Mediterranean, Central Asian, and Near Eastern Centers of Diversity.

Netherlands	1650 tons	$1,450,000
Australia	1600 tons	$1,450,000
Spain	625 tons	$500,000
Turkey	250 tons	$200,000
Total	**4200 tons**	**$4,000,000**

Cretans used opium medicinally as early as 3400 B.P., and it was in use by early Egyptians and Sumerians, apparently. By the time of Mohammed (A.D. 570–632), its medicinal and narcotic properties were appreciated by the Arabians. Its narcotic usage moved to India, thence China. Antagonistic roles of British smugglers, and Chinese officials, trying to curb the scourge of millions, ended up in the Opium Wars of 1840 and 1855. Then morphine and heroin reciprocated within a century, causing addiction in thousands of Caucasians, to be supplanted, at least in part, in the 1980s by cocaine.

Poppyseeds were used for food and oil two millennia before Christ in Egypt, and the plant now grows from 55° N in Russia to 40° S in Argentina. Here I see Australia challenging another new Dutch monopoly.

In 2000, the U.S. imported ~5300 MT of poppyseed worth nearly 4 million dollars (FAS, 2002).

ROSEMARY

Rosemary (1988 imports of 810 tons worth $682,800) is the aromatic foliage of the climatically Mediterranean shrub, *Rosmarinus officinalis*, of the Mediterranean Center of Diversity. On the last year of record (1988), imports were:

Spain	450 tons	$400,000
France	175 tons	$125,000
Yugoslavia	90 tons	$70,000
Portugal	60 tons	$55,000
Total	**800 tons**	**$675,000**

In the first century A.D., Pliny assigned all sorts of medicinal claims to rosemary, home of more than a dozen antioxidant and chemopreventive compounds. It is one of the main herbs in the NCI Designer Food Program to prevent cancer, with capsicum, flaxseed, and licorice.

Already in eleventh century English herbals, rosemary had moved fast, without Columbus.

SAFFRON

Saffron (34 tons worth $3.286 million) represents the stigmata of the flowers of Mediterranean *Crocus sativus*, assigned to the Mediterranean and Near Eastern Centers of Diversity. Imports in 1989, the first year recorded separately by FAS, were:

Pacific Islands	15 tons	$14,000
China	10 tons	$7200
Spain	8 tons	$3,000,000
Total	**34 tons**	**$3,000,000**

Obviously, there is something wrong with the accounting above. Spain is a classic supplier, and I believe the price there is correct. If anyone is buying saffron at less than $1000 a ton as the figures above indicate, it probably isn't saffron, possibly azafran or turmeric. While Azafran is the Spanish name for turmeric in Latin America, where turmeric is common and saffron is not, za'faran was the arabic word for yellow. Traditionally, saffron has been the western food colorant corresponding to turmeric in the east. Columbus may have changed all that, if I can believe all the FTEA statistics.

In 2000, the U.S. imported ~14.3 MT of saffron worth nearly 6 million dollars (FAS, 2002).

SAGE

Sage (2044 tons worth $6.833 million) represents leaves of temperate *Salvia officinalis,* another climatically Mediterranean herb assigned to the Mediterranean Center of Diversity and secondarily to the Indochina-Indonesia Center.

Yugoslavia	1000 tons	$3,500,000
Albania	700 tons	$2,500,000
Turkey	150 tons	$250,000
Greece	50 tons	$125,000
Total	**2000 tons**	**$7,000,000**

In 2000, the U.S. imported ~2500 MT sage worth ca. 4.5 million dollars (FAS, 2002).

SAVORY

Savory der bohnenkraut represents foliage of summer savory, *Satureja hortensis*, or winter savory, *Satureja montana*, both assigned to the Mediterranean Center of Diversity. On the last year of record (1988), I imported only from:

Yugoslavia	145 tons	$55,000
France	50 tons	$50,000

SESAME

Sesame (40,514 tons worth 39.962 million) is truly a seed of subtropical *Sesamum indicum*, assigned to the African and Hindustani Centers of Diversity and early in the China-Japan Center.

Much of our imported sesame seed end up in the hamburger rolls.

Mexico	27,000 tons	$27,000,000
Guatemala	3800 tons	$4,000,000
Salvador	3000 tons	$3,000,000
India	2700 tons	$2,500,000
Total	**40,000 tons**	**$40,000,000**

Herodotus tells us that sesame saved several innocent boys from becoming eunuchs. Corinthians were sending 300 boys to Ayates, who were hustled into a temple en route by sympathetic citizens. Forbidden from entering the temple, the Corinthians cut off the boys food supplies. But their saviors saved them with sesame cakes.

By 2000, sesame imports were 49,000 MT worth nearly 55 million dollars.

TARRAGON

Tarragon is the foliage of the warm temperate herb, *Artemisia dracunculus*, assigned to the Central Asian Center of Diversity. On the last year of record (1988), major U.S. imports were:

France	30 tons	$350,000
Netherlands	11 tons	$50,000
New Zealand	7 tons	$200,000
Israel	6 tons	$25,000
Total	**60 tons**	**$625,000**

Thyme

Crude thyme was aggregated, strangely, with the laurel or bay leaf above for 1989. Processed thyme (71 tons worth $123,000) is the leaf of climatically Mediterranean *Thymus vulgaris*, appropriately referred to the Mediterranean Center of Diversity. On the last year of record (1988), crude thyme imports to the U.S. were:

Spain	850 tons	$1,800,000
Morocco	50 tons	$40,000
Jamaica	30 tons	$200,000
France	20 tons	$55,000
Total	**950 tons**	**$2,000,000**

Advanced Thyme (1989):

Jamaica	23 tons	$84,000
Morocco	17 tons	$12,000
Lebanon	16 tons	$10,000
Jordan	9 tons	$6000
Total	**70 tons**	**$125,000**

In 2000, the U.S. imported ~1900 MT of thyme worth only about 500,000 dollars (FAS, 2002).

Turmeric

Turmeric (2147 tons worth $1.807 million) is the rhizome of tropical *Curcuma longa,* assigned to the Hindustani Center of Diversity.

India	1900 tons	$1,600,000
China	70 tons	$35,000
Costa Rica	45 tons	$50,000
Peru	45 tons	$25,000
Total	**2150 tons**	**$1,800,000**

Some scholars suggest that use of turmeric in magical rites intended to produce fertility became so entrenched that turmeric moved with early Hindus to the "Hinduized kingdoms of Southeast Asia," quoted in Marco Polo records, China in 1280. Turmeric reached East Africa in the eighth century and West Africa in the thirteenth. It reached Jamaica in 1783.

In 2000, the U.S. imported ~2400 MT of turmeric worth nearly three million dollars (FAS, 2002).

Vanilla

Vanilla (1107 tons worth $46.125 million) beans aren't beans, although they are seed pods of tropical *Vanilla planifolia,* assigned to the South American Center of Diversity.

Indonesia	525 tons	$11,000,000
Malagasy	425 tons	$25,000,000
Comores	100 tons	$7,500,000
Mexico	15 tons	$850,000
Total	**1100 tons**	**$46,000,000**

Truly an American Rain Forest species, vanilla might be suggested as an Extractive Reserve candidate. Vanilla can be harvested renewably from tropical agroforestry scenarios, although heavy shade reduces yields. In spite of the published threats of biotechnology, artificial vanillin as a byproduct of the forest industry (could be a tropical forest as well as a temperate forest), and vanilla from tissue culture, the natural vanilla has not been supplanted for some usages.

Native American vanilla is much produced abroad now, Indonesia recently replacing Madagascar as the "largest supplier of the U.S. market" (Dull, 1990). Recent surveys suggest that, of an estimated 400 million gallons of vanilla (still our most popular flavor) ice cream, 20–25% is all natural, 40–50% is vanilla flavored, and 25–35% is artificially flavored. Ice-cream continues to be the largest use of natural vanilla, at slightly less than half the market. I have heard people speculate that synthetic will replace the natural. But Dull (1990) says, "The continuing trend toward natural flavorings in food products is keeping demand for vanilla beans steady, despite strong competition from synthetic flavorings like vanillin." I, too, have heard it said that, more and more, Americans are demanding naturals rather than synthetics—for flavors, food colorants, antioxidants and preservatives, extenders and thickeners, and sometimes even medicines and pesticides. If this trend continues, it bodes well for Extractive Reserves.

On the other hand, if the 1988–1989 trend continues for vanilla, things are not so cheerful; biotechnology may be taking its toll. Here are the tonnage and dollar figures (excluding Belgium/Luxemburg middle men):

	1988		1989	
	MT	$1,000	MT	$1,000
Comoros Islands	184.9	12,694	107.2	7502
French Polynesia	3.1	215	9.0	430
Indonesia	423.3	8282	526.5	11,073
Madagascar	576.5	39,841	420.7	24,652
Mexico	10.2	460	15.6	825
Other Pacific Isles	7.7	426	28.4	1,643
Totals	**1214.7**	**61,918**	**1107.4**	**46,125**

The 1107 tons is just about 10% off the 1988 figure of 1214 tons. I should watch the figures in 1991. Yokoyama et al. (1988) give figures for U.S. imports for 1981–1986:

Year	Imports MT	Value (C.I.F.) (= Cost, Insurance, and Freight) $ million
1981	642	32
1982	886	47
1983	979	53
1984	839	51
1985	745	49
1986	1003	60

In 2000, the U.S. imported ~1300 MT of vanilla worth nearly 44 million dollars (FAS, 2002).

Abbreviations

Full reference citations, in their proper alphabetical (by author) sequences, will be found under References. Many of our primary reference citations follow the consistent system (abbreviation, volume, page) format. These are more meaningful to us, the compilers, than the PMID abstract number, e.g., EB, or JE, or PR followed by a number then a colon then another number, always means *Economic Botany, Journal of Ethnopharmacology,* or *Journal of Phytotherapy Research* respectively, followed by the volume number:page number.

The major references in the body of this book are indicated by concise and consistent three-letter abbreviations. You will find the short explanation in the alphabetical sequence for the oft-used three-letter abbreviations for our major references under Reference Abbreviations. Many primary sources are often cited via the PMID index, which in this book is indicated by X followed directly by the PubMed serial number. Even for the $3000 worth of journals to which I subscribe, I can usually find the PubMed citation in the same week that the journal gets my citation.

Conventional abbreviations are here under Abbreviations. So there are three types of citations, compactly squeezed into the all important Activities and Indications paragraphs and generously sprinkled elsewhere.

AA arachidonic acid
ABS abstract
ACAT Acyl-CoA: cholesterol acyltrans-ferase
ACE angiotensin converting enzyme
AChE antiacetylcholinesterase
ADD attention deficit disorder
ADR adverse drug reaction
AFG in Afghanistan, as based on KAB
AGE aged garlic extract
AHH arylhydrocarbon hydroxylase
AHP American Herbal Products Association
AIL Duke's computerized AILS file, source of *The Green Pharmacy*, etc.; soon to be online
AITC allylisothiocyanate
ALA alpha-linolenic acid
AP-1 activation protein-1
AMP adenosine monophosphate
APA American Pharmaceutical Association
APB as-purchased basis
APP amyloid precursor
ARC Aloe Research Council
ATP adenosine triphosphate
B[a]P benzo[a]pyrene
BAL Baluchistan, as based on KAB
BCG Bacillus Calmette-Guerin

BHA Butylated hydroxyanisole
BHT Butylated hydroxytoluene
BP before present
BPC British Pharmacopoeia
BPH benign prostatic hypertrophy
BUN blood urea nitrogen
CAM cell adhesion molecule
cAMP cyclic adenosine monophosphate
CDC Center for Disease Control
cf compare with
CFS chronic fatigue syndrome
CHD coronary heart disease
chd child
ckn chicken
CNS central nervous system
COM commercial
COMT catechol-O-methyl-transferase
COPD chronic obsessive pulmonary disorder
CORP corporation
COX cyclooxygenase
COX-I cyclooxygenase inhibitor (sometimes COX-1 or COX-2)
CP cyclophosphamide
cv cultivar
CVI chronic venous insufficiency
DGL deglycyrrhizinated licorice
DHT dihydrotestosterone

DIM Dithymoquinone
DMBA 7,12-dimethylbenz[a]anthracene (a carcinogen)
dml dermal
DOC Department of Commerce
EBV Epstein-Barr virus
ED50 effective dose at which 50% of subjects are "cured," "effected," "affected," or "altered"
e.g. for example
EO essential oil
EPA eicosapentaenoic acid
EPO Evening Primrose oil
ERT estrogen replacement therapy
etc. et cetera
ETP etoposide
ext extract
f folklore, not yet substantiated
FDA Food and Drug Administration
FT fitoterapia
frg frog
g gram
GA glycyrrhetinic acid
GABA gamma-amino-butyric acid
GC *Garcinia cambogia*
GERD gastroesophageal reflux disease
GFG green farmacy garden
GI gastrointestinal
GLA gamma-linolenic acid
GMO genetically modified organism
gpg guinea pig
GST glutathione S-transferase
GTF glucosyl-transferase
h (as a score for an activity or indication) homeopathic
H2O2 hydrogen peroxide
HCA hydroxycitric acid
HCN hydrocyanic acid
HDR *Herbal Desk Reference*; online version under my Medical Botany Syllabus (MBS)
HFR human fatality reported
HLE human leukocyte elastase
HMG hydroxymethylglutarate
hmn human
hr hour
HRT hormone replacement therapy
iar intraarterial
IBD inflammatory bowel disease
IBS irritable bowel syndrome
IC inhibitory concentration

ICMR Indian Council of Medical Research
ID50 inhibitory dose at which 50% of activity is inhibited
IgE immunoglobulin-E
igs intragastric
ihl inhalation
IL interleukin
ims intramuscular
inc incorporated
IND intradermal
inf infusion
iNOS inducible Nitric Oxide Synthase
ipr intraperitoneal
ith intrathecal
ivn intravenous
LD50 lethal dose at which 50% of experimental population is killed
LDlo lowest reported lethal dose
lf leaf
ltr long terminal repeat
l liter
MAOI monoamine oxidase inhibitor
MDR multidrug resistant
mg milligram
MIC used differently by various sources; minimum inhibiting concentration or mean inhibiting concentration
mky monkey
ml milliliter
MLC mean or minimal lethal concentration
MLD used differently by various sources; Merck meaning minimum lethal dose; some other sources meaning mean lethal dose, and some do not define it (with apologies to the reader from the compiler)
mM millimole
MMP-9 matrix metalloproteinase-9
MUFA monounsaturated fatty acid
mus mouse
NCI National Cancer Institute
NF-κB nuclear factor-kappa B
NH3 ammonia
NIDDM noninsulin-dependent diabetes mellitus
NKC natural killer cell
NO nitric oxide
NSAID nonsteroidal anti-inflamatory drug
NWP Northwest Province or Pushtu (dialect at border of northwestern Afghanistan)

OCD obsessive compulsive disorder

ODC ornithine-decarboxylase

OL Oleoresin

OPC oligomeric procyanidin

ORAC oxygen radical absorbance capacity

orl oral

OTC over the counter (or approved for sale in Europe)

oz ounce

PA pyrrolizidine alkaloids

PAF platelet aggregating factor

par parenteral

pc personal communication

PEITC phenethylisothiocyanate

pers. comm. personal communication

PG prostaglandin

pgn pigeon

PKC protein kinase C

PMS premenstrual syndrome

pp pages

ppm parts per million

PSA prostate-specific antigen

PTK protein tyrosine kinase

rbt rabbit

RSV respiratory syncytial virus

RT reverse transcriptase

SAC S-allylcysteine

SAD seasonal affective disorder

SAM S-adenosylmethionine

scu subcutaneous

SF Stephen Foster

SGPT serum glutamic pyruvic transaminase

SL sesquiterpene lactones

SLE systemic lupus erythematosus

SN serial number (when followed by a number)

SOD superoxide dismutase

SSRI selective serotonin reuptake inhibitor

sup suppository

TAM traditional Ayurvedic medicine

tbsp tablespoon

TCM traditional Chinese medicine

THC tetrahydrocannabinol

TNF tumor necrosis factor

TPA 12-O-tetradecanoylphorbol-13-acetate

TQ Thymoquinone

tsp teaspoon

unk unknown

uns unspecified

USDA United States Department of Agriculture

UTI urinary tract infection

UV ultraviolet

VD venereal disease

VEGF vascular endothelial growth factor

VOD veno-occlusive disease

Vol volume

wmn woman

WPW Wolff-Parkinson-White (syndrome)

X solitary X in the title line of the herb following the scientific name means do not take it without advice from an expert (think of it as a skull and cross-bones)

X followed by serial number PMID (PubMed ID number)

XO external use only

ZMB zero moisture basis

μg microgram

μl microliter

μM micromole

A

Aframomum melegueta K. Schum. (Zingiberaceae)
ALLIGATOR PEPPER, GRAINS OF PARADISE, GUINEA GRAINS, MELEGUETA PEPPER

Synonym — *Amomum melegueta* Roscoe.

Medicinal Uses (Grains of Paradise) — Viewed as an African panacea (UPW), the plant was introduced to the West Indies, probably during the slave trade days. Newly captured slaves were so dependent on the spice that slave ships had to carry an ample supply. Crushed seeds are rubbed on the skin as a counterirritant (WO2). Plant decoction taken as febrifuge. Root decoction given for constipation. Root used to expel tapeworms. Used with *Piper* in treating gonorrhea (UPW). Juice from fresh leaves staunches bleeding (WO2). Africans speculate that it has more synergistic power if given as an enema (UPW). Abreu and Noronha (1997) remind us that the pungent (spicy) principles of *A. melegueta* have antifeedant, antischistosomal, antiseptic, antitermite, and molluscicidal properties.

Used for tumors in Ghana and Nigeria (JLH). Lee and Surh (1998) note that the pungent vanilloids, (6)-gingerol and (6)-paradol, can induce apoptosis. (6)-gingerol and (6)-paradol have antitumor and antiproliferative effects (Surh, 1999). (6)-paradol, a pungent zingiberaceous phenolic, is antiseptic and analgesic. It tends to slow promotion of skin carcinogenesis and topical application inhibited TPA-induced ear edema. It may induce apoptosis in cultured human promyelocytic leukemia (HL-60) cells. It decreased the incidence and the multiplicity of skin tumors initiated by 7,12-dimethylbenz[a]anthracene (DMBA) and promoted by 12-O-tetradecanoylphorbol-13-acetate (TPA). (6)-paradol and its derivatives possess cancer chemopreventive potential (Chung et al., 2001).

Working with pungent principles, Eldershaw et al. (1992) showed that some gingerols and shogaols are thermogenic. The gingerols showed more molar potency than shogaols. (6)-Shogaol inhibits carrageenin-induced swelling of hind paw in rats and arachidonic acid (AA)-induced platelet aggregation in rabbits. Moreover, (6)-shogaol prevented prostaglandin I2 (PGI2) release from the aorta of rats when tested as an inhibitor of platelet aggregation. (6)-Shogaol inhibits cyclooxygenase and 5-lipoxygenase (X3098654).

And some of the pungent compounds have analgesic effects. Onogi et al. (1992) studied anesthetic effects of (6)-shogaol, structurally resembling capsaicin, and with similar but weaker effects on substance P. At 30 μM, (6)-shogaol dose-dependently increased immunoreactive substance P. The maximum effect of (6)-shogaol was observed at 100 μM, still less than half the effect of 10 μM capsaicin. Systemic administration of (6)-shogaol (160 mg/kg) was anesthetic in rats, with peak effects in 15 and 30 min (80 mg/kg was ineffective) (X1282221).

Indications (Grains of Paradise) — Allergy (1; FNF); Asthma (1; UPW; FNF); Backache (f; UPW); Bite (f; UPW); Bleeding (f; WO2); Cancer (f; JLH; UPW); Childbirth (f; UPW); Climacteric (f; UPW); Colic (f; UPW); Constipation (f; WO2); Earache (1; UPW; FNF); Fever (1; WO2; FNF); Fracture (f; UPW); Gastrosis (f; UPW); Gonorrhea (f; UPW); Headache (1; UPW; FNF); Impotence (f; UPW); Infection (1; ABS); Pain (f; UPW); Schistosomiasis (1; ABS); Snakebite (f; UPW); Sore (f; UPW); Tapeworm (f; WO2); Toothache (f; UPW; WO2); VD (f; UPW); Worm (f; WO2); Wound (f; UPW); Yaws (f; UPW).

Grains of Paradise for asthma:

- Antiallergic: 6-shogaol; gingerol; shogaol
- Antibronchitic: borneol
- Antihistaminic: 6-shogaol; 8-gingerol; gingerol; shogaol
- Antioxidant: 6-shogaol; gingerol; zingerone
- Antiprostaglandin: 6-shogaol; gingerol
- Antispasmodic: 6-shogaol; borneol
- Cyclooxygenase-Inhibitor: gingerol; shogaol; zingerone

Grains of Paradise for fever:

- Analgesic: 6-shogaol; borneol; gingerol; shogaol
- Antibacterial: alpha-terpineol
- Antiinflammatory: borneol; gingerol; shogaol; zingerone
- Antioxidant: 6-shogaol; gingerol; zingerone
- Antipyretic: 6-shogaol; borneol; gingerol; shogaol
- Antiseptic: alpha-terpineol; gingerol; oxalic-acid; paradol; shogaol
- Antispasmodic: 6-shogaol; borneol
- Cyclooxygenase-Inhibitor: gingerol; shogaol; zingerone
- Sedative: 6-shogaol; alpha-terpineol; borneol; gingerol; shogaol

Other Uses (Grains of Paradise) — "Melegueta pepper is the only major spice that is native to Africa" (Bown, 2001). In the Middle Ages, these grains, originally imported through Italy from Africa, across the Sahara through Tripoli, and finally to Italy, were ranked in Europe right along with the other hot spices, cinnamon, cloves, and ginger. One or all together were used to flavor wines and sauces. Portuguese eventually evolved sea routes from Guinea. Used during medieval times, e.g., ancient Rome, to flavor the spiced wine "hippocras" with cinnamon and ginger (BOW). It was used as a pepper substitute, but that use was banned in Britain by George III as injurious to the health. Burkill (1985) tells us that, in Britain, their use to strengthen beer, spirits, and wines, was viewed as a bit mischievous, and it was finally forbidden by law (IHB). In Europe and America, the spice is rarely used anymore except in veterinary preparations and for flavoring certain liqueurs and vinegars (GEO). Still, it seems to be better known as a spice than as a medicine or food. The aromatic seeds are used as a spice and condiment, with beer, bread, cordials, liqueurs, meats, and wines, and used in preparing perfumes (FAC, HHB, WO2). American food processors allegedly use the grains in candy, ice cream, and soft drinks (FAC). Seeds indispensable to the Moroccan spice mix called "ras el hanout" (AAR). Seeds may be substituted for black pepper, pounded or ground in a pepper mill. During wars, this was a frequent substitute for black pepper (HHB). In Brazil, under the name "malegueta pepper," it is critical in their hot chile/bean dish, feifoada (AAR). West Africans eat the pulp around the immature seeds and chew the pulp and/or seed as a stimulant (WO2). Congolese use the seeds in magic, usually in sevens or multiples of seven, in armlets, bracelets, and magic fetishes (UPW). Seeds sometimes used as a fish poison. Roots have the flavor of cardamom.

For more information on activities, dosages, and contraindications, see the *CRC Handbook of Medicinal Herbs,* ed. 2, Duke et al., 2002.

Cultivation (Grains of Paradise) — Widely cultivated in Africa and elsewhere in the tropics, both in sun and partial shade.

Chemistry (Grains of Paradise) — Here are a few of the more notable chemicals found in this species. For a complete listing of the phytochemicals and their activities, see the CRC phytochemical compendium, Duke and duCellier, 1993 (DAD) and the USDA database http://www.ars-grin.gov/duke/.

Gingerol — Analgesic; Antiaggregant 0.5–20 μM, 10–100 μM; Antiallergic; Anticancer; Antiemetic; Antihepatotoxic; Antihistaminic; Antiinflammatory; Antioxidant; Antiprostaglandin 0.5–20 μM; Antipyretic; Antischistosomic 5 ppm; Antiseptic; Antithromboxane 0.5–20 μM; Cardiotonic 1–30 μM; Cholagogue; Cyclooxygenase-Inhibitor; Fungicide; Gastrostimulant; Hepatoprotective; Hypotensive; Inotropic 1–30 μM; Molluscicide 5 ppm, LD20 = 12.5 ppm; Mutagenic; Nematicide; Positive Inotropic 1–30 μg/ml; Schistosomicide EC100 = 10 ppm; Sedative; Thermogenic.

6-Gingerol — Amphitensive 0.5–1 mg/kg; Analgesic 1.7–3.5 mg/kg ivn, 140 mg/kg orl mus; Anesthetic 1.75–3.5 mg/kg ipr mus, 140 mg/kg orl mus; Anti-5-HT; Antiallergic; Antiemetic 25 mg/kg; Antiinflammatory; Antileukemic; Antioxidant; Antiproliferative; Antiprostaglandin IC50 = 4.6–5.5 μM; Antipyretic 1.7–3.5 mg/kg ivn, 140 mg/kg orl mus; Antiseratoninergic; Antitussive; Antiulcer IC58 = 150 mg/kg, IC54 = 100 mg/kg; Apoptotic; Cardiodepressant; Cardiotonic; Chemopreventive; Cholagogue; Cyclooxygenase-Inhibitor; Cytotoxic; Depressor; Hepatoprotective; Hypertensive 0.5–1 mg/kg; Hypotensive 1–100 μg/kg; Larvicide LD90 = 62 μg /ml; LD100 = 250 μg/ml; 5-Lipoxygenase-Inhibitor IC50 = 3 μM; Mutagenic; Nematicide LD90 = 62 μg/ml; Positive Inotropic 10 μg/ml; Sedative; Stomachic; Thermogenic; LD50 = 25.5 inv mus; LD50 = 58.1 ipr mus; LD50 = 250 orl mus.

8-Gingerol — Anti-5-HT; Antiaggregant 0.5–20 μM; Anticathartic; Antiemetic; Antihistaminic; Antiprostaglandin IC50 = 2.5–5 μM; Cardiotonic; Enteromotility-Enhancer 5 mg/kg; 5-Lipoxygenase-Inhibitor IC50 = 0.36 μM; Positive Inotropic 1 μg/ml.

6-Paradol — Analgesic; Antiinflammatory; Antileukemic; Antiproliferative; Antitumor (Skin); Apoptotic; Chemopreventive; Cytotoxic.

8-Paradol — Antialzheimeran; Antiarthritic; Anticancer; Antiinflammatory; Chemopreventive; COX-2-Inhibitor IC50 = 3.4 μM.

Shogaol — Analgesic; Antiallergic; Antiemetic; Antihepatotoxic; Antihistaminic; Antiinflammatory; Antipyretic; Antiseptic; Cyclooxygenase-Inhibitor; Fungicide; Gastrostimulant; Hypotensive; 5-Lipoxygenase-Inhibitor; Molluscicide LC100 = 50 ppm, LC20 = 12.5 ppm; Mutagenic; Nematicide; Sedative; Thermogenic.

6-Shogaol — Analgesic 1.7–3.5 mg/kg ivn, 140 mg/kg orl mus; Anesthetic 1.75–3.5 mg/kg ipr mus, 140 mg/kg orl mus; Antiaggregant; Anti-5-HT; Antiallergic; Anticathartic; Antiedemic; Antiemetic; Antihistaminic; Antihypothermic 10 mg/kg; Antioxidant; Antiprostaglandin IC50 = 1.6–2.3 μM; Antipyretic 1.7–3.5 mg/kg ivn, 140 mg/kg orl mus; Antispasmodic; Antitussive; Antiulcer IC70 = 150 mg/kg; Bradycardic; Cardiodepressant 10–100 μg/kg; Cardiotonic; CNS-Depressant; COX-2-Inhibitor IC50 = 2.1 μM; Cyclooxygenase-Inhibitor; Enteromotility-Enhancer 2.5 mg/kg; Gastrostimulant; Hepatoprotective; Hypertensive 0.5–1 mg/kg; Hypotensive 1–100 μg/kg; Larvicide LD90 = 62 μg/ml; Molluscicide; Mutagenic; Nematicide LD90 = 62 μg/ml, MLC = 0.1 mg/ml; Positive Inotropic; Pressor; Sedative; Sprout-Inhibitor; Stomachic; Sympathomimetic; Vasoconstrictor; LD50 = 25–51 ivn mus; LD50 = 109 ipr mus; LD50 = 250–687 orl mus.

Zingerone — Adrenergic; Anesthetic; Antiatherosclerotic IC50 = 8.9 μM; Antiinflammatory; Antimutagenic; Antioxidant (LDL) IC50 = 8.9 μM; Antipyretic; Cardiotonic; Catecholaminigenic; Cyclooxygenase-Inhibitor; Hypotensive; Irritant; 5-Lipoxygenase-Inhibitor; Paralytic; Secretagogue; Tachyphylactic; Vasodilator; LD50 = 2560 orl rat.

A

Aframomum sceptrum (Oliv. and D. Hanb.) K. Schum.
(Zingiberaceae)
BLACK AMOMUM, GUINEA GRAINS

Medicinal Uses (Guinea Grains) — Seeds an ingredient in medicine for internal uses, for dysentery and gastrointestinal ailments. Seed is laxative. Seeds are chewed like kola, having a numbing effect, hence used for toothaches. They are used to disguise poison and drugs administered furtively, as aphrodisiacs and abortifacients. They are used as fish poison. For external use, seeds are crushed and rubbed on the body as a counterirritant, or applied as a paste for headache, earache, or pulverized and put on wounds and sores. Fruit chewed as a stimulant (CFR). Leaves are crushed with lime for spots of small-pox (Burkill, 2000). Juice of young leaves is styptic (CFR). Oil is a stimulating carminative. Roots are taken for constipation and as a vermifuge for tapeworm. In Mali, roots are taken for blennorrhagia and intestinal worms, whether dried and mixed with salt, ginger, or pine nuts, or crushed in cold water overnight and drunk in the morning on an empty stomach (Burkill, 2000).

Indications (Guinea Grains) — Blenorrhagia (f; IHB); Constipation (f; IHB); Earache (f; CFR); Headache (f; CFR); Smallpox (f; IHB); Toothache (f; CFR); Worm (f; IHB); Wound (f; CFR).

Other Uses (Guinea Grains) — Seeds are pungent and used for flavoring cordials and liquors, and in veterinary medicine. Seeds are also used in foods. Seeds are camphoraceous in taste (Burkill, 2000). Fruit-pulp around seeds is eaten, especially before maturity. Seeds yield an essential oil (EO) whose pungency is due to paradol, a yellow oily substance similar to gingerol, present in the fatty oil of the seeds. Oil has been used as a spice (CFR). Leaves are cooked in food as a flavoring. The Igala of Nigeria make costumes for masquerades from the leaves. Root is fibrous and used in SE Nigeria for weaving.

For more information on activities, dosages, and contraindications, see the *CRC Handbook of Medicinal Herbs* ed. 2, Duke et al., 2002.

Cultivation (Guinea Grains) — In most areas of tropical Africa, most fruits are gathered from wild plants growing in temporary clearings in the forest. Plants often grown in home or local gardens in the tropical forests, usually propagated by divisions of the rhizomes. Plants are easily grown under the forest conditions and do not have any special complications in cultivation. Methods of cultivation are similar to those used for ginger and cardamoms. Fruits and their seeds are harvested when ripe, mainly from plants in the forests, or from plants planted at edge of the forest. Seeds are removed from the fruit and dried until needed. Often, fruits are picked whenever they are desired as a chew, and there is no real time for harvest. Plants flower and fruit year round (CFR).

Chemistry (Guinea Grains) — Since I have found no chemical data, one could assume that it may share many of the same chemicals as *Aframomum melegueta*.

Allium cepa L. (Liliaceae)
BERMUDA ONION, ONION, SHALLOT

A

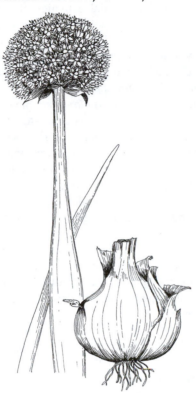

Medicinal Uses (Onion) — I think the whole chive/garlic/leek/onion/ramp/scallion alliance is as important for medicine as for food. Perhaps my near addiction to green beans and onions has spared me the adult onset diabetes. Oral extracts of onion and onion juice seem to lower blood sugar. The scales outside the onion bulb are one of the richer sources of quercetin, a very useful phytochemical also shared with evening primrose. Scales contain a heart stimulant that increases pulse volume, affects the uterus, promotes bile production, and reduces blood sugar. Onion bulbs are said to be aphrodisiac, diuretic, expectorant, emmenagogue, hypoglycemic, and stimulant (MPI). Onion juice and EO demonstrated antiaggregant and hypocholesterolemic activities in humans subjects (MPI). Onions are alleged to stimulate bile production, to speed healing of gunshot wounds, and to cure scorpion bites, freckles, and the common cold. Asian Indians eat raw onions, spiced up with lemon, pepper, and salt, for bronchitis, colic, edema, fever, and scurvy (Libster, 2002). Some people with parasites macerate an onion in white wine and drink it on an empty stomach in the morning. Or pediatric patients drink water in which onion has steeped overnight to kill parasites (Libster, 2002). Cooked onions are consumed by Japanese macrobioticists to calm the nervous system and alleviate irritability and sore muscles after heavy labor. A cut raw onion is placed under the pillow to aid insomnia (Libster, 2002). Reputed to be hypotensive, onions have recently been shown to contain the antihypertensive agent prostaglandin A_1, but only at ca. 1 ppm. With prostaglandins like this going at $10 per milligram, that means that a kilogram of onions (now costing about $1) contains $10 worth of prostaglandin. Add that to your chicken soup. Juice of the bulb is used for coughs and earache. Macerated in gin, the bulbs are used for dropsy and gravel (GMH). In India, onions are believed to be aphrodisiac, especially if retained in a cow dung year in a well-stoppered pot for four months (DEP). Even wilder, in an Indian formula for acute dysentery, one buries a grain of opium in an onion bulb and then roasts the onion (DEP). Most of the real and folk medicinal attributes of onion are shared with garlic and other lesser known members of the genus *Allium*. Garlic is popular with organic gardeners and naturopaths for its biological activities.

Libster (2002) comments on the bactericidal activities of onion, a close second to garlic, probably equipotent on a ZMB. Onion is indicated for its antibacterial actions, used for millennia for coughs and colds, whether bacterial or viral. Fresh onion extracts (not steamed or exposed to 37°C two days after grating) slow *Prevotella intermedia, Porphyromonas gingivalis, Streptococcus mutans,* and *S. sobrinus,* all implicated in tooth decay. Gram(+) bacteria are more sensitive than gram(–), at least to Egyptian onion oil. *In vitro,* onion oil significantly inhibited *Bacillus anthracis, B. cereus, Micrococcus luteus,* and *Staphylococcus aureus* among gram(+) and gram(–) *Escherichia coli* and *Klebsiella pneuomoniae.* But Libster reports other studies showing no bactericidal effect on *Moraxella catarrhalkis* and *Streptococcus pneumoniae* (Libster, 2002). Freshly expressed juice is said to be moderately bactericidal. Chen et al. (1985) showed onion was weaker as an antibacterial than garlic. Onion inhibited *E. coli, P. vulgaris, S. faecalis,* and *B. cereus* (X4064797). Onion proved antiseptic to *Bacillus subtilis, Escherichia coli,* and *Saccharomyces cerevisiae,* confirming traditional uses of spices as food preservatives, disinfectants, and antiseptics (De et al., 1999). The antibacterial activities of green onion are slightly weaker than regular onion. However, green onion inhibited *P. aeruginosa* and *M. luteus.*

Summarizing clinical applications of ayurvedic herbs, Khan and Balick (2001) note human studies on onion for colic, hyperlipidemia, scarring, and stomach cancer. They adduced evidence that onions fed to breast-feeding mothers increased colic symptoms in breast-feeding infants, "Garlic and onion can be ingested frequently in low doses without any side effects, and can still produce a significant antithrombic effect."

Antidiabetic (blood sugar regulating) properties of S-methyl-cysteine sulfoxide (200 mg/kg for 45 days in alloxan-diabetic rats) isolated from onion were comparable to those of glibenclamide and insulin (Khan and Balick, 2001). Al-Bekairi et al. (1985) fed aqueous extracts of onion to mice for three months (100 mg/kg added to drinking water) and found no estrogenic, antiestrogenic, or spermatotoxic activity. They did suggest aphrodisiac activity increased weight of testes and epididymus, and produced a significant increase in sperm count.

In human studies, onion and garlic significantly prevented fat-induced (dietary) increases in serum cholesterol and plasma fibrinogen. Briggs et al. (2001) showed that raw onion inhibited platelet-mediated thrombosis in dogs. A number of other studies suggest that dietary onion is good for the heart. Onion juice inhibits human platelet aggregation *in vitro.* Six dogs given 2.0 g/kg raw onion homogenate intragastrically showed significant antiaggregant activity, suggesting that raw onion may help prevent platelet-mediated cardiovascular disorders (X11584080).

Folklorically, roasted onion is applied to tumors. The seed, with honey, is a folk remedy for warts. An injection of the juice is said to help tumors of the breast and rectum, among others. The bulb, prepared in various manners, is said to help indurated glands and tumors. While garlic, at least on an as-purchased basis, is more often mentioned for cancer than onion, the onion is the better source of quercetin. And I can handle 10–20 g onion much more readily than a clove of raw garlic. Onion is by far my richest source of protocatechuic acid (PA) (0.45–1.7%) on a dry weight basis, if my sources are correct. At levels 1/1000th of the toxic dose, PA is a good chemopreventive, at least in mouse skin. Khan and Balick (2001) noted an inverse correlation between dietary onions and stomach cancer — antioxidants can be a doubled-edged sword for cancer control. As a silver bullet, isolated and concentrated phytochemical, PA can be a bad rather than a good guy, with allergenic, carcinogenic, glutathione-depleting, hepatotoxic, and nephrotoxic potential (JAF49:5674). It is my personal opinion that dietary onion is more liable to prevent than to cause cancer. According to Libster (2002), cooking onion with hamburger is supposed to reduce the carcinogenicity of the charbroil. Sugars released in cooking (and, to a lesser degree, quercetin) reduce the mutagenicity of ground beef (Libster, 2002). Dietary nutrients are almost always good guys when consumed in their dietary (evolutionary) context; it's when they are isolated and put in capsules that it becomes possible to overdose and overwhelm your homeostasis.

A

On December 10, 2001, the news broke on aromatase inhibitors for the prevention and containment of breast cancers. There was, expectedly, no mention of the natural aromatase inhibitors. Fiorellia et al. (1999) note that cells of colorectal adenocarcinoma also express aromatase, which was inhibited by 1 μM quercetin (abundant in evening primrose and onion), as by the synthetics tamoxifen (100 nM) and raloxifene (10 nM). That tells me that on a molar basis, quercetin was one tenth as potent as tamoxifen. Saarinen et al. (2001), aware of the *in vitro* antiaromatase activity, were unable to find such activity *in vivo*. In rats given doses of 50 mg/kg, neither apigenin, chrysin, nor naringenin inhibited aromatase. They suspect that this lack of inhibition may have been due to relatively poor absorption and/or bioavailability of the phytochemicals.

In a University of Hawaii study of 582 patients, those who ate the most onion had 50% less lung cancer, especially squamous cell carcinoma (cf 40% for apples). Boik (2001) cautiously praises the flavonoids, more specifically flavonols, quercetin, and kaempferol, if not onion, as a cancer preventive. Most of nearly 75 studies on the flavonol quercetin suggest that it inhibits proliferation of a wide array of tumors, usually at levels of 1–50 μM. Boik (2001) reports only four animal studies of quercetin, in two of which ipr quercetin inhibited leukemia, lymphoma, metastasis of melanoma, and reduction of human head and neck squamous cell carcinoma. The animal studies lead Boik to suggest human dosages of 1,200 to 4,000 mg/man/day. In the single human study, there was evidence of cancer inhibition. Quercetin may have some estrogenic activity, but only 10% of the estrogenic activity of genistein, which "appears capable of producing an estrogenic effect and stimulating cancer growth." Luteolin seems ca. 58% as estrogenic, resveratrol 22%, and apigenin ca. 16% as estrogenic. The flavones are more potent antioxidants than vitamin C, apigenin ca. 1.5 times more potent, resveratrol 2 times, with quercetin 4.7 times more potent, almost as potent as the oligomeric procyanidins (OPCs) at nearly 5 times as potent. In scavenging free radicals, these compounds can also prevent cancer. At moderate doses, *in vivo*, it's likely that quercetin and the other flavonoids serve as antioxidant rather than prooxidants. However, there are conditions under which quercetin can function as a mutagen and/or prooxidant. Boik (and/or my database) lists several mechanisms by which onion's quercetin might prevent cancer. So we have one flavonol with more than a dozen different activities that could reduce the incidence of cancer, and we have at least a half dozen other flavonoids doing many of the same things in slightly different levels and ways. Perhaps that enables Boik to predict synergies among these food farmaceuticals. Boldly, he calculates some tentative human dosages as 1.2–4.9 g/day (as scaled from animal antitumor studies), 1–4 g/day (as scaled from animal antiinflammatory studies), leading to a target dose of 3800 mg/day quercetin. Then, more speculatively, he suggests that synergies with other phytochemicals may reduce that minimum antitumor dose to 250 mg quercetin/day, paralleling his minimum anticancer dose 100 mg for genistein (e.g., in beans), and 170 mg luteolin through similar calculations. Looks like good reasoning for curried bean/celery soup with plenty of onion for its quercetin, and garlic for its DADS, etc. If you find this paragraph rough sledding, may I refer you to Boik's very interesting book (BO2).

You may find equally rough sledding on 450 induction, a detoxication mechanism for poisons, both natural and pharmaceutical, and p450 inhibition, a mechanism that potentiates some poisons, natural and pharmaceutical. More confusing, quercetin is both an inducer and inhibitor of p450 enzymes, reportedly inhibiting at IC50 = 50–100 μM, inducing at 5 μM/rat ipr. For more information, you should consult Dr. Flockhart's database at Georgetown (http://www.dml.georgetown.edu/depts/pharmacology/davetab.html).

Smoother sailing, I hope, with *The Color Code* (Joseph et al., 2002), where we read that all onions — purple, red, white, or yellow — contain the organosulfur phytocomedicinals (the hotter the onion, the more organosulfur). *The Color Code* stresses the red onions that contain anthocyanins, organosulfur compounds, and the quercetin praised above. More importantly, what *The Color Code* says moves my favorite antidiabetic string beans with raw onions into the lung cancer and osteoporosis column. With male rats fed 1 g powdered onion/day for a month, mineral content of bone increased >17% and bone thickness by 15%. Removing the ovaries from female rats resulted in

A

32% bone loss. But feeding the ovariectomized rats powdered onion, they lost 25% less bone. "Onion therapy was slightly more effective than the osteoporosis drug calcitonin" (JNU).

Allow me to get speculative; your body coevolved with phytochemicals like quercetin. Your genes have known these phytochemicals more than 5 million years. In those years, your body has evolved homeostatic mechanisms for keeping wholesome phytochemicals within bounds, grabbing a phytochemical when the body needs it, excluding a phytochemical, often as "expensive urine," when it is unneeded. It is the unnatural pharmaceutical (which may be inhibited or potentiated by hundreds of phytochemicals in our food chain) that is dangerous, not the natural phytochemical. In the opening years of the new millennium, the press, echoing biased interests, is frightening you from the safer foods and herbs, and into the waiting arms of the more dangerous pharmaceuticals. They are trying to sell you a pill a day for life instead of trying to make you well. Your body knows better than your physician, or pharmacist, or herbalist, which phytochemicals your body needs. Feed it a wide variety of plant foods and it will select from that phytochemical menu. Feed it a solitary pharmaceutical, and there is no menu — just the solitary silver bullet that may or may not be what the body needs. Feed it a fruit or vegetable, and it has hundreds of useful phytochemicals from which to choose. Can you body do this? Feed it, seven each, wholesome types of beans, fruits, grains, herbs, nuts, spices, and vegetables a day, and it has 49 menus from which to seek a remedy, homeostatically.

Indications (Onion) — Adenosis (f; JLH); Acne (f; JFM; WHO); Allergy (1; BRU; FNF); Ameba (1; X10594976); Anaphylaxis (1; X3932203); Angina (f; BGB; PHR); Anorexia (2; BGB; KOM; PH2; WHO); Apoplexy (f; DEP); Asthma (1; APA; BRU; JFM; PHR; PH2); Atherosclerosis (2; APA; FNF; JFM; KOM; PH2; SHT; WO2); Bacillus (1; X4064797); Bacteria (1; FNF; JFM; PH2); Bite (f; DEP); Biliousness (f; KAB); Bleeding (f; KAB); Blister (1; SKJ); Boil (1; SKJ); Bronchosis (2; BGB; PHR; PH2; WHO); Bruise (f; PHR; WHO); Bug Bite (f; PHR); Burn (f; JLH; PHR); Cancer (1; APA; FNF); Cancer, breast (1; FNF; JLH; JNU); Cancer, colon (1; JNU); Cancer, esophagus (1; JNU); Cancer, gland (1; FNF; JLH); Cancer, liver (1; FNF; JLH); Cancer, lung (1; JNU); Cancer, rectum (1; FNF; JLH); Cancer, stomach (1; APA; BRU; FNF; JLH; JAC7:405); Cancer, uterus (1; FNF; JLH); Candida (1; X10594976); Carbuncle (f; KAB); Cardiopathy (1; APA; JNU); Caries (1; X9354029); Catarrh (f; KAB); Chest Cold (f; JFM); Chilblain (f; KAP); Cholecystosis (f; JFM; PHR); Cholera (f; DEP; WHO); Cold (2; DEM; FNF; PHR; PH2); Colic (f; KAP; PHR; PH2; WHO); Colosis (f; KAP); Congestion (1; APA; BGB; JFM); Convulsion (f; KAB; LIB); Corn (f; JLH; LIB); Cough (2; BGB; JFM; PHR; PH2); Cramp (f; GHA); Deafness (f; GHA; JFM); Dermatosis (1; X7600010; SKJ); Diabetes (2; APA; FNF; WHO); Dropsy (f; BGB; DAD; GMH); Dysentery (f; BGB; DAD; JNU); Dysgeuzia (f; KAB); Dysmenorrhea (f; PHR); Dyspepsia (2; JFM; PHR; PH2); Dyspnea (f; BGB); Dysuria (f; KAB); Earache (1; APA; DEM; DEP); Edema (f; JFM; LIB); Epilepsy (f; JFM); Epistaxis (f; KAB; LIB); Escherichia (1; PH2; X4064797); Felon (f; JLH); Fever (2; DEM; PHR; PH2; WHO); Flu (1; DEM; FNF); Fungus (1; X10594976); Furuncle (f; PHR); Gas (f; DAD; JFM; SKJ); Gastrosis (f; GHA); Giardia (1; X10594976); Gingivosis (1; X9354029); Gravel (f; BGB; DAD; GMH); Headache (f; LIB); Hemorrhoid (f; LIB); Hepatosis (f; KAB); High Blood Pressure (2; FNF; PHR; PH2; SHT; WHO); High Cholesterol (2; APA; FNF; SHT; WHO); High Triglycerides (1; FNF; WHO); Hyperglycemia (1; FNF); Hyperlipidemia (f; SKJ); Hysteria (f; DEP; KAB); Immunodepression (2; FNF; PHR); Induration (f; JLH); Infection (2; FNF; JNU; PHR; PH2); Inflammation (2; FNF; PHR); Insomnia (f; KAB); Jaundice (f; LIB; WHO); Lymphangites (f; KAB); Malaria (f; DEP; JFM); Mange (f; JFM); Migraine (f; KAB); Mycosis (1; X10548758); Nephrosis (f; BGB; GHA); Neuralgia (f; JFM); NIDDM (2; WHO); Nyctalopia (f; KAB); Obesity (1; BGB; LIB; SKJ); Odontosis (f; KAB); Ophthalmia (f; SKJ); Osteoporosis (1; ACT5:330; JNU); Otosis (f; SKJ; WHO); Pain (1; FNF; JFM); Parasite (1; JFM; X10594976); Periodontosis (1; X9354029); Pertussis (f; PHR); Pharyngosis (2; PHR); Phthisis (f; DEP); Proctosis (f; KAB); Prolapse (f; KAB); Protozoa (1; X10594976); Rabies (f; KAB); Rheumatism (f; JFM); Salmonella (1; PH2); Scabies (f; JFM; LIB); Sinusosis (f; LIB); Sore (1; JNU;

WHO); Sore Throat (f; DEP; LIB); Splenosis (f; DEP; LIB); Sting (f; JFM); Stomachache (f; PHR); Stomatosis (2; PHR); Strangury (f; KAP); Streptococcus (1; X9354029; X4064797); Sunstroke (f; DEP); Swelling (f; JFM); Syncope (f; DEP; KAB); Tenesmus (f; BGB); Thrombosis (f; JFM); Tinnitus (f; JFM); Tonsilosis (f; JFM); Toothache (f; JNU); Tuberculosis (f; JFM; LIB); VD (f; LIB); Vertigo (f; KAB); Virus (1; X10594976); Wart (f; PHR); Whitlow (f; JLH); Worm (f; JFM); Wound (f; PHR); Yeast (1; WHO; X10594976).

Onion for colds:

- Analgesic: adenosine; caffeic-acid; ferulic-acid; p-cymene; quercetin
- Anesthetic: ethylene
- Antiallergic: cis-methylsulphinothioic-acid-s-1-propenylester; cis-n-propylsulphi-nothioic-acid-s-1-propenylester; ferulic-acid; kaempferol; n-propylsulphi-nothioic-acid-s-n-propylester; oleanolic-acid; quercetin; trans-5-ethyl-4,6,7-trithia- 2-decene-4-s-oxide; trans-methylsulphinothioic-acid-s-1-propenylester; trans-n-propylsulphi-nothioic-acid-s-1-propenylester; trans-trans-5-ethyl-4,6,7-trithia-2,8-decadiene-4-s-oxide; tuliposide-a
- Antibacterial: acetic-acid; allicin; alliin; benzyl-isothiocyanate; caffeic-acid; cycloartenol; cycloeucalenol; diallyl-disulfide; diallyl-sulfide; diallyl-trisulfide; endol-ysin; ferulic-acid; isoquercitrin; isorhamnetin; kaempferol; muramidase; oleanolic-acid; p-coumaric-acid; p-cymene; protocatechuic-acid; quercetin; rutin; silver; sinapic-acid; vanillic-acid
- Antiflu: allicin; caffeic-acid; diallyl-trisulfide; p-cymene; quercetin
- Antihistaminic: caffeic-acid; isorhamnetin; kaempferol; quercetin; rutin
- Antiinflammatory: alpha-amyrin; alpha-linolenic-acid; caffeic-acid; cycloartenol; feru-lic-acid; isorhamnetin; kaempferol; oleanolic-acid; protocatechuic-acid; quercetin; quer-cetin-3-o-beta-d-glucoside; rutin; salicylates; vanillic-acid
- Antioxidant: allicin; alliin; caffeic-acid; catechol; diallyl-trisulfide; ferulic-acid; fumaric-acid; glutathione; isoquercitrin; isorhamnetin; kaempferol; lutein; melatonin; oleanolic-acid; p-coumaric-acid; protocatechuic-acid; pyrocatechol; quercetin; rutin; sinapic-acid; spiraeoside; vanillic-acid
- Antipharyngitic: quercetin
- Antipyretic: salicylates
- Antiseptic: 5-hexyl-cyclopenta-1,3-dione; 5-octyl-cyclopenta-1,3-dione; allicin; alliin; benzyl-isothiocyanate; caffeic-acid; catechol; diallyl-sulfide; diallyl-trisulfide; kaempferol; oleanolic-acid; oxalic-acid; phloroglucinol; pyrocatechol; trigonelline
- Antitussive: protocatechuic-acid
- Antiviral: allicin; benzyl-isothiocyanate; caffeic-acid; catechol; diallyl-disulfide; dial-lyl-trisulfide; ferulic-acid; kaempferol; oleanolic-acid; p-cymene; protocatechuic-acid; quercetin; rutin
- COX-2-Inhibitor: kaempferol; melatonin; oleanolic-acid
- Cyclooxygenase-Inhibitor: allicin; cis-methylsulphinothioic-acid-s-1-propenylester; cis-n-propylsulphinothioic-acid-s-1-propenylester; kaempferol; melatonin; n-propylsulphi-nothioic-acid-s-n-propylester; oleanolic-acid; quercetin; trans-5-ethyl-4,6,7-trithia- 2-decene-4-s-oxide; trans-methylsulphinothioic-acid-s-1-propenylester; trans-n-propylsulphi-nothioic-acid-s-1-propenylester; trans-trans-5-ethyl-4,6,7-trithia-2,8-decadiene-4-s-oxide
- Expectorant: acetic-acid
- Immunostimulant: allicin; alliin; alpha-linolenic-acid; caffeic-acid; diallyl-disulfide; fer-ulic-acid; melatonin; protocatechuic-acid
- Phagocytotic: allicin; alliin; ferulic-acid; oleanolic-acid; protocatechuic-acid
- Viristat: benzyl-isothiocyanate

A

Onion for diabetes:

- Aldose-Reductase-Inhibitor: caffeic-acid; isoquercitrin; kaempferol; p-coumaric-acid; quercetin; rutin; spiraeoside; vanillic-acid
- Antiaggregant: adenosine; allicin; alliin; alpha-linolenic-acid; caffeic-acid; cycloalliin; ferulic-acid; kaempferol; melatonin; quercetin; salicylates
- Anticapillary-Fragility: quercetin; rutin
- Antidiabetic: allicin; alliin; diphenylamine; quercetin; rutin; s-methyl-cysteine-sulfoxide
- Antioxidant: allicin; alliin; caffeic-acid; catechol; diallyl-trisulfide; ferulic-acid; fumaric-acid; glutathione; isoquercitrin; isorhamnetin; kaempferol; lutein; melatonin; oleanolic-acid; p-coumaric-acid; protocatechuic-acid; pyrocatechol; quercetin; rutin; sinapic-acid; spiraeoside; vanillic-acid
- Antiperoxidant: caffeic-acid; p-coumaric-acid; protocatechuic-acid; quercetin
- Antiradicular: allicin; alliin; caffeic-acid; isoquercitrin; kaempferol; lutein; melatonin; protocatechuic-acid; quercetin; rutin; vanillic-acid
- Antithrombic: ferulic-acid; quercetin
- Hypocholesterolemic: 24-methylene-cycloartanol; adenosine; allicin; alliin; caffeic-acid; cycloartenol; diallyl-disulfide; diallyl-sulfide; diallyl-trisulfide; melatonin; rutin; trigonelline
- Hypoglycemic: adenosine; allicin; alliin; diallyl-disulfide; diallyl-trisulfide; diphenylamine; quercetin; salicylates; trigonelline
- Insulin-Sparing: allicin
- Insulinic: adenosine
- Insulinogenic: quercetin

Other Uses (Onion) — For millennia, onions have been famous for food, condiments, and medicine. Green onions are eaten raw with meats, fish, cheese, or as a vegetable, or chopped and added to cottage cheese, or cooked. Onions are eaten raw, boiled, baked, creamed, broiled, fried, french-fried, roasted, or pickled, and in soups, stews, dressings, or salads, but perhaps more importantly, added to other ingredients for innumerable dishes. Dry onions may be served as a vegetable dish or to flavor meat, fish, and poultry dishes and are also used in salt substitutes such as Spike, Mrs. Dash, and Vegit. A thick layer of cooked onion is used on the French dish pissaladière, sometimes called "Provençal pizza" (FAC). Onions are used in the Catalan sauces sofregit and samfaina (FAC). In Tunisia, a fermented onion paste called "hrous" is used to flavor couscous, soups, and stews (FAC). The papery outer skins, called "shuski" in slavic Macedonia, are used as a dye for coloring Easter eggs, and in Egypt they are used to color and flavor eggs called "hamine" (FAC). Shallots produce a dark, rich-hued dye (FAC). Onion skins are used as a smoke flavoring (FAC). The leaves of some cultivars are widely used as scallions. In Catalonia, the large shoots called "calçots" or "sprunzale," sprouted from bulbs planted in trenches, are blanched and eaten raw with bread, grilled, or used for flavoring beans and sauces (FAC). Sprouted seeds used in salads and on sandwiches.

Cooking onions changes some of the pungent sulfur compounds into sugars, making the cooked onion sweeter and blander. As with garlic, I personally believe the fresh is better medicine and spice. But in the kitchen, if not the medicine chest, 1 tbsp minced dry onion equals the flavor of a quarter cup minced fresh onion, and 1 tbsp onion powder moistened with water equals the flavor a medium sized onion (RIN). As with garlic, I treat the cooked and raw onion as two different medicines and two different spices, and often add both, raw and cooked, to soups and salads. Rinzler (1990) advises that it takes some 8 lb of fresh onion to make 1 lb of dry onions. Dried onions are available as onion flakes, onion powder, and onion salt.

What to do to avoid tears? Joseph et al. (2002) suggest holding a piece of bread between the teeth while dicing. Some of the rising fumes are absorbed by the bread before reaching your eyes.

You might also try chilling the onion before cutting, reducing the volatility of the lachrimatory compounds. And rinse your knife under cold water before chopping (JNU). I've also heard that breathing through your mouth, instead of your nose, helps, too.

For more information on activities, dosages, and contraindications, see the *CRC Handbook of Medicinal Herbs,* ed. 2, Duke et al. 2002.

Cultivation (Onion) — Propagated by seeds, dry sets, or transplants, the three methods being used for commercial onions in the United States. Soil preparations for growing onions consists of plowing, discing, bed-shaping, pre-plant irrigation, fertilization, and sometimes ditching. Bed-shaping and planting are usually done with a sled-type planter. When planting is shallow, irrigating after planting may be necessary. Pre-irrigation is usually practiced in fall-planted (October) onions to make soil most workable and to germinate weeds before final land preparation. In sandy loams of the Punjab, maximum bulb sizes and yields were recorded with a combination of 112 kg/ha N, 56 P_2O_5, and 25 k_2O. This return was also best from an economical point of view. In Nebraska studies, weeds allowed to grow in the row for 2, 4, 6, and 8 weeks after onion emergence reduced onion yields 20, 20, 40, and 65%, respectively. The weed check yielded 12 tons biomass compared to 40 MT onions per acre in the weeded plots. Weed biomass at 4 weeks was 690 kg/ha, but 3,530 kg/ha after 8 weeks. Development of herbicides for pre-emergence and post-emergence enables the grower to cultivate larger acreage, as much hand-weeding and hoeing are eliminated. Onion sets may be dropped by hand into shallow furrows, but on commercial onion farms, machines have been developed to make furrows, drop sets, and cover them in one operation. For mechanical planting, sets must be graded for uniform size. Sets are spaced 6.5–9 cm apart, in rows 30–40 cm apart. Onion growers are planting fewer sets and preferring onion transplants grown from seed. These are about 1 cm in diameter and about 15 cm tall when transplanted. Onions may also be propagated by bulbs, which are sliced across the top, thereby producing 6–10 shoots, which mature in about 3 months. Onion crops must be kept weeded and watered but well drained. Seed production, using male sterile selections, should be isolated from other onion crops by at least 1000 m (DAD).

From seeding to bulb maturity is 100–140 days, depending on cv and weather. Spring onions should be harvested much earlier (30–45 days). As they mature, onions cease to produce new leaves and roots. Food in the leaves moves to the bulbs, and the green tops weaken and fall over. Best time to harvest a field is when one quarter to one third of tops have fallen over. A small onion plow may be used to loosen the bulbs. In mucky land, bulbs are easily pulled by hand; in irrigated areas, water may be used to soften the soil a day or so before harvesting. Pulled onions are put in windrows with the bulbs shaded by the tops to minimize scalding from the sun. Onions are then transported to storage houses where they are cured by forced ventilation, being considered cured when they have lost 3–5% of their weight. High temperature and high humidity during the curing, with good air circulation, favor development of desirable skin color. Sprouting of stored bulbs can be reduced by sprout suppressants. Globe onions can be held for 6–8 months, Bermuda types for only 1–2 months, at 0ºC. A relative humidity of 65–70% is recommended for onion storage. Humidity up to 85% has been satisfactory. Green onions are harvested as needed. They are pulled and bunched, the roots sometimes pruned a bit, but the leaves left intact. Onion yields vary widely, e.g., 1900 kg/ha in Honduras, 27,650 kg/ha in Chile, and up to 29,000 kg/ha in U.S. The world average is about 12,000 kg/ha in 1979.

Chemistry (Onion) — According to Hager's Handbook (1969–1979), onion contains 0.005–0.015% EO, methylalliin, dihydroalliin (propylalliin), and cycloalliin. The colored epidermal layer may contain 4% quercetin, 1% spiraeoside, 0.45% protocatechuic acid, phloroglucin, protocatechuic methyl ester. Leaves contain quercetin, spiraeoside, ferulic acid ester, and caffeic acid. Cell walls contain mannan, pectin, pentosane, fructosane, myrosinase, and peroxidase. Most *Allium* species exhibit antioxidative activity in a linoleic-acid model system, and onion and garlic show it, even in minced pork. Rutin is said to be antiatherogenic, antiedemic, antiinflammatory, antithrombogenic, and hypotensive (DAD). Here are a few of the more notable chemicals found in onion. For a complete listing of the phytochemicals and their activities, see the CRC phytochemical compendium, Duke and duCellier, 1993 (DAD) and the USDA database http://www.ars-grin.gov/duke/.

Allicin — Alcohol-Dehydrogendase-Inhibitor 500 μM; Amebicide 30 $\mu g/mL$; Anthelminthic; Anti-aggregant 0.1–1 μM; Antiatherosclerotic 0.05–0.1 mg/kg orl hmn; Antibacterial MIC = 27 $\mu g/ml$, 500 $\mu g/ml$; Antidiabetic; Antiflu; Antiglaucomic; Antiherpetic; Antiinflammatory; Antileukemic 31+ μM; Antileukotriene; Antilymphoma; Antimutagenic; Antimycobacterial MIC = 1.67 mg/ml; Antineuralgic; Antioxidant 1.8 $\mu g/ml$; Antiproliferant; Antiprostaglandin IC67 = 50 μM; Antiradicular 1.8 $\mu g/ml$; Antisarcomic; Antiseptic; Antishigellic; Antistaphylococcic MIC = 27 $\mu g/ml$; Antitriglyceride 0.05–0.1 mg/kg orl hmn; Antitubercular MIC = 1.67 mg/ml; Antitumor; Antiviral; Apoptotic 31+ μM; Candidicide MIC 7 $\mu g/ml$; Cholinesterase-Inhibitor 500 μM; Cyclooxygenase-Inhibitor; Fungicide; Gram(+)-icide 8–12 ppm; Gram(–)-icide 8–12 ppm; Hepatotoxic 100 mg/kg/day (= 500 cloves a day); Hypocholesterolemic IC37–72 = 162 $\mu g/ml$, 0.05–0.1 mg/kg orl hmn, IC50 = 9 μM; Hypoglycemic 0.1 mg/kg; Hypolipidemic ID50 = 10 μM; Hypotensive; Immuno-stimulant; Insecticide; Insulin-Sparing 100 mg/kg/man; Larvicide; Lipolytic 4–6 mg/day; Lipoxy-genase-Inhibitor ED = 25 $\mu g/ml$; Mucokinetic; Nematicide; NO-Inhibitor IC50 = 2.5–5 μM; Papain-Inhibitor 500 μM; Phagocytotic; Prooxidant; Succinate-Dehydrogenase-Inhibitor 500 μM; Tri-chomonicide; Urease-Inhibitor 500 μM; Vibriocide; Xanthine-Oxidase-Inhibitor 500 μM; LD50 = 309 mg/kg orl mus (male); LD50 = 363 mg/kg orl mus (female); LD50 = 60mg/kg ivn mus; 120 mg/kg scu mus.

Alliin — Antiaggregant IC100 = 60 $\mu g/ml$; Antibacterial; Antidiabetic 200 mg/kg; Antihepatotoxic 0.5 $\mu g/ml$; Antineuralgic; Antioxidant; Antiradicular; Antisarcomic 1–3 mg ims rat; Antiseptic; Antithrombic; Antitumor; Hepatoprotective; Hypocholesterolemic; Hypoglycemic; Lipolytic 200 mg/kg orl rat; Lipolytic 7–11 mg/day.

Cepaene — Antiaggregant; Antiallergic; Antiasthmatic; Cyclooxygenase-Inhibitor; Lipoxygenase-Inhibitor.

Protocatechuic-Acid — Antiarrhythmic; Antiasthmatic; Antibacterial; Anticlastogenic; Anti-hepatotoxic; Antiherpetic; Antiinflammatory; Antiischemic; Antiophidic; Antioxidant; $^2/_3$ quer-cetin, 10 × alpha-tocopherol; Antiperoxidant IC50 = >100 μM; Antiradicular; Antispasmodic EC50 = 4.6–17 μM; Antitumor (colon); Antitumor (mouth); Antitumor (skin); Antitussive; Antiviral; Carcinogenic; Chemopreventive 100 ppm orl rat; Fungicide 500 $\mu g/ml$; Glutathione-Depleting; Hepatotoxic; Immunostimulant; Nephrotoxic; Phagocytotic; Prostaglandigenic; Secretagogue; Ubiquict.

Quercetin — See also *Alpinia officinarum*.

Allium sativum L. (Alliaceae)
AJO, GARLIC, ROCAMBOLE, SERPENT GARLIC

Medicinal Uses (Garlic) — I think this is the number one spice medicine, but closely rivaled by ginger, onion, red and black peppers, and turmeric. Garlic products occupy a >$1-billion-a-year category (Jain and Apitz-Castro, 1994). Amagase et al. (2001) give a table purportedly listing the top 15 supplements in the U.S., with garlic on top at 28%, ginseng at 14%, ginkgo at 13%, echinacea at 11%, chamomile and St. John's-wort at 8%, cayenne at 7%, and ginger at 6% (the others were chemicals or the like).

Strange that so many of the scientists at the sponsored garlic symposium warn about the hazards of garlic in its natural state, rather suggesting the aged garlic extract (AGE), which supported the symposium. Amagase et al. (2001), cite the following adverse effects associated with raw garlic and garlic powder (hinting but not saying that they are not associated with AGE): (1) diarrhea and other stomach disorders, (2) decrease of serum protein and calcium (onion apparently raises calcium (JAD), (3) anemia, (4) bronchial asthma, (5) contact dermatitis, and (6) inhibition of spermatogenesis. Additionally, they all seem rather concerted in their effort to bring down a formerly unquestioned truth. For example, in 1997, we read, "allicin is the most important substance in garlic in terms of both amount and medicinal power" (Fulder, 1997). But in the recent garlic symposium (Milner and Rivlin, 2001), "allicin can be an oxidizing agent that...can damage the intestinal lining and the stomach" (Amagase et al., 2001). I'll continue to take my garlic cooked in foods, powdered on toasts and salads and pizzas, and almost always accompanied by its milder cousin, the onion, and occasionally leeks and scallions. All possess various combinations of useful sulfureous phytochemicals. And it is clear that AGE contains many useful compounds in different proportions. The symposium stresses that. And I suspect that AGE, like garlic and onions, can decelerate the aging process, slowing oxidative damage, and boosting the immune system.

The great book by Koch and Lawson (1996) tabulates numerous clinical trials, making me think that garlic is great, both preventively and curatively, for early stages of cancer, cardiopathy,

A

diabetes, high blood pressure, high cholesterol, and infections like yeast. Read the book and the AGE symposium, and you'll know a lot more about garlic and its phytochemicals and their utility, and surprising variations and interpretations. But the folklore has been with us for years.

Asian Indians use the oil as eardrops and for atonic dyspepsia, colic, gas, and skin rashes. Russians apply raw garlic to corns for 12–18 hr (Libster, 2002). Both garlic and onion are reported to inhibit platelet aggregation. Garlic juice and EO demonstrated antiaggregant and antihypercholesterolemic activities in humans subjects (MPI). In *Science,* we read that garlic and onions have long been reputed to have such mystical powers as the ability to stimulate bile production, lower blood sugar, alleviate hypertension, speed healing of gunshot wounds, cure scorpion bites, freckles, and the common cold. Garlic is considered aphrodisiac, carminative, diaphoretic, diuretic, expectorant, stimulant, and stomachic. It acts as an analgesic in headaches, earaches, and rheumatic pains. The juice is rubefacient and, mixed with oil, is useful for curing skin diseases, ulcers, wounds, insect bites, and as eardrops for earache. As a bactericidal expectorant, it is useful in the treatment of tuberculosis. Many claim that deodorized garlic is as effective a medicine as the "stinking rose." Those who attribute the medicinal activities to malodorous sulfur compounds might disagree. "Processed garlic is not as effective as fresh garlic" (RIN). Other folk medicinal uses are reported in *Medicinal Plants of the Bible* (Duke, 1999), the CRC *Handbook of Alternative Cash Crops* (DAD), and the great Koch and Lawson book (1996).

Charak, the father of Ayurvedic medicine, claimed that garlic "maintains the fluidity of blood and strengthens the heart" (Rahman, 2001). And today, the *JAMA* warns us that it maintains the fluidity of the blood (might cause a bleeding incident). Yes, *JAMA* recommends aspirin for thinning the blood but warns about the perils of garlic. As McCaleb (2001) so aptly puts it, the *JAMA* authors warn about the "use of garlic supplement for surgery patients based on a single case that did not even involve the use of a garlic supplement, but extreme consumption of a food. One elderly man ate 15 grams of raw garlic — or about five medium sized cloves — per day for an extended period of time and subsequently experienced bleeding problems during surgery, possibly but not necessarily connected with the garlic. This one incident, more than a decade old, is the only case on record that supports the author's argument against garlic" (McCaleb, 2001). Garlic is cited in the *Egyptian Codex Ebers*, a 35-century-old document, as useful in cancer and heart disease (Rahman, 2001). And today, it is one of the most promising herbs for cancer and cardiopathy (JAD).

Rivlin (2001) comments, "According to the Bible, the Jewish slaves in Egypt were fed garlic and other allium vegetables, apparently to give them strength and increase their productivity, as it was believed to do for the indigenous Egyptian citizens. The Jewish people must have developed some fondness for garlic, because when they left from Egypt with Moses, it is written "that they missed…the onions and the garlic" (Num. 11:5) (Rivlin, 2001). This seems an apt place to quote Fulder's comment, "Garlic, with its sulfureous nature, is clearly both fierce and friendly…. Having brought about the Fall of Man, Satan stepped from the Garden of Eden; where his right foot first rested, the onion plant sprang up, and where his left foot met the ground, there grew garlic" (Fulder, 1997).

Even ancient medical texts from China, Egypt, Greece, India, and Rome each prescribed garlic as medicine. In many cultures, garlic was given to strengthen and increase work capacity for laborers. Hippocrates, the revered physician, prescribed garlic for several ailments. Garlic was given to the original Olympic athletes, perhaps one of the first ergogenic or "performance enhancing" herbs. Cultures that developed without contact with one another arrived at similar conclusions about the garlic and its efficacy. Modern science tends to confirm many of the beliefs of ancient cultures regarding garlic, defining mechanisms of action and exploring garlic's potential for disease prevention and treatment (Rivlin, 2001). Still, too much of the medical establishment in the U.S. writes it off as folk superstition, dangerous at causing bleeding incidents and interfering with hard core pharmaceuticals. I think the pharmaceuticals are more dangerous, more expensive, and kill more Americans (JAD).

Louis Pasteur described the antibacterial effect of garlic and onion juices. Garlic has been used worldwide to fight bacteria. *Allium* vegetables, particularly garlic, are broadly antibiotic against gram(+) and gram(–) bacteria; (1) raw garlic juice is effective against many intestinal bacteria responsible for diarrhea in humans and animals; (2) garlic is effective against strains that are resistant to antibiotics; (3) garlic is synergistic with antibiotics; (4) garlic does not generate resistance; (5) garlic may prevent toxin production by microorganisms (Sivam, 2001). The classical assumption of modern pharmacy and pharmacognosy (what I term the "silver bullet monochemical philosophy") is that single entities or simple mixes of bioactive compounds are assumed to be the principal bioactive component(s). Rosen et al. (2001) state that, "In a nutraceutical or herbal philosophy, garlic or AGE is important as the whole herb," which contains many sulfureous compounds and other phytochemicals all contributing to the overall antioxidant and other bioactivities. I'm more inclined to the herbal philosophy, which I term the holistic whole herbal "shotgun." And the more aged I get, the more my aged stomach appreciates the AGE. When crushed whole garlic enters the stomach acid, it decomposes into several volatile compounds including DAS and DADS. I share Rosen's surprise that consumption of raw garlic, not known to contain p-cymene and limonene (a breast cancer preventive), quickly leads to their appearance in the breath. It's the holistic whole herb working with the whole holistic human that results in biological activities (Rosen et al., 2001).

And Ohnishi et al. (2001), after *in vivo* and *ex vivo* studies, propose that a cocktail of antioxidants could lessen the incidence and severity of crisis and reduce anemia in sickle cell disease. A clinical herbalist I know concurred that "use of antioxidants slows permanent sickling, and the permanently sickled cells cause microhemorrhage in the spleen, and so antioxidants definitely slow progression of the disease, and garlic is a good antioxidant" (A. K. Tillotsen, pers/comm 2001). And AGE inhibits the activation of the oxidant-induced transcription factor, nuclear factor (NF)-kappa B, which has clinical significance in human immunodeficiency virus gene expression (Borek, 2001).

And AGE may have a role in protecting against loss of brain function in aging and possess other antiaging effects, as suggested by its ability to increase cognitive functions, memory, and longevity in age-accelerated mice models (Borek, 2001). And it may help in loss of penile function as well. Garlic is just one of many tonic spices, with equally malodorous asafetida, black pepper, cardamom, cayenne, cinnamon, cloves, cumin, fennel, and ginger considered to gently stimulate the endocrine system. Even in the Arabic or Unani Tibb system, garlic is suggested as an aphrodisiac for low sex drive, and even as a tonic for the elderly (Bergner, 1996). There may be a rationale for the garlic's aphrodisiac activity in Koch and Lawson's rationale for garlic's hypotensive activity. Aqueous and alcoholic extracts of garlic increase the production of nitric oxide, which is associated with decreased hypertension. *In vivo*, 4 g oral fresh garlic (ca. one clove) can double the nitric oxide synthetase activity of blood platelets in 3 hr (Koch and Lawson, 1996). Nitric oxide is said to be necessary to achieve and maintain an erection. Some people even resort to nitroglycerine patches that contain nitric acid to help achieve erections. In 1988, Lawson adds that among the unique features of garlic is its high content of free amino acids "strongly dominated by arginine" (Lawson, 1998). Arginine is used by the cells that line the artery walls to manufacture nitric oxide, which facilitates blood flow to the penis. Without nitric oxide, erections are impossible.

The Codex Ebers, ca. 3500 B.P., is one of the earliest sources indicating prescription of garlic to treat abnormal growths that probably represented malignancies of one kind or another. The Codex also suggested garlic, as we do today, for circulatory ailments, general malaise, and infestations with insects and parasites (Rivlin, 2001). Hippocrates prescribed eating garlic for uterine tumors. I'd be quicker to use garlic than to use taxol, were I of the female gender. The Bower manuscript, dating about 450 A.D. in India, suggested garlic for abdominal tumors. So my brother and I, genetically targeted for colon cancer, take a lot of food famacy garlic. AGE protects DNA against free radical-mediated damage and mutations, inhibits multi-step carcinogenesis, and defends against ionizing radiation and UV-induced damage, including protection against some forms of UV-induced immunosuppression. NCI files (Hartwell, 1982) report that cancer incidence in France is supposedly lowest where garlic consumption is greatest, that garlic eaters in Bulgaria do not have cancer, and that a physician in Victoria, British Columbia, reportedly treated malignancies by prescribing garlic eating. Just inhaling the stalk is said to help uterine tumors, fibroids, polyps, and neoplasms. A poultice of the bulb is said to help tumors (bladder and uterus), the root ointment is said to help tumors and corns, the juice to help hard swellings and skin cancer (BIB, DAD, JLH).

There is strong animal and *in vitro* evidence for anticarcinogenic effects of garlic and/or its active ingredients. Epidemiology addresses cancers of the stomach, colon, head and neck, lung, breast, and prostate. Nineteen studies were reviewed regarding garlic consumption and cancer incidence. Site-specific case-control studies of stomach and colorectal cancer, suggest a protective effect for raw and/or cooked garlic (Fleischauer and Arab, 2001). *Helicobacter pylori* is considered responsible for some stomach cancers and ulcers. The incidence of stomach cancer is lower where Allium uptake is high. *In vitro*, *H. pylori* is susceptible to garlic extract. Even some antibiotic-resistant *H. pylori* strains are reduced by garlic. Garlic should be studied as a low-cost remedy for *H. pylori* (Sivam, 2001).

Horie et al. (2001) show how garlic preparations can ease some of the side effects of chemotherapy. More importantly to me, Lamm and Riggs (2001) hinted that garlic immunotherapy might be clinically compared to bacillus Calmette-Guerin (BCG) immunotherapy. They said BCG was superior to chemotherapy (doxorubicin, mitomycin, and thiotepa) and comparable to other immunotherapies (interleukin-2, interferon-alpha, and the interferon inducer, broprimine) for bladder carcinoma *in situ*. They did not necessarily say that garlic is as good as BCG at boosting the immune system but did list some nice garlic immunostimulating credentials. Garlic stimulates proliferation of lymphocytes and macrophage phagocytosis; induces the infiltration of macrophages and lymphocytes in transplanted tumors; induces splenic hypertrophy; stimulates release of interleukin-2, tumor necrosis factor-alpha, and interferon-gamma; and enhances natural killer cell, killer cell, and lymphokine-activated killer cell activity (Lamm and Riggs, 2001).

Say "cancer" and "cardioprotection," and many Americans think of garlic. Hypoglycemic, hypolipidemic, and hypocholesterolemic effects, with antiandrogenic side effects, have been documented. "Even a modest reduction of cardiovascular disease with garlic could save billions of dollars annually in the United States." Heart attacks run $25,000 to $75,000 each (Bergner, 1996). Negative numbers in the *Southern Medical Journal* below indicate that dietary therapy saves money by reducing other medical problems as well. Garlic is one of the best for cholesterol, according to Bergner and myself. Sharing my opinion that fresh garlic is best, especially as an antibiotic, Bergner adds that every form of garlic tested has cardiovascular benefits and anti-cancer properties. Powders can vary fourfold in allicin content, and garlic oil ingredients can vary twentyfold. If you want a commercial product, however, you might select a name brand, such as Garlicin, Kwai, or Kyolic. These have all been tested in clinical trials or other studies and have therapeutic levels of active ingredients (Bergner, 1996).

Ajoene has been identified as one active ingredient in garlic. Ajoene may be as effective as aspirin at preventing blood clots, heart attacks, and strokes (RIN). Jain and Apitz-Castro (1994) cite studies showing that ajoene is absent in "any of the commercially available 'garlic preparations' including the fermented 'aged' garlic extract (e.g., Kyolic), freeze dried garlic, or even fresh garlic bulbs." They suggest that for "clearing arteries," only aged ether extracts or healthier alcoholic extracts exhibit antiplatelet activity, which can be ascribed to the potent ajoene (SPI). Ajoene can also interfere with absorption of dietary fat by inhibiting gastric lipase. Ajoene antiplatelet activity might even underlie the anticancer activity. Aggregation of host platelets by circulating tumor cells could be important in metastasis (Fukushima et al., 2001). Four studied garlic preparations significantly enhanced natural killer (NK) and killer cell activities of the spleen cells of tumor-bearing mice. Different types of garlic preparations have different pharmacologic properties (Kasuga et al., 2001, X11238821).

Relative Expense of Saving a Year of Life in Patients with Elevated Serum Cholesterol	
Dietary advice (e.g., garlic)	–$2500
Niacin	–$1250
Psyllium husk	–$650
Lovastatin	$50,000
Colestipol	$75,000
Colestyramine	$90,000
Gemfribrozol	$110,000

Source: Adapted from Bergner, P., *The Healing Power of Garlic*, Prima Publishing, Rocklin, CA, 1996.

The antibiotic activity of the garlic is as good as that of many synthetics. But here in the U.S., we give our cattle antibiotics, worsening antibiotic resistance in our country. In Italy, they give garlic to the pigs as an antibiotic, stimulating growth without affecting the taste of the pork negatively. Garlic proved antiseptic to *Bacillus subtilis, Escherichia coli,* and *Saccharomyces cerevisiae,* confirming traditional uses of spices as food preservatives, disinfectants, and antiseptics (De et al., 1999). Garlic extracts are fungicidal against *Candida albicans,* and low concentrations of garlic extract are lethal and/or inhibitory to numerous strains of *Cryptococcus neoformans*. As an vermifuge for tapeworms, garlic is eaten along with prescribed medicine. Libster (2002) reports 79% of 34 men symptom free of *Tinea pedis* after 7 days on Acuagel (a cream with ajoene, the organic trisulfur compound in garlic). The other 21% were symptom free after 7 more days (Libster, 2002).

GARLIC VS. CIPRO

Can "Russian penicillin" (garlic) prevent or reduce the likelihood of getting anthrax if you have been hit with 8000 spores? I may be the only herbalist in America who says yes, for at least four reasons:

1. It boosts the immune system (e.g., Libster notes that one fraction, "a protein isolated from aged garlic extract, is an efficient immunopotentiator"). Even the CDC and the FDA and the newscasters note that aged or sick people with depressed immune systems are more liable to get anthrax than healthier citizens. As noted elsewhere, garlic boosts the immune system in several ways (Garlic Symposium, 2001).
2. The sulfur-containing compounds have been proven antiseptic and bactericidal or bacteriostatic to gram(+) bacteria and other species of *Bacillus*.
3. The strong garlic aroma repels germ-laden people, if not germs themselves.
4. Whole polychemical garlic is less likely to generate bacterial resistance than a monochemical like Cipro. Libster (2002) notes that garlic, maybe even just its monochemical allicin, is effective against certain multi-drug-resistant bacteria.

Libster (2002) adduces some more evidence that I here recycle, hoping to convince CDC and FDA to at least examine the garlic in comparative trials before they dismiss it as useless. With the big mix of a dozen sulphur drugs, possibly synergistic, garlic may be 1/100th, maybe 1/10th, or possibly half as potent as cipro or penicillin, neither of which has been proven clinically against anthrax. Neither the pharmacophilic CDC, nor FDA, nor the herbophilic ABC, AHPA, nor HRF can be sure which is more potent until they have been clinically compared. Why isn't affordable, safe garlic being investigated as the herbal alternative to cipro, doxicyclin, or penicillin instead of being relegated to the allopathic dumpster? I like Ralph Moss' answer, "If you are looking for why alternative therapies are relegated to the 'junk heap' by allopathic medicine, it is because these substances cannot generate megaprofits the way patented pharmaceuticals can" (Moss, 2001). Centuries ago, before synthetics and antibiotics, Russians believed that garlic amulets protected them from the evil spirits that cause infectious diseases. It is still respected in Russia for its antibacterial, antiparasitic, and antiviral properties. It was during World War II, when they used garlic in the field in treating wounds, reportedly preventing gangrene and sepsis, that the name "Russian penicillin" evolved. Current Russian medicine uses garlic for colic, dyspepsia, enterosis, and putrefactive bacteria in the intestines. Traditional Russian herbalists use garlic to prevent cold, flu, and other viruses. As to bacteria, garlic or its allicin is apparently effective against certain MDR-bacteria. I speculate that this is because it contains dozens of antiseptic compounds, several working in different directions.

Libster (2002) notes that, in England, garlic vinegar was taken "internally to prevent being infected with the plague." I had heard it was also applied topically to thieves looting bodies of the dead. Libster notes that, historically, it was also used for infections due to polluted drinking water, and specifically for leprosy. For smallpox, it was chopped and applied to the feet bound in linen cloth. Purple-skinned garlic (probably with anthocyanins) is believed most effective in TCM for ameba and microbes. Garlic extracts have been shown effective against *Helicobacter pylori in vivo*, whereas garlic oil (often lacks the sulfur compounds) was shown ineffective clinically (Libster, 2002).

Indications (Garlic) — Abscess (1; DAA; PNC); Acne (f; FAD); Adenosis (f; JLH); Aegilops (f; JLH); Aging (1; PH2); Allergy (1; AKT; FNF); Alopecia (1; WHO; WO2); Altitude Sickness (f; KAL); Alzheimer's (1; JN131:1010); Amebiasis (2; FAY; PNC); Amenorrhea (1; BGB; LIB); Anemia (f; DAD); Anorexia (f; FAY); Appendicitis (1; FAY; PNC); Aphtha (1; KAL); Arrhythmia (1; FNF); Arthrosis (1; FAD; KAL; PHR; PH2); Asthma (1; PNC; FNF; WHO); Atherosclerosis (3; AKT; APA; BGB; BIS; FAD; FNF; KAL; PHR; PH2; SHT; WHO); Athlete's Foot (2; FNF; LIB; TGP); Bacillus (1; LAW; X10548758); Bacteria (1; FNF; JFM; PH2); Bite (f; FAY; JFM); Boil (1; DAA); Bronchiectasis (1; KAL); Bronchosis (2; FAD; FNF; PHR; PH2; WHO); Burn (2; KAL); Callus (f; JFM; PH2); Cancer (2; AKT; FAD; FNF; PH2); Cancer, abdomen (1; AKT; FNF; JLH); Cancer, bladder (1; FNF; JLH; X11341051; X11238818); Cancer, breast (1; BRU; JN131:989); Cancer, colon (1; AKT; FNF; JLH; X11238811); Cancer, esophagus (1; JN131:1075); Cancer, gland (1; FNF; JLH); Cancer, liver (1; BO2); Cancer, lung (1; BRU; FNF; JLH; JN131:989); Cancer, prostate (1; X11102955); Cancer, skin (1; FNF; JLH); Cancer, stomach (1; AKT; X11238811); Cancer, uterus (1; FNF; JLH); Candida (2; CAN; FNF; KAL); Carbuncle (f; FAY); Cardiopathy (3; BGB; FAD; FNF; SKY); Caries (1; FNF; KAB); Catarrh (1; AKT; BGB); Celiac (1; KAL); Childbirth (f; JFM; KAB); Cholecystosis (f; APA); Cholera (1; PNC); Chronic Fatigue (f; JFM); Coccidiosis (1; KAL); Cold (2; AKT; FAD; FNF; PHR; PNC); Colic (1; WHO); Colosis (1; KAL; LAW); Congestion (1; FAY); Constipation (f; JFM; PH2); Convulsion (f; KAB; PHR); Corn (f; JLH; LIB; PHR); Cough (2; APA; FAD; PHR); Cramp (f; PH2); Cryptococcus (1; DAA); Cystosis (f; JFM); Cytomegalovirus (1; KAL); Deafness (f; LAW); Debility (f; PH2); Dementia (1; X11238823); Dermatosis (1; AKT; DAA; DAD; KAL; PNC); Diabetes (1; MAM; PH2; PNC); Diarrhea (1; AKT; PNC); Diphtheria (f; DAA; DAD); Dropsy (f; KAB); Dysentery (2; AKT; DAD; FAD; PNC); Dysmenorrhea (f; PHR; PH2); Dyspepsia (1; AKT; BIS; JFM; KAL; PNC; WHO); Dyspnea (1; FAD; FAY); Earache (1; FAD); Edema (1; FNF; JFM; PNC); Enterosis (2; AKT; APA; FAD; PH2; WHO); Epigastrosis (2; WHO); Epilepsy (f; AKT; FAY); Escherichia (1; LAW; WO2); Felon (f; JLH); Fever (2; FAD; PHR; PH2); Fibroid (f; DAD; JLH); Filaria (1; KAL); Flu (1; AKT; APA; FNF; KAL; PNC); Fungus (1; AKT; FNF; JFM); Gangrene (f; KAP); Gas (1; DAD; JFM; PH2; WHO); Gastroenterosis (2; BIS; DAD; FAD); Gastrosis (2; AKT; FAD; FAY; PH2; WHO); Giardia (1; KAL; X11101670); Gout (f; DEP; FAD; JFM); Headache (f; JFM); Helicobacter (1; AKT; X11238826); Hemorrhoid (f; JFM); Hepatosis (1; APA; FNF); Hepatotoxicity (Acetaminophen) (2; MAM); Herpes (1; KAL); High Blood Pressure (2; AKT; FAD; FNF; PH2; SHT; WHO); High Cholesterol (3; AKT; APA; FNF; KAL; PH2;

A

SHT); High Triglycerides (3; AKT; APA; KAL; SHT); HIV (1; FNF; KAL); Hookworm (1; AKT; KAL; WHO); Hyperlipidemia (3; SHT; WHO); Hyperperistalsis (2; WHO); Hypoglycemia (f; FAY); Hypotension (f; DAD); Hysteria (f; JFM); Immunodepression (2; FNF; PHR; SKY); Impotence (1; AKT; X11238821); Induration (f; JLH); Infection (2; AKT; FNF; JFM; SHT); Inflammation (1; FNF; JFM); Insanity (f; AKT); Insomnia (f; JFM); Intermittent Claudication (2; BGB; SHT; TGP); Keratosis (1; KAL); Lambliasis (1; KAL); Laryngosis (1; KAL; KAP); Lead Poisoning (1; PNC); Leishmaniasis (1; X11119248) Leprosy (f; JFM); Leukemia (f; JLH); Leukoderma (f; KAB); Lumbago (f; PH2); Lupus (f; KAL); Lymphoma (1; BO2; JLH); Malaria (f; DAD; JFM); Mange (f; JFM); Melancholy (f; JFM); Melanoma (1; JN131:1027); Meningosis (f; DAA); Menopause (f; JFM); Mucososis (1; KAL); Mycosis (1; AKT; FNF; PNC); Myofascitis (f; DAA); Myosis (f; PHR; PH2); Nausea (1; WHO); Nephrosis (1; KAL); Neuralgia (1; KAL; PHR); Neuroblastoma (1; JN131:1027); Nicotinism (1; KAL); Odontosis (f; KAB); Otosis (1; FAD; SKY); Pain (1; FNF; JFM; PH2); Palpitation (f; JFM); Paradentosis (1; KAL); Paralysis (f; KAB); Parasite (1; AKT); Paratyphoid (f; KAP); Paraty-phus (f; LAW); Periodontosis (1; LAW); Pertussis (2; DAD; FAD; FAY; PNC); Pharyngosis (2; PHR); Pinworm (1; AKT; FAY); Pneumonia (1; DAD; LAW); Poliomyelitis (1; LAW); Polyp (f; JLH); Pulmonosis (f; KAP); Pulposis (1; LAW); Raynaud's (2; TGP); Respirosis (1; AKT; BGB; LAW; PH2; WHO); Rheumatism (1; FAD; LAW; PH2); Rhinosis (2; BGB); Ringworm (1; APA; DAA; WHO); Roundworm (1; KAL; WHO); Salmonella (1; WO2); Scabies (1; DAA; JFM); Sciatica (f; PHR; PH2); Senile Dementia (1; LAW; X11238823); Sepsis (1; LAW); Shigella (1; LAW; WO2); Sinusosis (1; FAY); Snakebite (f; FAD; FAY); Sore (1; FAD; JFM); Sore Throat (1; LAW); Splenosis (f; KAB); Sporotrichosis (1; KAL); Staphylococcus (1; LAW); Stomachache (f; FAY); Stomatosis (2; PHR); Streptococcus (2; X9354029); Stroke (1; JN131:1010); Swelling (f; AKT; FAD; FAY; JFM); Syncope (f; KAB); Tapeworm (f; JFM); Thirst (f; KAB); Thrombosis (1; FAY); Tonsilosis (1; LAW); Trachoma (f; DAA); Trichomonaisis (1; DAA); Trypanosomiasis (1; KAL); Tuberculosis (1; APA; JFM; KAL); Typhoid (f; DAA); Typhus (1; DAD; KAL); Tumor (1; FNF); Ulcer (1; AKT; X11238826); Ulcus cruris (2; KAL); UTI (1; WHO); Vaginosis (2; APA; DAA; KAL); Varicosis (f; JFM); Virus (1; FNF; PH2); Wart (f; PHR; PH2); Water Retention (1; FNF); Wen (f; JLH); Whitlow (f; JLH); Worm (1; AKT; APA; JFM); Wound (f; PHR); Yeast (2; APA; CAN; FNF; JAD; WO2).

Garlic for cardiopathy:

- ACE-Inhibitor: glutathione
- Antiaggregant: (-)-n-(1'-deoxy-1'-d-fructopyranosyl)-s-allyl-l-cysteine-sulfoxide; 2-vinyl-4h-1,3-dithiin; adenosine; ajoene; allicin; alliin; allyl-methyl-trisulfide; allyl-trisul-fide; alpha-linolenic-acid; apigenin; caffeic-acid; cycloalliin; ferulic-acid; kaempferol; methyl-allyl-trisulfide; phytic-acid; quercetin; salicylates
- Antiarrhythmic: adenosine; apigenin; ferulic-acid
- Antiatherogenic: rutin; s-allyl-cysteine-sulfoxide
- Antiatherosclerotic: allicin; oleanolic-acid; quercetin
- Anticardiospasmic: allithiamin
- Anticoronary: lignin
- Antiedemic: caffeic-acid; oleanolic-acid; rutin
- Antihemorrhoidal: rutin
- Antioxidant: allicin; alliin; allyl-mercaptan; apigenin; caffeic-acid; chlorogenic-acid; diallyl-pentasulfide; diallyl-trisulfide; ferulic-acid; glutathione; kaempferol; lignin; myricetin; oleanolic-acid; p-coumaric-acid; p-hydroxy-benzoic-acid; phytic-acid; quer-cetin; rutin; s-allyl-cysteine-sulfoxide; s-allyl-l-cysteine; salicylic-acid; sinapic-acid; tau-rine; vanillic-acid
- Antitachycardic: adenosine
- Antithrombic: ferulic-acid; quercetin
- Arteriodilator: adenosine; ferulic-acid

- COX-2-Inhibitor: apigenin; kaempferol; oleanolic-acid; quercetin; salicylic-acid
- Calcium-Antagonist: allicin; caffeic-acid; trans-ajoene
- Cardiotonic: oleanolic-acid
- Cyclooxygenase-Inhibitor: ajoene; allicin; apigenin; kaempferol; oleanolic-acid; quercetin; salicylic-acid
- Diuretic: apigenin; asparagine; caffeic-acid; chlorogenic-acid; citrulline; kaempferol; myricetin; oleanolic-acid
- Hypocholesterolemic: 2-vinyl-4h-1,3-dithiin; adenosine; ajoene; allicin; alliin; caffeic-acid; chlorogenic-acid; diallyl-disulfide; diallyl-sulfide; diallyl-trisulfide; inulin; lignin; methyl-ajoene; nicotinic-acid; phytic-acid; rutin; s-allyl-cysteine-sulfoxide; s-allyl-l-cysteine; s-methyl-l-cysteine-sulfoxide; taurine; trigonelline
- Hypotensive: adenosine; allicin; alpha-linolenic-acid; apigenin; kaempferol; prostaglandin-a-1; quercetin; rutin; tyrosinase
- Sedative: adenosine; apigenin; caffeic-acid; citral; geraniol; linalool; oleanolic-acid
- Vasodilator: adenosine; apigenin; kaempferol; myricetin; prostaglandin-e-1; quercetin

Garlic for infection:

- Amebicide: allicin; diallyl-trisulfide
- Analgesic: adenosine; allithiamin; caffeic-acid; chlorogenic-acid; ferulic-acid; quercetin; salicylic-acid
- Anesthetic: linalool
- Antibacterial: ajoene; allicin; alliin; allistatin-i; allistatin-ii; apigenin; caffeic-acid; chlorogenic-acid; citral; diallyl-disulfide; diallyl-sulfide; diallyl-tetrasulfide; diallyl-trisulfide; endolysin; ferulic-acid; geraniol; kaempferol; lignin; linalool; muramidase; myricetin; oleanolic-acid; p-coumaric-acid; p-hydroxy-benzoic-acid; quercetin; rutin; salicylic-acid; sinapic-acid; vanillic-acid
- Antiedemic: caffeic-acid; oleanolic-acid; rutin
- Antiinflammatory: ajoene; allicin; alpha-linolenic-acid; apigenin; caffeic-acid; chlorogenic-acid; ferulic-acid; kaempferol; myricetin; oleanolic-acid; quercetin; quercetin-3-o-beta-d-glucoside; rutin; salicylates; salicylic-acid; vanillic-acid
- Antiseptic: 2-propene-1-sulfinothioic-acids-2-propenyl-ester; allicin; alliin; caffeic-acid; chlorogenic-acid; citral; diallyl-sulfide; diallyl-tetrasulfide; diallyl-trisulfide; geraniol; kaempferol; linalool; myricetin; oleanolic-acid; phloroglucinol; salicylic-acid; trigonelline
- Antiviral: allicin; allyl-alcohol; apigenin; caffeic-acid; chlorogenic-acid; diallyl-disulfide; diallyl-trisulfide; ferulic-acid; kaempferol; lignin; linalool; myricetin; oleanolic-acid; quercetin; rutin
- Bacteristat: quercetin
- COX-2-Inhibitor: apigenin; kaempferol; oleanolic-acid; quercetin; salicylic-acid
- Cyclooxygenase-Inhibitor: ajoene; allicin; apigenin; kaempferol; oleanolic-acid; quercetin; salicylic-acid
- Fungicide: ajoene; allicin; alpha-phellandrene; beta-phellandrene; caffeic-acid; chlorogenic-acid; citral; diallyl-disulfide; eruboside-b; ferulic-acid; geraniol; linalool; p-coumaric-acid; phloroglucinol; phytic-acid; quercetin; salicylic-acid; sinapic-acid
- Fungistat: p-hydroxy-benzoic-acid
- Immunostimulant: allicin; alliin; alpha-linolenic-acid; caffeic-acid; chlorogenic-acid; diallyl-disulfide; ferulic-acid; inulin
- Lipoxygenase-Inhibitor: ajoene; allicin; caffeic-acid; chlorogenic-acid; kaempferol; myricetin; p-coumaric-acid; quercetin; rutin

Other Uses (Garlic) — Cultivated for the pungent bulb, it is used fresh, dried, or powdered as a seasoning, rather than as a vegetable. It is best crushed finely and used in moderation. According to Rinzler (1990), $\frac{1}{4}$ tsp garlic powder equals two small, fresh garlic cloves. Fried in too hot fat or oil, it develops an acrid flavor. Bulbs may be baked, boiled, broiled, roasted, or sautéed. Allioli, aioli, bagna cauda, skordaliá, and tarator are garlic sauces widely used in Mediterranean and Middle Eastern cookery (FAC). Fleurs d'ail is a sauce made of the flowers. Mild young leaves are considered a delicacy in the Orient and are used in salads, soups, egg-dishes, etc. Flowering stalks, sometimes called "garlic chives," are also used for flavoring and are occasionally sold in bunches in oriental stores (FAC). The seeds and sprouted seeds are also eaten (FAC). Dehydrated garlic is a common ingredient of herbal salt substitutes like Vegit, Spike, and Mrs. Dash. Garlic oil is occasionally used in cooking. Bulbs yield 0.06–0.1% EO, containing allyl propyl disulfide, diallyl disulfide, and two other sulfur compounds, allicin and allisatin (I and II). Italians feed garlic to their pigs in lieu of zinc bacitracin. The antibiotic activity of the garlic is as good, hence it stimulates growth without affecting the taste of the pork — maybe even improving it. Garlic and its active principles, whether natural or synthetic, constitute relatively safe pesticides, shown to help control or destroy aphids, cabbage-white butterfly caterpillars, Colorado beetle larvae, mosquito larvae, pulse beetles, root knot nematodes, horseflies, armyworms, bacteria, ticks, and several fungi. It is reported that garlic extracts contain a powerful bactericide allylthiosulfinic allyl ester or allicin, formed by the interaction of a garlic enzyme alliinase and the substrate S ethyl L-cysteine sulfoxide. When enzyme or substrate was inoculated into mice with sarcoma, all animals died within 16 days; when enzyme was allowed to react with substrate, followed by administration to the tumor-bearing animals, no tumor growth occurred, and the animals remained alive during a 6 month observation period. A nutritional supplement known as "garlic balls" contains odorless garlic mixed with ginseng and honey. Can be grown in indoor window sill gardens (FAC).

For more information on activities, dosages, and contraindications, see the *CRC Handbook of Medicinal Herbs,* ed. 2, Duke et al. 2002.

Cultivation (Garlic) — Cloves are separated and individual cloves planted on ridges or raised beds to provide drainage. Cloves should be planted to a depth of 2.5–5 cm, 10–15 cm apart, in rows 30–90 cm apart. Plantings are made in the spring in the North and in the late fall in the South, being planted at a rate of 500–1100 kg cloves per hectare. Irrigation is very beneficial, especially at the beginning of the season, but should be discontinued as plants approach maturity. A green manure crop or other organic matter plus supplement of commercial fertilizer (5–10–5) at rate of 900–1359 kg/ha is recommended. A surface dressing of N as bulbing begins may be helpful. Extra potash may be needed in some areas. In India, garlic is rotated with ragi (*Eleusine*), chile (*Capsicum*), corn (*Zea*), potatoes (*Solanum*), and beans, perhaps helping curb pest and disease problems. Garlic requires about 4 months (sometimes to 8 months) to mature. It is usually harvested in May in the South and June or July in the North, and in tropical areas during the dry season. Maturity is often indicated by the leaves browning and drying out. Bulbs are carefully lifted, freed from soil, and either allowed to dry in the field for a few days or thoroughly dried in the sun elsewhere. When dry, the bulbs are delivered to a commercial processing plant, where they are packaged or ground into garlic powder or chips. Garlic will keep 6–8 months at 0.5°C with 70% relative humidity. Most commercially grown garlic yields about 4.5 MT/ha. Usually, the yield is ca. 10 times the bulbs planted (DAD).

Chemistry (Garlic) — A garlic bulb (whole) can contain up to 1.8% alliin (fresh weight; 4% dry weight) (but no allicin). On crushing, that same bulb can attain 0.37% allicin (3700 ppm). Allicin can be an oxidizing agent that not only impedes bacterial growth but may possibly damage the intestinal lining and the stomach. Enteric-coated garlic products designed to generate allicin in the delicate intestine may damage the intestinal tract. The symposium (Amagaze et al., 2001) convinced me that AGE might be a relatively safer product, but I am not so positive about the relative proportionality of efficacy with safety. Alliin is one (85%) of three s-alkylcysteine sulfoxides able to make thiosulfinates, with 5% isoalliin and 10% methiin. Reporting on 22 strains of garlic, Koch and Lawson (1996) show that these vary almost threefold from variety to variety, from 3500–11,800 ppm, but alliin itself may vary fivefold, from 3000–15,000 ppm. Alliin is rare outside the genus, but allicin has been reported from roots of my garlic mustard *Alliaria officinalis*, 100 ppm in fresh roots, along with some sulforaphane and Amazonian *Adenocalymna alliaceum*, at 1000 ppm, which is one of the adjuvants to some ayahuasca formulae (LAW).

List and Hohammer (1969–1979) add several enzymes (alliinase, myrosinase, peroxidase, desoxyribonuclease, tyrosinase), choline, iodine, traces of uranium, 20% inulin-containing polyoses, saponin, methyl cysteine, methyl cysteine sulfoxide, etc. (HHB). Three compounds in garlic inhibit platelet aggregation (1) diallyl trisulfide, (2) 2-vinyl-4H-1,3-dithiin, and (3) ajoene. Other compounds identified were allicin, allyl methyl trisulfide, diallyl disulfide, diallyl tetrasulfide, and 3-vinyl-4H-1,2-dithiin mg Fe, 4 mg Na, 326 mg K, 920 mg beta-carotene equivalent, 0.11 mg thiamine, 0.14 mg.

Here are a few of the more notable chemicals found in garlic. For a complete listing of the phytochemicals and their activities, see the CRC phytochemical compendium, Duke and duCellier, 1993 (DAD) and the USDA database http://www.ars-grin.gov/duke/.

Ajoene — Antiaggregant 90–100 μ*M*; Antiallergic; Antibacterial MIC = 55–150 μg/ml; Antiherpetic; AntiHIV; Antiinflammatory; Antileukemic 40 μ*M*; Antileukotriene; Antilymphomic; Antimalarial 50 mg/kg; Antimetastatic; Antimutagenic; Antimycotic; Antiproliferant; Antiprostaglandin; Antiseptic; Antistaphylococcic MIC = 55 μg/ml; Antistomatitic; Antithrombic; Antitumor; Antitumor (colon); Antiviral; Apoptotic 40 μ*M*; Candidicide MIC 70 μg/ml; Candidistat < 20 μg/ml; COX-2-Inhibitor; Cytotoxic 2–50 μg/ml; Cyclooxygenase-Inhibitor; Cytostatic; Fungicide IC100 = 100 μg/ml; Fungistat < 20 μg/ml; Gram(+)-icide; Gram(–)-icide; Hypocholesterolemic IC37–72 = 234 μg/ml, IC50 = 9 μ*M*; Lipolytic; Lipoxygenase-Inhibitor; NF-kB-Inducer; NO-Inhibitor IC50 = 2.5–5 μ*M*; Protisticide; Tineacide; Trypanosomicide.

Allicin — See also *Allium cepa.*

Alliin — See also *Allium cepa.*

Arginine — Antidiabetic?; Antiencephalopathic; Antihepatosis; Antiinfertility 4 g/day; Antioxidant?; Aphrodisiac 3 g/day; Diuretic; Hypoammonemic; Pituitary-Stimulant; Spermigenic 4 g/day.

Cycloalliin — Antiaggregant; Fibrinolytic; Lachrimatory.

Tryptophan — Analgesic 750 mg/4 × day/orl/man/; Antianxiety 500–1000 mg/meal; Antidementic 3 g/day; Antidepressant 1–3 g/3 × day/orl/man/; Antidyskinetic 2–8 g/orl/wmn/day/; Antiinsomniac 1–3 g/day; Antimanic 12 g/man/day/orl; Antimenopausal 6 g/day; Antimigraine 500 mg/man/4 × day; Antioxidant?; Antiparkinsonian 2 g 3 × day; Antiphenylketonuric; Antipsychotic 12 g/man/ day; Antirheumatic; Carcinogenic; Essential; Hypoglycemic; Hypnotic; Hypotensive 3 g/day; Insulinase-Inhibitor; Insulinotonic; Monoamine-Precursor; Prolactinogenic; Sedative 3–10 g/man/ day; Serotoninergic 6–12 g/day/orl/man; Tumor-Promoter; RDA = 300–1200 mg/day.

Alpinia galanga (L.) Sw. (Zingiberaceae)
GALANGAL, GREATER GALANGAL, LANGUAS, SIAMESE GINGER

Synonyms — *Languas galanga* (L.) Stuntz, *Maranta galanga* L.

Medicinal Uses (Greater Galangal) — Popular folk remedy for cancer, especially of the mouth and stomach. Rootstocks are considered aphrodisiac, aromatic, carminative, stimulant, stomachic, and tonic. Its expectorant activity has been compared to anise and dill, useful in pediatric respiratory problems. I would not hesitate to mix it with those for my grandchildren during flu season. I would hesitate to give them a brand new synthetic. Roots, flayed on one end until brush-shaped, are dipped in vinegar and rubbed on spots caused by "panu," a common skin disease in Java. As a paste, with a little garlic and vinegar (red wine vinegar is better), it is a last resort drastic remedy for herpes. Rhizomes are also recommended for use in dyspepsia, diabetes, impotence, nervous debility, and food poisoning. They also have disinfectant properties and are used as a deodorant of foul smells in the mouth and other parts of the body, as well as being used as a fragrant adjunct in various prescriptions. Aromatic parts, made into a paste, are applied in acne and other skin diseases. Seeds are alterative, calefacient, sternutatory, and stomachic and are useful in diarrhea and vomiting as well as for medicinal uses same as the rhizome. In some parts of the world, its main use is in clearing the voice.

In one clinical trial, 261 osteoarthritics with moderate to severe pain in the knee were enrolled in a randomized, double-blind, placebo-controlled, multicenter, parallel-group, 6-week study. After washout, patients received the herb extract or placebo twice daily, with acetaminophen allowed as rescue medication. An extract of ginger and greater galangal significantly (but modestly) reduced symptoms of osteoarthritis. There was a good safety profile, with mild GI adverse events reported in the herbal group compared to controls (Altman and Marcussen, 2001).

Plants contain an EO that is carminative and reduces spasms of involuntary muscle tissue, diminishing excessive peristalsis in the intestines (WO2). Al-Yahya et al. (1990) reported antisecretory, antiulcer, and cytoprotective activities of ethanolic rhizome extracts at 500 mg/kg in rats (PM58:124). 1'-acetoxychavicol-acetate and 1'-acetoxyeugenol-acetate are reportedly powerful antiulcer agents (WO2).

Rhizome EO inhibited Mycobacterium tuberculosis at 25 mg/ml. It was more potent against gram(–) bacteria (at 0.4–0.6 mg/ml). LD50 of the oil in guinea pigs was 0.68 ml/kg (mode of administration not stated) (MPI). Steam volatile portion of the rhizome extracts stimulated the bronchial glands; the nonvolatile portion acted reflexly through the gastric mucosa (MPI). And as another example of synergy, (E)-8 beta,17-epoxylabd-12-ene-15,16-dial synergistically enhanced the antifungal activity of quercetin and chalcone against *Candida albicans*. The antifungal activity may be due to changes of membrane permeability arising from membrane lipid alteration (X8792660).

In a study of clinical applications of ayurvedic herbs, Khan and Balick (2001) note human studies on greater galanga, showing that the ethanolic extract (at 125 mg/kg significantly decreased effects of induced micronucleated polychromatic erythrocytes without modifying cytotoxicity.

Looking at potential toxicity, Qureshi et al. (1992) found no spermatotoxic activity at acute doses of 0.5, 1, and 3 g/kg ethanolic extracts or chronic dosages of 100 mg/kg. They noted weight gain in sex organs and increased sperm motility and sperm counts. The alcoholic extract of the rhizome lowered temperature in mice and potentiated amphetamine toxicity (MPI).

Indications (Greater Galangal) — Bacteria (1; FNF; HHB; MPI); Bronchosis (1; FNF; HHB); Cancer (1; X3575509); Candida (1; FNF; X8792660); Catarrh (f; HHB; MPI); Childbirth (f; DAA); Cold (1; FNF); Colic (f; DAA; WO2); Cough (f; WO2); Cramp (1; FNF; WO2); Diabetes (f; HHB); Diarrhea (1; DAA; FNF); Dyspnea (f; HHB); Earache (f; DAA); Enterosis (f; WO2); Fever (1; MPI); Fungus (1; FNF; X8792660); Gastrosis (f; WO2); Infection (1; FNF; HHB); Inflammation (1; FNF); Mycosis (1; FNF; X8792660); Nausea (f; DAA); Pain (1; FNF; JPP42:877); Protozoa (1; HHB); Pulmonosis (f; HHB); Rheumatism (1; FNF; HHB); Tuberculosis (1; MPI; WO2); Ulcer (1; FNF; WO2; X1017082); Water Retention (1; WO2); Yeast (1; X8792660).

Greater Galangal for infection:

- Analgesic: borneol; camphor; eugenol; myrcene; p-cymene
- Anesthetic: 1,8-cineole; camphor; eugenol; linalool; myrcene
- Antibacterial: 1,8-cineole; alpha-pinene; alpha-terpineol; bornyl-acetate; eugenol; limonene; linalool; myrcene; p-cymene; terpinen-4-ol
- Antiedemic: caryophyllene-oxide; eugenol
- Antiinflammatory: alpha-pinene; beta-pinene; borneol; caryophyllene-oxide; eugenol; galangin; kaempferide
- Antiseptic: 1,8-cineole; alpha-terpineol; beta-pinene; camphor; eugenol; limonene; linalool; terpinen-4-ol
- Antiviral: alpha-pinene; ar-curcumene; beta-bisabolene; bornyl-acetate; galangin; limonene; linalool; p-cymene
- COX-2-Inhibitor: eugenol
- Cyclooxygenase-Inhibitor: galangin
- Fungicide: 1,8-cineole; camphor; caryophyllene-oxide; chavicol; eugenol; linalool; myrcene; p-cymene; terpinen-4-ol; terpinolene
- Fungistat: limonene
- Lipoxygenase-Inhibitor: galangin

Greater Galangal for ulcer:

- Analgesic: borneol; camphor; eugenol; myrcene; p-cymene
- Anesthetic: 1,8-cineole; camphor; eugenol; linalool; myrcene
- Antibacterial: 1,8-cineole; alpha-pinene; alpha-terpineol; bornyl-acetate; eugenol; limonene; linalool; myrcene; p-cymene; terpinen-4-ol
- Antiinflammatory: alpha-pinene; beta-pinene; borneol; caryophyllene-oxide; eugenol; galangin; kaempferide
- Antioxidant: camphene; eugenol; gamma-terpinene; myrcene
- Antiprostaglandin: eugenol
- Antiseptic: 1,8-cineole; alpha-terpineol; beta-pinene; camphor; eugenol; limonene; linalool; terpinen-4-ol

A

- Antispasmodic: 1,8-cineole; borneol; bornyl-acetate; camphor; eugenol; limonene; linalool; myrcene
- Antiulcer: 1′-acetoxy-eugenol-acetate; 1′-acetoxychavicol-acetate; ar-curcumene; beta-bisabolene; beta-sesquiphellandrene; eugenol
- Antiviral: alpha-pinene; ar-curcumene; beta-bisabolene; bornyl-acetate; galangin; limonene; linalool; p-cymene
- COX-2-Inhibitor: eugenol
- Cyclooxygenase-Inhibitor: galangin
- Fungicide: 1,8-cineole; camphor; caryophyllene-oxide; chavicol; eugenol; linalool; myrcene; p-cymene; terpinen-4-ol; terpinolene
- Lipoxygenase-Inhibitor: galangin
- Vulnerary: terpinen-4-ol

Other Uses (Greater Galangal) — Known in cultivation for at least 14 centuries, it is grown for the aromatic rootstock used in liqueurs, curries, medicine, and as a flavoring for foods. In Kerala, rhizomes are used to flavor fish and in pickling. Used to flavor bean curd and curries in Java, sauces in Malaya. Also used as a flavoring in fish dishes, meat, soups, and in masaman and other curries (FAC). Slices of young rhizome are added as side dishes or spices to rice dishes, or eaten raw or steamed. Underdeveloped lateral shoots are eaten, but only when cooked, as they are very hot before preparation. Rhizome is too hot to be eaten raw. Rhizome dyes wool yellow. Flower buds and flowers may be pickled, eaten raw, steamed, used in soups and salads, or mixed with chili paste (FAC). The red fruits are edible (FAC). The EO is used to flavor liqueurs such as *Chartreuse, Angostura* and other bitters, and soft drinks (FAC). The EO is sometime used in perfumery. The same EO is said to be good at knocking down house flies. Furthering the old adage that spices were classically used as both antioxidants and antiseptics is sound evidence that Galangal extract may inhibit lipid oxidation and increase microbial stability of minced meat (X10716573).

For more information on activities, dosages, and contraindications, see the *CRC Handbook of Medicinal Herbs,* ed. 2, Duke et al. 2002.

Cultivation (Greater Galangal) — Plants are mainly cultivated by rhizome divisions or rhizome cuttings. These are planted in rich, well-cultivated soil in partial shade, often at distances of 0.5–1 m square. As Ochse (1980) puts it picturesquely, the plant is thankful of well-tilled fertile habitat, which should be humid but not swampy. Once planted, the roots are earthed up. Relatively high moisture in soil is necessary for successful cultivation. Rhizomes are collected in late summer or early autumn (WO2). Earliest harvest could be at 2.5–3 months, but there is more to be had later. Don't wait too long. Bown (2001) says four- to six-year-old rhizomes are harvested at the end of the growing season. The rhizomes get fibrous when allowed to grow too long. Bown (2001) notes that spider mites can be a problem under glass.

Chemistry (Greater Galangal) — Steam distillation of fresh rhizomes yields ca. 0.04% of a spicy EO, with ca. 48% methyl cinnamate, 20–30% cineole, and some camphor and pinene. Our star for myrcene is greater galangal; its rhizomes can attain 4.5% (ZMB) of the anesthetic myrcene, higher than any spice in my database. Leaves of the bayrum tree and/or allspice (up to 2.4% on a calculated dry weight basis), nutmeg (0.59), rosemary (0.56), eucalyptus (to 0.5), cardamom (0.3), fennel (0.3) cornmint (0.25), wild bergamot (0.19), parsley seed (0.17), caraway seed (0.16), spearmint (0.14), tarragon (0.1), dill seed (0.09), and mountain dittany (0.07% on a dry weight basis). Here are a few of the more notable chemicals found in greater galangal. For a complete listing of the phytochemicals and their activities, see the CRC phytochemical compendium, Duke and duCellier, 1993 (DAD) and the USDA database http://www.ars-grin.gov/duke/.

1′-Acetoxyeugenol–Acetate — Antisarcoma; Antitumor; Antiulcer.

Beta–Bisabolene — Abortifacient; Antirhinoviral IC50 = 1,800?; Antiulcer IC57 = 100 mg/kg; Antiviral IC50 = 1,800?; Stomachic.

1,8-Cineole — See also *Elettaria cardamomum.*

Galangin — Antiaflatoxic IC50 = 1.19 μ*M*, IC50 = 0.32 ppm; Anticancer; Antigenotoxic; Antiinflammatory IC50 = 5.5 μ*M*; Antimutagenic; Antioxidant; Antiperoxidant IC50 = 39 μ*M*; Antiradicular; Antiviral; Aromatase-Inhibitor IC20 = 1 μ*M*/L; Copper-Chelator; Cyclooxygenase-Inhibitor IC50 = 5.5 μ*M*; COX-2-Inhibitor; Hepatoprotective IC50 = 1.19 μ*M*, IC50 = 0.32 ppm; Inotropic; Mutagenic; NO-Inhibitor; Quinone-Reductase-Inducer 4 μ*M*; Topoisomerase-I-Inhibitor; Tyrosinase-Inhibitor.

Myrcene — Analgesic; Anesthetic 10–20 mg/kg ipr mus, 20–40 mg/kg scu mus; Antibacterial; Anticonvulsant; Antimutagenic; Antinitrosaminic; Antioxidant; Antipyretic; Antispasmodic; Fungicide; Insectifuge; Irritant.

Alpinia officinarum Hance (Zingiberaceae)
CHINESE GINGER, LESSER GALANGAL

Synonym — *Languas officinarum* (Hance) Farw.

Medicinal Uses (Lesser Galangal) — It is a favorite spice and medicine in Estonia and Lithuania (GMH). A salve, prepared from the root, is said to be a folk remedy for cancer in Louisiana and Oklahoma. Prescribed for gastralgia and chronic enteritis. Rhizomes are considered aphrodisiac, aromatic, carminative, stimulant, and stomachic, being especially useful in dyspepsia, and in preventing fermentation and flatulence. In India, it is considered a nervine tonic and an aphrodisiac. Reportedly, it clears halitosis when chewed, and sore throat when swallowed. According to India folklore, if given to children, they learn to talk earlier. Powdered with oil or water, it is said to remove freckles. Supposedly reduces the urine flow in diabetics (DEP). Powder is used as a snuff, especially for catarrh (GMH), and sometimes used as medicine for cattle.

Demonstrating synergies between natural compounds and antibiotics, Liu et al. (2001) were able to reverse vancomycin resistance in enterococci with certain flavonoids, including galangin. Combining galangin or 3,7-dihydroxyflavone with vancomycin sensitizes resistant strains of *Enterococcus faecalis* and *Enterococcus faecium*. MICs of vancomycin against 67% of resistant clinical isolates and a type strain of enterococci were lowered from >250 μg/ml to <4 μg/ml with galangin (12.5 μg/ml) or with 3,7-dihydroxyflavone (6.25 μg/ml) (Liu et al., 2001).

Though working with an unrelated species, Kubo et al. (2001) reported tyrosinase-inhibitory activity in three galangal flavonols, galangin, kaempferol, and quercetin (also COX-Inhibitors). The activity stemmed from their ability to chelate copper in the enzyme. The corresponding flavones, chrysin, apigenin, and luteolin, did NOT chelate copper in the enzyme. Galangin inhibits monophenolase activity and all three inhibit diphenolase (X10976523). The COX-2-Inhibitors in this zingiberaceous plant, like those in ginger and turmeric, might render it a useful food pharmaceutical for alleviating, decelerating, or preventing Alzheimer's, arthritis, and colon cancer. Galangin, with antioxidant and radical scavenging activities can modulate enzyme activities and suppress the genotoxicity of chemicals. Galangin may be a promising cancer chemopreventive (Kajiya, 2001).

Indications (Lesser Galangal) — Adenosis (f; HHB; MAD); Ague (f; DAA); Allergy (1; FNF); Alzheimer's (f; MAD); Anemia (f; MAD); Anorexia (2; DAA; KOM; MAD; PH2); Arthrosis (1; COX; FNF); Bacteria (1; BOW; FNF); Bronchosis (2; FNF; PHR; PH2); Cancer (1; FNF; JLH; X11344041); Cancer, bladder (f; JLH); Cancer, colon (1; COX; FNF); Cancer, penis (f; JLH); Cancer, skin (f; BOW); Catarrh (f; GMH); Cholecystosis (2; MAD; PHR; PH2); Cholera (f; DAA); Cold (2; FNF; PHR; PH2); Cramp (1; FNF); Dermatosis (f; BOW); Diarrhea (f; DAA; MAD); Dysmenorrhea (f; DAA; HHB; MAD); Dyspepsia (2; DAA; FNF; GMH; KOM; PH2); Enterosis (f; DAA; PH2; PHR); Epigastrosis (f; BOW); Fever (2; DAA; GMH; PHR; PH2); Flu (1; FNF); Freckle (f; DEP); Fungus (1; BOW; FNF; X1025003); Gas (f; MAD); Gastrosis (f; GMH); Gingivosis (f; BOW); Halitosis (f; DEP); Hepatosis (2; DAA; FNF; PHR; PH2); Hypochondria (f; DAA); Infection (1; FNF; PH2; X1025003); Inflammation (2; FNF; PHR); Malaria (f; EFS); Pain (1; FNF; PH2); Pharyngosis (2; PHR; PH2); Polyuria (f; DEP); Pulmonosis (f; MAD); Rheumatism (f; BOW; MAD); Roemheld Syndrome (f; PH2); Seasickness (f; DAA; GMH; MAD); Sore Throat (f; DEP); Stomachache (f; DAA; MAD; PH2); Stomatosis (2; FNF; PHR; PH2); Stone (f; MAD); Swelling (f; HHB); Syncope (f; DAA; HHB); Toothache (f; DAA); vertigo (f; HHB); vomiting (f; GMH).

Lesser Galangal for cancer:

- AntiHIV: quercetin
- Antiaggregant: eugenol; kaempferol; quercetin
- Antiarachidonate: eugenol
- Anticancer: camphor; eugenol; galangin; isorhamnetin; kaempferol; limonene; linalool; quercetin
- Antifibrosarcomic: quercetin
- Antihepatotoxic: quercetin
- Antiinflammatory: beta-pinene; eugenol; galangin; isorhamnetin; kaempferide; kaempferol; quercetin
- Antileukemic: kaempferol; quercetin
- Antileukotriene: quercetin
- Antilipoperoxidant: quercetin
- Antimelanomic: quercetin
- Antimutagenic: eugenol; galangin; kaempferide; kaempferol; limonene; quercetin
- Antinitrosaminic: quercetin
- Antioxidant: camphene; eugenol; isorhamnetin; kaempferol; quercetin
- Antiperoxidant: galangin; quercetin
- Antiproliferant: quercetin
- Antiprostaglandin: eugenol; phenyl-alkyl-ketones
- Antithromboxane: eugenol
- Antitumor: eugenol; kaempferol; limonene; quercetin
- Antiviral: galangin; kaempferol; limonene; linalool; p-cymene; quercetin
- Apoptotic: kaempferol; quercetin
- COX-2-Inhibitor: eugenol; kaempferol; quercetin
- Chemopreventive: limonene
- Cyclooxygenase-Inhibitor: galangin; kaempferol; quercetin

A

- Cytochrome-p450-Inducer: 1,8-cineole; delta-cadinene
- Cytotoxic: eugenol; quercetin
- Hepatoprotective: eugenol; isorhamnetin; quercetin
- Hepatotonic: 1,8-cineole
- Lipoxygenase-Inhibitor: galangin; kaempferol; quercetin
- Mast-Cell-Stabilizer: quercetin
- Ornithine-Decarboxylase-Inhibitor: limonene; quercetin
- p450-Inducer: 1,8-cineole; delta-cadinene; quercetin
- PTK-Inhibitor: quercetin
- Protein-Kinase-C-Inhibitor: quercetin
- Topoisomerase-II-Inhibitor: kaempferol; quercetin
- Tyrosine-Kinase-Inhibitor: quercetin

Lesser Galangal for colds:

- Analgesic: camphor; eugenol; p-cymene; quercetin
- Anesthetic: 1,8-cineole; camphor; eugenol; linalool
- Antiallergic: 1,8-cineole; kaempferol; linalool; quercetin; terpinen-4-ol
- Antibacterial: 1,8-cineole; delta-cadinene; eugenol; isorhamnetin; kaempferol; limonene; linalool; p-cymene; quercetin; terpinen-4-ol
- Antibronchitic: 1,8-cineole
- Antiflu: limonene; p-cymene; quercetin
- Antihistaminic: isorhamnetin; kaempferol; linalool; quercetin
- Antiinflammatory: beta-pinene; eugenol; galangin; isorhamnetin; kaempferide; kaempferol; quercetin
- Antioxidant: camphene; eugenol; isorhamnetin; kaempferol; quercetin
- Antipharyngitic: 1,8-cineole; quercetin
- Antipyretic: eugenol
- Antiseptic: 1,8-cineole; beta-pinene; camphor; eugenol; kaempferol; limonene; linalool; terpinen-4-ol
- Antitussive: 1,8-cineole; terpinen-4-ol
- Antiviral: galangin; kaempferol; limonene; linalool; p-cymene; quercetin
- Bronchorelaxant: linalool
- COX-2-Inhibitor: eugenol; kaempferol; quercetin
- Cyclooxygenase-Inhibitor: galangin; kaempferol; quercetin
- Decongestant: camphor
- Expectorant: 1,8-cineole; camphene; camphor; limonene; linalool

Other Uses (Lesser Galangal) — Cultivated primarily as a spice, used for over 1000 years in Europe, probably introduced by Arabian or Greek physicians. Arabs feed it to their horses to make them fiery (GMH). The reddish-brown aromatic and pungent rhizomes taste somewhere between pepper and ginger. They have been used to flavor vinegar and the liqueur called "nastoika." Rhizomes are used as a tea or with tea by the Tartars (DEP). It is also used in making curries. Reddish brown powder used as a snuff. Leaves are also edible (FAC). In India, the oil is favored in perfumes. Guenther's *Essential Oils* says it imparts a "warm, unique, and somewhat spicy note." Alcohol "freely extracts all the properties" (GEO).

For more information on activities, dosages, and contraindications, see the *CRC Handbook of Medicinal Herbs,* ed. 2, Duke et al. 2002.

Cultivation (Lesser Galangal) — Sometimes propagated from seed, but usually grown by divisions of the rhizome. Bown (2001) suggests rich, well-drained soils in partial shade with high humidity.

Four- to six-year-old rhizomes are harvested at the end of the growing season. The rhizomes get fibrous when allowed to grow too long. Rhizomes can be used fresh or dried. Bown (2001) notes that spider mites can be a problem under glass.

Chemistry (Lesser Galangal) — Here are a few of the more notable chemicals found in lesser galangal. For a complete listing of the phytochemicals and their activities, see the CRC phytochemical compendium, Duke and duCellier, 1993 (DAD) and the USDA database http://www.ars-grin.gov/duke/.

Galangin — See also *Alpinia galanga.*

Kaempferol — Antiaflatoxic IC50 = 8.73 μ*M*, IC50 = 3.28 ppm; Antiaggregant 30 μ*M*; Antiallergic; Antibacterial 20 μg/ml; Anticancer; Antifertility 250 mg/kg day/60 days/orl rat; Antigingivitic 20 μg/ml; Antihistaminic; Antiimplantation; Antiinflammatory 20 mg/kg, 200 mg/kg ipr rat; Antileukemic IC50 = 3.1 μg/ml; Antilymphocytic; Antimutagenic ID50 = 10–40 nM; Antioxidant IC50 = 40 μ*M*; ¾ quercetin; Antiperiodontic 20 μg/ml; Antiplaque 20 μg/ml; Antiradicular (7 × quercetin); Antiseptic 20 μg/ml; Antiserotonin 200 mg/kg ipr rat; Antispasmodic; Antistaphylococcic; Antitumor; Antiulcer 50–200 mg/kg ipr rat; Antiviral; Apoptotic 60 μ*M*; Aromatase-Inhibitor IC12 = 1 μ*M*/l; 11B-HSD-Inhibitor; cAMP-Phosphodiesterase-Inhibitor; Carcinogenic; Choleretic; Copper-Chelator; COX-2-Inhibitor; Diaphoretic?; Diuretic; Estrogenic EC50 = 0.1–25 μ*M*/l, EC50 = 0.56 μ*M*; Hepatoprotective IC50 = 5.46 μ*M*, IC50 = 1.30 ppm; HIV-RT-Inhibitor IC50 = 50–150 μg/ml; Hypotensive; Inotropic; Iodothyronine-deiodinase-Inhibitor; Lipoxygenase-inhibitor; 5-Lipoxygenase-Inhibitor IC50 (μ*M*) = 20; MAOI; Mutagenic; Natriuretic; Neuroprotective; Protisticide; Quinone-Reductase-Inducer 3 μ*M*; Teratologic; Topoisomerase-I-Inhibitor; Topoisomerase-II-Inhibitor IC50 = 8.1 μg/ml; Tyrosinase-Inhibitor; Uterotropic EC50 = 0.1–25 μ*M*/L; Vasodilator.

Quercetin — Aldose-Reductase-Inhibitor; Allelochemic IC82 = 1 mM; Analgesic; Anesthetic; Antiaflatoxic IC50 = 25 μ*M*, IC50 = 7.5 ppm; Antiaggregant IC50 = 55 μ*M*, 30 μ*M*; Antiallergic IC50 = 14 μ*M*; Antianaphylactic; Antiasthmatic IC50 = 14 μ*M*; Antiatherosclerotic; Antibacterial; Anticancer; Anticapillary Fragility; Anticarcinomic (Breast) IC50 = 1.5 μ*M*; Anticariogenic ID50 = 120 μg/ml; Anticataract; AntiCrohn's 400 mg/man/3×/day; Anticolitic 400 mg/man/3×/day; Antidermatitic; Antidiabetic; Antiencephalitic; Antiescherichic; Antielastase IC50 = 0.8 μg/ml; Antiestrogenic; Antifeedant (IC52 = <1000 ppm diet); Antifibrosarcomic; Antiflu; Antigastric; Antigonadotropic; AntiGTF ID50 = 120 μg/ml; Antihepatotoxic; Antiherpetic; Antihistaminic IC50 = <10 μ*M*; AntiHIV; Antihydrophobic; Antiinflammatory (20 mg/kg) 150 mg/kg; Antileukemic IC50 = >10 μg/ml, 5.5–60 μ*M*, IC50 = 10 μ*M*; Antileukotrienic; Antilipoperoxidant IC67 = 50; Antimalarial IC50 = 1–6.4 μg/ml; Antimelanomic; Antimetastatic; Antimutagenic ID50 = 2–5 nM; Antimyocarditic; Antinitrosaminic; Antioxidant IC96 = 300 ppm, 4.7 × Vit. E; Antiperiodontal; Antipermeability; Antiperoxidant; Antipancreatotic; Antipharyngitic; Antiplaque; AntiPMS 500 mg/2×/day/wmn; Antipodriac; Antipolio; Antiproliferant; Antiprostanoid; Antiprostatic; Antipsoriac; Antiradicular IC50 = 4.6 μ*M*; Antispasmodic; Antistreptococcic ID50 = 120 μg/ml; Antithiamin; Antithrombic; Antitumor; Antitumor (Bladder); Antitumor (Breast); Antitumor (Colon); Antitumor (lung); Antitumor (Ovary); Antitumor (Skin) 20 μ*M*; Antiulcer; Antiviral IC50 = 10 μ*M*; Apoptotic 20–60 μ*M*; ATPase-Inhibitor; Bacteristat 10 mg/ml; 11B-HSD-Inhibitor; Bradycardiac; Calmodulin-Antagonist; cAMP-Phosphodiesterase-Inhibitor; Candidicide; Carcinogenic 40,000 ppm (diet) mus; Catabolic; COMT-Inhibitor; Copper-Chelator; COX-2-Inhibitor <40 μ*M*; Cyclo-oxygenase-Inhibitor; Cytotoxic ED50 = 70 μg/ml, IC82 = 100 μg/ml; Deiodinase-Inhibitor; Diaphoretic?; Differentiator 5.5 μ*M*; Estrogenic (10% genistein); Fungicide; Hemostat; Hepatomagenic 5000 ppm (diet) rat; Hepatoprotective; HIV-RT-Inhibitor IC50 = <1 μg/ml; Hypoglycemic 100 mg/kg orl rats; Hypotensive; Inotropic; Insulinogenic; Juvabional; Larvistat (8,000 ppm diet); Lipoxygenase-Inhibitor IC11 = 1.25 mM, IC50 = 0.1–5 μ*M*; MAOI-Inhibitor; Mast-Cell-Stabilizer; Metalloproteinase-Inhibitor IC50 = >42 μ*M*; MMP-9-Inhibitor 20 μ*M*; Mutagenic; NADH-Oxidase-Inhibitor; NEP-

Inhibitor IC50 = >42 μ*M*; Neuroprotective 5–25 μ*M*; NO-Inhibitor 5–50 μ*M*; ODC-Inhibitor <10 μ*M*; p450-Inducer 5 μ*M*; p450-Inhibitor 50–100 μ*M*; Phospholipase-Inhibitor; Protein-Kinase-C-Inhibitor; PTK-Inhibitor 0.4–24 μ*M*; Quinone-Reductase-Inducer 6 μ*M*, 13 μ*M*; Teratologic; Topoisomerase-I-Inhibitor IC50 = 42 μ*M*, IC50 = 12.8 μg/ml; Topoisomerase-II-Inhibitor IC50 = 23–40 μ*M*, IC50 = 1–6.9 μg/ml; Tumorigenic (0.1% diet orl rat/yr); Tyrosinase-Inhibitor; Tyrosine-Kinase-Inhibitor; Vasodilator; Antinitrosaminic; Xanthine-Oxidase-Inhibitor IC50 = >0.4 μg/ml; LD50 = 160 (orl mus) (may have been contaminated with podophyllin); LD50 = >2000 orl rat PAM.

Amomum aromaticum Roxb. and *Amomum subulatum* Roxb. (Zingiberaceae)

Bengal Cardamom, Black Cardamom, Brown Cardamom, Greater Cardamom, Indian Cardamom, Jalpaiguri Cardamom, Nepalese Cardamom

Hopelessly (taxonomically) inextricable cardamom species, which we will call Nepalese Cardamom.

Medicinal Uses (Nepalese Cardamom) — *The Wealth of India* says that the seeds are used for spice and medicine, much as those of *A. subulatum*. Indeed, they are so alike that Hager's Handbook (1972) treats them under the same entry. I'm inclined to do the same thing here in this spice book, knowing how difficult members of the ginger family are to separate taxonomically. Ayurvedics use the pungent seeds for abdominal pains, biliousness, enlarged spleen, indigestion, itch, and other ailments of the head, mouth, and rectum (KAB). The herbal PDR notes that TCM uses the species for diarrhea, digestive upsets, malaria, and vomiting. The seeds are credited as being alexeteric, astringent, stimulant, and stomachic, having been prescribed for abdominal diseases, biliousness, dyspepsia, rectal diseases, and vomiting. In large doses (30 grains), the seeds are taken with quinine for neuralgia. The seed decoction is gargled for gum and tooth problems. The seeds, with those of melon, are used as diuretics in kidney stones (WOI). Seeds promote elimination of bile, hence useful in liver problems. Seeds also used in gonorrhea (WOI). Unani regard the seeds as astringent, cardiotonic, hepatotonic, hypnotic, orexigenic and stomachic (KAB). The husk of the fruit (pericarp) is used for headache and "heals stomatitis" (WOI). While I'd be reluctant myself, Indians apply the aromatic oil from the seeds to their eyes to soothe inflammation.

Jafri et al. (2001) validated the use of large cardamom (fruit of *A. subulatum*), commonly known as "Heel kalan" or "Bari Ilaichi," in the Unani system of medicine in gastrointestinal disorders. A crude methanolic extract and its different fractions, viz. EO, petroleum ether (60 to 80°), ethyl acetate, and methanolic fractions, were studied in rats for their ability to inhibit the gastric lesions induced by aspirin, ethanol, and pylorus ligature. In addition, their effects on wall mucus, output of gastric acid, and pepsin concentration were recorded. The extract and its fractions of *A. subulatum*, inhibited gastric lesions induced by ethanol significantly, but not those induced by pylorus ligation and aspirin. However, ethyl acetate fraction increased the wall mucus in pylorus ligated rats. The results suggest a direct protective effect of ethyl acetate fraction on gastric mucosal barrier. The observation of decrease in gastric motility by EO and petroleum ether fractions suggests the gastroprotective action of the test drug (X11297839). And there are four antioxidant compounds reported from the fruits, two more potent than tocopherol, two comparable, all with strong radical scavenging activities (X11508709).

Indications (Nepalese Cardamom) — Alzheimer's (1; FNF); Anorexia (f; HH2); Bacteria (1; FNF); Biliousness (f; WOI); Bronchosis (1; FNF); Catarrh (1; FNF); Chill (f; HH2); Cholera (f; KAB); Cold (f; HH2); Conjunctivosis (f; WOI); Cramp (1; FNF); Diarrhea (f; PH2); Dyspepsia (f; PH2; WOI); Enterosis (f; WOI); Fatigue (1; FNF); Fever (f; PH2); Gastrosis (f; KAB); Gingivosis (f; WOI); Gonorrhea (f; KAB); Gravel (f; WOI); Headache (f; WOI); Hepatosis (f; WOI); Impotence (f; KAB); Infection (1; FNF); Inflammation (1; FNF); Malaria (f; PH2); Nephrosis (f; WOI); Neuralgia (f; KAB; WOI); Odontosis (f; WOI); Pain (1; FNF; WOI); Proctosis (f; WOI); Snakebite (f; HH2); Staphylococcus (1; FNF); Sting (f; HH2); Stomatosis (f; WOI); Trichomonas (1; FNF); Ulcer (1; FNF); VD (f; WOI); Vomiting (f; PH2).

Other Uses (Nepalese Cardamom) — Seeds widely used as a spice. *A. aromaticum* is probably not as popular in India as the large cardamom, *A. subulatum*. Both are apparently used interchangeably as spice and as ingredients in masticatories and snuffs. In South India, it is a major constituent in "agarbatties." It is an important constituent in Afghan "char marsala," a culinary spice mix. Whole fruits may be ground and powdered into rice pilafs (FAC). The pungently aromatic seeds are often substituted for true cardamom, the husks often powdered and added to cattle feed (WOI).

For more information on activities, dosages, and contraindications, see the *CRC Handbook of Medicinal Herbs,* ed. 2, Duke et al. 2002.

Cultivation (Nepalese Cardamom) — Plants are probably cultivated like other ginger relatives, by rhizome divisions or rhizome cuttings, as new growth resumes. These are planted in rich, well-cultivated, moisture-rich soils in partial shade. Seeds probably harvested at the end of the growing season. Bown (2001) notes that spider mites can be a problem under glass.

Chemistry (Nepalese Cardamom) — Here are a few of the more notable chemicals found in nepalese cardamom. For a complete listing of the phytochemicals and their activities, see the CRC phytochemical compendium, Duke and duCellier, 1993 (DAD) and the USDA database http://www.ars-grin.gov/duke/.

Beta-Pinene — Antiinflammatory; Antiseptic; Antispasmodic; Candidicide; Insectifuge; Herbicide; Irritant; Spasmogenic; LD50 = 4700 mg/kg orl rat.

Cardamomin — AntiEBV IC50 = 3.1 μM; Antiviral IC50 = 3.14 μM.

1,8-Cineole — See also *Elettaria cardamomum.*

Limonene — See also *Carum carvi.*

Amomum compactum Soland. ex Maton (Zingiberaceae)
CLUSTER CARDAMOM, JAVA CARDAMOM, ROUND CARDAMOM, SIAM CARDAMOM

Synonyms — *Amomum kepulaga* Sprague & Burk.

Medicinal Uses (Round Cardamom) — According to Hartwell (1982), the plants are used in folk remedies for indurations of the liver and uterus, and for cancer. Reported to be antitoxic, antiemetic, carminative, and stomachic. Rarely used alone in China, more frequently used in combinations (e.g., mixed with fresh egg yolks), it is used during parturition. Used, along with other cosmetic fragrances, in a Malayan recipe for madness (DAD).

Indications (Round Cardamom) — Ague (f; DAD); Alzheimer's (1; FNF); Bacteria (1; FNF); Bronchosis (1; FNF); Cachexia (1; DAD; FNF); Cancer (1; DAD; FNF; JLH); Cancer, liver (f; DAD; JLH); Cancer, uterus (f; DAD; JLH); Catarrh (f; DAD); Childbirth (f; DAD); Cold (1; DAD; FNF); Cough (1; FNF); Cramp (1; DAD; FNF); Dyspepsia (f; DAD); Fatigue (1; FNF); Flu (1; FNF); Gout (1; DAD; FNF); Heartburn (f; DAD); Hepatosis (f; DAD; JLH); Induration (f; JLH); Infection (1; FNF); Inflammation (1; FNF); Madness (f; DAD); Nausea (f; DAD); Ophthalmia (f; DAD); Pain (1; FNF); Rheumatism (1; DAD; FNF); Staphylococcus (1; FNF); Trichomonas (1; FNF); Ulcer (1; FNF); Uterosis (f; JLH); Vomiting (f; DAD).

Round Cardamom for Alzheimer's:

- ACE-Inhibitor: alpha-terpineol
- AChE-Inhibitor: limonene

- Antiacetylcholinesterase: 1,8-cineole; carvone; limonene; p-cymene; terpinen-4-ol
- Anticholinesterase: 1,8-cineole
- Antiinflammatory: alpha-pinene; beta-pinene; borneol; caryophyllene
- CNS-Stimulant: 1,8-cineole; borneol; camphor; carvone

Round Cardamom for cold/flu:

- Analgesic: borneol; camphor; p-cymene
- Anesthetic: 1,8-cineole; camphor
- Antiallergic: 1,8-cineole; terpinen-4-ol
- Antibacterial: 1,8-cineole; alpha-pinene; alpha-terpineol; caryophyllene; limonene; p-cymene; terpinen-4-ol
- Antibronchitic: 1,8-cineole; borneol
- Antiflu: alpha-pinene; limonene; p-cymene
- Antiinflammatory: alpha-pinene; beta-pinene; borneol; caryophyllene
- Antipharyngitic: 1,8-cineole
- Antipyretic: borneol
- Antiseptic: 1,8-cineole; alpha-terpineol; beta-pinene; camphor; carvone; limonene; terpinen-4-ol
- Antitussive: 1,8-cineole; terpinen-4-ol
- Antiviral: alpha-pinene; limonene; p-cymene
- Decongestant: camphor
- Expectorant: 1,8-cineole; alpha-pinene; camphor; limonene

Round Cardamom for cramp:

- Analgesic: borneol; camphor; p-cymene
- Anesthetic: 1,8-cineole; camphor
- Antiinflammatory: alpha-pinene; beta-pinene; borneol; caryophyllene
- Antispasmodic: 1,8-cineole; borneol; camphor; caryophyllene; limonene
- Carminative: camphor; carvone
- Myorelaxant: 1,8-cineole; borneol
- Sedative: 1,8-cineole; alpha-pinene; alpha-terpineol; borneol; carvone; caryophyllene; limonene; p-cymene
- Tranquilizer: alpha-pinene

Other Uses (Round Cardamom) — One of the two more important cardamoms of Indonesia. Rumpf preferred them for culinary purposes to *Elettaria cardamomum*. At one time, the fruit was official in the French Codex. Grieve's Herbal speculates that the round cardamoms of Dioscorides are those called *Amomi uva* by Pliny (perhaps *A. globosum*). The fruits serve as a warm aromatic spice. Fruits have a sweet, turpentine aroma and flavor and are used as a spice or chewed to sweeten the breath. Seeds are used in cakes in Malaysia (FAC). In Java, hearts of the young shoots of closely related *Amomum maximum*, also called Java Cardamom, are eaten as lablab. Young inflorescences and young fruits are also cooked with rice and eaten. The fresh juicy aril of ripe seeds of that species are a relished delicacy (DAD). Young, pungent shoots are eaten raw, roasted, or cooked and served with rice (FAC).

For more information on activities, dosages, and contraindications, see the *CRC Handbook of Medicinal Herbs,* ed. 2, Duke et al. 2002.

Cultivation (Round Cardamom) — Planted from small segments of the rhizome, usually in prepared ground under trees, e.g., coconut. Probably cultivated like the true cardamom or other ginger relatives, by rhizome divisions or rhizome cuttings, as new growth resumes. These are

A

planted in rich, well-cultivated, relatively moisture rich soils, with partial shade. Seeds probably harvested at the end of the growing season. Bown (2001) notes that spider mites can be a problem under glass.

Chemistry (Round Cardamom) — Roughly a kilogram of roots and rhizome will give a cubic centimeter of EO. Seeds yield 2.4% oil with the aroma of borneol and camphor, both of which have been identified in the oil, also terpineol. The fruit oil contains 12% cineole. Purseglove et al. (1981), casting doubt on earlier analyses, report 1,8-cineole as the major constituent (67%), with 16% beta-pinene, 4% alpha-pinene, 5% alpha-terpineol, 3% humulene, some p-cymene, limonene, myrcene, d-camphor, carvone, myrtenal, d-borneol, alpha- and gamma-terpineol, terpinen-4-ol, caryophyllene, and humulene epoxiide II, but no alpha-terpinyl acetate. Here are a few of the more notable chemicals found in round cardamom. For a complete listing of the phytochemicals and their activities, see the CRC phytochemical compendium, Duke and duCellier, 1993 (DAD) and the USDA database http://www.ars-grin.gov/duke/.

Alpha-Pinene — Allelochemic; Antibacterial; Anticancer; Antifedant; Antiflu; Antiinflammatory; Antispasmodic; Antiviral; Coleoptiphile; Expectorant; Herbicide IC50 = 30 μ*M*; Insectifuge (50 ppm); Insectiphile; Irritant; p450(2B1)-Inhibitor IC50 = 0.087 μ*M*; Sedative; Spasmogenic; Tranquilizer; Transdermal.

Beta-Pinene — Antiinflammatory; Antiseptic; Antispasmodic; Candidicide; Insectifuge; Herbicide; Irritant; Spasmogenic; LD50 = 4700 mg/kg orl rat.

Borneol — Allelochemic; Analgesic; Antiacetytlcholine; Antibacterial MIC = 125–250 μg/ml; Antibronchitic; Antiescherichic MIC = 125 μg/ml; Antifeedant; Antiinflammatory; Antiotitic; Antipyretic; Antisalmonella MIC = 125 μg/ml; Antispasmodic ED50 = 0.008 mg/ml; Antistaphylococcic MIC = 250 μg/ml; Candidicide MIC = 250 μg/ml; Choleretic; CNS-Stimulant; CNS-Toxic; Fungicide; Hepatoprotectant; Herbicide IC50 = 470 mM, IC50 = 470 μ*M*; Inhalant; Insectifuge; Irritant; Myorelaxant; Negative Chronotropic 29 μg/ml; Negative Inotropic 29 μg/ml; Nematicide MLC = 1 mg/ml; Sedative; Tranquilizer; LDlo = 2000 (orl rbt).

D-Borneol — Convulsant; Emetic; Fragrance; Irritant; LD50 = 1059 mg/kg orl rat.

Bornyl-Acetate — Antibacterial; Antifeedant; Antispasmodic ED50 = 0.09 mg/ml; Antiviral; Expectorant; Insectifuge; Myorelaxant; Negative Chronotropic 933 nl/ml; Negative Inotropic 933 nl/ml; Sedative.

Camphor — Allelopathic; Analgesic; Anesthetic; Antiacne; Anticancer; Antidiarrhea 500 μg/ml; Antidysenteric 500 μg/ml; Antiemetic 100–200 mg man orl; Antifeedant IC50 = 5000 ppm diet; Antifibrositic; Antiitch; Antineuralgic; Antiseptic; Antispasmodic ED50 = 0.075 mg/ml; CNS-Stimulant; Carminative; Convulsant; Cosmetic; Counterirritant; Decongestant; Deliriant; Ecbolic; Emetic; Epileptigenic; Expectorant; Fungicide ED50 = 2.7 mM; Herbicide IC50 = 3.3–180 mM; Insectifuge; Irritant; Nematicide MLC = 1 mg/ml; Oculoirritant; p450(2B1)-Inhibitor IC50 = 7.89 μ*M*; Pesticide; Respirainhibitor; Respirastimulant; Rubefacient; Stimulant; Transdermal; Verrucolytic; Vibriocide 500 μg/ml; LD = 1200 mg/kg ipr rat.

1,8-Cineole — See also *Elettaria cardamomum*.

Apium graveolens L. (Apiaceae)
CELERY

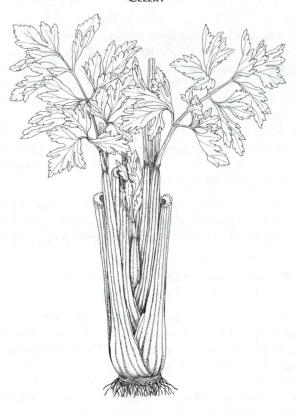

Medicinal Uses (Celery) — Regarded as aperient, carminative, diuretic, emmenagogue, nervine, sedative, stimulant, stomachic, and tonic. Eclectics suggested the herbal or root infusion for bronchosis, intermittent fever, neurosis, and rheumatism. The bruised plant is applied topically as an analgesic (FEL). Celeryseed extract is the only "medicine" I take every day, and either celery seed, old age, or serendipity has prevented gout crises for 6 years, since I discontinued the pharmaceutical allopurinol. Ayurvedic scientists said in 1976 that celery prevents rheumatism and gout. Hildegarde von Bingen reported its use for gout ("gicht") closer to 1000 years ago. They could not have known then that apigenin was a COX-2 Inhibitor. After chamomile (to 0.6%), celery stalks (to 0.2%) are the best source of apigenin cited in my USDA database. Maybe chamomile would be as useful for gout as celeryseed and celery stalks are.

Weed (2002) suggests heavy consumption of certain spices when estrogen levels are down. Seeds like anise, caraway, celery, coriander, cumin, and fennel contain phytoestrogens, as do their oils, says Weed. She suggests using these seeds "lavishly" when cooking or making tea with any one of them, drinking 3 to 4 cups a day "for best results" (Weed, 2002). A yellowish oil derived from the roots is said to repair sexual impotence brought about by illness. A rheumatism remedy recommended in the old herbals sounds pretty potent. It combines the "aphrodisiac" celery with the "aphrodisiac" damiana and the narcotic coca, source of cocaine. That should at least take one's mind off the rheumatism.

The Herbal PDR (PH2) notes that animal tests demonstrated anticonvulsive and sedative, but not diuretic effects, for the roots, seeds, and stalks. The EO inhibits bacteria and fungi but not dramatically (PH2), and it has anticonvulsant and tranquilizing activities in rats (MPI). Celery juice and stalks have been reported to lower blood pressure in hypertensive humans (HAD). This may seem contradictory, since celery stalks are rather rich in sodium (about 50 mg per average serving).

Of course, the role of salt in hypertension has flip-flopped recently anyhow. Is it bad or not? Cardiac and hypertensive patients are still often advised to avoid celery (but remember, the equivalent of four stalks a day has been shown to lower high blood pressure). Rinzler (1990) tells us that a teaspoon of celery seed on a salad adds only 3 mg sodium, an amount considered low. In a study of clinical applications of Ayurvedic herbs, Khan and Balick (2001) note that celery extracts (ipr injected in rats) significantly decreased serum cholesterol concentrations. In Wistar rats fed high-cholesterol diets, those rats also fed aqueous extracts had significantly lower total serum cholesterol, LDL cholesterol, and triglycerides (PM61:18). Extracts of the seeds or phytochemicals therein showed antiinflammatory, antitumor, hepatoprotective, and hypocholesterolemic effects (Khan and Balick, 2001).

As a natural COX-2 Inhibitor, apigenin may just deserve as much hype as the synthetic COX-2-Inhibitors, alleviating arthritis, gout, inflammation, rheumatism, and possibly preventing Alzheimer's and cancer. Jeong et al. (1999) screened 28 flavonoids for inhibitory effects against aromatase (prepared from human placenta). Over 50% of the flavonoids significantly inhibited aromatase activity, with apigenin strongest at IC50 = 0.9 µg/ml, hesperetin next at 1.0 µg/ml, and chrysin at 1.1 µg/ml (Jeong et al., 1999). Dec. 10, 2001, NBC national news highly praised synthetic aromatase inhibitors for the prevention and containment of breast cancers. There was, expectedly, no mention of the natural aromatase inhibitors. But a quick look at PubMed abstracts showed that several natural flavonoids were also active, some at levels 1/10th as potent as the possibly more dangerous synthetics. Here's just one more line of evidence that a variety of vegetables can contribute to the prevention and hence reduction of cancer. Celery and chamomile are my top sources of apigenin which is one fifth to one tenth as potent as the synthetics. On the negative side, Saarinen et al. (2001), aware of the *in vitro* antiaromatase activity, were unable to find such activity *in vivo*. In rats given doses of 50 mg/kg, neither apigenin, chrysin, nor naringenin inhibited aromatase. They suspected this lack of inhibition may have been due to relatively poor absorption and/or bioavailability of the phytochemicals (Saarinen et al., 2001). Turmeric and tamarind might possibly improve availability. Boik (2001) certainly praises apigenin, if not celery or celeryseed, as a cancer preventive. More than 26 studies on the flavones apigenin and/or luteolin show that they inhibit a wide variety of cancer cell lines usually at levels of 1–50 μM. Boik (2001) reports only one animal study of apigenin, wherein ipr apigenin inhibited growth and metastasis of melanoma in mice at 25–50 mg/kg, roughly equivalent to human doses of 1.2–2.5 apigenin a day in man. Apigenin may have some estrogenic activity (BO2), but only 16% the estrogenic activity of genistein, which "appears capable of producing an estrogenic effect and stimulating cancer growth." Luteolin seems ca. 58% as estrogenic, resveratrol 22%, and quercetin ca. 10% as estrogenic. The flavones are more potent antioxidants than vitamin C, apigenin ca. 1.5 times more potent, resveratrol 2 times, quercetin 4.7 times, and OPCs nearly 5 times as potent. In scavenging free radicals, these compounds can also prevent cancer. My database lists several mechanisms by which apigenin might prevent cancer and improve cell-to-cell communication: antiangiogenic (at 4 μM), antiaromatase (IC50 = 0.9 µg/ml), antihistaminic (IC50 = 10–35 μM), antileukemic (20–50 μM), antimelanomic (1–50 μM), antimetastatic (1–50 μM), antimutagenic (IC50 = 10–40), antiproliferant (1–50 μM), apoptotic, COX-2-Inhibitor (<40 μM), cytotoxic (1–50 μM), beta-Glucuronidase-Inhibitor (IC50 = ~40 μM), hyaluronidase-inhibitor (IC50+ = 50–250 μM), p21-inducer (10–70 μM), p450-inducer PKC-inhibitor (IC50 = 10–40 μM), Polyamine-Synthesis-Inhibitor, PTK-Inhibitor (10–100 μM), Quinone-Reductase-Inducer 20 μM, ras-cascade inhibitor, Topoisomerase-I-Inhibitor; Topoisomerase-II-Inhibitor (IC28 = 18 μM). So we have one flavone with more than a different activities that could reduce the incidence of cancer, and we have at least half a dozen other flavonoids doing many of the same things in slightly different levels. I presume that is what emboldens Boik to predict synergies among these food farmaceuticals. Boldly he calculates some tentative human dosages as 1.2–2.5 g/day (as scaled from animal antitumor studies), 0.66–4.1 g day (as scaled from animal antiinflammatory studies), leading to a target dose of 1,500 mg/day apigenin. Even more speculatively, he suggests that synergies with other phytochemicals may reduce that minimum antitumor dose to 100 mg apigenin/day, paralleling his minimum anticancer dose for genistein (e.g., in beans). He comes out with a minimum antitumor dose of 170 mg luteolin and 250 mg quercetin through similar calculations. Looks like good reasoning for curried bean/celery soup with plenty of onion for its quercetin and garlic for its DADS, etc. Apigenin is reportedly, like quercetin, both an inducer and inhibitor of p450 enzymes, reportedly inhibiting at (IC50 = 1.35 μM, IIIA-4; 100 μg and IC50 = 0.1 μM, inducing at 5 μM/rat ipr). For more information, you should consult Dr. Flockhart's database at Georgetown (http://www.dml.georgetown.edu/depts/pharmacology/davetab.html).

Celery seed oil is bacteriostatic against *Bacillus pumilus*, *Bacillus subtilis*, *Corynebacterium diptheriae*, *Pseudomonas solanacearum*, *Salmonella typhi*, *Shigella dysenteriae*, *Staphylococcus albus*, *Staphylococcus aureus*, *Streptococcus faecalis*, *Streptococcus pyogenes*, and *Vibrio cholerae*. Celery showed potent antimicrobial activities against *Bacillus subtilis*, *Escherichia coli*, and *Saccharomyces cerevisiae*, confirming traditional uses of spices as food preservatives, disinfectants, and antiseptics (De et al., 1999). The oil shows a chemotactic effect and cercaricidal activity of the cercaria of *Schistosoma mansoni* (SPI).

Indications (Celery) — Alzheimer's (1; COX; FNF); Amenorrhea (f; CRC; DEP; KAB); Anasarca (f; CRC; DEP; KAB; WO2); Anorexia (f; KAB; PHR; PH2); Anxiety (1; APA); Arthrosis (1; APA; FNF; PNC); Ascites (f; KAB); Asthma (f; DEP; JFM; KAB); Bacillus (1; X10548758); Bacteria (1; PH2); Bronchosis (f; DEP; KAB); Cancer (1; CRC; FNF); Cancer, breast (1; CRC; FNF); Cancer, eye (1; CRC; FNF); Cancer, foot (1; CRC; FNF); Cancer, liver (1; CRC; FNF); Cancer, penis (1; CRC; FNF); Cancer, spleen (1; CRC; FNF); Cancer, stomach (1; CRC; FNF; JAC7:405); Cancer, testes (1; CRC; FNF); Cancer, uterus (1; CRC; FNF); Cancer, vulva (1; CRC; FNF); Cardiopathy (1; APA; KAB); Catarrh (f; KAB); Cholecystosis (f; PH2); Colic (f; DEP; MBB; WO2); Condyloma (f; JLH); Congestion (f; JFM); Convulsion (f; KAP); Corn (f; CRC; JLH); Cough (f; KAB; PH2); Cramp (1; FNF); Cystosis (1; APA; CAN; FNF; MBB); Depression (f; PED); Diabetes (f; APA; MAM); Dysmenorrhea (f; APA; JFM); Dyspepsia (f; APA); Dysuria (f; KAB); Edema (1; FNF; JFM); Enterosis (f; KAB); Epilepsy (1; PNC; WO2); Escherichia (1; X10548758); Exhaustion (f; PH2); Felon (f; CRC; JLH); Fever (f; FEL; KAB); Fungus (1; PH2; X10548758); Gallstone (f; PHR); Gas (1; CRC; JFM); Gastrosis (1; JAC7:405); Gout (1; CAN; FNF; MBB; MPI; PH2); Hepatosis (f; APA; CRC; DEP; JLH); Hiccup (f; KAB); High Blood Pressure (2; APA; CRC; FNF; PNC); High Cholesterol (1; APA; PM61:18); High Triglycerides (1; PM61:18); Impostume (f; JLH); Induration (f; CRC; JLH); Infection (1; X10548758); Inflammation (1; FNF; KAB); Insomnia (1; APA; FNF; PNC); Jaundice (f; JFM); Kidney Stone (f; PHR); Lumbago (f; CRC); Malaria (f; FEL); Nausea (f; KAB); Nephrosis (f; APA; PH2); Nervous Anxiety (f; APA; PHR); Obesity (f; APA); Ophthalmia (f; KAB); Ovary (f; PH2); Pain (1; FNF; KAB); Proctosis (f; KAB); Pulmonosis (f; JFM); Rheumatism (1; CAN; CRC; FEL; FNF; MPI; PH2); Rhinosis (f; KAB); Scabies (f; KAB); Schistosoma (1; SPI); Scirrhus (f; JLH); Sore (f; CRC); Splenosis (f; CRC; DEP; JLH; KAB; WO2); Sting (f; KAB); Stomachache (f; CRC; JFM); Stone (f; DEP; PHR; PH2); Stress (1; APA); Swelling (1; FNF; MBB); Toothache (f; KAB); Tumor (1; CRC; FNF; JLH); Uterosis (f; JFM); UTI (1; CAN; FNF); Water Retention (1; FNF); Wen (f; JLH); Whitlow (f; CRC; JLH); Worm (1; X11305251); Yeast (1; X10548758).

Celery for high blood pressure:

- Antiaggregant: adenosine; alpha-linolenic-acid; apigenin; bergapten; caffeic-acid; cinnamaldehyde; eugenol; falcarindiol; ferulic-acid; imperatorin; kaempferol; menthone; myristicin; nodakenin; quercetin; salicylates; thymol
- Antioxidant: apigenin; caffeic-acid; camphene; chlorogenic-acid; eugenol; ferulic-acid; fumaric-acid; gamma-terpinene; isoquercitrin; kaempferol; linalyl-acetate; luteolin; mannitol; myrcene; myristicin; p-coumaric-acid; p-hydroxycinnamic-acid; protocatechuic-acid; quercetin; rutin; scopoletin; sinapic-acid; thymol

- Antistress: myristicin
- Antithromboxane: eugenol
- Anxiolytic: apigenin
- Arteriodilator: adenosine; ferulic-acid
- Cardiodepressant: guaiacol
- Cardiotonic: guaiacol
- Diuretic: adenine; apigenin; apiole; asparagine; caffeic-acid; chlorogenic-acid; glycolic-acid; isopimpinellin; isoquercitrin; kaempferol; luteolin; mannitol; myristicin; terpinen-4-ol
- Hypotensive: adenosine; alpha-linolenic-acid; angelicin; apigenin; benzyl-benzoate; bergapten; cinnamaldehyde; isoquercitrin; kaempferol; myristicin; psoralen; quercetin; rutin; scoparone; scopoletin; valeric-acid
- Myocardiotonic: adenine
- Sedative: adenosine; alpha-pinene; alpha-terpineol; angelic-acid; angelicin; apigenin; caffeic-acid; carvone; caryophyllene; cinnamaldehyde; citronellal; eugenol; geranyl-acetate; imperatorin; isovaleric-acid; limonene; linalool; linalyl-acetate; menthone; methyl-phthalides; myristicin; p-cymene; perillaldehyde; sedanenolide; thymol; valeric-acid
- Vasodilator: adenine; adenosine; apigenin; apiole; cinnamaldehyde; kaempferol; luteolin; quercetin; scoparone

Celery for rheumatism:

- Analgesic: adenosine; caffeic-acid; chlorogenic-acid; eugenol; falcarindiol; ferulic-acid; gentisic-acid; menthone; myrcene; nicotine; osthol; p-cymene; quercetin; scoparone; scopoletin; thymol
- Anesthetic: cinnamaldehyde; eugenol; formaldehyde; guaiacol; linalool; linalyl-acetate; myrcene; myristicin; thymol
- Antiarthritic: thymol
- Antidermatitic: apigenin; fumaric-acid; guaiacol; quercetin; rutin; xanthotoxin
- Antiedemic: caffeic-acid; caryophyllene; caryophyllene-oxide; eugenol; rutin; scopoletin
- Antiinflammatory: alpha-linolenic-acid; alpha-pinene; apigenin; bergapten; beta-pinene; butylidene-phthalide; caffeic-acid; caryophyllene; caryophyllene-oxide; chlorogenic-acid; cinnamaldehyde; cnidilide; eugenol; ferulic-acid; gentisic-acid; imperatorin; isopimpinellin; kaempferol; luteolin; mannitol; myristicin; osthol; protocatechuic-acid; quercetin; quercetin-3-o-galactoside; rutin; salicylates; scoparone; scopoletin; thymol; umbelliferone; xanthotoxin
- Antiprostaglandin: caffeic-acid; eugenol; ligustilide; scopoletin; umbelliferone
- Antirheumatic: gentisic-acid; p-cymene; thymol
- Antispasmodic: adenosine; angelicin; apigenin; apiin; apiole; benzyl-benzoate; bergapten; butylidene-phthalide; caffeic-acid; caryophyllene; cinnamaldehyde; cnidilide; eugenol; ferulic-acid; herniarin; kaempferol; ligustilide; limonene; linalool; linalyl-acetate; luteolin; mannitol; menthone; methyl-phthalides; myrcene; myristicin; p-coumaric-acid; protocatechuic-acid; psoralen; quercetin; rutin; scopoletin; thymol; umbelliferone; valeric-acid; xanthotoxin; z-ligustilide
- COX-2-Inhibitor: apigenin; eugenol; kaempferol; quercetin
- Counterirritant: thymol
- Cyclooxygenase-Inhibitor: apigenin; cinnamaldehyde; kaempferol; quercetin; thymol
- Lipoxygenase-Inhibitor: caffeic-acid; chlorogenic-acid; cinnamaldehyde; kaempferol; luteolin; p-coumaric-acid; quercetin; rutin; umbelliferone
- Myorelaxant: adenosine; angelicin; apigenin; benzyl-benzoate; ligustilide; luteolin; rutin; scoparone; scopoletin; thymol; valeric-acid

Other Uses (Celery) — The leaf stalk is the better known part of the plant in America. They are blanched and eaten raw in salads, or braised, fried or steamed, or stuffed with cream cheese, peanut butter, or pimenta cheese. They are added to casseroles, juices, soups, stews, stuffings, and casseroles. Leaves and seeds are both often used in continental cookery, where the leaves are added at the last moment to broth, soups, stews, and stuffing. Dried leaves, occasionally marketed as "celery flakes," are stronger flavored than seeds. They too are added to salads and sandwiches (FAC). Celery seed, seed extract, and oil are used for flavoring baked goods, beverages, herbal salts and peppers (celery seed with black pepper), juices, omelettes, pickles, soups, stews, etc. Celery juice is often blended with carrot, parsley, and spinach juices, at my house more often with carrot, garlic, and onion, which I consider great for preventing hypertension (FAC, JAD). Rinzler adds that celery seeds are the smallest seeds used as a spice or flavoring, 760,000 seed to the pound. The seed or the oil derived therefrom are used in celery salt, celery tonic, liqueurs, perfumes, soups, stews, and toilet waters. The French prepare from celeryseed, the liqueur called Creme d'Celery, which does not differ greatly in flavor from kummel. The recipe may contain celery seed, plus or minus caraway, cumin or fennel, crushed and steeped in sugared vodka. Anise may be added but with care, or it will be the dominant flavor (FAC, JAD). The seed oil lends a floral-like odor to oriental perfumes, imparting a warm and clinging note (Vernin and Parkanyi, 1994). Nakatani's summary of Mori et al.'s study (1974) "Essential oils of celery, cinnamon, coriander and cumin" were comparable to sorbic acid at preventing the slimy spoilage of Vienna sausage (Nakatani, 1994, SPI).

For more information on activities, dosages, and contraindications, see the *CRC Handbook of Medicinal Herbs,* ed. 2, Duke et al. 2002.

Cultivation (Celery) — Slow starting and slow growing, this well-known biennial vegetable is not so well known as an herb. Start seeds indoors, quite early, for transplant to the field. The slowly germinating seedlings require hardening for frost and drought. Originally, the plant was rather tolerant of saline marshy soils. It is difficult to grow the vegetable, easier to grow the herbage. Leaves, as opposed to celery stalks, can be harvested anytime and used fresh if dried slowly. Stalk yields of 20–30 tons/a are possible (Duke, 1985).

Chemistry (Celery) — Here are a few of the more notable chemicals found in celeryseed. For a complete listing of the phytochemicals and their activities, see the CRC phytochemical compendium, Duke and duCellier, 1993 (DAD) and the USDA database http://www.ars-grin.gov/duke/.

Apigenin — Antiaflatoxic IC50 = 9.52 μM, IC50 = 2.57 ppm; Antiaggregant; Antiallergic; Antiangiogenic 4 μM; Antiarrhythmic; Antibacterial; Anticancer; Anticomplementary; Antidermatitic; Antiestrogenic; Antiherpetic; Antihistaminic IC50 = 10–35 μM; AntiHIV IC72 = 200 μg/ml, IC50 = 143 μg/ml; Antihyaluronidase IC50+ = 50–250 μM; Antiinflammatory (= indomethacin), IC~65 = 1,000 μM; Antileukemic 20–50 μM; Antimelanomic 1–50 μM; Antimetastatic 1–50 μM; Antimutagenic ID50 = 10–40 nM; Antioxidant 1.5 × Vit. E; Antiproliferant 1–50 μM; Antispasmodic EC50 = 1–5 μM; Antithyroid; Antitumor 1–50 μM; Antitumor (Breast) 1–50 μM; Antitumor (Lung); Antitumor (Skin); Antiviral; Anxiolytic 10 mg/kg; Apoptotic 12–60 μM; Aromatase-Inhibitor IC65 = 1 μM/l; 11B-HSD-Inhibitor; Calcium-blocker?; Choleretic; CNS-Depressant; COX-1-Inhibitor IC65 = 1,000 μM; COX-2-Inhibitor IC>65 = 1,000 μM, <40 μM; Cytotoxic IC88 = 10 μg/ml, 1–50 μM; Deiodinase-Inhibitor; Differentiator MIC 30 μM, IC40 = 40 μM; Diuretic; Estrogenic EC50 = 0.1–25 μM/L, (16% genistein), EC50 = 1 μM; Beta-Glucuronidase-Inhibitor IC50 = ~40 μM; Hypotensive; Inotropic; MAOI; Musculotropic; Mutagenic; Myorelaxant; NADH-Oxidase-Inhibitor; NO-Inhibitor 5–50 μM; Nodulation-Signal; ODC-Inhibitor; p21-Inducer 10–70 μM; PKC-Inhibitor IC50 = 10 μM; Polyamine-Synthesis-Inhibitor; Progestational; Protein-Kinase-C-inhibitor IC50 = 10–40 μM; PTK-Inhibitor 10–100 μM; Quinone-Reductase-Inducer 20 μM; Radioprotective; Sedative 30–100 mg/kg; Sunscreen; Topoisomerase-I-Inhibitor; Topoisomerase-II-Inhibitor IC28 = 18 μM, IC45 = 180 μM, 50 μg/ml; Uterotropic EC50 = 0.1–25 μM/l; Vasodilator.

Apiin — Aldose-Reductase-Inhibitor; Antiarrhythmic; Antibradykinic; Anatispasmodic.

Apiole — Abortifacient 5,000 orl rbt; Antidysmenorrheic; Antimalarial; Antineuralgic; Antipyretic; Antireproductive 0.013 man; Antispasmodic; Calcium-Antagonist IC50 = 29.2 μM; CNS-Stimulant (= caffeine); Diuretic; Emmenagogue; Hepatotoxic; Insecticide; Intoxicant; Irritant; Secretolytic; Synergist; Uterotonic; Vasodilator; LDlo = 500 (scu dog); LDlo 25,000–35,000/man; LD 2,000 orl gpg; LD50 = 50 mg/kg ivn mus.

Sedanenolide — Sedative.

Sedanolide — Antitumor 60 mg/mus/6 days.

Armoracia rusticana P. Gaertn. et al. (Brassicaceae)
HORSERADISH

Synonyms — *A. lapathifolia* Gilib. ex Usteri, *Cochlearia armoracia* L., *Nasturtium armoracia* (L.) Fr., *Radicula armoracia* (L.) B. L. Rob., *Rorippa armoracia* (L.) Hitchc.

Medicinal Uses (Horseradish) — Could horseradish be one of the bitter herbs of the Bible? Well, horseradish is reportedly one of the five bitter herbs, with coriander, horehound, lettuce, and nettle, during the Jewish Passover (Libster, 2002). None of my Bible books suggest that horseradish is one of the bitter herb(s). Tucker and DeBaggio (2000) tell us that horseradish was cultivated prior to the Exodus of the Hebrew slaves. But Bown (2001) suggests it may have been brought in to cultivation less than 2000 years ago, perhaps first medicinal, secondarily spice, not "becoming popular as a flavoring until the late 16th century." It is bitter and so important, traditionally, in our modern Passover feasts. Considered antiscorbutic, antiseptic, digestive, diuretic, expectorant, rubefacient, stimulant, and vermifuge, horseradish is about as useful in the medicine chest as it is in the spice rack. In ancient times, it was used as a preventative or cure for many ailments. Eclectics said picturesquely, "It promotes all the secretions, the urinary in particular, and stimulates the stomach when this organ is enfeebled." The hot cider infusion has been recommended for dropsy (FEL). "Locally, the vinegar infusion is said to remove tan and freckles" (FEL). The herb is approved, internally, by Germany's Commission E for bronchosis, coughs, respiratory catarrh, and UTIs, externally in rubefacient liniments for respiratory ailments and myalgia (BGB, PH2). In the U.S., horseradish is an ingredient in the proprietary Rasapen, a UTI (BGB). Native Americans have even incorporated the imported species into their pharmacopoeia as well (DEM), also using it for digestive and urinary disorders. Reportedly, they used it for inducing abortions as well, though probably not early on with the eclectic prescription. Eclectics made a saturated whiskey infusion of new or recent roots, of which they recommended 4 oz, 3 or 4 times a day (that's a pint a day in my book) and continuing treatment until the abortion was obtained (FEL). The root can be chewed for toothache. Horseradish extracts are used to treat gout in Europe, where it is said to compare favorably with synthetics (WOI). Extracts are also used for hepatosis. Cooked in milk and honey, it is used folklorically for hoarseness. The juice, in vinegar, diluted with water and sweetened with glycerine, was once given to children with whooping cough. Horseradish was once given to children with worms. Also used for facial neuralgia (GMH). The pulp is good for skin cancer (WOI). Extracts inhibit growth of ascites carcinoma in mice and Jensen sarcoma in rat (WO2). The dose offered by the herbal PDR (PH2) seems a wee bit high at 20 g fresh root, that is ca. $^2/_3$ oz, more than I could comfortably ingest at a setting. The oil is one of the most hazardous of all EOs and is not recommended for either external or internal use (CAN). Excessive doses may lead to diarrhea or night sweats. One case of a heart attack has been recorded; the patient survived (TAD).

Horseradish is one of the better sources of allylisothiocyanate (AITC), along with the mustards and other members of the cabbage family, and garlic mustard. Nielsen and Rios (2000) showed that volatiles containing AITC were most effective at inhibiting various bread molds: *Penicillium commune, P. roqueforti, Aspergillus flavus, Endomyces fibuliger,* and *E. fibuliger.* Using volatile EOs and oleoresins from various spices and herbs, they found mustard EO most efficacious, with cinnamon, clove, and garlic also highly active, oregano only slightly active. They did not analyze horseradish, but it is almost as pungent as the mustard and loaded with AITC. The minimal inhibitory concentration (MIC) for AITC was 1.8–3.5 µg/ml for the various fungi and yeast. Whether AITC was fungistatic or fungicidal depended on its concentration and the concentration of spores. When the gas phase contained at least 3.5 µg/ml, AITC was fungicidal to all tested fungi (X11016611). Horseradish proved antiseptic to *Bacillus subtilis, Escherichia coli*, and *Saccharomyces cerevisiae,* confirming traditional uses of spices as food preservatives, disinfectants, and antiseptics (De et al., 1999).

Fearing cancer, I might seek isothiocyanates, I'd go for a cruci-fix, a solitary dish or mix of the crucifers I had on hand. A series of new studies reported on Medline indicate that feeding isothiocyanates to experimental animals protects them from cancers of the breast, esophagus, liver, lungs, mammaries, and stomach. Dr. Paul Talalay, MD, Johns Hopkins School of Medicine, reviews the mechanisms by which isothiocyanates block carcinogenesis and suggests that they are ideal for chemoprevention of cancers. The list of well-known crucifers is long and tasty: broccoli, brussels sprouts, cabbage, collards, cress, kale, kohlrabi, mustard greens, pak choy, radishes, turnip greens, and watercress. Both for flavor and other sulfur-containing cancer-preventing compounds, I'd spice up my cruci-fix with the likes of chives, garlic, leek, onions, and ramps. And to increase the flavor and the heat, I'd add a dash of cayenne and tabasco.

Indications (Horseradish) — Allergy (1; FNF; LIB; PED); Anorexia (f; APA; DEM); Arthrosis (1; APA; BGB; CAN); Asthma (1; BGB; DEM; FNF); Atony (f; FEL); Bacillus (1; X10548758); Bacteria (2; FNF; HHB; KOM); Bronchosis (2; APA; FNF; PHR; PH2; SKY); Cancer (1; FNF; JLH); Cancer, abdomen (1; FNF; JLH); Cancer, breast (1; FNF); Cancer, colon (1; FNF; JLH); Cancer, liver (1; FNF; JLH); Cancer, nose (1; FNF; JLH); Cancer, spleen (1; FNF; JLH); Cancer, stomach (1; FNF; JLH); Cancer, skin (1; FNF; JLH; WO2); Catarrh (2; KOM; PHR); Chilblain (f; GMH); Cholecystosis (f; PHR; PH2); Cold (1; DEM; FNF; SKY); Colic (f; APA; PH2); Congestion (1; APA); Cough (2; GMH; PHR; PH2); Cramp (1; HHB); Cystosis (1; LIB; PHR); Debility (f; BOW); Dental Plaque (f; FAD); Diabetes (f; DEM; LIB); Dropsy (f; FEL; HHB); Dysmenorrhea (f; DEM); Dyspepsia (f; PHR; PH2; SKY); Dysuria (2; CAN; PED; PHR); Edema (f; BGB; CAN); Enterosis (1; PH2; WO2); Escherichia (1; X10548758); Fever (f; BOW); Flu (f; GMH; PHR; PH2); Freckle (f; FEL); Fungus (1; FNF; HHB; X10548758); Gastrosis (f; LIB); Glossosis (f; DEM);

Gout (f; BGB; GMH; HHB; PHR; WO2); Gravel (f; DEM); Hepatosis (f; HHB; PHR; PH2); High Blood Pressure (1; LIB); Hoarseness (f; FEL; GMH; WO2); Induration (f; JLH); Infection (2; FNF; PH2); Inflammation (1; CAN; FNF; PH2); Mycosis (1; FNF; HHB; X10548758); Myosis (2; BGB; KOM; PH2); Neuralgia (f; DEM; GMH); Pain (1; DEM; FNF; PH2); Pericardosis (f; BOW); Pertussis (f; GMH; LIB); Pleurisy (f; BOW); Respirosis (2; APA; DEM; KOM; PHR; PH2); Rheumatism (f; DEM; HHB; PHR); Rhinosis (1; JLH; PED); Sciatica (f; APA; BGB; GMH); Sinusosis (1; LIB; SKY); Sore (f; LIB); Sore Throat (f; LIB; SKY); Splenosis (f; GMH; WO2); Stomachache (f; LIB); Stomatosis (f; DEM); Stone (1; CAN; LIB); Swelling (f; BGB; JLH); Toothache (f; DEM; LIB); Typhoid (1; WO2); Urethrosis (2; KOM; PH2); UTI (2; APA; BGB; KOM; PH2); Worm (f; APA; GMH); Wound (f; APA; BOW); Yeast (1; X10548758).

Horseradish for bronchosis:

- Antibacterial: aesculetin; allyl-sulfide; caffeic-acid; gentisic-acid; kaempferol; limonene; p-hydroxy-benzoic-acid; quercetin; raphanin; sinapic-acid; vanillic-acid
- Antihistaminic: caffeic-acid; kaempferol; quercetin
- Antiinflammatory: aesculetin; caffeic-acid; gentisic-acid; kaempferol; quercetin; salicylates; vanillic-acid
- Antioxidant: caffeic-acid; kaempferol; p-hydroxy-benzoic-acid; quercetin; sinapic-acid; vanillic-acid
- Antipharyngitic: quercetin
- Antipyretic: aesculetin; salicylates
- Antispasmodic: caffeic-acid; kaempferol; limonene; quercetin
- Antiviral: caffeic-acid; gentisic-acid; kaempferol; limonene; quercetin
- COX-2-Inhibitor: kaempferol; quercetin
- Candidicide: quercetin
- Candidistat: limonene
- Cyclooxygenase-Inhibitor: kaempferol; quercetin
- Decongestant: allyl-isothiocyanate
- Expectorant: limonene
- Immunostimulant: caffeic-acid
- Phagocytotic: sinigrin

Horseradish for cancer:

- AntiHIV: caffeic-acid; quercetin
- Antiaggregant: allyl-sulfide; caffeic-acid; kaempferol; quercetin; rutoside; salicylates
- Anticancer: aesculetin; allyl-isothiocyanate; caffeic-acid; kaempferol; limonene; p-hydroxy-benzoic-acid; quercetin; sinapic-acid; sinigrin; vanillic-acid
- Anticarcinogenic: caffeic-acid
- Antifibrosarcomic: quercetin
- Antihepatotoxic: caffeic-acid; quercetin; sinapic-acid
- Antiinflammatory: aesculetin; caffeic-acid; gentisic-acid; kaempferol; quercetin; salicylates; vanillic-acid
- Antileukemic: kaempferol; quercetin
- Antileukotriene: caffeic-acid; quercetin
- Antilipoperoxidant: quercetin
- Antimelanomic: quercetin
- Antimutagenic: 4-pentenyl-isothiocyanate; aesculetin; allyl-isothiocyanate; caffeic-acid; kaempferol; limonene; p-hydroxy-benzoic-acid; quercetin
- Antinitrosaminic: caffeic-acid; quercetin

A

- Antioxidant: caffeic-acid; kaempferol; p-hydroxy-benzoic-acid; quercetin; sinapic-acid; vanillic-acid
- Antiperoxidant: caffeic-acid; quercetin
- Antiproliferant: quercetin
- Antiprostaglandin: caffeic-acid
- Antitumor: caffeic-acid; kaempferol; limonene; quercetin; vanillic-acid
- Antiviral: caffeic-acid; gentisic-acid; kaempferol; limonene; quercetin
- Apoptotic: kaempferol; quercetin
- COX-2-Inhibitor: kaempferol; quercetin
- Chemopreventive: limonene
- Cyclooxygenase-Inhibitor: kaempferol; quercetin
- Cytoprotective: caffeic-acid
- Cytotoxic: aesculetin; caffeic-acid; quercetin
- Hepatoprotective: caffeic-acid; quercetin
- Immunostimulant: caffeic-acid
- Lipoxygenase-Inhibitor: aesculetin; caffeic-acid; kaempferol; quercetin
- Mast-Cell-Stabilizer: quercetin
- Ornithine-Decarboxylase-Inhibitor: caffeic-acid; limonene; quercetin
- p450-Inducer: quercetin
- PTK-Inhibitor: quercetin
- Prostaglandigenic: caffeic-acid; p-hydroxy-benzoic-acid
- Protein-Kinase-C-Inhibitor: quercetin
- Sunscreen: aesculetin; caffeic-acid
- Topoisomerase-II-Inhibitor: kaempferol; quercetin
- Tyrosine-Kinase-Inhibitor: quercetin
- UV-Screen: aesculetin

Other Uses (Horseradish) — According to the Oracle at Delphi, "the radish was worth its weight in lead, the beet its weight in silver, and the horseradish its weight in gold." It is hard to imagine Passover, or hot roast beef, or shrimp cocktail sauce without grated horseradish. Back around 1600, John Gerarde said, "The Horse Radish stamped with a little vinegar put thereto, is commonly used among the Germans for sauce to eat fish" (Bown, 2001). The spice is cultivated for its thick, fleshy white root which has a delicious, intense pungent and cooling taste. It is primarily used in cocktail sauces, vinegars, mustards, and mixed flavorings. In 2000, the U.S. was producing some 6 million gallons of horseradish sauce, enough to season a line of sandwiches estimated to encircle the Earth 12 times (TAD). Today, Horseradish preparations may be red (preserved in beet juice) or white (preserved in vinegar) (FAC). Polish serve a special beet/horseradish condiment at Easter with ham and/or sausage (FAC). In Germany, where horseradish is one of the most important cultivated medicinal spice, leaves are eaten as a vegetable and in salads. Leaves can be cooked like spinach. Roots brought indoors to a dark place can give etiolated white sprouts by Christmas or better yet, Easter. Germans also cook the sliced roots as a vegetable, like parsnip. Grated roots, mixed with vinegar and salt, enhance flavor of meats and seafoods. Ethanol extracts (tinctures) of fresh or partially dried roots are more liable to be used medicinally than culinarily. Sliced horseradish in milk is said to be an excellent cosmetic; in vinegar, it helps to remove freckles (FAC, WOI, WO2).

For more information on activities, dosages, and contraindications, see the *CRC Handbook of Medicinal Herbs,* ed. 2, Duke et al. 2002.

Cultivation (Horseradish) — The plant rarely produces viable seed. Propagation is by root division with a section of the crown attached, which is cut off when the roots are harvested in late fall. These are overwintered in a cool humid cellar. The crop seems to fare better in average garden loams, moist but not constantly wet, pH 5.0–7.5 (average 6.5) in full sun. It is hardy to zone 5

(TAD). Fresh roots can be temporarily buried in sand or soil. Cuttings (pieces of roots ca. 1.5 in. or 4 cm long) are outplanted as soon as the soil can be worked, in furrows 8–30 cm deep, at intervals of 60 cm in rows 75 cm apart, to give a rate of about 22,000/ha. Root cuttings increase in diameter but usually not in length. However, by clearing away the soil from the upper part of the root and removing fibrous side-roots, tap roots become thicker and longer. Roots are ready for use 4–5 months after outplanting (WOI, WO2). Here at the Green Farmacy Garden, the plant comes back even after you harvest it, the deeper lateral roots and/or severed taproots, keep coming back like a spicy song.

Chemistry (Horseradish) — Volatile isothiocyanates, locked up temporarily as the glycosides nasturtiin and/or sinigrin, are released when hydrolyzed by the enzyme myrosinase or thioglucosidase, also present in the root. This enzymatic oxidation of the hot compounds in horseradish (and other crucifers) rather parallels what happens with some hot sulfur-containing compounds in the garlic family as well. The enzymes go into action only when the root is scratched. And these sulfur compounds seem to be very important in cancer prevention when used in moderation. Here are a few of the more notable chemicals found in horseradish. For a complete listing of the phytochemicals and their activities, see the CRC phytochemical compendium, Duke and duCellier, 1993 (DAD) and the USDA database http://www.ars-grin.gov/duke/.

Allyl – Isothiocyanate—Antiasthmic; Anticancer; Antifeedant; Antimutagenic; Antiseptic; Counterirritant; Decongestant; Embryotoxic; Fungicide MIC = 1.8–3.5 µg/ml; Herbicide IC100 = 0.4 mM; Insectiphile; Mutagenic; Nematiovistat 50 µg/ml; Spice FEMA 1–80 ppm; LD50 = 339 orl rat.

Isothiocyanate — Antibacterial; Anticancer; Antiseptic; Antithyroid; Antitumor; Chemopreventive; Fungicide; Hypotensive; Goitrogenic; p450-Inhibitor; Mucolytic; Nematicide; Respiradepressant; LD50 = 120.

Rutoside — Antiaggregant; Anticapillary-Fragility; Choleretic (50 mg/kg ipr); COMT-Inhibitor.

Sinigrin — Anticancer; Antifeedant; Larvicide; Mutagenic; Phagostimulant; Quinone-Reductase-Inducer 15 µM.

B

Bixa orellana L. (Bixaceae)
Achiote, Annatto, Annoto, Arnato, Bija, Lipstick pod, Lipstick Tree

Medicinal Uses (Annatto) — Seeds and latex used for tumors, cancer, and condyloma. Seeds gargled with vinegar and rice water for cancer of the mouth. Leaf infusion used in Costa Rica to prevent baldness. Leaf infusion gargled for tonsilitis. Bolivians press leaves on aching body parts. Seeds are reportedly aphrodisiac, astringent, cordial, expectorant (laxative and vermifuge), and febrifugal. Astringent febrifugal fruit pulp is used for dysentery and kidney disease. The reddish paste is applied as an unguent to burns. Considered a cosmetic, dye, food, hair dressing, medicine, ornamental tree, and vitamin source. In India, where the useful ornamental weed tree has established itself, as it has pantropically, leaves are used for jaundice and snakebite, the root bark for fevers, including malaria. Fruits are considered astringent and laxative. The plant is also recommended for gonorrhea (MPI). The hot water extracts potently inhibit lens aldose reductase, perhaps due to isoscutellarein (X1814628).

Here's what Rodolfo Vasquez and I had to say about the plant a few years back in our *Amazonian Ethnobotanical Dictionary*. The plant is thriving in our ReNuPeRu Ethnobotanical Garden there. "*Bixa orellana*" L. Bixaceae. Achote, Achiote amarillo. Cultivated. Natives mainly use it for food coloring and to decorate their bodies. There are experimental plots for the extraction of bixin. In Piura, the shoot decoction is considered antidysenteric, antipyretic, antiseptic, antivenereal, aphrodisiac, and astringent (GEO). The foliage is used to treat skin problems and hepatitis; also used as aphrodisiac, antidysenteric, and antipyretic. Considered good for the digestive system and for treatment of liver disease. Very effective as a gargle for tonsilitis (DAV). Chinatecas poultice leaves on cuts to avoid scars (DAV). People from Cojedes use the flower infusion as purge and to avoid phlegm in newborn babies. Kayapo massage stomachs of women in labor with the leaves. Waunana use to dye demijohns and baskets. Bark yields a gum similar to gum arabic. Fiber used as cordage. Kayapo use to tint to the body (DAV). Dye said to be an antidote for HCN (SAR). Seeds believed to be expectorant, the roots digestive (SAR) and antitussive (BDS). Around Explorama, fresh leaf stalks, devoid of blades, are inserted into a glass of water; the mucilage that forms is applied in conjunctivitis (DAV). It is recommended in the Amazon for pink-eye (conjunctivitis), which more often than not in Latin America is viral. Mustalish notes, perhaps thankfully, that some shamans "extract the vascular fluids from young stems as a treatment for conjunctivitis and to relieve the discomfort of corneal abrasions" (Mustalish and Baxter, 2001). Yes, I saw our shaman Antonio Montero Pisco treating Mustalish successfully for corneal abrasions. Annatto dye is a potential rare cause of anaphylaxis, causing angiodema, hypotension, and urticaria (X1994783, DAD).

Note that nowhere do we mention prostate. But by January 2001, Peruvians were manufacturing processed supplements, dominated by *Bixa* for BPH. The label on one product read "Contiene achiote 'C,' Uña de gato, Chanca piedras, etc." In Spanish, it also says, "Keep out of the reach of children" and "Store in a cool dry place." There's even an expiration date. But it doesn't specifically tell you that you are taking it for prostate health ("Prostasana;" the title alone tells you), nor how much to take for your prostate. The first three herbs mentioned are common around my Peruvian haunts, annatto, *Bixa orellana*; cat's claw, *Uncaria* spp.; stonebreaker, *Phyllanthus niruri*. None of my early literature suggested achiote for the prostate. But it does contain lycopene and several carotenoid derivatives of lycopene.

Some questions regarding the safety of ingesting the tea for diabetes mellitus have been raised, having found in fact a hyperglycemic activity for trans-bixin. Perhaps annatto should be counter-indicated for diabetics. The toxicity is "low grade," especially in well nourished experimental animals. Toxicity commences in dogs given 60 mg/kg trans-bixin.

Indications (Annatto) — Acne (f; 60P); Alopecia (f; DAD); Asthma (f; JFM); Bacteria (1; 60P; FNF; TRA); Bite (f; MPI); Bleeding (1; WO2); Blister (f; WO2); Burn (f; DAD; TRA; WO2); Cancer (f; JLH); Cancer, mouth (f; BOW; DAD); Childbirth (f; BOW; DAV; IED);Colic (f; BOW); Condylomata (f; JLH); Conjunctivosis (1; DAV; FNF); Dermatosis (f; JFM); Diabetes (1; FNF; JFM; WO2); Diarrhea (f; IED; JFM); Dysentery (f; 60P; DEP; JFM; WO2); Dyspnea (f; JFM); Epilepsy (f; WO2); Eruption (f; JFM); Fever (f; 60P; JFM; MPI; WO2); Flu (f; JFM); Gastrosis (f; DAV); Glaucoma (1; X1814628); Gonorrhea (1; JFM; MPI; TRA); Headache (f; JFM); Hemorrhoid (f; JFM); Hepatosis (f; DAV; JFM); High Blood Pressure (1; 60P; WO2); High Blood Sugar (1; 60P); Infection (1; FNF; WO2); Inflammation (1; FNF); Jaundice (f; JFM; WO2); Leprosy (f; JFM; WO2); Malaria (f; MPI; WO2); Measles (f; JFM); Nausea (f; JFM); Nephrosis (f; DAD; DEP); Pain (f; DAD); Parasite (1; DAD); Pleurisy (f; JFM); Prostatosis (1; FNF); Pulmonosis (f; IED); Respirosis (f; IED); Salmonella (1; TRA); Scar (f; WO2); Snakebite (f; MPI); Sore (f; JLH; WO2); Sore Throat (f; JFM; WO2); Stomachache (f; JFM); Stomatosis (f; DAD; JFM); Tonsilosis (f; DAV); Tumor (f; JLH); Uterosis (f; JFM); VD (f; DAV; JFM; MPI); Vomiting (f; JFM); Worm (f; DAD); Wound (f; JFM); Yeast (1; FNF; TRA).

Annatto for conjuctivosis:

- Antibacterial: ellagic-acid; lignin; salicylic-acid
- Antiinflammatory: ellagic-acid; luteolin-7-glucoside; salicylic-acid
- Antiviral: ellagic-acid; lignin; luteolin-7-glucoside
- COX-2-Inhibitor: salicylic-acid
- Cyclooxygenase-Inhibitor: salicylic-acid

Annatto for diabetes:

- Aldose-Reductase-Inhibitor: ellagic-acid; luteolin-7-glucoside; salicylic-acid
- Antioxidant: bixin; crocetin; cyanidin; ellagic-acid; lignin; lutein; salicylic-acid
- Antiperoxidant: ellagic-acid

- Antiradicular: lutein
- Hypocholesterolemic: crocetin; lignin
- Hypoglycemic: salicyclic-acid
- Thermogenic: salicylic-acid

B

Other Uses (Annatto) — First imported into Europe in the 16th century, annatto was first cultivated in India in 1787. Annatto is cultivated for its seeds, the bright crimson seed coats of which yield the red dye bixin (80%) and the yellow dye orellin (20%). This is one of the few FDA approved natural colorants. Bixin is used to impart color to certain foods (e.g., butter, cheese, margarine), foodstuffs, and wax polishes. Mercadente et al. (1997) remind us that annatto is second in importance among natural colorants. The seeds of achiote, alias roucou, are briefly steeped in hot oil or lard, which is then strained, cooled, and used as coloring and flavoring for fish, meat, poultry, rice, and vegetables (FAC). Also used to tint baked goods, drinks, fats, ice-cream, salad dressing, snacks, and yoghurts. As a colorant, it may still have antioxidant properties. Martinez-Tome et al. (2001) reported that at 5% concentration > annatto > BHA > sweet paprika > cumin > hot paprika > saffron > BHT at 100 μg/g as an antioxidant. In Yucatan, the whole seeds are ground with various spices into a paste, giving a more pronounced color and flavor. Mexicans stain poultry and suckling pigs with a paste of the seeds ground with other spices. Widely used in Latin America to color rice, soups, and others, including meats. Aztecs mixed ground annatto seed to a chocolate beverage. Annatto is used as body paint and hair dressing by Amerindians, serving for lipstick and rouge. In the Philippines, seeds are ground for a condiment. Jamaicans use annatto, with chiles and onions, with their national dish, salt cod and akee. Sausages colored with annatto do not lose their color. One tsp powdered seed to a cup of water can nicely color a dish of rice (AAR). Yes, this dirt cheap colorant, especially in the tropics where *Bixa* is almost a weed tree, is incredibly cheaper, and perhaps healthier, than the world's most expensive spice, saffron. Bown (2001) notes that annatto is used to stain maggots red to make them more appealing as fish bait. Bixin has been used as a textile dye for cotton and silk. For printing purposes, the dye is dissolved in caustic soda and developed with acid, alum, or stannous chloride. Brown shoe polish is made from the seeds, and floor wax is made by dissolving the dye in kerosene. A fairly good fiber may be obtained from the bark, which also contains small amounts of dye. A gum somewhat similar to gum arabic can be extracted from the trunk. Wood used for firesticks and the bark for cordage. Roots said to impart the taste and color of saffron to meats. It is said to be one of the few shrubs that can grow through and kill the dreaded lalang grass. Used as a wind break in coffee and tea plantations. (AAR, BOW, DAD, FAC, WO2).

For more information on activities, dosages, and contraindications, see the *CRC Handbook of Medicinal Herbs*, ed. 2, Duke et al., 2002.

Cultivation (Annatto) — Soil should be prepared in the same manner as for cotton. Seeds, previously softened by soaking in water, are planted in holes or furrows 2.5–3.5 m apart in shaded nurseries. Seeds germinate in 8–10 days. As the young plants develop, they should be protected by artificial shade or intercrops, with increased light as they get older. When 15–25 inches high, they are ready for outplanting, spaced for final distribution at 4.5–6 m apart. Plantings fare well if in 60 cm cubed plots filled with well aged farm manure. Water well at planting. Can be intercropped with cassava, corn, and malanga. After 3 months, plantation should be weeded and superfluous plants removed. Except for periodical weeding, the plantation needs little attention. ANAI recommends growing medicinal or culinary herbs, like lemongrass, between the shrubs (DAD, WO2). Collection of seed may begin as early as 18 months, a full crop expected 3–4 years after sowing. Trees remain productive for 10–15 years. Capsules are gathered, usually by hand, when they are reddish and beginning to break open. It is wise to prune branches rather vigorously when harvesting the capsules; pruned plants yield better. Capsules are dried in the shade for about 10 days then exposed to the sun until all have opened. Clusters and seeds are then placed in a bag and beaten with a stick to loosen the seed. Thus, seeds are easily removed from the capsule, and little dye is lost. Seeds are then sifted to separate seed from trash. Seed is again sun-dried 4–6 hours before bagging. For home extraction of dye/spice, pouring hot water over the pulp and seeds to macerate and separate them by pounding with a wooden pestle. Remove seeds, letting the pulp settle, pouring off excess water; dry pulp gradually in the shade. In India, the plant produces throughout the year, with two main crops in March and September. In Hawaii, harvests are in May, September, and the best yields are obtained in January, with about 44 kg/ha for round pod variety, and 939 kg/ha for pointed-pod variety. A tree should yield 4.5–5 kg dried seed per year. An average yield of 500–2000 kg/ha per year is satisfactory, but up to 4500 kg/ha have been reported in five-year old fields; 100 kg of seed yield about 5–6 kg of material which contains 12–30% bixin.

Chemistry (Annatto) — Mercadente et al. (1997) note that, of the 500–600 known carotenoids, many have not yet proven to have pro-vitamin-A activity, but many have other useful health-giving properties. They found five apocarotenoids, all new to *Bixa* and three new to science, and recapitulated others, e.g., bixin (estimated at >80% of carotenoids in the seed coat). Carotenoids are frequently added to foods as colorants, presumed health-giving as well. The following are permitted by the European Union: annatto, beta-apo-8'-carotenal, beta-apo-8'-carotenic ethyl ester, bixin, canthaxanthin, capsanthin, capsorubin, beta-carotene, carotene mixes, lutein, lycopein, norbixin, paprika extract (Haila et al., 1996). Here are a few of the more notable chemicals found in annatto. For a complete listing of the phytochemicals and their activities, see the CRC phytochemical compendium, Duke and duCellier, 1993 (DAD) and the USDA database http://www.ars-grin.gov/duke/.

Arginine — See also *Allium sativum.*

Bixin — ADI = 1.25 mg/kg; Antioxidant 30–60 µg/g; Colorant; Dye.

Histidine — ADI = 15 g/day/orl; Antiatherosclerotic; Antinephrotic; Antioxidant; Antiulcer; Anti-uremic; Essential; Oxidant.

Boswellia sacra Flueck. (Burseraceae)
FRANKINCENSE, OLIBANUM TREE

Synonym — *Boswellia carteri* Birdw.

Medicinal Uses (Frankincense) — For over a century, the true identity of frankincense has been illusive. Here's a quote from the *Dictionary of the Economic Products of India*, cerca 1889. "It is probable that several species yield (Frankincense) of which *B. carterii* is perhaps one of the most important.... The Arabs, as early as the tenth century, carried Olibanum to India, and the Indian names for it have, through the lapse of time, become almost hopelessly mixed up with those given to the Indian species.... Many centuries before Christ, the drug was one of the most important articles of trade which the Phoenicians and Egyptians carried on with Arabia (DEP). But all seem to agree that the frankincense or olibanum, whichever species they be, seems to share many aromatic and medicinal principles with pitches and turpentines emanating from the firs and pines and their relatives, some of which were approved by Germany's Commission E. And over 100 years ago, in India, olibanum was recommended in chronic lung ailments like bronchorrhea and laryngitis, both as an internal concoction, and as an inhalant. And ten centuries ago, Avicennia recommended frankincense for dysentery, fever, tumors, ulcers, and vomiting (GMH). Ointments of olibanum were suggested for boils, carbuncles, and sores. The resin is used to stimulate digestion, to treat mastitis, strengthen the teeth, and mixed into hair products. Soot collected from burning the resin is used as kohl for soothing sore eyes. Other authors suggest it is more cosmetic than medicine. Pregnant Yemenis chew the gum; also chewed for emotional and psychological problems. Arabians often chew it as a masticatory, believed to improve the memory, or add it to coffee. The resin is presumed to be diuretic and laxative. Thieret (1996) adds that, in Graeco-Roman medicine, frankincense was prescribed for abscesses, bruises, chest ache, hemorrhage, hemorrhoids, paralysis, and ulcers. In northern Africa, it is used for back problems, chest congestion, chronic coughs, poliomyelitis, and venereal ailments (Thieret, 1996).

As a natural COX-2 Inhibitor, boswellic-acid may alleviate arthritis, gout, inflammation, and rheumatism, and possibly prevent Alzheimer's and cancer (Newmark and Schulick, 2000). And I am specifically watching the press for COX-2-Inhibitors and colon cancer. Boik (2001) suggest boswellic-acid, if not frankincense, as a cancer preventive. Seven *in vitro* studies suggest it inhibits proliferation or induces differentiation in leukemia and CNS-cancer cell lines, usually at levels of 2–40 μM. Two human studies suggest that frankincense extracts can alleviate brain cancer pathologies, perhaps also reducing brain edema. Alcoholic extracts of Boswellic are antiinflammatory in rats at oral doses of 50 mg/kg, with antiarthritic mitigation (including a reduction in collagen degradation) at 100 mg/kg. The whole extract was more effective than pure boswellic acid, suggesting that a frankincense liqueur might be better than a silver bullet boswellic-acid. Boswellic acid induced apoptosis in leukemia cells ($IC50 = 30$ μM), induced differentiation of human leukemia cells at 11–22 μM, and inhibited proliferation on four human brain cancer cell lines ($IC50 = 30$ to 40 μM), of 11 lines of meningiomas ($IC50 = 2$–8 μM), of leukemia cells ($IC72 = 4$ μM). Boik (and/or my database) lists several mechanisms by which boswellic-acid might prevent cancer: antiangiogenic, antiinflammatory, antileukemic, antimeningiomic, antimetastatic, antiproliferant, apoptotic, COX-2-Inhibitor, cytotoxic, beta-glucuronidase-inhibitor, hyaluronidase-inhibitor (10–70 μM), and topoisomerase-inhibitor. So we have one phytochemical in the Biblical frankincense with more than a dozen different activities that could reduce the incidence of cancer. Boik (2001) calculates tentative human dosages as 340–2400 mg/day (as scaled from animal antitumor studies), 730–3200 mg/day (as scaled from animal antiinflammatory studies), 3600–5400 mg/day used in human anticancer studies, suggesting a target dose of 1800 mg/day boswellic-acid. He suggests that fifteenfold synergies with other phytochemicals may reduce that minimum antitumor dose to 120 mg boswellic-acid/day. "Since the target dose is achievable, synergetic interventions may not be required for boswellic acid to produce an anticancer effect in humans. Still, it may greatly benefit from synergistic interactions and is best tested in combinations" (Boik, 2001).

Indications (Frankincense) — Abscess (f; HAD); Alzheimer's (1; COX; FNF); Anxiety (f; BOW); Arthrosis (1; COX; FNF); Asthma (f; HHB); Backache (f; HAD); Bilharzia (f; BIB); Bleeding (f; HAD); Boil (f; DEP); Bronchosis (f; BIB; DEP); Bruise (f; HAD); Callus (f; BIB); Cancer (1; COX; FNF; JLH); Cancer, anus (1; BIB; COX); Cancer, breast (1; BIB; COX); Cancer, eye (1; BIB; COX); Cancer, penis (1; BIB; COX); Cancer, spleen (1; BIB; COX); Cancer, teat (1; BIB; COX); Cancer, testicle (1; BIB; COX); Carbuncle (f; DEP; JLH); Colitis (1; FNF); Corn (f; JLH); Cough (f; HAD); Cramp (1; FNF); Crohn's Disease (1; FNF); Dermatosis (f; GMH); Dysentery (f; BIB); Dysmenorrhea (f; BOW); Dyspepsia (f; HAD); Edema (1; FNF); Fever (f; BIB); Gingivosis (f; BOW); Gonorrhea (f; BIB); Hemorrhoid (f; HAD); Infection (f; BOW); Inflammation (1; FNF); Laryngosis (f; BIB; DEP); Leprosy (f; BIB); Leukemia (1; FNF); Mastosis (f; JLH); Meningioma (1; FNF); Myelosis (f; HAD); Neurosis (f; BIB; HAD); Ophthalmia (f; JLH); Orchosis (f; JLH); Pain (1; FNF; HHB); Polio (f; HAD); Polyp (f; JLH); Proctosis (f; JLH); Psychosis (f; HAD); Respirosis (f; PH2); Rheumatism (1; BIB; FNF); Sore (f; DEP); Spermatorrhea (f; BIB); Splenosis (f; JLH); Stomachache (f; BIB); Stomatosis (f; BOW); Swelling (1; BIB; FNF); Syphilis (f; BIB); Tumor (1; FNF); Ulcer (f; HAD); Urogenitosis (f; BIB); Uterosis (f; HHB); UTI (f; BOW); Vaginosis (f; BOW); VD (f; BIB); Vomiting (f; HAD); Water Retention (1; FNF); Wound (f; PH2).

Frankincense for Alzheimer's:

- ACE-Inhibitor: alpha-terpinene; alpha-terpineol; gamma-terpinene; myrcene
- AChE-Inhibitor: limonene
- Antiacetylcholinesterase: alpha-terpinene; carvone; gamma-terpinene; limonene; p-cymene; terpinen-4-ol
- Antiinflammatory: acetyl-11-keto-beta-boswellic-acid; acetyl-beta-boswellic-acid; alpha-boswellic-acid; alpha-pinene; beta-boswellic-acid; beta-pinene; borneol; caryophyllene; caryophyllene-oxide
- Antileukotriene: acetyl-11-keto-beta-boswellic-acid; acetyl-beta-boswellic-acid; alpha-boswellic-acid; beta-boswellic-acid
- Antioxidant: alpha-boswellic-acid; camphene; gamma-terpinene; myrcene
- CNS-Stimulant: borneol; carvone

Frankincense for cancer:

- Antiadenomic: farnesol
- Anticancer: alpha-pinene; alpha-terpineol; aromadendrene; carvone; limonene; linalool; mucilage
- Antiinflammatory: acetyl-11-keto-beta-boswellic-acid; acetyl-beta-boswellic-acid; alpha-boswellic acid; alpha-pinene; beta-boswellic-acid; beta-pinene; borneol; caryophyllene; caryophyllene-oxide
- Antileukemic: acetyl-beta-boswellic-acid; farnesol
- Antileukotriene: acetyl-11-keto-beta-boswellic-acid; acetyl-beta-boswellic-acid; alpha-boswellic-acid; beta-boswellic-acid
- Antimelanomic: farnesol
- Antimutagenic: anisaldehyde; limonene; myrcene
- Antioxidant: alpha-boswellic-acid; camphene; gamma-terpinene; myrcene
- Antitumor: alpha-humulene; caryophyllene; caryophyllene-oxide; limonene
- Antiviral: alpha-pinene; dipentene; limonene; linalool; p-cymene
- Apoptotic: farnesol
- Chemopreventive: limonene
- Cytochrome-p450-Inducer: delta-cadinene
- Hepatoprotective: borneol
- Ornithine-Decarboxylase-Inhibitor: limonene
- p450-Inducer: delta-cadinene

Frankincense for rheumatism:

- Analgesic: borneol; myrcene; p-cymene
- Anesthetic: linalool; myrcene
- Antiedemic: acetyl-11-keto-beta-boswellic-acid; beta-boswellic-acid; caryophyllene; caryophyllene-oxide
- Antiinflammatory: acetyl-11-keto-beta-boswellic-acid; acetyl-beta-boswellic-acid; alpha-boswellic-acid; alpha-pinene; beta-boswellic-acid; beta-pinene; borneol; caryophyllene; caryophyllene-oxide
- Antirheumatalgic: p-cymene
- Antispasmodic: borneol; caryophyllene; farnesol; limonene; linalool; myrcene
- Myorelaxant: borneol

Other Uses (Frankincense) — Frankincense came to the Holy Land via the famous spice route across southern Arabia and some of the littoral stations of East Africa, the same caravan highway used also for goods from India and points farther east (Zohary, 1982). Today, the Catholic Church may be a major consumer, often using frankincense in ceremonial incenses. Botanical historian John W. Thieret (1996) seems to agree with Zohary, noting that a main source of frankincense is *Boswellia sacra*. "Herodotus (born 484 B.C.) wrote that the frankincense trees were guarded by vast numbers of small winged serpents;" he was wrong. Most frankincense comes from Somalia (following bananas and cattle as leading export), where it provides work for some 10,000 Somali families. Herodotus also said that Arabs every year brought to Darius 1000 talents of Frankincense as tribute. Modern Parsis in western India still maintain the same incense ritual (GMH). Some is gathered in Arabia. Most goes to Saudi Arabia, Yemen, and Egypt, the major markets. Early botanist Theophrastus, some three centuries before Christ, said that most frankincense came from Saba (southwestern Arabia, once ruled by the famed Queen Sheba). That ancient country got rich in the incense trade. Circa 2335 B.P., Alexander the Great's army captured Gaza, plundering its frankincense and sending it to Greece. Tons of incense were buried in the temples of Babylon and Nineveh. And in King Tut's tomb, 3000 year old balls of frankincense were recovered. One Roman Church formula had 10 oz olibanum (frankincense), 4 oz benzoin, and 1 oz Storax. "In today's churches, frankincense is an ingredient in the incense that sometimes nearly suffocates the faithful.... Because frankincense and myrrh no longer enjoy the esteem that they did two millennia ago, I wonder what the Wise Men would bring today. Perhaps gold, dates, and oil" (Thieret, 1996). Import statistics are hard to come by. Thieret (1996) suggests total yearly production of myrrh is perhaps 500 tons, frankincense 1000 tons. Recent U.S. imports run 5–20 tons. The United Kingdom imports ca. 30 tons frankincense each year, one perfume manufacturer alone consuming 5 tons annually (Thieret, 1996).

The aromatic resin is chewed or manufactured into an EO used in baked goods, candies, gelatins, ice creams, puddings, and soft drinks (FAC). Flowers and seed of the Indian variety are consumed as foods. Leaves yield an EO. The rosin is used in making balsam substitutes, inks, lacquers, paints, and varnishes. The volatile oils from the resin are suitable for the soap and perfume industry, paints, and varnishes. The resin is, of course, often used as incense. The "kohl" or black used by Egyptian women to stain their eyelids is made of charred frankincense, with or without other odorants added. It is also melted to make a depilatory and made into a paste to perfume the hands. In cold snaps, Egyptians warm their rooms with a brazier on which incense is burned.

For more information on activities, dosages, and contraindications, see the *CRC Handbook of Medicinal Herbs*, ed. 2, Duke et al., 2002.

Cultivation (Frankincense) — Not cultivated, but harvested from the wild. Ghazanfar (1994) notes that, in southern Arabia, luban trees occur in wadis extending to the coast on the lower slopes of the gullies and runoffs. Bown (2001) suggests propagation with semi-ripe cuttings; grow in well-drained to dry soils in full sun, with light pruning in spring. The gum exuding from cuts is the major medicinal incense, being burned to give a perfumed smoke, used to improve the aroma of clothing, hair, and residences. Of the Indian Olibanum, WO2 reports it demands light, tolerates fire, and makes gregarious open forests, coppicing readily and producing root suckers as well. It regenerates nicely from seed and root-suckers, and cuttings strike well even during drought. This being the only non-coniferous source of "turpentine" in India, it can be tapped so as not to injure the tree; trees yield 0.9–2.5 kg gum per tree. Healthy trees, girth 90 cm or more (obviously much bigger than the true desert frankincenses), are tapped by shaving off a thin band of bark ca. 20 cm broad, 30 cm high, about 15 cm above the soil line. The cut is made about half the thickness of the bark. In India, tapping begins in November and is stopped before the monsoon. Bown (2001) says gum can be collected all year, though the best is that from the driest hottest months in the driest hottest areas. For more details, see WO2.

Chemistry (Frankincense) — Here are a few of the more notable chemicals found in frankincense. For a complete listing of the phytochemicals and their activities, see the CRC phytochemical compendium, Duke and duCellier, 1993 (DAD) and the USDA database http://www.ars-grin.gov/duke/.

Boswellic-Acid — Analgesic 20–55 mg/kg ipr rat; Antiallergic; Antiarthritic 100 mg/kg orl rat; Antiasthmatic; Anticomplement IC100 = 0.1 μM; Antiedemic; Antieicosanoid; Antihyaluronidase; Antiinflammatory 50 mg/kg orl rat, IC100 = 0.1 μM; Antileukemic 11–30 μM, IC69 = 25 mg/kg ipr mus, IC82 = 50 mg/kg ipr mus; Antimeningiomic IC50 = 11–22 μM; Antioxidant; Antiproliferant — 40 μM, IC69 = 25 mg/kg ipr mus, IC82 = 50 mg/kg ipr mus; Antitumor (Brain) IC50 = 30–40 μM; AntiVEGF; Apoptotic 30–40 μM; COX-2-Inhibitor; Differentiator 11–22 μM; Elastase-Inhibitor IC50 = 15 μM; Beta-Glucuronidase-Inhibitor 100 mg/kg; 5-HETE-Inhibitor; 5-Lipoxygenase-Inhibitor IC50 = 1.5–33 μM; Sedative 20–55 mg/kg ipr rat; Topoisomerase-Inhibitor.

alpha-Boswellic-Acid — Antiallergic; Antiarthritic; Antiasthmatic; AntiCrohn's; Anticolitic; Antiinflammatory; Antileukotriene; COX-2-Inhibitor; 5-Lipoxygenase-Inhibitor.

Beta-Boswellic-Acid — Antiallergic; Antiarthritic; Antiasthmatic; AntiCrohns; Anticolitic; Antiedemic 50–200 mg/kg; Antiinflammatory 5 μM; Antileukotriene 5 μM; 5-Lipoxygenase-Inhibitor 5 μM.

C

Capparis spinosa L. (Capparaceae)
CAPER, CAPERBUSH

Synonym — *Capparis rupestris* Sm.

Medicinal Uses (Caper) — The caper contains some of the same isothiocyanates found in the mustard family and may serve a similar role medicinally. Root and root bark of the Biblical caper, prepared variously (malagmas, cataplasms, drunk with wine or vinegar, etc.), are folk remedies for indurations (of the bladder, kidney, liver, spleen, and uterus), tumors in general, and warts (JLH). The root bark is viewed as alterative, analgesic, aperient, aphrodisiac, astringent, diuretic, emmenagogue, expectorant, stimulant, tonic, and vermifue, and is used in rheumatism, scurvy, enlarged spleen, sclerosis (spleen), tubercular glands, and toothache. Unani use the juice to kill worms; they also consider the root bark aperient, analgesic, emmenagogue, expectorant, and vermifuge, and use it in adenopathy, paralysis, rheumatism, splenomegaly, and toothache. Broken leaves are used as a poultice in gout. Tender stems are used for dysentery. According to Biblical scholars Moldenke and Moldenke (1952), the caper is a stimulant, exciting both hunger and thirst and thus strengthening the appetite when it becomes a bit sluggish. Bedouins are said to use caper with *Teucrium pilosum* as an inhalant for colds. Bedouins boil the chopped or powdered leaves in water and inhale the vapors for headache. For arthritic-like pains of the back, joints and limbs, they boil the ground leaves and poultice them, wrapped in thin cloth, to the ache, even as they sleep. Barren women are covered with a mixture of ground leaves of *Capparis* and *Tamarix* to inhale the vapors and sweat to correct their barrenness. Lebanese boil the root and plant for dengue, malaria, and Malta fever. Lebanese regard the roots as specific for malaria or splenomegaly following malaria. Iranians use it for intermittent fever and rheumatism. Algerians boil the whole plant in oil as a puerperal hydragogue. Crushed seeds have been suggested for dysmenorrhea, female sterility, ganglions, scrofula, and ulcers. The capers themselves have been suggested for atherosclerosis, chills, oph-thalmia, and sciatica, especially in North Africa. Fruits, considered antiscorbutic, are used for colds, dropsy, and sciatica (BIB). Capers are eaten as food farmacy, for dry skin.

Shirwaikar et al. (1996) found antihepatotoxic activity in alcoholic, ether, ethyl acetate, and petrol extracts of the root bark of the caper, all of which reduced the elevated serum transaminases. The various extracts were administered orally to rats up to 2000 mg/kg, with no evident toxicity or mortality (Shirwaikar et al., 1996).

Indications (Caper) — Adenosis (f; BIB; JLH); Aging (f; BIB); Arthrosis (f; BIB); Atherosclerosis (f; BIB); Bleeding (f; BOW); Cancer (1; BIB; FNF); Cancer, abdomen (1; FNF; JLH); Cancer, bladder (1; FNF; JLH); Cancer, colon (1; FNF; JLH); Cancer, groin (1; FNF; JLH); Cancer, head (1; FNF; JLH); Cancer, kidney (1; FNF; JLH); Cancer, liver (1; FNF; JLH); Cancer, neck (1; FNF; JLH); Cancer, spleen (1; FNF; JLH); Cancer, uterus (1; FNF; JLH); Cataract (f; BIB); Chill (f; BIB); Cirrhosis (f; WO2); Cold (f; BIB); Conjunctivosis (f; BOW); Cough (f; BOW); Cramp (1; FNF); Cystosis (f; JLF); Dengue (f; BIB); Diarrhea (f; BOW); Dropsy (f; BIB); Dysentery (f; BIB); Dysmenorrhea (f; BIB); Enterosis (f; BOW); Fracture (f; BIB); Gastrosis (f; BOW); Gout (f; SKJ; WO2); Headache (f; BIB); Hepatosis (1; FNF; JLH; WO2); Induration (f; JLH); Infection (f; BOW); Inflammation (1; FNF); Infertility (f; BIB); Malaria (f; BIB); Malta Fever (f; BIB); Nephrosis (f; JLH; WO2); Ophthalmia (f; BIB); Otosis (f; BIB); Pain (f; BIB); Paralysis (f; HAD); Rheumatism (f; WO2); Sclerosis (f; BIB); Sciatica (f; BIB); Scurvy (1; WO2); Scrofula (f; BIB); Snakebite (f; BIB); Splenomegaly (f; BIB); Splenosis (f; BIB; WO2); Toothache (f; BIB); Tuberculosis (1; BIB; WO2); Tumor (f; BIB); Ulcer (f; BIB); Uterosis (f; JLH); Wart (f; BIB; JLH).

Caper for cancer:

- AntiHIV: quercetin
- Antiaggregant: coumarin; quercetin; vitamin-e
- Antiaging: vitamin-e
- Antiandrogenic: coumarin
- Anticancer: coumarin; mucilage; quercetin; rutin; sinigrin; vitamin-e
- Antifibrosarcomic: quercetin
- Antihepatotoxic: glucuronic-acid; quercetin; rutin
- Antiinflammatory: coumarin; quercetin; rutin; vitamin-e
- Antileukemic: beta-sitosterol-beta-d-glucoside; quercetin; vitamin-e
- Antileukotriene: quercetin; vitamin-e
- Antilipoperoxidant: quercetin
- Antimelanomic: coumarin; quercetin
- Antimetastatic: coumarin
- Antimutagenic: coumarin; quercetin; rutin
- Antinitrosaminic: quercetin; vitamin-e
- Antioxidant: quercetin; rutin; vitamin-e
- Antiperoxidant: quercetin
- Antiproliferant: quercetin; vitamin-e
- Antitumor: beta-sitosterol-beta-d-glucoside; coumarin; quercetin; rutin; vitamin-e
- Antiviral: quercetin; rutin
- Apoptotic: quercetin; vitamin-e
- COX-2-Inhibitor: quercetin
- Chemopreventive: coumarin
- Cyclooxygenase-Inhibitor: quercetin
- Cytotoxic: quercetin
- Hepatoprotective: quercetin; vitamin-e

C

- Immunostimulant: coumarin; vitamin-e
- Lipoxygenase-Inhibitor: quercetin; rutin; vitamin-e
- Lymphocytogenic: coumarin
- Lymphokinetic: coumarin
- Mast-Cell-Stabilizer: quercetin
- Ornithine-Decarboxylase-Inhibitor: quercetin; vitamin-e
- p450-Inducer: quercetin
- PTK-Inhibitor: quercetin
- Protein-Kinase-C-Inhibitor: quercetin; vitamin-e
- Sunscreen: rutin
- Topoisomerase-II-Inhibitor: quercetin; rutin
- Tyrosine-Kinase-Inhibitor: quercetin

Caper for hepatosis:

- Antiedemic: coumarin; rutin
- Antihepatotoxic: glucuronic-acid; quercetin; rutin
- Antiherpetic: quercetin; rutin; vitamin-e
- Antiinflammatory: coumarin; quercetin; rutin; vitamin-e
- Antileukotriene: quercetin; vitamin-e
- Antilipoperoxidant: quercetin
- Antioxidant: quercetin; rutin; vitamin-e
- Antiperoxidant: quercetin
- Antiradicular: quercetin; rutin; vitamin-e
- Antivaricose: rutin
- Antiviral: quercetin; rutin
- COX-2-Inhibitor: quercetin
- Cyclooxygenase-Inhibitor: quercetin
- Detoxicant: glucuronic-acid
- Hepatoprotective: quercetin; vitamin-e
- Immunostimulant: coumarin; vitamin-e
- Lipoxygenase-Inhibitor: quercetin; rutin; vitamin-e
- Phagocytotic: coumarin; sinigrin

Other Uses (Caper) — The young pickled buds of the caper give the Biblical "desire" or relish to food; "…fears shall be in the way, and the almond tree shall flourish, and the grasshopper shall be a burden, and desire shall fail…because man goeth to his long home…" (Ecclesiastes 12). Today, in the Mediterranean islands, capers are gathered and steeped in vinegar for an appetizer. Capers are unusual in that they are the flower buds used as the condiment in cooking. They are used to flavor canapes, gravies, salads, and sauces, after being cooked and pickled. Raw capers are all but unpalatable and much improved in the pickling process. Bay leaves, black pepper, and/or tarragon are good in the pickling vinegar (TAD). Some favor capers pickled in sea salt. Some French sauces graced with capers include ravigote, remoulade, tartare, and some vinaigrettes (TAD). Rinzler (1990) notes they are particularly good in a sour cream sauce (RIN). For pickling, one kilo of buds is steeped in one kilo of brine vinegar for a month. Capers can be substituted for oriental fermented black beans (toushi). Sprouts are sometimes eaten like asparagus, as well as the buds and shoots. In California, where they prune the shrubs back to the ground, the new shoots are eaten as a delicacy in spring. Pickled fruits are eaten in the Punjab and Arabia, as well as in Cyprus, where branch tips are also pickled. They are also grown for the large ornamental white and purple flowers. The timber is said to resist termites (BIB, FAC, RIN).

For more information on activities, dosages, and contraindications, see the *CRC Handbook of Medicinal Herbs*, ed. 2, Duke et al., 2002.

Cultivation (Caper) — Hardy only to zone 9, capers are best grown in full sun, well-drained sandy soils of pH 6.3–8.3. Plants will freeze back at least to the ground at 32°F (0°C). Treated seed sown in sandy soil under glass in the spring; plants then transplanted outdoors during warm, settled weather. Seeds should first be treated with a dilute sulfuric acid or permanganate. May be propagated also from cuttings of short shoots planted in a sandy soil in frames in a glasshouse. In the North, it is grown by training the long slender shoots on a trellis in a glasshouse, or by treating the plant as a tender annual and planting it outdoors when weather permits. The shrub produces a profusion of flowers in the summer and is often grown as an ornamental. In the South, it can be grown outdoors. In southern California, where hardy, rooted cuttings are outplanted in February-March, spaced 16 × 16 feet (5 × 5 m). During the next two years, they may require two or three irrigations. Each plant might be fertilized with $^1/_2$ lb ($^1/_4$–$^1/_5$ kg) 16–16–16 fertilizer. Plants are pruned back to the ground each year. Flower buds are harvested from May to August. In Spain, intercropping with cereals is recommended and said to increase the yield of capers. Harvest manually the flower buds before they show any color. A three-year-old bush can yield more than 2 lb of buds per year, four-year-olds and older >20 pounds/year (CFR, TAD).

Chemistry (Caper) — Small wonder the plant has so much anticancer folklore, containing such antitumor compounds as beta-sitosterol, beta-sitosterol-beta-d-glucoside, citric-acid, coumarin, quercetin, and rutin, as well as indole glucosinolates and isothiocyanates. Here are a few of the more notable chemicals found in caper. For a complete listing of the phytochemicals and their activities, see the CRC phytochemical compendium, Duke and duCellier, 1993 (DAD) and the USDA database http://www.ars-grin.gov/duke/.

Beta-Sitosterol — ADI = 9–30 g/day/man; Androgenic; Anorectic; Antiadenomic; Antiandrogenic; Antibacterial; Anticancer; Antiestrogenic; Antiedemic IC54 = 320 mg/kg orl; Antifeedant; Antifertility; Antigonadotropic; Antiinflammatory; Antileukemic; Antilymphomic; Antimutagenic 250 µg/mL; Antiophidic 2.3 mg mus; Antiprogestational; Antiprostaglandin (30 mg/day/12 wk); Antiprostatadenomic; Antiprostatitic 10–20 mg/3×/day/orl man; Antitumor (Breast); Antitumor (Cervix); Antitumor (Lung); Antiviral; Artemicide LC50 = 110 ppm; Candidicide; Estrogenic; Gonadotropic; Hepatoprotective; Hypocholesterolemic 2–6 g/man/day/orl; Hypoglycemic; Hypolipidemic 2–6 g/day; Hypolipoproteinaemic; Spermicide; Ubiquict; Ulcerogenic 500 mg/kg ipr rat; LD50 = 3000 mg/kg ipr mus; LDlo = >10,000 inj rat.

Beta-Sitosterol-Beta-D-Glucoside — Antileukemic; Antispasmodic (20 mg/kg); Antitumor; Hypoglycemic.

Citric-Acid—See also *Hibiscus sabdariffa.*

Coumarin — See also *Dipteryx odorata.*

Quercetin — See also *Alpinia officinarum.*

Rutin — Aldose-Reductase-Inhibitor; Allelochemic; Anesthetic; Antiapoplectic; Antiatherogenic; Antibacterial; Anticancer; Anticapillary-Fragility 20–100 mg orl man; Anticataract; Anticlastogenic; Anticonvulsant; AntiCVI 270 mg/man/day; Antidementic; Antidermatitic; Antidiabetic; Antiedemic 270 mg/day/orl man; Antierythemic; Antifeedant; Antiglaucomic 60 mg/day; Antihematuric; Antihemorrhoidal; Antihepatotoxic; Antiherpetic; Antihistaminic; Antiinflammatory 20 mg/kg; Antimalarial IC50 = >100 µg/ml; Antimutagenic ID50 = 2–5 nM; Antinephrotic; Antioxidant IC28 = 30 ppm, IC50 = 120 µ*M*; Antipurpuric; Antiradicular (9 × quercetin); Antispasmodic; Antisunburn; Antithrombogenic; Antitumor-Promoter; Antitrypanosomic 100 mg/kg; Antiulcer; Antivaricose; Antiviral; cAMP-Phosphodiesterase-Inhibitor; Catabolic; Estrogenic?; Hemostat; Hepatomagenic 20,000 ppm (diet) rat; Hypocholesterolemic; Hypotensive; Insecticide; Insectiphile; Juvabional; Larvistat IC95 = 4000–8000 ppm diet; Lipoxygenase-Inhibitor IC75 = 2.5 mM; Mutagenic; Myorelaxant; Oviposition-stimulant; Radioprotective; Sunscreen; Topoisomerase-II-Inhibitor IC50 = 1 µg/ml; Vasoconstrictor; LD50 = 950 (ivn mus).

Capsicum spp. L. (Solanaceae)
Bird Chilli, Bird Pepper, Chili Pepper, Hot Pepper, Red Chili, Spur Pepper, Tabasco Pepper

Synonyms — *Capsicum minimum* Blanco, *Capsicum fastigiatum* Bl.

Medicinal Uses (Hot Pepper) — Tabascos are highly regarded as disease preventives, like garlic and onion, with which it is often mixed. Finely powdered seed or fruit given in delirium tremens. Cayenne pepper used externally as stimulant, counterirritant, and rubefacient; internally as a digestive, and to dispel gas and rouse the appetite. Navajo are said to use cayenne in weaning children from breast feeding (Libster, 2002). Remember that milk will dissolve some capsaicin. Experimentally, red pepper is hypoglycemic.

Many medicinal properties depend on amount of capsaicin (0.2–1%) present. Fruit from Africa is highest in capsaicin. Capsaicin, the major pungent principle, stimulates salivation and sweating. Some people suggest capsicum for alopecia. Weed (2002) says re alopecia, "Avoid cayenne. Heroic herbalists say it increases hair growth by improving blood circulation to the scalp. But when there is hair loss, says Janet Roberts, MD, specialist in women's hair loss and member of the Oregon Menopause Network, there are inflamed follicles. Cayenne increases inflammation, ultimately increasing hair loss" (Weed, 2002). Warming herbs like cayenne, cinnamon, and ginger may increase energy but may increase hot flashes, too (Weed, 2002).

Andrews (1995) cites the work of Irwin Ziment, UCLA School of Medicine, who routinely prescribes hot spices such as black pepper, ginger, horseradish, hot pepper, and mustard for asthma, bronchitis, and other chronic respiratory tract problems. But this was all probably part of the Mayan tradition. Ziment recommends a chilli a day, ca. one three-inch cayenne-type, with meals. Capsicum may be helpful in colds, coughs, or hay fever. Capsaicin irritates the mucous membranes lining the nose and throat, causing them to weep a watery secretion, "making it easier for you to cough up mucus or clear your nose when you blow" (RIN). And heating up that chicken soup with hot pepper may be a pleasant decongestant for colds (HAD, RIN). The oral LD50 is 97–294 mg/kg in mice, meaning that to ingest a lethal dose, I, a 220 lb (100 kg) rat, would need ingest some 135–415 oz of hot pepper. No way (TAD).

Alcoholics seem to have an affinity for hot peppers. Today, there's the Bloody Mary; tomorrow, the hot pepper for the hangover. According to Libster (2002), in the wild, wild west, cayenne was believed the best remedy for the DTs. People ingested capsicum in soup to help alleviate the craving for alcohol and for the nausea of too much alcohol. Hot peppers were believed to prevent vomiting, tonify the stomach, and promote digestion in alcoholics (Libster, 2002). Mandram is a West Indian stomachic preparation of mashed chilies in Madeira wine, with cucumber, shallot, chives or onions, and lime juice. Said to aid alcoholics by reducing dilated blood vessels, thus relieving chronic congestion.

Lavishing antiseptic praise on Louisiana hot sauce, Andrews (1995) cites LSU studies that showed that straight Louisiana hot sauce killed all the bacteria in a test tube within a minute. Even at a dilution of 1:16, it killed them in 5 min. For ostreaphiles, the hot sauce killed *Vibrio vulnificus*, a germ that makes oyster-eating dangerous. Sharma et al. (2000) showed that, though bactericidal on their own, extracts of black pepper, *Capsicum*, and turmeric partially protected *Bacillus megaterium, Bacillus pumilus,* and *Escherichia* from radiation, probably protecting their DNA. Chile was strongest. While there is no record to my knowledge of *Capsicum* combating *Bacillus anthracis*, cayenne inhibits *Bacillus cereus* and *B. subtilis, Clostridium* spp, and *Streptococcus pyogenes*, at least *in vitro* (Libster, 2002). Chile was antiseptic to *Bacillus subtilis, Escherichia coli,* and *Saccharomyces cerevisiae* confirming traditional uses of spices as food preservatives, disinfectants, and antiseptics (De et al., 1999). And, rightly or wrongly, Andrews (1995) reports that hot pepper repels some big critters too—dogs from garbage cans, squirrels from bulb plantings and birdseed in bird feeders, cats from flowerbeds, burrowing armadillos, and children who suck coins out of slot machines, not to mention weaning children off the breast.

The liniment made from the fruit is said to be a folk remedy for indolent tumors and indurated mammae. The whole plant, steeped in milk, is said to remedy hardened tumors (JLH). In a review paper, Buchanan (1978) reported that feeding rats a protein-deficient diet containing 10% chili peppers produced a 54% incidence of hepatomas, suggesting that capsaicin may contribute to the etiology of human liver cancer, particularly in those regions where dietary protein is minimal. Authors have reported that red peppers are carcinogenic or co-carcinogenic. Still, the low incidence of gastric cancers in Latin America has suggested to others that the hot pepper might be anticarcinogenic. Even the American Cancer Institute, an agency that also recommends tamoxifen, a carcinogen, for the prevention of cancer, also admits that diets rich in foods well endowed with beta-carotene may lower the risks of some types of cancer. Red peppers are rich in carotenoids and richer in ascorbic acid than citrus (TAD). But capsaicin may be a double-edged sword. Studies on the toxicity, mutagenic, and carcinogenic/co-carcinogenic activities of capsaicin had conflicting results, most depending on the dose. Most studies have shown that low consumption of chiles is beneficial, while high consumption may be deleterious. Chile consumption may create a risk for gastric cancer, but it also protects against aspirin-induced injury of the gastroduodenal mucosa in humans and exhibits a protective factor against peptic ulcers. Orally administered capsaicin also exhibits chemoprotective activities against some chemical carcinogens and mutagens, but one study in Chile found that gallbladder carcinoma was correlated with the high intake of both green and red chiles (TAD). Translation: all things in moderation. Rats fed 10% chili had a higher incidence of liver tumors than controls. The latest study available to me showed that rats fed diets with 500 ppm capsaicin diets had 60% fewer colon cancers (Yoshitani et al., 2001). Andrews (1995) evokes a personal communication from Peter Gannett, one of the earlier scientists who showed that, in absurd quantities (corresponding to several pounds of hot pepper a day for life), capsaicin might be mutagenic or tumorigenic. Gannet told Andrews that the protective effects of capsaicin seem to outweigh any carcinogenic or mutagenic potential.

According to Joseph et al. (2002), green peppers are better at blocking nitrosamine than tomatoes (JNU), all the more reason to add green pepper with your red pepper to those barbecues (JAD). As one of the best sources of ORAC antioxidants in general, containing ascorbic acid, beta-carotene, beta-cryptoxanthin, lutein, and zeaxanthin, peppers are colorful additions to your medicine chest as well, when you consider all the diseases those nutrients are reported to prevent or alleviate.

Red pepper contains twice the vitamin C and nine times the vitamin A equivalent of green peppers. Mix and match! Enjoy!

Wood (1993) edited a whole book on capsaicin and pain. Andrews (1995) presents a beautifully illustrated book with 34 color plates, all on *Capsicum*. She too has a great section on pain. Although giving us a formula for homemade capsaicin cream, Andrews confesses that she herself indulges in the store-bought analgesic, Zostrix, which costs more than $38 per oz. She was, like so many writers, suffering from writer's arthritis, if not carpal tunnel syndrome. She rubbed her arthritic fingers slowly four times a day and experienced "blessed relief in less than three weeks." She also found that it abated the pain of elbow tendinitis and carpal tunnel syndrome in the thumb. To make 16 oz "capsaid," you'll need 8 oz of ripe Habaneros, a pint of glycerine or inert mineral oil, and an ounce of broken beeswax or paraffin. Microwave the peppers ca. 3 min with 1–2 tbsp water in a sealed polyethylene bag, until the stems are easy to remove. Strain the water off the peppers, remove the stem, and macerate or chop, with the seeds. Put peppers with oil in pot; bring slowly to a boil, lower heat, and simmer 4 hr, then cool for 4 hr. Repeat two times for a total of 12 hr cooking, 12 hr cooling. Strain the pulp, expressing the juice to go through the strain. Discard the residue. Blend in electric blender the pureed pulp and oil. Strain one more time through fine metal strainer lined with one-ply tissue paper. Return to clean pan, adding the beeswax or paraffin and melting carefully over low heat, stirring until well blended. Cool slightly and pour into plastic vials or flat container jars in which your salve can harden, hopefully not too much (Andrews, 1995, adapted from Jeanne Rose's *The Aromatherapy Book*).

Indications (Hot Pepper) — Adenoma (1; X11604990); Ague (f; IED); Alcoholism (1; PHR; PH2; WO2); Allergy (1; FNF); Angina (f; LIB); Anorexia (1; APA; PHR; WBB; WO2); Anorexia nervosa (f; PH2); Arrhythmia (1; FNF); Arthrosis (Pain) (1; APA; BGB; FNF; SKY); Asthma (1; JFM; JNU); Atherosclerosis (1; PHR; PH2); Bacillus (1; X10548758); Backache (1; APA; WBB); Bacteria (1; FNF; X10548758); Bleeding (f; DAD); Boil (f; IED; JFM); Bronchosis (1; APA); Burn (f; LIB); Bursitis (1; SKY); Cancer (f; JLH); Cancer, breast (f; JLH); Cancer, colon (1; X11604990); Cancer, nose (f; JLH); Cancer, skin (f; JLH); Cardiopathy (1; FNF; PHR; PH2); Cataract (1; DAD); Chickenpox (1; APA); Chilblains (1; BGB; PNC; WO2); Childbirth (1; 60P); Chill (f; APA); Cholera (f; IED; PH2; JAF49:31010); Chromhidrosis (1; LIB); Circulosis (1; WAM); Cluster Headache (1; APA); Cold (1; APA; FNF; JFM; RIN); Colic (1; APA; FNF; JFM; PNC); Congestion (1; DAD; FNF; JFM); Cough (f; JFM; PH2); Cramp (2; FNF; KOM; PH2); Cystosis (f; LIB); Delirium (f; LIB); Diabetes (1; APA); Diabetic Neuropathy (1; SKY); Diarrhea (f; PHR; PH2); Diphtheria (f; LIB); Dropsy (f; IED); Dyspepsia (1; APA; BGB; IED; PH2; WO2); Dyspnea (f; DAV); Earache (f; IED); Edema (f; PH2); Enterosis (f; PH2); Epithelioma (f; JLH); Escherichia (1; X10548758); Fever (1; IED; PHR; PH2; TAD); Flu (1; DAV; FNF); Frostbite (f; BGB; PHR; PH2; SPI); Fungus (1; X10548758); Gangrene (f; LIB); Gas (1; APA; DAV); Gastrosis (1; JFM; PH2; TRA; WO2); Giddiness (f; IED); Gout (f; IED; PH2); Hay Fever (1; RIN); Headache (1; APA; WAM); Head Cold (1; RIN); Hemorrhoid (f; IED; JFM; WBB); Hepatosis (f; WBB; WO2); Herpes Zoster (1; DAV; SKY); High Blood Pressure (1; FNF); High Cholesterol (1; APA; FNF; TRA); High Triglycerides (1; APA); Hoarseness (f; PHR); Impotence (f; LIB; PHR); Induration (f; JLH); Infection (1; FNF; IED; PH2); Inflammation (1; FNF; TRA; WO2); Inorgasmia (f; PHR); Ischemia (1; FNF); Itch (2; ABS); Kernel (f; JLH); Laryngosis (f; PNC); Lumbago (1; APA; PHR; PH2; PNC); Malaria (f; IED; PHR; PH2); Mastosis (f; JLH); Mycosis (1; X10548758); Myosis (2; APA; KOM; PHR; PH2; PNC); Nephrosis (f; LIB); Neuralgia (1; APA; SKY; WO2); Neuropathy (1; TAD); Obesity (1; FNF; HAD); Osteoarthrosis (1; LIB; TAD); Otosis (f; PH2); Pain (2; APA; BGB; FNF; PH2; WBB); Pharyngosis (1; DAD; PH2); Plague (f; WBB); Pneumonia (f; LIB); Proctosis (f; LIB); Prurigo (2; ABS); Psoriasis (1; APA; FNF; SKY); Pulmonosis (f; IED; 60P); Respirosis (f; IED); Rheumatism (2; APA; FNF; PHR; PH2; TRA); Rhinosis (f; JLH); Scarlet Fever (f; PH2); Sciatica (1; PH2); Seasickness (f; PH2); Shingles (1; APA); Snakebite (f; IED; 60P); Sore (f; LIB); Sore Throat (1; JFM; PHR; PH2); Sprain (1; APA); Strain (1; APA); Stomachache (f; JAF49:3101); Stomatosis (f; LIB); Stroke (1; PHR; PH2); Streptococcus (1; LIB);

Swelling (f; DAD; WBB); Tachycardia (1; FNF); Tennis Elbow (1; JAD); Tension (2; PH2); Thumb-Sucking (1; APA; BGB); Thyrosis (f; PED); Tonsilosis (f; LIB); Toothache (1; DAV; 60P); Typhoid (f; IED); Typhus (f; JAF49:3101); Ulcer (f; BGB; LIB); UTI (f; PH2); Varicosis (1; JAD; WBB; WO2); Virus (1; WO2); Water Retention (1; FNF); Wound (1; JFM; WO2); Xerostoma (1; FNF); Yeast (1; X10548758); Yellow Fever (f; JAF49:3101; PH2).

Hot Pepper for cardiopathy:

- ACE-Inhibitor: alpha-terpineol; myrcene
- Antiaggregant: alpha-linolenic-acid; caffeic-acid; capsaicin; ferulic-acid; quercetin; salicylates
- Antiarrhythmic: apiin; capsaicin; ferulic-acid; scopoletin
- Antiatherogenic: rutin
- Antiatherosclerotic: lutein; quercetin
- Antiedemic: caffeic-acid; caryophyllene; rutin; scopoletin
- Antihemorrhagic: phylloquinone
- Antihemorrhoidal: rutin
- Antiischemic: capsaicin
- Antioxidant: caffeic-acid; capsaicin; chlorogenic-acid; ferulic-acid; hesperidin; lutein; myrcene; p-coumaric-acid; quercetin; rutin; scopoletin
- Antitachycardic: capsaicin; solasodine
- Antithrombic: ferulic-acid; quercetin
- Arteriodilator: ferulic-acid
- COX-2-Inhibitor: quercetin
- Calcium-Antagonist: caffeic-acid; capsaicin; hesperidin; solanine
- Cardiotonic: capsaicin; solanine
- Cyclooxygenase-Inhibitor: capsaicin; hesperidin; quercetin
- Diuretic: asparagine; betaine; caffeic-acid; chlorogenic-acid; terpinen-4-ol
- Hypocholesterolemic: beta-ionone; caffeic-acid; chlorogenic-acid; dihydrocapsaicin; rutin
- Hypotensive: 1,8-cineole; alpha-linolenic-acid; phylloquinone; quercetin; rutin; scopoletin; valeric-acid
- Sedative: 1,8-cineole; alpha-terpineol; benzaldehyde; caffeic-acid; carvone; caryophyllene; isovaleric-acid; limonene; pulegone; solanine; valeric-acid
- Vasodilator: capsaicin; quercetin

Hot Pepper for cold/flu:

- Analgesic: caffeic-acid; camphor; capsaicin; chlorogenic-acid; ferulic-acid; hesperidin; myrcene; quercetin; scopoletin; solanine
- Anesthetic: 1,8-cineole; benzaldehyde; camphor; capsaicin; cinnamic-acid; myrcene
- Antiallergic: 1,8-cineole; ferulic-acid; quercetin; terpinen-4-ol
- Antibacterial: 1,8-cineole; acetic-acid; alpha-terpineol; benzaldehyde; beta-ionone; caffeic-acid; caryophyllene; chlorogenic-acid; cinnamic-acid; delta-3-carene; ferulic-acid; limonene; myrcene; p-coumaric-acid; pulegone; quercetin; rutin; scopoletin; terpinen-4-ol; thujone
- Antibronchitic: 1,8-cineole; solanine
- Antiflu: caffeic-acid; hesperidin; limonene; quercetin
- Antihistaminic: caffeic-acid; chlorogenic-acid; hesperidin; pulegone; quercetin; rutin
- Antiinflammatory: alpha-linolenic-acid; beta-pinene; caffeic-acid; capsaicin; caryophyllene; chlorogenic-acid; cinnamic-acid; delta-3-carene; ferulic-acid; hesperidin; quercetin; rutin; salicylates; scopoletin; solasodine

C

- Antioxidant: caffeic-acid; capsaicin; chlorogenic-acid; ferulic-acid; hesperidin; lutein; myrcene; p-coumaric-acid; quercetin; rutin; scopoletin
- Antipharyngitic: 1,8-cineole; quercetin
- Antipyretic: pulegone; salicylates
- Antiseptic: 1,8-cineole; alpha-terpineol; benzaldehyde; beta-pinene; caffeic-acid; camphor; capsaicin; carvone; chlorogenic-acid; limonene; oxalic-acid; scopoletin; terpinen-4-ol; thujone
- Antitussive: 1,8-cineole; terpinen-4-ol
- Antiviral: caffeic-acid; chlorogenic-acid; ferulic-acid; hesperidin; limonene; quercetin; rutin
- Bronchorelaxant: scopoletin
- COX-2-Inhibitor: quercetin
- Cyclooxygenase-Inhibitor: capsaicin; hesperidin; quercetin
- Decongestant: camphor
- Expectorant: 1,8-cineole; acetic-acid; camphor; limonene
- Immunostimulant: alpha-linolenic-acid; benzaldehyde; caffeic-acid; chlorogenic-acid; ferulic-acid
- Interferonogenic: chlorogenic-acid
- Phagocytotic: ferulic-acid

Hot Pepper for rheumatism:

- Analgesic: caffeic-acid; camphor; capsaicin; chlorogenic-acid; ferulic-acid; hesperidin; myrcene; quercetin; scopoletin; solanine
- Anesthetic: 1,8-cineole; benzaldehyde; camphor; capsaicin; cinnamic-acid; myrcene
- Antidermatitic: quercetin; rutin
- Antiedemic: caffeic-acid; caryophyllene; rutin; scopoletin
- Antiinflammatory: alpha-linolenic-acid; beta-pinene; caffeic-acid; capsaicin; caryophyllene; chlorogenic-acid; cinnamic-acid; delta-3-carene; ferulic-acid; hesperidin; quercetin; rutin; salicylates; scopoletin; solasodine
- Antiprostaglandin: caffeic-acid; hesperidin; scopoletin
- Antispasmodic: 1,8-cineole; apiin; benzaldehyde; caffeic-acid; camphor; capsaicin; caryophyllene; cinnamic-acid; ferulic-acid; limonene; myrcene; p-coumaric-acid; quercetin; rutin; scopoletin; thujone; valeric-acid
- COX-2-Inhibitor: quercetin
- Counterirritant: 1,8-cineole; camphor; thujone
- Cyclooxygenase-Inhibitor: capsaicin; hesperidin; quercetin
- Lipoxygenase-Inhibitor: caffeic-acid; chlorogenic-acid; cinnamic-acid; hesperidin; p-coumaric-acid; quercetin; rutin
- Myorelaxant: 1,8-cineole; rutin; scopoletin; valeric-acid

Other Uses (Hot Pepper) — Source of cayenne pepper and chili powder, used in tabasco sauce and other preparations. Ground red pepper, or cayenne pepper, is a sharp, hot powder made from dried ripe chiles. Ground red pepper is an ingredient in many spice blends, particularly chili powder (red cayenne, 83%; cumin (or substitute ground caraway), 9%; oregano, 4%; salt, 2.5%; and garlic powder, 1.5%) (RIN). Cayenne pepper is incorporated in laying mixtures for poultry. Extract used in manufacture of ginger beer and other beverages. The hot piquant fruits are pickled, and used as a spice, or made into hot pepper sauce. Young leaves and tops of the stems are occasionally steamed and eaten as vegetables. In Thailand, juice from the leaves adds color and that hot, spicy flavor to curries (FAC). For a hot/cold treat, try a sprinkling of candied green jalapeño over chocolate ice cream (Arndt, 1999).

And don't get the stuff on your fingers and then touch your sensitive mucous membranes. Water will not cut capsaicin — not even soapy water. Alkaline materials like chlorine or ammonia will ionize the capsaicins, permitting them to go into solution. Tucker and Debaggio (2000) suggest, after handling hot peppers, washing the hands with ammonia or chlorine bleach, which turns capsaicin into water soluble salts. Rinzler (1990) mentions that capsaicin does dissolve in alcohol, milk fat, and vinegar (RIN). For burning in the mouth, cheap vodka may be the best alternative. Some people drink tomato juice or eat a fresh lemon or lime to stop the burning, the theory being that the acid counteracts the alkalinity of the capsaicin. If you overdo a hot pepper, the best coolants are milk, yogurt, sour cream, and other dairy products — say, ice cream. They are also useful in removing residue from the hands. Casein in dairy products breaks the bond of capsaicin with the pain receptors in the mouth. So milk works, too, if you don't like or have vodka (TAD). Beer, bread, rice, and tortillas can also help (AAR).

For more information on activities, dosages, and contraindications, see the *CRC Handbook of Medicinal Herbs,* ed. 2, Duke et al., 2002.

Cultivation (Hot Pepper) — Cultivation is same as for *C. annuum*, except that it takes longer for *C. frutescens* peppers to mature. Peppers, perennial and hardy only in frost-free situations, treated as annuals in frost country. They fare well in full sun, friable and porous soils, moist but not constantly wet, pH 4.3–8.7 (average 6.1). Germination takes 6–10 days. One kilogram of seed produces 15,000–17,000 plants, enough to plant a hectare. Seed are sown in seedbeds early in spring (mid March) and transplanted to fields in 7–8 weeks, spaced about 60 cm apart, in rows 90–120 cm apart. Seedlings from field sown seed are thinned to 30–45 cm apart. Grown monoculture, can be transplanted and cultivated by machine. Irrigation may be necessary. Frequent shallow cultivation is necessary to control weeds until harvesting begins. Do not plant peppers where tomatoes or potatoes have grown within 3 years or where peppers were grown the previous season. Irrigation, used when necessary, can increase yields sixfold. Soil should have adequate nutrients to keep peppers growing well. Stable manure or complete fertilizer should be used on field before planting, with side dressings in late May or June. Inorganic gardeners use sprays as soon as insects are evident. In warm weather, early cultivars (cvs) mature in 3–4 months; other cvs may require 4–5 months. Fruits are harvested manually, every 7 days. Fruit of paprika is dried with artificial heat in barns or sundried. Harvest continues until frost or adversity. In India, peppers are grown under dry conditions in rotation with sorghums, groundnut, and millet, and under irrigation with sugarcane and corn. Plants degenerate under cultivation in 1–2 years. Re *C. annuum*, the seeds remain viable for 2–3 years. Like tomatoes, green or red peppers may be ripened artificially, between 21.5°C and 25°C being best for ripening. At 0°C, peppers may be kept in good condition for about 30 days or more, with 95–98% relative humidity advised. Maximum yields rarely exceed 10 tons/ha. In spacing studies in Georgia, yields of 7.5 to 15 tons/ha of pimentos are reported. (See DAD for more.)

Chemistry (Hot Pepper) — The active ingredient capsaicin is very heavily studied as an analgesic (Wood, 1993). Patel and Srinivasan (1985) noted that dietary capsaicin significantly increased lipase, maltase, and sucrase activities. Capsaicin causes contact dermatitis. Here are a few of the more notable chemicals found in hot pepper. For a complete listing of the phytochemicals and their activities, see the CRC phytochemical compendium, Duke and duCellier, 1993 (DAD) and the USDA database http://www.ars-grin.gov/duke/.

Beta-Carotene — Androgenic?; Antiacne; Antiaging; Antiasthmatic; Anticancer 22 ppm; Anticarcinomic; Anticoronary 50 mg/man/2 days; Antiichythyotic; Antileukoplakic; Antilupus 150 mg/man/day/2 mo; Antimastitic; Antimutagenic; Antioxidant; Antiozenic; Antiphotophobic 30–300 mg man/day; AntiPMS; Antiporphyric 30–300 mg/man/day; Antipityriasic; Antiproliferant; Antipsoriac; Antiradicular; Antistress; Antitumor; Antiulcer 12 mg/3×/day/man/orl; Antixerophthalmic; Colorant; Hypokeratotic; Immunostimulant 180 mg/man/day/orl; Interferon-Synergist; Mucogenic;

Phagocytotic; Prooxidant 20 µg/g; Thymoprotective; Ubiquict; RDA = 2.25–7.8 mg/day; PTD = 15–30 mg/day; LD50 = >1000 mg/kg orl rat.

Beta-Crypotxanthin — Antiproliferant; Antitumor (5 × carotene); Antitumor (Breast); Antitumor (Cervix); Colorant; Vitamin-A-Activity.

Capsaicin — Adrenergic; Analgesic; Anaphylactic; Anesthetic; Antiaggregant; Antiarrhythmic 100 µ*M*; Anticancer; Anticolonospasmic; Antiinflammatory; Antiischemic 100 µ*M*; Antimastalgic; Antineuralgic; Antineuritic; Antinitrosaminic; Antiodontalgic; Antioxidant; Antipsoriatic; Antipyretic 7 mg/kg; Antiseptic; Antispasmodic; AntiSubstance-P; Antitachycardic 100 µ*M*; Antitumor (Lung); Antiulcer; ATPase-Inhibitor; Bronchoconstrictor; Calcium-Antagonist 100 µ*M*; Carcinogenic; Cardiotonic; Catabolic; Catecholaminigenic; Cyclooxygenase-Inhibitor; Cytochrome-p450-Inhibitor; Diaphoretic; Digestive; Endocrinactive 50 mg/kg scu rat; Irritant; Lactase-Promoter; Laxative; 5-Lipoxygenase-Inhibitor; Maltase-Promoter; Mutagenic; Neurotoxic; Repellent; Radioprotective; Respirasensitizer; Rubefacient; Secretagogue; Sialagogue; Sucrase-Promoter; Tachyphylactic; Thermogenic; Vasodilator; LDlo = 1.6 ivn cat; LD50 = 0.56 ivn mus; LD50 = 7.56 ipr mus; LD50 = 10 ipr rat; LD50 = 47 mg/kg orl rat; LD50 = 190 orl mus; LD50 97–294 orl mus; LD50 = 9.00 scu mus.

Capsidol — Fungicide; Phytoalexin.

Lutein — Antiatherosclerotic; Antimaculitic; Antinyctalopic; Antiproliferant; Antitumor (Breast); Antitumor (Colon); Antioxidant; Antiradicular; Cardioprotctive; Colorant; Prooxidant 5–40 µg/g; Quinone-Reductase-Inducer 2.5 µg; Retinoprotectant Optometry; Ubiquict.

Zeaxanthin — Antitumor (Breast); Colorant; Hepatoprotective; Quinone-Reductase-Inducer.

Carum carvi L. (Apiaceae)
CARAWAY

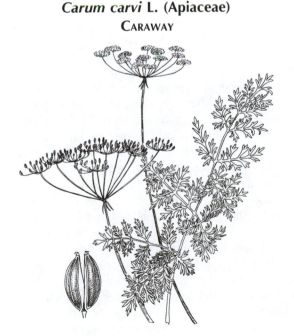

Synonym — *Carum velenovskyi* Rohlena

Medicinal Uses (Caraway) — Antispasmodic, bactericidal, carminative, diaphoretic, digestive, diuretic, emmenagogue, expectorant, fungicidal, lactagogue, laxative, stimulant, and stomachic, caraway has been used to alleviate alopecia, bruises, cholera, earache, fistula, halitosis, headache, hookworm, hysteria, impotence, prolapse, sores, and syphilis (LIL). Chinese use the roots for arthralgia, convulsions, fever, flu, headache, and tetanus. Enteric-coated and non-enteric-coated caraway-peppermint oil combinations are safe preparations which act locally causing smooth muscle relaxation. I think even my colonoscopist uses it, surely the peppermint, to keep my colon from cramping down on his scope. This myorelaxant effect may explain results observed in clinical trials with patients suffering functional dyspepsia (Micklefield et al., 2000).

Weed (2002) suggests heavy consumption of certain spices when estrogen levels are down. Seeds like caraway, celery, coriander, cumin, poppy and sesame, mustard and anise, fennel and fenugreek all contain phytoestrogens, as do their oils, says Weed. She suggests using these seeds "lavishly" when cooking or making tea with any one of them, drinking 3–4 cups a day "for best results" (Weed, 2002). One nursing formula recommends any one of the aromatic umbelliferous seeds, caraway, coriander, cumin, dill, fennel, or anise, with raspberry or nettle leaves and blessed thistle or borage leaves. The caraway seeds "increase milk production and tone the digestive system; their powers are carried through the breast milk and into the child, curtailing colic and indigestion" (Weed, 1985).

Caraway seed oil is both antibacterial and antispasmodic, and inhibits skin tumors (TAD). The oil also has antihistaminic, fungicidal, and larvicidal properties. EOs of caraway are extremely active relative to the tuberculosis bacterium.

The monoterpenoid limonene has received a lot of favorable press lately, especially relative to breast cancer. Limonene not only inhibits the initiation of cancer in the lung, mammaries, and forestomach, it can even lead to regression in mammary cancers (in rats). And I find that those things proving useful for breast cancers usually prove useful in other hormone-dependent cancers, like colon and prostate. Shwaireb et al. (1995) showed that rats whose diet consisted of 20% caraway and watercress, both good soup ingredients that could be spiced up with turmeric and other anticancer spices, might prevent breast cancer. Caraway afforded nearly 50% protection, watercress 28.5%. Huang et al. (1994) indicated that d-carvone, at 200 μM was equivalent to d-limonene at 200 μM at least at inhibiting lung and forestomach cancer. Our caraway seed, richest source of limonene (up to 3%), is also one of the richest sources of carvone (up to nearly 4.5%). It also contains perillyl-alcohol.

Boik (2001) mentions limonene first among monoterpenes he deems valuable against cancer. He cites 15 *in vitro* and 7 animal studies suggesting that monoterpenes inhibit a variety of and induce regressions in tumor lines. The anticancer effects of limonene have been studied intensively at the University of Wisconsin. While Boik specifies orange oil as a major source, he admits that some commercial orange oils contain no detectable limonene. But caraway and celeryseed are the best whole plant sources in my database, attaining levels potentially as high as 3% and 2.5%, respectively. At levels lower than 1% in rat diet, limonene inhibited experimental breast cancer development. One of the major modes of monoterpene action may be in detoxification. The related terpene, perillyl alcohol, is 5–10 times more potent in inhibiting breast cancer. Both limonene and perillyl alcohol are rapidly metabolized to perillic acid. At 2% of diet (ca. 1,500 mg/kg), perillyl alcohol reduced chemically induced liver tumor masses tenfold. Apoptosis was markedly increased. Much lower doses (75–150 mg/kg) prevented the incidence of chemically induced rodent colon cancer. Boik notes that, as so often, the chemopreventive dose is lower than the antitumor effect, in this case the preventive dose is 1/10th to 1/20th the tumor reduction level. Boik assumes, and I find this a reasonable assumption after pondering synergy for two decades, a full fifteenfold increase in potency due to synergistic interactions. "Synergistic interactions are probably necessary for limonene and perillyl alcohol to produce an anticancer effect" (BO2). (For more details see the Synergy Principle chapter in Kaufman et al., 1998.) Like other compounds Boik discusses, monoterpenes "will be most effective when taken in at least three divided doses per day" (BO2). Boik warns that, as with other drugs and pure solitary phytochemicals, tumor cells may evolve resistance to monoterpenes (BO2). But I am not arguing for isolated phytochemicals; I argue only for phytochemicals in their natural context. If Boik is correct, we can roughly convert rat dosage data (in mg/kg) to estimated human dosage data simply by dividing by 4 (or 4.319 for the purist), mouse data dividing by 7 (or 7.28 for the purist). But oversimplifying, it took a hypothetical 730 mg/kg for a mouse, and 430 mg/kg for a rat; it would only take a manageable 100 mg/kg for me, which if I calculate correctly translates to 10 g for my 100 kg body.

Remember, it is caraway seed, not rye seed, that adorn most rye breads. So be sure and eat your crust, if like me, you are genetically targeted for colon cancer. Rye bran, even better than soy, prevented colonic crypts in animals (Davies et al., 1999).

Indications (Caraway) — Alactea (f; EFS); Allergy (1; FNF); Alzheimer's (1; FNF); Anemia (1; APA); Anorexia (2; HHB; KOM; PIP); Bacteria (1; FNF; HH2; PH2; WO2); Bronchosis (2; FNF; PHR); Cancer (1; BO2; FNF); Cancer, breast (1; BO2; FNF); Cancer, colon (1; BO2; FNF); Cancer, liver (1; BO2; FNF); Cancer, pancreas (1; BO2; FNF); Cancer, prostate (1; BO2; FNF); Cancer, stomach (1; BO2; FNF); Candida (1; FNF; PHR); Cardiopathy (f; PHR; PH2); Cholecystosis (2; PHR); Cold (2; APA; FNF; PHR); Colic (1; DEP; KOM; PIP; WO2); Congestion (1; FNF); Cough (2; APA; FNF; PHR); Cramp (1; DEP; FNF; PHR; PH2; SHT; WO2); Dermatosis (f; PH2); Diarrhea (f; BOW); Dysmenorrhea (f; APA); Dyspepsia (1; APA; DEP; HHB; KOM; PHR; SHT); Ectoparasite (f; HHB); Enterosis (f; DEP; PH2); Fever (2; PHR); Flu (1; FNF); Fungus (1; FNF; HH2; WO2); Gas (1; HHB; KOM; SHT); Gastrosis (1; PH2; PIP; PNC WO2); Hemorrhoid (f; DEP); Hepatosis (2; FNF; PHR); Hernia (f; BOW); Hiatal Hernia (f; BOW); Immunodepression (1; FNF); Incontinence (f; APA); Infection (1; FNF; HH2; PHR; PH2); Inflammation (1; FNF); Laryngosis (f; BOW); Lumbago (f; WO2); Melanoma (1; FNF); Meteorism (f; PHR); Myosis (1; APA) Nausea (f; APA); Nervousness (f; PHR); Neurosis (f; PH2); Ophthalmia (f; DEP); Pain (1; FNF); Pharyngosis (2; PHR); Pleurosis (1; HHB); Rheumatism (1; HHB; WO2); Scabies (1; WO2); Stomachache (1; PNC); Stomatosis (2; FNF; PHR); Stone (1; FNF); Trichomonas (1; FNF); Ulcer (f; BOW); Uterosis (f; DEP); Water Retention (f; EFS); Worm (1; DEP; EFS; FNF); Yeast (1; HH2; PH2).

Caraway for cancer:

- 5-Alpha-Reductase-Inhibitor: alpha-linolenic-acid
- AntiEBV: geranial
- AntiHIV: caffeic-acid; methanol; quercetin
- Antiaggregant: alpha-linolenic-acid; caffeic-acid; falcarindiol; kaempferol; myristicin; quercetin; salicylates
- Anticancer: alpha-linolenic-acid; alpha-pinene; alpha-terpineol; beta-myrcene; caffeic-acid; carvone; decan-1-al; hyperoside; isoquercitrin; kaempferol; limonene; linalool; myristicin; p-coumaric-acid; quercetin; quercetin-3-o-beta-d-glucoside; scopoletin; umbelliferone; xanthotoxin
- Anticarcinogenic: caffeic-acid
- Antifibrosarcomic: quercetin
- Antihepatotoxic: caffeic-acid; hyperoside; p-coumaric-acid; quercetin; scopoletin
- Antiinflammatory: alpha-linolenic-acid; alpha-pinene; beta-pinene; caffeic-acid; carvacrol; delta-3-carene; hyperoside; kaempferol; myristicin; quercetin; quercetin-3-o-beta-d-glucoside; quercetin-3-o-beta-d-glucuronide; salicylates; scopoletin; umbelliferone; xanthotoxin
- Antileukemic: astragalin; kaempferol; quercetin
- Antileukotriene: caffeic-acid; quercetin
- Antilipoperoxidant: quercetin
- Antilymphomic: xanthotoxin
- Antimelanomic: carvacrol; perillaldehyde; perillyl-alcohol; quercetin

C

- Antimetastatic: alpha-linolenic-acid
- Antimutagenic: caffeic-acid; falcarindiol; kaempferol; limonene; myrcene; quercetin; scopoletin; umbelliferone; xanthotoxin
- Antinitrosaminic: caffeic-acid; p-coumaric-acid; quercetin
- Antioxidant: caffeic-acid; camphene; carvacrol; gamma-terpinene; hyperoside; isoquercitrin; kaempferol; myrcene; myristicin; p-coumaric-acid; quercetin; scopoletin
- Antiperoxidant: caffeic-acid; p-coumaric-acid; quercetin
- Antiproliferant: perillyl-alcohol; quercetin
- Antiprostaglandin: caffeic-acid; carvacrol; scopoletin; umbelliferone
- Antistress: myristicin
- Antitumor: caffeic-acid; isoquercitrin; kaempferol; limonene; p-coumaric-acid; quercetin; scopoletin; xanthotoxin
- Antiviral: alpha-pinene; caffeic-acid; geranial; hyperoside; kaempferol; limonene; linalool; p-cymene; quercetin
- Apoptotic: kaempferol; perillyl-alcohol; quercetin
- COX-2-Inhibitor: kaempferol; quercetin
- Chemopreventive: limonene; perillyl-alcohol
- Cyclooxygenase-Inhibitor: carvacrol; kaempferol; quercetin
- Cytochrome-p450-Inducer: 1,8-cineole
- Cytoprotective: caffeic-acid
- Cytotoxic: caffeic-acid; falcarindiol; p-coumaric-acid; quercetin; scopoletin; xanthotoxin
- Hepatoprotective: caffeic-acid; herniarin; isorhamnetin-3-galactoside; isorhamnetin-3-glucoside; quercetin; scopoletin
- Hepatotonic: 1,8-cineole
- Immunostimulant: alpha-linolenic-acid; astragalin; caffeic-acid
- Lipoxygenase-Inhibitor: caffeic-acid; kaempferol; p-coumaric-acid; quercetin; umbelliferone
- Lymphocytogenic: alpha-linolenic-acid
- Mast-Cell-Stabilizer: quercetin
- Ornithine-Decarboxylase-Inhibitor: caffeic-acid; limonene; quercetin
- p450-Inducer: 1,8-cineole; quercetin
- PTK-Inhibitor: quercetin
- Phytohormonal: scopoletin
- Prostaglandigenic: caffeic-acid; p-coumaric-acid
- Protein-Kinase-C-Inhibitor: quercetin
- Sunscreen: caffeic-acid; umbelliferone
- Topoisomerase-II-Inhibitor: isoquercitrin; kaempferol; quercetin
- Tyrosine-Kinase-Inhibitor: quercetin

Caraway for cold/flu:

- Analgesic: caffeic-acid; falcarindiol; myrcene; p-cymene; quercetin; scopoletin
- Anesthetic: 1,8-cineole; carvacrol; linalool; myrcene; myristicin
- Antiallergic: 1,8-cineole; herniarin; kaempferol; linalool; quercetin; terpinen-4-ol
- Antibacterial: 1,8-cineole; alpha-pinene; alpha-terpineol; caffeic-acid; carvacrol; citronellol; cuminaldehyde; delta-3-carene; falcarindiol; geranial; herniarin; hyperoside; isoquercitrin; kaempferol; limonene; linalool; myrcene; p-coumaric-acid; p-cymene; perillaldehyde; perillyl-alcohol; quercetin; scopoletin; terpinen-4-ol; thujone; umbelliferone; xanthotoxin
- Antibronchitic: 1,8-cineole
- Antiflu: alpha-pinene; caffeic-acid; hyperoside; limonene; p-cymene; quercetin

- Antihistaminic: caffeic-acid; kaempferol; linalool; quercetin; umbelliferone; xanthotoxin
- Antiinflammatory: alpha-linolenic-acid; alpha-pinene; beta-pinene; caffeic-acid; carvacrol; delta-3-carene; hyperoside; kaempferol; myristicin; quercetin; quercetin-3-o-beta-d-glucoside; quercetin-3-o-beta-d-glucuronide; salicylates; scopoletin; umbelliferone; xanthotoxin
- Antioxidant: caffeic-acid; camphene; carvacrol; gamma-terpinene; hyperoside; isoquercitrin; kaempferol; myrcene; myristicin; p-coumaric-acid; quercetin; scopoletin
- Antipharyngitic: 1,8-cineole; quercetin
- Antipyretic: salicylates
- Antiseptic: 1,8-cineole; alpha-terpineol; beta-pinene; caffeic-acid; carvacrol; carvone; citronellol; furfural; kaempferol; limonene; linalool; n-nonanal; scopoletin; terpinen-4-ol; thujone; umbelliferone
- Antistress: myristicin
- Antitussive: 1,8-cineole; carvacrol; terpinen-4-ol
- Antiviral: alpha-pinene; caffeic-acid; geranial; hyperoside; kaempferol; limonene; linalool; p-cymene; quercetin
- Bronchorelaxant: linalool; scopoletin
- COX-2-Inhibitor: kaempferol; quercetin
- Cyclooxygenase-Inhibitor: carvacrol; kaempferol; quercetin
- Expectorant: 1,8-cineole; alpha-pinene; astragalin; beta-phellandrene; camphene; carvacrol; limonene; linalool
- Immunostimulant: alpha-linolenic-acid; astragalin; caffeic-acid

Caraway for hepatosis:

- AntiEBV: geranial
- Antiedemic: caffeic-acid; scopoletin
- Antihepatotoxic: caffeic-acid; hyperoside; p-coumaric-acid; quercetin; scopoletin
- Antiherpetic: caffeic-acid; quercetin
- Antiinflammatory: alpha-linolenic-acid; alpha-pinene; beta-pinene; caffeic-acid; carvacrol; delta-3-carene; hyperoside; kaempferol; myristicin; quercetin; quercetin-3-o-beta-d-glucoside; quercetin-3-o-beta-d-glucuronide; salicylates; scopoletin; umbelliferone; xanthotoxin
- Antileukotriene: caffeic-acid; quercetin
- Antileukotrienogenic: scopoletin
- Antilipoperoxidant: quercetin
- Antioxidant: caffeic-acid; camphene; carvacrol; gamma-terpinene; hyperoside; isoquercitrin; kaempferol; myrcene; myristicin; p-coumaric-acid; quercetin; scopoletin
- Antiperoxidant: caffeic-acid; p-coumaric-acid; quercetin
- Antiprostaglandin: caffeic-acid; carvacrol; scopoletin; umbelliferone
- Antiradicular: caffeic-acid; carvacrol; isoquercitrin; kaempferol; quercetin
- Antiviral: alpha-pinene; caffeic-acid; geranial; hyperoside; kaempferol; limonene; linalool; p-cymene; quercetin
- COX-2-Inhibitor: kaempferol; quercetin
- Cholagogue: caffeic-acid; herniarin; scopoletin
- Choleretic: 1,8-cineole; caffeic-acid; kaempferol; p-coumaric-acid; scopoletin; umbelliferone
- Cyclooxygenase-Inhibitor: carvacrol; kaempferol; quercetin
- Cytoprotective: caffeic-acid
- Hepatoprotective: caffeic-acid; herniarin; isorhamnetin-3-galactoside; isorhamnetin-3 glucoside; quercetin; scopoletin

- Hepatotonic: 1,8-cineole
- Hepatotropic: caffeic-acid
- Immunostimulant: alpha-linolenic-acid; astragalin; caffeic-acid
- Lipoxygenase-Inhibitor: caffeic-acid; kaempferol; p-coumaric-acid; quercetin; umbelliferone

Other Uses (Caraway) — Lovers who ate caraway seed were said to always remain faithful, and articles containing the seed were said never to be stolen, in the old folklore. The seed on American rye breads, also essential on Kosher Rye, caraway has many culinary uses. Traditional with beet, cabbage, and carrot, it is often served with baked apples, other fruits, and breadstuffs. It is used with goulash and roast pork. Cheeses, soups, and vinegars are also flavored with caraway. Seeds should be added after the dish is almost ready, as prolonged cooking renders them bitter (RIN). Several liqueurs utilize caraway, e.g., Aquavit, Danzig, Danzigerwasser, Goldvasser, Kummel, L 'huile de Venus, and certain types of Schnapps and Brandies. One lusty liqueur called "Layaway" (a Jim Duke original) consists of 1 oz each seeds of caraway, anise, and fennel, steeped in one pint vodka and strained. In Scotland, buttered bread is dipped into a saucer of caraway seed. Unbroken seed retain their aroma if stored in air tight container and protected from light. Ground or mashed caraway seed can be substituted for cummin in homemade chili powders and curries. Seeds are also used with sugar-coated plums and taken as a breath freshener and digestive aid after spicy meals. Seeds are crushed and brewed into a tea. The seed oil is used commercially in ice cream, candy, pickles, soft drinks, etc. Caraway leaves may be chopped up and used as a parsley substitute. The young leaves form a good salad, while older ones may be boiled and served like spinach or added to soups and stews. Roots are boiled and eaten or chopped and used in soups, especially in northern Europe. (DEP, FAC, LIL).

Potential Carawade (alias limonenade) Ingredients (all herbs or spices, if you count the citrus peel as spice. I enjoy beverages combining any of these that I have on hand)

PPM		
Apium graveolens	Celeryseed	530–24,700 SD
Carum carvi	Caraway	7860–30,180 SD
Citrus aurantiifolia	Lime	2795–6400 FR
Citrus aurantium	Sour orange	1000–8000 FR
Citrus limon	Lemon	2796–8000 EO
Citrus reticulata	Tangerine	6500–9400 FR
Citrus sinensis	Orange	8300–9700 FR
Elettaria cardamomum	Cardamom	595–9480 FR
Foeniculum vulgare	Fennel	200–9420 FR
Illicium verum	Star-anise	100–5220 FR
Mentha spicata	Spearmint	200–5725 FR
Myristica fragrans	Nutmeg	720–5760 FR
Thymus vulgaris	Thyme	15–5200 PL

For more information on activities, dosages, and contraindications, see the *CRC Handbook of Medicinal Herbs*, ed. 2, Duke et al., 2002.

Cultivation (Caraway) — Hardy zones 3–9, caraway fares well in full sun, in light or humus-rich soils, wet but not constantly wet, pH 4.8–7.8. (average 6.4). Flowering and dying in its second season, the biennial caraway does well on well-prepared, upland fertile soil. Seeds (11,000/oz; 390/g) are sown (ca. 4–6 lb/a) in spring or autumn, and thinned to 8–12 in apart in rows. Seed germinate best at 59°F (15°C) for 8 hr and 50°F (10°C) for 16 hr. Germination is optimal after soaking the seed 3–6 days and drying for 4 hr before planting. Israeli's like ca. 30 plants per square yard (37 per square meter) (TAD). There should be two to three plants per foot in rows ca. 30 in apart. Seed will not mature until the second season; hence caraway is good for intercropping with such annuals as beans, coriander, dill, mustard, and peas. After intercrops are harvested the first year, caraway need only be weeded until it matures the second year. Large growers are mechanized, from mechanical drilling of the seed, and mechanized cultivation, on to mechanized harvesting and threshing. The small herbalist may cut the plants to dry, using a sickle or scythe, or may pull up the plants or cut off the tops one by one. One can expect about 1.3 cups of seed per plant. Seed yields usually run about 500–2000 lb/acre. There is a residual ton of straw, which can serve as animal food (Rosengarten, 1973). Seed must be dried and stored in a dry place. Ripe seed can be dried in the sun or over low heat, stirring occasionally. Some herbalists suggest the seeds be sterilized with scalding hot water to get rid of insect pests. Seed can also be salt-cured, frozen, or steeped in vinegar (LIL, TAD, Duke, 1987).

Chemistry (Caraway) — Rinzler (1990) seems to seek out the negative in her useful book, so she mentions that carvone and limonene are irritants. "Limonene is also a photosensitizer, a chemical that makes your skin more sensitive to sunlight" (RIN). I cannot tell where she got her data, but she acknowledges talking with Walter Lewis, Varro Tyler, and myself. Hence, I have added her "photosensitizer" to my database for limonene and cited RIN as the source. Rinzler fails to mention the potential of limonene in breast cancer, and the chemopreventive nature of carvone. My database presents a more balanced (negative and positive) account of the biological activities of limonene, including the newly reported photosensitizer. Here are a few of the more notable chemicals found in caraway. For a complete listing of the phytochemicals and their activities, see the CRC phytochemical compendium, Duke and duCellier, 1993 (DAD) and the USDA database http://www.ars-grin.gov/duke/.

Carvacrol — See also *Cunila origanoides*.

Carveol — CNS-Stimulant.

Carvone — Antiacetylcholinesterase IC50 = 1.4–1.8 mM; Anticancer; Antiseptic (1.5 × phenol); Carminative; CNS-Stimulant; Insecticide; Insectifuge; Motor-Depressant; Nematicide MLC = 1 mg/ml; Sedative; Trichomonicide LD100 = 300 μg/ml; Vermicide; LD50 = 1640 (orl rat).

Limonene — ACHe-Inhibitor; Allelochemic; Antiacetylcholinesterase IC22–26 = 1.2 mM; Antibacterial; Anticancer; Antifeedant; Antiflu; Antilithic; Antimutagenic; Antiseptic; Antispasmodic ED50 = 0.197 mg/ml; Antitumor; Antitumor (Breast); Antitumor (Pancreas); Antitumor (Prostate); Antiviral; Candidistat; Chemopreventive; Detoxicant; Enterocontractant; Expectorant; Fungistat; Fungiphilic; Herbicide IC50 = 45 μM; Insectifuge; Insecticide; Irritant; Nematicide IC = 100 μg/ml; ODC-Inhibitor ~750 mg/kg (diet); p450-Inducer; Photosensitizer; Sedative ED = 1–32 mg/kg; Transdermal; LD50 = 4600 (orl rat).

Ceratonia siliqua L. (Fabaceae)
Carob, Locust Bean, St. John's-Bread

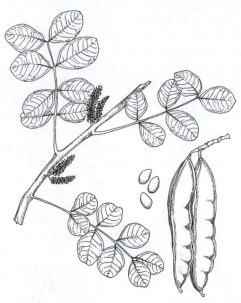

Medicinal Uses (Carob) — Seeds were once sold at a high price by pharmacists, with singers imagining they cleared the voice. The pod is used as an anticatarrhal, demulcent, and resolvent, the leaf as astringent (LEG). In southern Europe, the pods are used for asthma and cough. Various portions of the plant are used as antitussive, astringent, pectoral, and laxative. Since the pods contain gallic acid, a reported antitumor compound, and seeds contain tannins, it seems logical that carob has a folk reputation against cancer (specifically verruca and indurations). Medicinally, the bean germ flour is used for diabetic foods, and the bean pulp is used to treat infantile diarrhea. Leung (1980) mentions folk usage of carob flour for infantile diarrhea and a pod decoction for catarrhal infection. Suffering from salmonella diarrhea in Panama, I was put on a diet of carob flour and was subsequently cured, either by carob or coincidence. Leung adds that the flour is considered binding, the flowers laxative (LAF). Still today, Dr. Linda White suggests for diarrhea adding carob to applesauce or other pectin containing gentle-on-the-bowels foods (White et al., 2000).

Like pectin, fiber-laden carob, fed to rats given a high cholesterol diet, controlled elevations in blood and liver cholesterol. Rats fed a fiber-free diet containing 1% cholesterol exhibited a small increase in serum cholesterol and a fivefold increase in liver cholesterol. Addition of 10% pectin or 10% locust bean gum kept the increase in liver cholesterol down to 100% (Kritchevsky, 1982).

Indications (Carob) — Asthma (f; BIB); Bacteria (1; FNF); Candida (1; FNF); Catarrh (f; BIB; PNC); Celiac (f; PH2); Childbirth (f; PH2); Colitis (f; PH2); Cough (f; BIB; HHB; PH2; WO2); Cramp (1; FNF); Dehydration (f; WO2); Diabetes (1; FNF; LAF; WO2); Diarrhea (2; DAW; FNF; HHB; SKY); Dyspepsia (2; FNF; SKY; WO2); Enterosis (f; PH2); Heartburn (1; SKY); High Cholesterol (1; BIB; FNF; LAF); Hyperglycemia (1; LAF); Hyperperistalsis (f; WO2); Induration (f; JLH); Inflammation (1; FNF); Obesity (1; LAF); Sprue (f; PH2); Steatorrhea (f; HHB); Voice (f; PNC); Vomiting (f; PH2); Wart (f; JLH).

Carob for diarrhea:

- Antibacterial: (–)-epicatechin; benzoic-acid; ferulic-acid; gallic-acid; lignin; myricetin; pyrogallol; quercetin
- Antidiarrheic: lignin

- Antispasmodic: ferulic-acid; phloroglucinol; quercetin
- Antiviral: (–)-epicatechin; ferulic-acid; gallic-acid; lignin; myricetin; quercetin
- Astringent: formic-acid; gallic-acid
- Candidicide: ferulic-acid; myricetin; pyrogallol; quercetin
- Demulcent: mucilage
- Hepatoprotective: (+)-catechin; ferulic-acid; quercetin

C

Carob for high cholesterol:

- Antiaggregant: (+)-catechin; (–)-epicatechin; ferulic-acid; quercetin; salicylates
- Antiatherosclerotic: quercetin
- Antidiabetic: (–)-epicatechin; maltose; pinitol; quercetin
- Antilipoperoxidant: (–)-epicatechin; quercetin
- Antioxidant: (+)-catechin; (–)-epicatechin; delta-5-avenasterol; ferulic-acid; gallic-acid; l-alanine; lignin; myricetin; pyrogallol; quercetin
- Cholagogue: ferulic-acid
- Choleretic: benzoic-acid; ferulic-acid; gallic-acid
- Diuretic: myricetin
- Hepatoprotective: (+)-catechin; ferulic-acid; quercetin
- Hypocholesterolemic: (–)-epicatechin; lignin
- Hypolipidemic: ferulic-acid
- Hypotensive: quercetin

Other Uses (Carob) — Primarily cultivated for its fruit (pod) and seeds, both are high in sugar and calcium content and low in protein and fats. Carob seeds are said to be the ancient and original weight used by goldsmiths, instituted early as the carat weight. The fruit, source of food for the poor in Jewish folklore, also appears in the Christian tradition as "St. John's Bread," eaten by St. John the Baptist in the wilderness. The locust fruit is spread out and dried as food, sustaining for cattle as well as for people. In the Prodigal Son story, the younger son asked for his inheritance, which he spent quickly and unwisely. When he had no money or food left, he hired himself out as a swine tender. And since he had no bread, he longed to eat the carob pods fed to swine. The pods were said to have been the only food available for cavalry horses in Wellington's Peninsular Campaign and in Allenby's Campaign in old Palestine. In the U.K., the pods have been used to make dog biscuits. Now carob has been elevated from cattle, dog, horse, and swine feed, to health food as a chocolate substitute. It is especially recommended to those sensitive to chocolate. On April 25, 1982, in Burtonsville, Maryland, carob-coated walnuts sold for $2.89 a pound, more fitting for a profligate son. In Palestine, a molasses named "dibs" is prepared from the ripe fruits. In Cyprus, a brittle candy known as "pasteli" is made from the pods. Alcoholic beverages have been made from infusions of the pod, probably like the algarobinya I enjoyed in Cusco, Peru, made from the related legume, *Prosopis*. Matter of fact, I welcomed in the New Year 2000 with algarobinya at Machu Picchu. Each harvest, one carob tree may carry eight hundred pounds of husks. Seeds are equal to other grains in nutritive value. From the ground endosperm, gum is made, known as locust bean gum, gum tragon, tragasol, carubin, and carob flour. The gum has great water-absorbing qualities and makes an excellent stabilizer, used in many food products [such as ice cream, pickles, salad dressings, and sauces (FAC)], in the manufacture of jellies, in the manufacture of paper, in cosmetics and drugs, and in the chemical industry. The pods are of equal importance, being even higher in sugar content than the seeds, and are popularly used in foodstuffs, juices, and flour for bread and macaroni. The sugar can be extracted from the pods by alcohol, and the molasses obtained as a by-product can be fermented to ethanol. Roast seeds have been used as a coffee substitute. Wood is hard and heavy and is used locally to make furniture and wheels. Bark contains ca. 50% tannin. Carob is also used in textile printing, synthetic resins, insecticides, and fungicides. American imports are mostly used in tobacco flavorings, and cosmetics.

For more information on activities, dosages, and contraindications, see the *CRC Handbook of Medicinal Herbs*, ed. 2, Duke et al. 2002.

Cultivation (Carob) — Can be propagated by seed, grafting, or cuttings. Seeds, removed from the pods soon after maturity, before hardening, are mixed with coarse sand and kept moist in a greenhouse or between layers of sterilized sacking or burlap. When they start swelling they are planted in a propagating bed, composed of clay rolled and packed hard, upon which the seeds are placed and covered lightly with sand and good soil to the depth of 1.2 cm. When seedlings show two sets of leaves, they are transferred to 6.5 cm pots with clay soil, the transplants being allowed to stand 24 hr without water. Then they are watered freely. When ca. 12–15 cm high, they are transferred to good potting soil in one-gallon cans or boxes 15 × 20 × 20 cm. When 1–1.8 m tall, saplings are outplanted 7 m each way in poor rocky soil, or 12–14 m each way in good fertile soil. On hillsides, the trees can best be set in terraces constructed on the contour. Seedlings are slow bearers, and sex is indeterminable for a long time; hence, budding or grafting on known stocks is better. Cuttings may be started with bottom heat, careful treatment, and hormones. Grafting the third year is best, only onto the healthiest plants. It seems best to graft onto the branches, not the stem, leaving the smaller branches to utilize the winter deposit of sap. Such branches may be cut off the following year. Seedlings may be budded the second year in the field when the stalk is about 0.8 cm in diameter. The tree is usually grown as a rain-fed crop, but irrigated crops bear better. Shallow cultivation can disturb the roots. Hence, cultivation is not recommended. Oil or chemical herbicides may be used to control insect problems and weeds (LEG).

Chemistry (Carob) — Here are a few of the more notable chemicals found in carob. For a complete listing of the phytochemicals and their activities, see the CRC phytochemical compendium, Duke and duCellier, 1993 (DAD) and the USDA database http://www.ars-grin.gov/duke/.

Arginine — Antidiabetic?; Antiencephalopathic; Antihepatosis; Antiinfertility 4 g/day; Antioxidant?; Aphrodisiac 3 g/day; Diuretic; Hypoammonemic; Pituitary-stimulant; Spermigenic 4 g/day.

Gallic-Acid — ACE-Inhibitor; Analgesic; Antiadenovirus; Antiallergic; Antianaphylactic; Antiasthmatic; Antibacterial MIC = 1000 µg/ml; Antibronchitic; Anticancer; Anticarcinomic ED50 = 3; Antiescherichic; Antifibrinolytic; Antiflu; Antihepatotoxic; Antiherpetic EC50 = >10 µg/ml; Anti-HIV; Antiiflammatory; Antileishmanic EC50 = 4.4 µg/ml; Antimutagenic; Antinitrosaminic; Antioxidant IC44 = 33 ppm; Antiperoxidant IC50 = 69 µ*M*; Antipolio; Antioxidant (7 × quercetin); Antiradicular (7 × quercetin); Antiseptic; Antistaphylococcic MIC = 1000 µg/ml; Antitumor; Antiviral; Apoptotic; Astringent; Bacteristat; Bronchodilator; Candidicide; Carcinogenic; Choleretic; Cyclooxygenase-Inhibitor; Floral-Inhibitor; Gram(–)icide MIC = 1000 µg/ml; Gram(+)icide MIC = 1000 µg/ml; Hemostat; Immunomodulator; Immunosuppressant; Insulin-Sparer; Myorelaxant; Nephrotoxic; Styptic; Topoisomerase-I-Inhibitor; Xanthine-Oxidase-Inhibitor IC50 = 24 µ*M*; LD50 = 5000 orl rbt; LD50 = 4500 scu rat.

Glutamic-Acid — ADI = 120 mg/kg, 5–12/g/man/day; Antalkali? 500–1000 mg/day/orl/man; Antiepileptic; Antilithic; Antiprostatitic 125–250 mg 3 × day; Antiretardation; Anxiolytic; Hypoammonemic; Neurotoxic.

Linoleic-Acid — 5-Alpha-Reductase-Inhibitor; Antianaphylactic; Antiarthritic; Antiatherosclerotic; Anticancer; Anticoronary; Antieczemic; Antifibrinolytic; Antigranular; Antihistaminic; Antiiflammatory; Antileukotriene-D4 IC50 = 31 µ*M*; Antimenorrhagic; Anti-MS; Antiprostatitic; Carcinogenic; Hepatoprotective; Hypocholesterolemic; Immunomodulatory; Insectifuge; Metastatic; Nematicide.

Pectin — Antiatheromic 15 g/man/day; Antibacterial; Anticancer; Antidiabetic 10 g/man/day/orl; Antidiarrheic; Antimutagenic; Antiobesity; Antitumor (Colon); Antitussive; Antiulcer; Chemopreventive; Demulcent; Fungicide; Hemostat; Hypocholesterolemic; Hypoglycemic; Peristaltic; LDlo = 1800 unk mus.

Tyrosine — Anticancer; Antidepressant?; Antiencephalopathic; Antiparkinsonian 100 mg/kg day; Antiphenylketonuric; Antiulcer 256 mg scu, 400 mg/kg ipr rat; Monoamine-Precursor.

Cinchona sp. (Rubiaceae), *C. pubescens* (Vahl), and *C. calisaya* (Wedd.), formerly *C. officinalis* (Auct.)
REDBARK QUININE, RED CINCHONA, AND YELLOWBARK QUININE, YELLOW CINCHONA

As with cinnamon and cassia, I doubt many studies are vouchered and hence doubt whether anyone knows which species they were studying. Lamentable but almost always true.

Medicinal Uses (Cinchona) — Reportedly analgesic, anesthetic, antiarrhythmic, antimalarial, antiperiodic, antipyretic, astringent, bitter, contraceptive, and tonic properties. The bark and/or its extracts are used in folk remedies for cancer carcinomata, and tumors. Bark is used as a bitter and stomachic; in small doses, it is a mild irritant and stimulant of the gastric mucosa. Powdered bark can be used as a dentifrice.

Quinine is reported to be anesthetic, antiperiodic, antipyretic, antiseptic, astringent, contraceptive, insecticide, schizonticide, stomachic, tonic, and uterotonic. Quinine is used in collyria as an anesthetic, astringent, bactericidal eyewash. Quinine has been used to treat hemorrhoids and varicose veins. Quinine sulfate is used to treat colds and leg cramps. Quinine and moreso quinidine regulate atrial fibrillation and flutter. Quinidine will suppress abnormal rhythms in any heart chamber. Serendipitously, malaria patients treated with *Cinchona* bark were found to be free of arrhythmia. Quinidine, with antiarrhythmic activity, increases the activity of another plant-derived compound, vincristine, the antileukemic principle from the Madagascar periwinkle. Quinidine is also used to treat hiccups. In 1973, 0.18% (2,758,000) of all prescriptions in the U.S. (1.532 billion) contained quinidine. As often echoed in the popular press, more than 25% of all prescriptions contained one or more active constituents or derivatives now or once obtained from seed plants. Lately, the senior author has enjoyed a triple-antimalarial gin tonic, with a spray of foliage of *Artemisia annua*, the Chinese "qinghao" with antimalarial artemisine, to commingle with the juniper and cinchona antimalarials in an attractive and tantalizing beverage "gin hao tonic" (DAD). Cold water infusions will not extract the antimalarial alkaloids.

Cinchonidine, cinchonine, quinidine, and quinine are some of the antipyretic compounds in this growing medicine chest called quinine (MPI). From an amebicidal point of view, compounds modified from quinine and quinidine were not as potent as emetine and derivatives. I might mix my ipecac and quinine in Amazonia. Chloro-9-desoxy-quinine and quinidine were least active while 9-desoxy-dihydro derivatives showed fair activity. More compounds which showed slight *in vitro* activity were also active *in vivo*, exceptions being 9-desoxy-dihydro quinine and quinidine which were active *in vivo* (MPI). Morton notes that a mixture of 59% quinine and urea hydrochloride is injected as a sclerosing agent for internal hemorrhoids, hydrocele, varicose veins, and pleural vacities after thoracoplasty. Quinine has recently proved active against the causative agent of "Chagas Disease," *Trypanosoma cruzi*.

Indications (Cinchona) — Adenosis (f; CRC; JLH); Ague (f; FEL); Alcoholism (1; PH2); Alopecia (f; CRC); Amebiasis (1; CRC; FNF; WO2); Anemia (f; FEL; HHB; PH2); Anorexia (2; KOM; PHR; PH2); Arrhythmia (1; FNF; WO2); Arthrosis (f; 60P); Asthma (f; HHB); Bacteria (1; FNF; WO2); Bleeding (1; HHB); Cachexia (f; FEL); Cancer (f; CRC; JLH; PHR; PH2); Cancer, breast (f; CRC); Cancer, gland (f; CRC; JLH); Cancer, liver (f; CRC); Cancer, mesentery (f; CRC); Cancer, spleen (f; CRC); Carcinomata (f; CRC); Cardopathy (1; CRC;

FNF); Catarrh (f; MPI); Childbirth (f; FEL); Chill (f; 60P); Chlorosis (f; FEL); Cold (1; CRC; FNF; PNC; 60P); Conjunctivosis (f; FEL); Cough (f; WO2); Cramp (1; CRC; PH2; PNC; 60P); Debility (f; FEL; GMH; PH2); Dermatosis (f; PH2); Diabetes (1; FNF; WO2); Diarrhea (f; CRC; PH2; 60P); Diphtheria (f; FEL); Dysentery (1; CRC; FNF; MPI; WO2); Dyspepsia (2; FEL; KOM; PHR; PH2; 60P); Elephantiasis (f; WO2); Embolism (1; FNF); Erysipelas (f; FEL); Felons (f; CRC; JLH); Fever (1; CRC; FNF; JAD; PH2); Flu (1; CRC; FNF; PH2; PNC; WO2); Gas (2; PHR; PH2); Gastrosis (f; PHR); Gonorrhea (f; FEL); Hangover (f; CRC); Hay Fever (f; FEL; MPI); Headache (f; FEL; WO2); Heart (1; FNF; MPG); Hemicrania (f; MPI); Hemorrhoid (1; CRC; WO2); Hepatosis (f; JLH); Herpes (1; FNF); Hiccup (f; CRC); High Blood Pressure (1; WO2); Hydrocele (f; CRC); Immunodepression (1; FNF); Infection (1; FNF; MPI); Inflammation (1; FNF; PH2); Lumbago (1; CRC; FNF); Malaria (2; FNF; PHR; PH2; 60P); Metastasis (f; JLH); Myosis (f; FEL); Myotonia (1; WO2); Neuralgia (f; CRC; FEL; HHB; MPI; PH2); Neurosis (f; CRC; PH2); Pain (1; FEL; FNF PH2); Palpitation (1; MPG); Paludism (f; 60P); Pertussis (f; CRC; HHB; MPI); Piles (f; CRC); Pinworm (f; CRC; WO2); Pneumonia (f; FEL; CRC; MPI); Pyemia (f; MPI); Rash (f; PH2); Respirosis (f; PH2); Rheumatism (f; FEL; MPI); Rhinosis (f; MPI); Sciatica (1; CRC; FNF; PH2); Septicemia (f; CRC); Sore (f; JLH; PHR; PH2); Sore Throat (f; CRC; MPI); Splenomegaly (f; PHR; PH2); Splenosis (f; JLH; MPI); Stomatosis (f; CRC); Sunstroke (f; FEL); Tachycardia (1; 60P); Tonsilosis (1; MPI); Trachoma (f; FEL); Trypanosomiasis (1; FNF); Tumor (f; CRC); Typhoid (f; CRC; FEL); Ulcer (f; JLH); Varicosis (f; CRC; WO2); Virus (1; FNF); Water Retention (1; FNF); Wen (f; JLH); Wound (f; PHR; PH2); Yeast (1; FNF).

Cinchona for cardiopathy:

- ACE-Inhibitor: (–)-epicatechin
- Antiaggregant: (–)-epicatechin; caffeic-acid; cinchonine
- Antiarrhythmic: hydroquinidine; protocatechuic-acid; quinidine
- Antiedemic: caffeic-acid
- Antihemorrhoidal: quinine
- Antiischemic: protocatechuic-acid
- Antioxidant: (–)-epicatechin; alizarin; caffeic-acid; chlorogenic-acid; hyperoside; proto-catechuic-acid
- Antitachycardic: quinidine
- Beta-Adrenergic Receptor Blocker: (–)-epicatechin
- Calcium-Antagonist: alizarin; caffeic-acid; chrysazin
- Cardiotonic: (–)-epicatechin
- Diuretic: caffeic-acid; chlorogenic-acid; hyperoside; avicularin
- Hypocholesterolemic: caffeic-acid; chlorogenic-acid; (–)-epicatechin
- Hypotensive: hyperoside; quinidine
- Sedative: caffeic-acid
- Vasodilator: (–)-epicatechin

Cinchona for malaria:

- Amebicide: cinchonamine; quinidine; quinine
- Antibacterial: (–)-epicatechin; alizarin; alizarin-2-methyl-ether; caffeic-acid; chloro-genic-acid; hyperoside; protocatechuic-acid; purpurin-1-methyl-ether; quinine
- Antihepatotoxic: caffeic-acid; chlorogenic-acid; hyperoside; protocatechuic-acid
- Antiinflammatory: (–)-epicatechin; caffeic-acid; chlorogenic-acid; hyperoside; protocat-echuic-acid; quinovic-acid

- Antimalarial: cinchonidine; cinchonine; dihydroquinidine; hydroquinidine; quinidine; quinidinone; quinine
- Antipyretic: cinchonidine; cinchonine; quinidine; quinine
- Antiseptic: caffeic-acid; chlorogenic-acid
- Antitrypanosomic: quinine
- Antiviral: (–)-epicatechin; caffeic-acid; chlorogenic-acid; cinchonain-III-b; cinchonidine; hyperoside; protocatechuic-acid
- Hepatoprotective: alizarin; caffeic-acid; chlorogenic-acid
- Immunostimulant: (–)-epicatechin; caffeic-acid; chlorogenic-acid; protocatechuic-acid
- MDR-Inhibitor: cinchonine; quinidine
- Protisticide: quinine

Other Uses (Cinchona) — Quinine inhibits meat decay and yeast fermentation. *Cinchona* extracts in hair tonics are said to promote and stimulate hair growth. Quinine and cinchona extracts are used in quinine and tonic waters, bitters, and liqueurs (up to 278 ppm red cinchona extract in alcoholic beverages); also in baked goods, candies, condiments, frozen dairy desserts, and relishes. Red cinchona bark extracts are used to flavor baked goods, bitters, candies, condiments, ice creams, liqueurs (like Campari and Dubonnet), and relishes (FAC). *Cinchona* barks approved in the U.S. for use in beverages only, not to exceed 83 ppm in the finished beverage (§172.510; §172.575). *Cinchona* bark is still used for tanning after extraction for alkaloids (DAD).

For more information on activities, dosages, and contraindications, see the *CRC Handbook of Medicinal Herbs*, ed. 2, Duke et al., 2002.

Cultivation (Cinchona) — Can be propagated by seed, cuttings, or grafting, the method used depending upon economic factors and purpose for which plants are desired. Freshly harvested seed give 90–98% germination. Seed remain viable 5–48 months, depending on storage method; best stored in the dark at 7–24°C at 33–66% relative humidity. Seed are small, about 2500/g. Seeds do not germinate in darkness, even though all other factors are favorable. They require an optimal light intensity for germination between 30–75 footcandles for 6 hr during the brightest part of the day; optimal germination temperature is 24–29.5°C with a rapid loss above this. Seeds planted in nursery in raised beds with overhead protection from sun and rain. Beds usually located on lower slopes of hills, not on valley bottoms. In Philippines, 2.5 g of seed sown per meter of bed; in Uganda, 1.5 g/m of bed. Use of sphagnum moss seems to prevent fungal disease, but forest soil topped with 2.5 cm of sphagnum works well. Seedlings are very susceptible to damping-off disease. Keep seedbeds evenly moist and seeds germinate in 3–4 weeks. Seedlings are moved to shaded transplant beds when 5–7.5 cm tall, at about 4–6 months of age, where they remain until about 2 m tall and have become gradually accustomed to sunlight. When trees have reached proper height, they are side grafted. In Philippines, seedlings are field planted when 20–30 cm tall; in Java, they are set out when 45–60 cm tall about 2 years after seeding. Seedlings should be moved with a ball of soil and set about 2 m apart, slightly deeper than in seedbed; water to settle soil, and then as needed. Cover crops such as *Crotalaria* and *Calopogonium* spp. may be grown between trees. Noxious weeds and grasses should be removed as they appear. Keep weed-free, particularly in seedling stage. Plantation should be cleared twice a year for first 2–3 years, then hoeing each tree is sufficient. Rogue all diseased and poor seedlings regularly. Usually no fertilizer is needed, although bone meal at rate of 30 g/m² is helpful. In Guatemala, composted manure added to nursery beds benefits growth, as plants require rather high levels of N for more rapid growth. Watering must be done carefully to avoid washing seed out or encouraging damp-off disease. In the absence of rain, watering in field is necessary, especially when young trees are first set out. Cuttings are too slow and uncertain for commercial use. Graftings are made during the rainy season, but not when rains are particularly heavy. About a year after grafting, trees may be set out in a grove,

C

spaced 1.4 m each way, with the plan to thin them, the ultimate spacing being about 2 m each way. Usually, rootstock of *C. pubescens* is used for grafting (DAD). Trees begin to flower after 3–4 years, and seed may be taken in 4–5 years onward. The entire plantation should be cut by the time trees are 10–12 years old. For seed production, different species should be isolated from each other in forest clearings or separated by a belt of tall growing trees. Capsules collected by hand from trees, using bamboo ladders. Must be harvested before capsule splits releasing seeds, but seed must be well ripened. Capsules are sorted, put into linen bags, and taken to a darkened room for threshing. Capsules opened by hand or gently cracked with pestle and mortar. In Java, a mechanical winnower with a slow current of air is found satisfactory. *Cinchona* bark is gathered during the rainy season when it is easy to remove from tree. Sometimes some bark is shaved off, being careful not to girdle the tree. Often the entire tree is uprooted and all the bark used — trunk, branches, and roots. The plantation is thinned this way as the trees grow. In Java, trees are cut into short lengths and bark removed by pounding with wooden mallets. After being dried in the sun or in ovens, bark is crushed in a mill and packed in gunny sacks. Fine quality bark is marketed as quills. It is obtained from trees that acquire a bush-like growth by being coppiced close to the roots. Sprouts are clipped and stripped of bark, which, when dried, assume the form of quills. These are carefully packed and shipped in boxes (DAD). Modern plantations of good trees should yield bark containing 10% of quinine or more. At 8 years, a plantation yields from 500–600 kg/ha of dried bark, capable of yielding ca. 31 kg of quinine sulfate. Higher yields, 9–16 MT dry bark/ha, have been projected. Good trees yield about 1000 capsules containing about 1000 seeds each. In 1976, the major producers of *Cinchona* were Java and other Indonesian areas, eastern slope of Andes, and Guatemala. India consumes most of what it raises. Around 1974, world consumption of cinchona alkaloids was estimated at 200–500 tons, which at an average alkaloid content of 5% would imply 4000–10,000 tons bark (DAD).

Chemistry (Cinchona) — The bark contains up to 16% (mostly 6–10%) total quinoline alkaloids (quinine, quinidine, cinchonine, cinchonidine). Other alkaloids include epiquinine, epiquinamine, hydroquinidine, hydroquinine, quinamine, etc. Tannins, quinovin, quinic acid, starch, resin, wax, and other items are also reported. According to Hager's Handbook cuscamine, cuscanoidine, homocinchonine, javanine, dicinchonine, dicinchoninine, and pericine are dubious names from the old literature (HHB). The Handbook devotes more than 20 fine-print pages to just the alkaloids of *Cinchona*. Here are a few of the more notable chemicals found in cinchona. For a complete listing of the phytochemicals and their activities, see the CRC phytochemical compendium, Duke and duCellier, 1993 (DAD) and the USDA database http://www.ars-grin.gov/duke/.

Cinchonidine — Antiflu; Antiherpetic; Antimalarial 150–500 mg/man/4 × day; Antipyretic; Antiviral; LD50 = 206 (ipr rat).

Cinchonine — Antiaggregant IC50 = 125–180 µ*M*; Antimalarial; LD50 = 152 ipr rat; Antipyretic; MDR-Inhibitor 250 mg/kg igs rat.

Quinidine — Amebicide IC50 = 16.6 µg/ml; Antiarrhythmnic 72 mg/kg ipr mus; Anticholinergic; Antihiccups; Antimalarial 10 mg/kg/orl/man/8 hr; Antipyretic; Antitachycardic; Antitumor; Cardiodepressant; Dermatitigenic; Emetic; Hypotensive; Immunosuppressive; Laxative; MDR-Inhibitor 100 µ*M*; Oxytocic; LD50 = 190 ipr mus; LD50 = 594 orl mus; LD50 = 1000 orl rat.

Quinine — Abortifacient; Amebicide IC50 = 14.8 µg/ml; Analgesic; Anesthetic; Antibacterial; Antidysenteric; Antifeedant; Antifibrillatory; Antihemorrhoidal; Antilumbagic; Antimalarial IC50 = 0.18 µg/ml; Antineuralgic; Antineuritic; Antipyretic; Antisciatic; Aperitif; Astringent; Candidicide; Cardiodepressant; Cytotoxic IC50 = 183 µg/ml; Dermatitigenic; Oxytocic; Phototoxic; Protisticide; Spermicide; Stomachic; Trypanosomicide; LDlo = 8000 mg/man; LDlo = 300 orl gpg.

Cinnamomum aromaticum Nees (Lauraceae)
Cassia, Cassia Bark, Cassia Lignea, China Junk Cassia, Chinese Cassia, Chinese Cinnamon, Saigon Cinnamon

Synonym — *Cinnamomum cassia* auct.

Cinnamon and cassia are often combined in the spice trade, so who knows which has been studied when it is reduced to powdered bark.

Medicinal Uses (Cassia) — Regarded as alexeteric, aperient, balsamic, lactafuge, stimulant, tonic, and vermicidal, cassia is a folk remedy for arthritis, chills, dizziness, dysmenorrhea, goiter, headache, jaundice, lumbago, menorrhagia, nausea, phymata, postpartum, rheumatism, and snakebite. Prolonged use of Cassia is thought to improve the complexion, giving one a more youthful aspect. Constant use is said (I'm not convinced) to prevent gray hair. The bark is prepared as a tea for excessive salivation. The leaves are taken internally for rheumatism. Unani consider the bark carminative, emmenagogue, and tonic, using it for headache, inflammation, piles, and pregnancy. The bark is antiseptic, astringent, and carminative, and the EO has demonstrated cardiovascular, hypotensive, and antiviral activities. Extracts are antibacterial and fungicidal. The bark extracts have shown anesthetic (2% bark solution anesthetizing nerve fiber) and antiallergic activity. Bark is a component of a Chinese proprietary drug used in epilepsy and where sedative or tranquilizing activities are needed in neurological disorders (WO2).

The aqueous extracts prevent increases in urinary protein levels when given orally to nephritic rats (WO2). Its antiulcer activity has been compared to Cimetidine. Sharma et al. (2001) proved that cassia has antimutagenic activity against two mutagens, viz. benzo[a]pyrene (B[a]P) and cyclophosphamide (CP). The antimutagenic potential was probably due to its effects on xenobiotic bioactivation and detoxification processes (X11506812). Trans-cinnamaldehyde is antimutagenic in *Escherichia coli*.

Two glucosides, 2 epicatechin derivatives, 2 cinnamic aldehyde, 1,3-acerals, and (dl)-syrigaresinol in the bark extract exhibited antiallergic activity. Cinncassiol D4 and cinncassiol D4 glucoside from a bark fraction also exhibited antiallergic activity. The aqueous extract slightly inhibits the production of hemolytic plaque-forming cells, evidently with an anticomplement action, inhibiting complement-dependent allergic reactions. The major component, cinnamaldehyde, is sedative and antipyretic. The eugenol found in the oil from the bark (12.5%) has an antiseptic, irritant, and local anesthetic properties, as well as weak tumor-promoting activity on mouse skin. It enhances trypsin activity *in vitro*. Cassia contains the antitumor (and probably oncogenic) agents benzaldehyde, coumarins, and tannins. The bark EO contained 70.5% cinnamaldehyde and 12.5% eugenol but the leaf EO contained 92.2% benzyl benzoate and only 4.2% cinnamaldehyde.

Indications (Cassia) — Allergy (1; FNF; WO2); Amenorrhea (1; PH2; WO2); Anesthetic (1; WO2); Anorexia (2; BGB; KOM; PH2); Ascites (f; WO2); Asthenia (f; BGB); Asthma (1; BGB; WO2); Bacteria (1; BGB; FNF; LAF; PH2); Bloating (2; BGB; KOM); Bronchosis (1; BGB); Cancer (1; CAN; FNF; JLH); Cancer, bladder (f; JLH); Cancer, diaphragm (f; JLH); Cancer, kidney (f; JLH); Cancer, liver (f; JLH); Cancer, rectum (f; JLH); Cancer, spleen (f; JLH); Cancer, stomach (f; JLH); Cancer, uterus (f; JLH); Cancer, vagina (f; JLH); Cold (1; BGB; CAN; FNF); Colic (1; BGB; CAN; PH2); Condyloma (f; JLH); Constipation (f; WO2); Cough (f; BGB); Cramp (1; BGB; CAN; FNF); Cystosis (f; JLH); Diaphragmosis (f; JLH); Diarrhea (1; BGB; CAN; FNF; PH2); Dyspepsia (2; BGB; CAN; FNF; KOM; PH2); Dysuria (f; WO2); Edema (f; WO2); Enterosis (f; BGB; PH2; WO2); Enuresis (f; PH2); Epilepsy (f; WO2); Fatigue (f; PH2); Fever (1; AHP; BGB; FNF; WO2); Flu (1; FNF); Fungus (1; BGB; LAF; PH2); Gas (1; BGB; CAN; PH2; WO2); Gastrosis (f; BGB; PH2; WO2); Gray Hair (f; WO2); Hepatosis (f; JLH); Hernia (f; PH2); High Blood Pressure (1; WO2); HIV (1; FNF); Immunodepression (1; PH2); Impotence (f; PH2); Induration (f; JLH); Infection (1; BGB; FNF; LAF; PH2); Inflammation (1; FNF); Insomnia (f; WO2); Menopause (f; PH2); Mycosis (1; BGB; LAF; PH2); Nephrosis (1; BGB; WO2); Nervousness (1; FNF; WO2); Neuralgia (1; WO2); Neurasthenia (f; PH2); Ophthalmia (1; WO2); Orchosis (f; PH2); Pain (1; FNF; WO2); Pharyngosis (f; WO2); Sore (f; JLH); Splenosis (f; JLH); Tracheosis (1; WO2); Tumor (1; CAN); Ulcer (1; BGB; CAN; FNF; PH2; WO2); Urethrosis (f; WO2); Uterosis (f; WO2); Vaginosis (f; JLH); Virus (1; BGB; FNF; LAF); Vomiting (1; CAN; PH2); Wart (f; JLH); Water Retention (f; WO2).

Cassia for cancer:

- AntiEBV: (–)-epicatechin; geranial
- AntiHIV: (+)-catechin; (–)-epicatechin; opc; procyanidins
- Antiadenomic: farnesol
- Antiaggregant: (+)-catechin; (–)-epicatechin; cinnamaldehyde; coumarin; eugenol; isoeugenol; safrole
- Antiandrogenic: coumarin
- Antiarachidonate: eugenol
- Anticancer: (+)-catechin; (–)-epicatechin; alpha-pinene; alpha-terpineol; benzaldehyde; cinnamaldehyde; cinnamic-acid; coumarin; eugenol; isoeugenol; limonene; linalool; methyl-eugenol; methyl-salicylate; mucilage; opc; safrole; salicylic-acid; trans-cinnamaldehyde
- Antihyaluronidase: opc; procyanidins
- Antiinflammatory: (+)-catechin; (–)-epicatechin; alpha-pinene; beta-pinene; cinnamaldehyde; cinnamic-acid; coumarin; eugenol; methyl-salicylate; opc; salicylic-acid
- Antileukemic: (–)-epicatechin; cinnamaldehyde; farnesol
- Antilipoperoxidant: (–)-epicatechin
- Antimelanomic: coumarin; farnesol
- Antimetastatic: coumarin
- Antimutagenic: (+)-catechin; (–)-epicatechin; cinnamaldehyde; cinnamic-acid; cinnamyl-alcohol; coumarin; eugenol; limonene; trans-cinnamaldehyde
- Antioxidant: (+)-catechin; (–)-epicatechin; camphene; eugenol; gamma-terpinene; isoeugenol; methyl-eugenol; opc; procyanidins; salicylic-acid

- Antiperoxidant: (+)-catechin; (–)-epicatechin
- Antiprostaglandin: (+)-catechin; eugenol
- Antithromboxane: eugenol
- Antitumor: benzaldehyde; benzyl-benzoate; coumarin; eugenol; limonene; salicylic-acid
- Antiviral: (–)-epicatechin; alpha-pinene; cinnamaldehyde; geranial; limonene; linalool; opc; p-cymene; procyanidins
- Apoptotic: farnesol
- Beta-Glucuronidase-Inhibitor: procyanidins
- COX-2-Inhibitor: (+)-catechin; eugenol; salicylic-acid
- Chemopreventive: coumarin; limonene
- Cyclooxygenase-Inhibitor: (+)-catechin; cinnamaldehyde; salicylic-acid
- Cytochrome-p450-Inducer: 1,8-cineole; safrole
- Cytotoxic: (+)-catechin; (–)-epicatechin; cinnamaldehyde; eugenol; isoeugenol
- DNA-Binder: methyl-chavicol; safrole
- Hepatoprotective: (+)-catechin; eugenol
- Hepatotonic: 1,8-cineole
- Immunostimulant: (+)-catechin; (–)-epicatechin; benzaldehyde; coumarin
- Lipoxygenase-Inhibitor: (–)-epicatechin; cinnamaldehyde; cinnamic-acid
- Lymphocytogenic: coumarin
- Lymphokinetic: coumarin
- Ornithine-Decarboxylase-Inhibitor: limonene
- p450-Inducer: 1,8-cineole
- Reverse-Transcriptase-Inhibitor: (–)-epicatechin
- Sunscreen: opc; procyanidins

Cassia for cold/flu:

- Analgesic: coumarin; eugenol; methyl-salicylate; p-cymene; salicylic-acid
- Anesthetic: 1,8-cineole; benzaldehyde; benzoic-acid; cinnamaldehyde; cinnamic-acid; eugenol; guaiacol; linalool; methyl-eugenol; safrole
- Antiallergic: 1,8-cineole; linalool; opc; procyanidins; terpinen-4-ol
- Antibacterial: (–)-epicatechin; 1,8-cineole; alpha-pinene; alpha-terpineol; benzaldehyde; benzoic-acid; cinnamaldehyde; cinnamic-acid; cuminaldehyde; eugenol; geranial; guaiacol; limonene; linalool; methyl-eugenol; opc; p-cymene; procyanidins; safrole; salicylic-acid; terpinen-4-ol
- Antibronchitic: 1,8-cineole
- Antiflu: alpha-pinene; limonene; p-cymene
- Antihistaminic: linalool; opc; procyanidins
- Antiinflammatory: (+)-catechin; (–)-epicatechin; alpha-pinene; beta-pinene; cinnamaldehyde; cinnamic-acid; coumarin; eugenol; methyl-salicylate; opc; salicylic-acid
- Antioxidant: (+)-catechin; (–)-epicatechin; camphene; eugenol; gamma-terpinene; isoeugenol; methyl-eugenol; opc; procyanidins; salicylic-acid
- Antipharyngitic: 1,8-cineole
- Antipyretic: benzoic-acid; cinnamaldehyde; eugenol; methyl-salicylate; salicylic-acid
- Antiseptic: 1,8-cineole; alpha-terpineol; benzaldehyde; benzoic-acid; beta-pinene; cresol; eugenol; furfural; guaiacol; hexanol; limonene; linalool; methyl-benzoate; methyl-eugenol; methyl-salicylate; o-methoxycinnamaldehyde; opc; procyanidins; safrole; salicylic-acid; terpinen-4-ol
- Antitussive: 1,8-cineole; terpinen-4-ol
- Antiviral: (–)-epicatechin; alpha-pinene; cinnamaldehyde; geranial; limonene; linalool; opc; p-cymene; procyanidins

- Bronchorelaxant: linalool
- COX-2-Inhibitor: (+)-catechin; eugenol; salicylic-acid
- Cyclooxygenase-Inhibitor: (+)-catechin; cinnamaldehyde; salicylic-acid
- Expectorant: 1,8-cineole; alpha-pinene; benzoic-acid; camphene; guaiacol; limonene; linalool
- Immunostimulant: (+)-catechin; (–)-epicatechin; benzaldehyde; coumarin
- Phagocytotic: (+)-catechin; coumarin

Cassia for dyspepsia:

- Analgesic: coumarin; eugenol; methyl-salicylate; p-cymene; salicylic-acid
- Anesthetic: 1,8-cineole; benzaldehyde; benzoic-acid; cinnamaldehyde; cinnamic-acid; eugenol; guaiacol; linalool; methyl-eugenol; safrole
- Antigastritic: opc; procyanidins
- Antiinflammatory: (+)-catechin; (–)-epicatechin; alpha-pinene; beta-pinene; cinnamalde-hyde; cinnamic-acid; coumarin; eugenol; methyl-salicylate; opc; salicylic-acid
- Antioxidant: (+)-catechin; (–)-epicatechin; camphene; eugenol; gamma-terpinene; isoeu-genol; methyl-eugenol; opc; procyanidins; salicylic-acid
- Antipeptic: benzaldehyde
- Antiulcer: (+)-catechin; cinnamaldehyde; eugenol
- Antiulcerogenic: melilotic-acid
- Carminative: eugenol; methyl-salicylate; safrole
- Circulostimulant: cinnamaldehyde
- Demulcent: mucilage
- Secretagogue: 1,8-cineole
- Sedative: 1,8-cineole; alpha-pinene; alpha-terpineol; benzaldehyde; cinnamaldehyde; coumarin; eugenol; farnesol; isoeugenol; limonene; linalool; methyl-eugenol; p-cymene
- Tranquilizer: alpha-pinene; cinnamaldehyde

Other Uses (Cassia) — Other than medicines, three products are obtained from Cassia: (1) the bark, equivalent and interchangeable with cinnamon, (2) oil from bark, leaves, or twigs, and (3) cassia "buds," which are in fact dried unripe fruits. These dried green fruits are the Cassia "buds" of commerce, which rather resemble cloves. True cloves are in fact also flower buds, unopened buds in the tightly fitted calyx. Cassia bark is an important spice. Exodus suggests that cassia and cinnamon, as distinct spices, were both incorporated in Biblical holy oil (Exodus 30:23–25). Cassia is essentially equivalent to cinnamon as medicine and spice, and the two are aggregated in the spice trade. Arndt (1999) laments that aggregating the two in the U.S. spice trade (and many other places) denies us a useful refinement in flavoring our foods; some dishes are better with cassia, others with cinnamon. Arndt (1999) says cassia, not cinnamon, is the authentic flavor for dishes in Burma, China, Indonesia, Korea, Malaysia, and Vietnam. She speculates that this is why it is called "Chinese cinnamon," and why it is one of the ingredients in Chinese five-spice. Arndt (1999) tells us how to distinguish cassia and cinnamon. Ground cassia bark is darker and redder brown than the medium tan ground cinnamon. The bark in cassia sticks is thicker than that in cinnamon sticks. Cassia sticks usually consist of a single piece rolled inward from both ends; cinnamon sticks may be multiple layers rolled around each other in "quills." If your sample is not too old, chewed cassia does not give you a gelatinous or slippery feel in the mouth; cinnamon does (AAR). Sticks (cassia or cinnamon) are preferable to ground spice in any simmerings, in chocolate, cocoa, or coffee, mulled ciders and wines, fruit syrups, and custard creams (AAR). Cassia is used in spice blends and pickling spices (where cracked cinnamon is used) and to flavor baked goods, candy (even chocolates), curry powders, fruit dishes, pickles, sauces. The bark, and an EO derived therefrom, are used to flavor baked goods, beverages, chewing gum, condiments, confectionaries, and curries. The cassia "buds" are used in flavoring breads, cakes, chocolate, and pickles (FAC). In Chinese cooking, cassia is often used with meats, especially pork, even marbled Chinese tea eggs. The bark, and probably all plant parts, contains cinnamaldehyde, a biologically active compound, which is distilled for export. Oil of Cassia can be distilled from the leaves, chips of wood, and bark. The oil is also used as a natural disinfectant. Though the main use of cassia bark is culinary in the West, especially in flavoring processed foods, the oil also finds its way into China's "Tiger Balm" and into soaps, perfumes, spice essences, and beverages. Quills of cassia and/or cinnamon can perfume your drawers; or boil 1/2 tsp cassia in 2 cups of water in the kitchen to freshen aromas there. (AFL, DAD, RIN, WO2).

For more information on activities, dosages, and contraindications, see the *CRC Handbook of Medicinal Herbs*, ed. 2, Duke et al., 2002.

Cultivation (Cassia) — In South China, it is grown in a moderately cool climate, annual rainfall ca. 1200–1300 mm, elevation mostly 90–300 m, usually propagated by cutting but sometimes grown from seed. Young plants may be grown from seed in nurseries 20–25 cm apart; when a few centimeters tall, are transplanted to the field, spaced at intervals of ca. 3 m apart each way. More frequently, cassia is propagated from cuttings. Plant produces flushes of growth, with 2–10 leaves on a branch, leaves standing 1–2 cm apart at maturity. Propagation should be delayed until leaves of a flush are horizontal and of firm texture. Entire flush may be used or two-node cuttings with both leaves attached are also very satisfactory. Rooting chemicals induce more and larger roots per cutting; rooting takes place within 2–4 months; usually 2–4 months with mist propagation and about 27°C bottom heat. Some cuttings take longer but eventually root. Rooted cuttings should be potted in a clean medium such as peat or sphagnum. Sterilized soil may be used if care is taken to avoid puddling. Only the uppermost axillary bud should be planted so that it is exposed to full light. Cuttings rooted by mist propagation should be conditioned to drier conditions after potting. Reduced misting over a 2 week period is usually sufficient. Keep free of weeds. Care must be given to young, tender growth to avoid leaf damage, drying winds, chemical sprays, dry soil, wilting, and rough handing. Callus formation occurs readily on most cuttings and usually precedes rooting. If, after 1 year, no roots have appeared, part of the callus should be removed and the area examined for signs of roots; one root is sufficient to start the cutting for normal plant development. In harvesting, the fragrant inner bark is peeled from slender canes and rolled like a scroll, with both edges towards the middle. Bits of outer bark may adhere. These scrolls of cassia are sold as "cinnamon sticks" in the U.S. (AAR, DAD, WO2).

Once, older trees were destructively harvested to produce cassia or camphor. Then time of planting to harvest was cut to 10–15 years. Lately, trees have been grown as a coppiced bush. At 5–9 years of age, bushes are cut off close to the ground, and new shoots spring up from the roots, so plantations may last indefinitely. From March to May, cuts are made in the bark of trees at least six years old. Branches about 3 cm thick are cleared of twigs and leaves; two longitudinal slits are then made and circumferential slits are made through the bark at about 40 cm apart to yield semicylindric quills about 40 cm long. The bitter-tasting epidermis is exfoliated, by planing or scraping, and the quills are dried in the sun. The bark is then distilled to extract cassia oil. Cassia buds are gathered when fruits are ripe on trees 10 years or older. Average trees in Sumatra yield 8 kg of best quality bark and 3 kg each of second- and third-grade bark. Cinnamaldehyde may be distilled from the bark, which yields 0.3–0.8% EO, dominated by cinnamaldehyde. Cinnamaldehyde is now also produced synthetically on a large scale from coal tar bases, particularly toluol. Cassia buds, which contain the fleshy ovaries, especially those that are plump and fresh with a fine cinnamon flavor and free from stalks and dirt, are the best. If the buds are packed with the bark, the flavor of both is improved.

Chemistry (Cassia) — See also *Cinnamomum verum*, with many of the same chemicals. Here are a few of the more notable chemicals found in cassia. For a complete listing of the phytochemicals and their activities, see the CRC phytochemical compendium, Duke and duCellier, 1993 (DAD) and the USDA database http://www.ars-grin.gov/duke/.

Cincassiol-D4 — Antiallergic.

Cinnamaldehyde — See also *Cinnamomum verum.*

Cinnamic-ACID — See also *Cinnamomum verum.*

Trans-Cinnamaldehyde — See also *Cinnamomum verum.*

Cinnamomum verum J. Presl (Lauraceae)
CEYLON CINNAMON, CINNAMON

Synonyms — *Cinnamomum zeylanicum* Blume, *Laurus cinnamomum* L.

Cinnamon and cassia are often combined in the spice trade, so who knows which has been studied when it is reduced to powdered bark.

Medicinal Uses (Cinnamon) — Well known for millennia, and oft mentioned in the Bible, cinnamon was reportedly one of the ingredients in Moses' holy ointment (Libster, 2002). Libster (2002) suggests that cinnamon tea might be useful for elderly bedridden patients with cold extremities, or in preparing such a patient for a recuperative walk, or cast removal from a fractured extremity. Regarded as antipyretic, antiseptic, astringent, balsamic, carminative, diaphoretic, fungicide, stimulant, and stomachic, it is a fragrant cordial, useful for weakness of stomach and diarrhea, checking nausea and vomiting, and used in other medicinal mixtures. Powdered bark in water (or EO, or tiger balm containing many of the same chemicals) is applied to the temple in headaches and neuralgia. Cinnamon bark prevents platelet agglutination and shows antithrombic and antitumor activity. Lebanese use cinnamon as a stimulant, for colds, rheumatism, halitosis, and to check slobbering in young and elderly people. It is also used to loosen coughs. Ayurvedics consider the bark aphrodisiac and tonic, using it for biliousness, bronchitis, diarrhea, itch, parched mouth, worms, and cardiac, rectal, and urinary diseases. They use the oil for "eructations," gas, loss of appetite, nausea, and toothache. Unani consider the oil carminative, emmenagogue, and tonic to the liver, using it for abdominal pains, bronchitis, head colds, and inflammation. They consider the bark alexeteric, aphrodisiac, carminative, expectorant, sialagogue, and tonic, using it for gas, headache, hiccup, hydrocele, liver ailments, piles, and scorpion stings. One of our Belizean ecotourists was complaining about a snoring roommate. Famed Belizean herbalist Rosita Aruigo suggests 1 cup of cinnamon tea with 2 tsp grated ginger, adding honey and milk to taste. Drink at bedtime each night until cured. Since I have mentioned one spicy Ayurvedic triad, trikatu (ginger, long pepper, and pepper), I may as well mention another spicy triad, trijataka [cardamom, cinnamon, and "tejapatra," which I am told is cassia (unidentified in DEP and WOI)], three more aromatics often used together for lengual paralysis, stomach cramps, and toothache (BIB, DAD, DEP, JLH, WOI, WO2).

In massive doses, it aided in the treatment of cancer. It has been regarded as a folk remedy for indurations (of spleen, breast, uterus, liver, and stomach) and tumors (especially of the abdomen, liver, and sinews). Cinnamon contains the antitumor agent benzaldehyde.

Cinnamon invigorates the blood, helps regulate the menstrual cycle, and checks flooding (Weed, 2002). The usual dose is a cup of cinnamon tea, 5–10 drops of tincture once or twice a day. Cautioning her readers not to use straight cinnamon oil internally, she suggests a strong cinnamon tea or using 5–10 drops of tincture every 15 min or so to slow flooding and relieve uterine cramping. But you can chew on a cinnamon stick or sprinkle powdered cinnamon on toast or tea or what-have-you. Mothers are given hot cinnamon tea, with ginger and caraway, to prevent blood clotting. Cinnamon is used as a stimulant of the uterine muscular fiber in menorrhagia and in tedious labor due to defective uterine contractions. It is also used for chills and menstrual cramps. Ayurvedics consider the bark aphrodisiac.

Cinnamon oil is antifungal, antiviral, antibacterial, and larvicidal. The cinnamon leaf oil, distilled from dried green leaves, is a powerful antiseptic and is also used in perfumes, spices, and in the synthesis of vanillin. The EOs are antiseptic, the ether-soluble fraction antioxidant. Libster (2002) also suggests experimentation with topical and internal cinnamon tea for patients with candida. Cinnamon killed *Bacillus subtilis, Escherichia coli,* and *Saccharomyces cerevisiae,* confirming traditional uses of spices as food preservatives, disinfectants, and antiseptics (De et al., 1999). Vapors of several EOs are active against dermatophytes, like *Tinea* and *Trichophyton.* Of those seven studied by Inouye et al. (2001), cinnamon was most potent, followed by lemongrass, thyme, and perilla (chiso) oils, which killed the conidia and inhibited germination and hyphal elongation at 1–4 μg/ml air. Eugenol is antiseptic. At a 0.1% concentration, a liquid CO_2 extraction of cinnamon bark completely suppressed growth of *candida, escherichia,* and *staphylococcus.* Libster (2002) reviews Chinese studies in the last decade where three of five patients with oral HIV and candida significantly improved with commercial cinnamon lozenges. The scientists speculated that it is improbable that systemic candida can be challenged, but therapeutic levels can be obtained in oropharynx (Libster, 2002).

USDA studies show that cinnamon (ca. $\frac{1}{8}$ tsp) can treble insulin efficiency and may be useful in adult onset diabetes (Khan et al., 1990). But recall that Hikino (1985) ascribes hyperglycemic activity to the labile and volatile cinnamaldehyde, main ingredient of the cinnamon.

Indications (Cinnamon) — Amenorrhea (1; CRC; WHO; WO2); Amnesia (f; ZUL); Anorexia (2; CAN; KOM; PH2; WHO); Arthrosis (f; CRC); Asthenia (1; BGB); Asthma (f; CRC; LIB); Bacillus (1; X10548758); Bacteria (1; FNF; WO2); Bleeding (1; APA); Bloating (1; BGB); Bronchosis (2; CRC; PHR); Cancer (f; CRC); Cancer, abdomen (f; JLH); Cancer, bladder (f; JLH); Cancer, breast (f; JLH); Cancer, colon (f; JLH); Cancer, diaphragm (f; JLH); Cancer, ear (f; JLH); Cancer, gum (f; JLH); Cancer, kidney (f; JLH); Cancer, liver (f; JLH); Cancer, mouth (f; JLH); Cancer, neck (f; JLH); Cancer, rectum (f; JLH; WO2); Cancer, sinus (f; JLH); Cancer, spleen (f; JLH); Cancer, stomach (f; JLH); Cancer, uterus (f; JLH); Cancer, vagina (f; JLH); Candida (1; CRC; LIB); Cardiopathy (f; LIB); Childbirth (f; LIB); Chill (f; PHR; PH2); Cholera (1; CRC; WO2); Cold (2; CAN; FNF; PHR; ZUL); Colic (1; APA; CAN; TRA); Condylomata (f; JLH); Congestion (1; FNF); Conjunctivosis (f; WHO); Convulsion (f; LIB); Cough (2; CRC; FNF; PHR); Cramp (1; APA; DEP; ZUL); Debility (f; LIB); Depression (f; LIB); Diarrhea (1; DEP; PHR; TRA; WHO); Dysentery (f; CRC; DEP; WO2); Dysmenorrhea (1; APA; WHO); Dyspepsia (2; CAN; FNF; IED; KOM; PH2; WHO); Dyspnea (f; WHO); Earache (f; LIB); Enterosis (1; JLH; WHO); Enterospasm (2; KOM; WHO); Escherichia (1;

CRC; X10548758); Exhaustion (f; LIB); Fever (2; AHP; FNF; PHR; TRA); Fistula (f; CRC); Flu (1; FNF; PHR; PH2); Frigidity (f; LIB; WHO); Fungus (1; FNF; LIB; X10548758); Gas (2; APA; DEP; KOM; TRA; WHO); Gastrosis (f; DEP; WO2); Gastrospasm (2; KOM); Gingivosis (f; JLH); Glossosis (f; DEP; WO2); Gonorrhea (f; LIB); Halitosis (f; PH2); Headache (1; DEP; WO2; ZUL); Heart (f; CRC); Hepatosis (f; JLH); High Blood Pressure (f; LIB; ZUL); Immunodepression (1; FNF); Impotence (f; LIB; WHO); Infection (2; FNF; PHR; WO2); Inflammation (1; FNF; LIB); Leukemia (1; TRA; WO2); Leukorrhea (f; WHO); Lumbago (f; CRC); Lungs (f; CRC); Lupus (f; LIB); Lymphoma (1; WO2); Mastosis (f; JLH); Menorrhagia (f; CRC; LIB); Mycosis (1; FNF; ZUL); Nausea (1; CRC; FNF; TRA; ZUL); Nephrosis (f; CRC; LIB); Nervousness (1; FNF); Neuralgia (f; DEP; WHO; WO2); Oketsu Syndrome (f; LIB); Otosis (f; LIB); Pain (1; FNF; WHO; WO2); Paralysis (f; DEP; WO2); Pharyngosis (2; PHR); Phthisis (f; CRC); Phymata (f; JLH); Proctosis (f; CRC; JLH); Prolapse (f; CRC); Psoriasis (f; CRC); Rheumatism (f; APA; WHO; WO2; ZUL); Salmonella (1; WO2); Sinusosis (f; JLH); Sore (f; JLH); Spasm (f; CRC); Splenosis (f; JLH); Staphylococcus (1; CRC); Stomatosis (2; CRC; JLH; PHR); Stress (f; LIB); Syncope (f; WO2); Tension (f; LIB); Toothache (f; DEP; PH2; WHO); Tuberculosis (f; LIB); Tumor (f; CRC; JLH); Typhoid (f; LIB); Ulcer (1; WHO); Vaginosis (f; CRC; JLH; WHO); VD (f; LIB); Virus (f; LIB); Vomiting (f; CRC; PH2); Wart (f; CRC; JLH); Wen (f; JLH); Worm (f; PHR; PH2); Wound (f; PHR; PH2; WHO); Yeast (1; APA; WO2; X10548758).

Cinnamon for dyspepsia:

- Analgesic: borneol; caffeic-acid; camphor; coumarin; eugenol; myrcene; p-cymene; phenol
- Anesthetic: 1,8-cineole; benzaldehyde; benzyl-alcohol; camphor; cinnamaldehyde; cinnamic-acid; eugenol; linalool; linalyl-acetate; methyl-eugenol; myrcene; phenol; safrole
- Antiemetic: camphor
- Antiinflammatory: (–)-epicatechin; alpha-pinene; beta-pinene; borneol; caffeic-acid; caryophyllene; caryophyllene-oxide; cinnamaldehyde; cinnamic-acid; coumarin; delta-3-carene; eugenol; eugenyl-acetate; mannitol; salicylates
- Antioxidant: (–)-epicatechin; caffeic-acid; camphene; eugenol; gamma-terpinene; isoeugenol; linalyl-acetate; mannitol; methyl-eugenol; myrcene; p-coumaric-acid; phenol; proanthocyanidins; vanillin
- Antipeptic: benzaldehyde
- Antiulcer: cinnamaldehyde; eugenol
- Antiulcerogenic: caffeic-acid
- Carminative: camphor; eugenol; safrole
- Circulostimulant: cinnamaldehyde
- Demulcent: mucilage
- Secretagogue: 1,8-cineole
- Sedative: 1,8-cineole; alpha-pinene; alpha-terpineol; benzaldehyde; benzyl-alcohol; borneol; bornyl-acetate; caffeic-acid; caryophyllene; cinnamaldehyde; citronellal; coumarin; eugenol; farnesol; geraniol; geranyl-acetate; isoeugenol; limonene; linalool; linalyl-acetate; methyl-eugenol; nerol; p-cymene
- Tranquilizer: alpha-pinene; cinnamaldehyde

Cinnamon for infection:

- Analgesic: borneol; caffeic-acid; camphor; coumarin; eugenol; myrcene; p-cymene; phenol
- Anesthetic: 1,8-cineole; benzaldehyde; benzyl-alcohol; camphor; cinnamaldehyde; cinnamic-acid; eugenol; linalool; linalyl-acetate; methyl-eugenol; myrcene; phenol; safrole
- Antibacterial: (–)-epicatechin; 1,8-cineole; alpha-pinene; alpha-terpineol; benzaldehyde; bornyl-acetate; caffeic-acid; caryophyllene; cinnamaldehyde; cinnamic-acid; citronellal;

cuminaldehyde; delta-3-carene; eugenol; geranial; geraniol; limonene; linalool; methyl-eugenol; myrcene; nerol; p-coumaric-acid; p-cymene; phenol; safrole; terpinen-4-ol
- Antiedemic: caffeic-acid; caryophyllene; caryophyllene-oxide; coumarin; eugenol; proanthocyanidins
- Antiinflammatory: (–)-epicatechin; alpha-pinene; beta-pinene; borneol; caffeic-acid; caryophyllene; caryophyllene-oxide; cinnamaldehyde; cinnamic-acid; coumarin; delta-3-carene; eugenol; eugenyl-acetate; mannitol; salicylates
- Antilymphedemic: coumarin
- Antiseptic: 1,8-cineole; alpha-terpineol; benzaldehyde; benzyl-alcohol; beta-pinene; caffeic-acid; camphor; citronellal; eugenol; furfural; geraniol; hexanol; limonene; linalool; methyl-eugenol; nerol; o-methoxycinnamaldehyde; oxalic-acid; phenol; proanthocyanidins; safrole; terpinen-4-ol
- Antiviral: (–)-epicatechin; alpha-pinene; bornyl-acetate; caffeic-acid; cinnamaldehyde; geranial; limonene; linalool; p-cymene; phenol; proanthocyanidins; vanillin
- Bacteristat: coumarin; isoeugenol
- COX-2-Inhibitor: eugenol
- Circulostimulant: cinnamaldehyde
- Cyclooxygenase-Inhibitor: cinnamaldehyde
- Fungicide: 1,8-cineole; alpha-phellandrene; beta-phellandrene; caffeic-acid; camphor; caryophyllene; caryophyllene-oxide; cinnamaldehyde; cinnamic-acid; coumarin; cuminaldehyde; eugenol; furfural; geraniol; linalool; methyl-eugenol; myrcene; o-methoxy-cinnamaldehyde; p-coumaric-acid; p-cymene; phenol; terpinen-4-ol; terpinolene; vanillin
- Fungistat: isoeugenol; limonene; methyl-eugenol
- Immunostimulant: (–)-epicatechin; benzaldehyde; caffeic-acid; coumarin
- Lipoxygenase-Inhibitor: (–)-epicatechin; caffeic-acid; cinnamaldehyde; cinnamic-acid; p-coumaric-acid

Other Uses (Cinnamon) — In Biblical times, spices like cinnamon were used to prepare incense and holy oils used in religious rites, medicines, and perfumes. Exodus suggests that cassia and cinnamon, as distinct spices, were both incorporated in Biblical holy oil (Exodus 30:23–25). Arndt (1999) laments that aggregating the two in the U.S. spice trade (and many other places) denies us a useful refinement in flavoring our foods; some dishes are better with cassia, others with cinnamon. Arndt tells us how to distinguish cassia and cinnamon. Ground cassia bark is darker and redder brown than the medium tan ground cinnamon. The bark in cassia sticks is thicker than that in cinnamon sticks. Cassia sticks usually consist of a single piece rolled inward from both ends; cinnamon sticks may be multiple layers rolled around each other in "quills." If your sample is not too old, chewed cassia does not give you a gelatinous or slippery feel in the mouth; cinnamon does (AAR). The bark, as the condiment cinnamon, is used in food, dentifrices, incenses, and perfumes. The bark is commonly used to flavor curries, buns, rolls, apple butter, puddings, beverages, etc. (FAC). EOs from the bark and leaves are similarly used. Cinnamon leaves are one of the ingredients used for flavoring jerked pork in Jamaica, where they are also used in place of cinnamon bark in puddings and hominy dishes. Cinnamon sugar, a fragrant mixture of cinnamon and white sugar, is a popular topping for French toast and cappuccino (FAC). Cinnamon bark oil, distilled from chips and bark of inferior quality, is used in foods, perfumes, soaps, cordials, and in drug and dental preparations. Early Egyptians apparently used cinnamon in embalming. Fat from the fruits has been used in candle making. Cinnamon sticks were once used to stake vanilla vines and some of the stakes took root.

Recently, it has been shown that compounds from cinnamon prevent stored potatoes from sprouting, probably due to aromatic aldehydes, like cinnamaldehyde, destroying the potato eye. "These natural sprouting inhibitors also kill a fungus that causes dry rot in potatoes…including a strain that has become resistant to fungicides" (BIB, DAD). Nakatani (1994) notes that EOs of celery, cinnamon, coriander, and cumin are comparable to sorbic acid at preventing the slimy spoilage of Vienna sausage (Nakatani, 1994, SPI).

For more information on activities, dosages, and contraindications, see the *CRC Handbook of Medicinal Herbs*, ed. 2, Duke et al., 2002.

Cultivation (Cinnamon) — Cinnamon is propagated by seed, three-leaved cuttings, layering, and division of old rootstocks, the latter resulting in earlier harvests. The short-lived seed germinate in 2–3 weeks. Pulp should be removed from seed before planting. Seed may be planted, four or five to *in situ* spots. They are probably better planted in a nursery. Shade, furnished to the young seedlings, is removed when they are about 15 cm high. After about 4 months, seedlings transplanted to containers. Four months later, they are outplanted, spaced ca. 2–3 m apart. Weeds should be removed, perhaps used as green manure or mulch. After about 2 years, stems are cut back close to ground to encourage coppice. Four to six of the coppice shoots are allowed to grow about 2 more years, pruning them to maintain straight canes. When 2–3 m high and 2–5 cm in diameter, these canes are harvested. Only a few canes are taken, a full crop being made when tree is 10 years old, after which the yield diminishes (DAD, WOI). Two crops are taken annually, usually in May and November, depending upon rains. Bark is more easily separated from the wood following the first rain. Semi-cylindrical quills of bark are manually cut. Quillings are pieces broken off quills in handling and shipping; featherings are bark from twigs; chips are shavings of outer bark sometimes with attached bits of inner bark, as happens when peeling crooked canes. Quills, quillings, and featherings, all true good cinnamon, are exported in cylindrical bales of 100 lb net, while chips are exported in pressed cakes of 2–3 cwt. net. Quills are fermented for 24 hr or so, and the epidermis and green cortical tissues and cork scraped off. After drying, quills contract. These are rolled up within each other, further dried in shade and compacted by rolling by hand daily. After 4 or 5 years, first cuttings yield about 65 kg quills/ha/year, increasing to 200 kg/ha and starting to decline after about 10 years. Large trees, ca. 60 cm in diameter, may yield 45 kg bark. But one should not expect yields of higher than 600 kg every five years. Leaf harvest has been estimated at 1.9–2.5 MT/ha. Leaves contain 0.6–0.8% EO.

Chemistry (Cinnamon) — Orally, cinnamaldehyde stimulates the CNS at low doses and inhibits it at high doses. It accelerates release of catecholamines (mainly adrenaline) from the adrenal glands into the blood. According to Sambaiah and Srinivasan (1989), cinnamon stimulated liver microsomal cytochrome p450 dependent aryl hydroxylase. Of the various types of cinnamon bark oils, that of *C. verum* may have the largest amount of eugenol. Eugenol is reportedly absent in cassia bark oil. Here are a few of the more notable chemicals found in cinnamon. For a complete listing of the phytochemicals and their activities, see the CRC phytochemical compendium, Duke and duCellier, 1993 (DAD) and the USDA database http://www.ars-grin.gov/duke/.

Cinnamaldehyde — ADI = 700 µg/kg; Adrenergic; Allelochemic IC100 = 2.5 mM; Anesthetic; Antiaggregant 200 µM; Antibacterial; Anticancer; Antiherpetic; Antiinflammatory; Antileukemic; Antimutagenic; Antipyretic; Antisalmonellic; Antispasmodic; Antiulcer 500 mg/kg/orl; Antiurease; Antiviral; Candidicide MIC = 16–40 ppm; Choleretic 500 mg/kg/orl; Chronotropic; Circulostimulant; CNS-Depressant; CNS-Stimulant; Cyclooxygenase-Inhibitor; Cytotoxic ED50 5–60 µg/ml; Fungicide 1.7–250 ppm; MIC = 16–40 ppm; Herbicide; Histaminic; Hypoglycemic; Hypotensive; Inotropic; Insecticide; Lipoxygenase-Inhibitor; Monoaminergic; Mutagenic; Nematicide 100 µg/ml; Sedative; Sprout-Inhibitor; Teratogenic; Tranquilizer; Tyrosinase-Inhibitor ID50 = 129 µg/ml; Vasodilator; LD50 = 2225 orl mus; LD50 = 610 ipr mus; LD50 = 132 ivn mus.

Cinnamic-Acid — Anesthetic; Antibacterial; Anticancer; Antiinflammatory; Antimutagenic; Antispasmodic; Choleretic; Dermatitigenic; Fungicide; Herbicide; Laxative; Lipoxygenase-Inhibitor; Vermifuge.

Eugenol — See also *Pimenta*.

Trans-Cinnamaldehyde — Allelochemic (IC50 = 0.38 mM); Anticancer; Antifeedant; Antimutagenic; Sweetner (50 × sucrose).

Cocos nucifera L. (Arecaceae)
COCONUT, COCONUT PALM, COPRA, NARIYAL

C

Medicinal Uses (Coconut) — Most of us think of coconut as food; fewer think of medicine or spice. Yet coconut water is diuretic, at least in dogs (MPI), and the water is also regarded as aperient, cooling, and demulcent. Coconuts are used in folk remedies for tumors (JLH). Reported to be antibacterial, antidotal, antipyretic, antiseptic, aperient, aphrodisiac, astringent, depurative, diuretic, hemostat, pediculicide, laxative, stomachic, styptic, suppurative, and vermifuge (NUT).

Soy interests have made saturated tropical fats look bad, but DeRoos et al. (2001) show that ingestion of a solid fat rich in lauric acid gives a more favorable serum lipid profile in healthy men and women than consumption of a solid fat rich in trans-fatty acids (partially hydrogenated soybean oil rich in trans-fatty acids). Solid fats rich in lauric acids appear to trans-fats in food manufacturing, where hard fats are indispensable (X11160540). The hemicelulose fiber (unlike the cellulose) decreased total cholesterol, LDL, VLDL while increasing HDL.

Coconut fiber reduced mutagenic and carcinogenic effects of chile and 1,2-dimethylhydrazine, respectively (Khan and Balick, 2001). And oil from the shell has proven antiseptic against *Aspergillus niger, Candida albicans, Penicillium chrysogenum, Rhizoctonia bataticola, R. solani* and *Trichophyton mentagrophytes* (MPI).

Indications (Coconut) — Abscess (f; DAD); Alactea (f; DAV); Alopecia (f; DAD); Amenorrhea (f; DAD); Asthma (f; DAD; DAV); Bacteria (1; FNF); Blennorrhagia (f; DAD); Bronchosis (f; DAD; PH2); Bruise (f; DAD); Burn (f; DAD); Cachexia (f; DAD); Calculus (f; DAD); Cancer (1; FNF; JLH; PH2); Cancer, breast (1; DAV); Cancer, colon (1; JAC7:405); Candida (1; FNF; JFM); Caries (f; WO2); Childbirth (f; DAV); Cold (f; DAD; PH2); Constipation (f; DAD); Cough (f; DAD; PH2); Debility (f; DAD); Dermatosis (f; DAD; PH2); Diabetes (f; IED); Dropsy (f; DAD); Dysentery (f; DAD; SKJ); Dysmenorrhea (f; DAD); Dysuria (f; SKJ; WO2); Earache (f; DAD); Erysipelas (f; DAD); Fever (f; DAD; SKJ); Flu (f; DAD); Fungus (1; FNF; MPI); Gingivosis (f; DAD); Gonorrhea (f; DAD); Gray Hair (f; PH2); Headache (f; IED); Hematemesis (f; DAD); Hemoptysis (f; DAD); Hepatosis (f; SKJ); High Blood Pressure (f; IED); High Cholesterol (1; JAC7:405); Impotence (f; DAD); Infection (1; FNF; MPI); Inflammation (1; FNF; PH2); Jaundice (f; DAD); Mastosis (f; JFM); Menorrhagia (f; DAD); Miscarriage (f; DAV); Mycosis (1; FNF; MPI); Nausea (f; DAD; IED); Parasite (f; IED); Pharyngosis (f; PH2); Phthisis (f; DAD); Pregnancy (f; DAD); Rash (f; DAD); Scabies (f; DAD); Scurvy (f; DAD); Sore (f; PH2); Sore Throat (f; DAD; PH2); Swelling (f; DAD); Syphilis (f; DAD); Toothache (f; DAD; JFM); Tuberculosis (1; DAD; MPI); Tumor (1; DAD; FNF); Typhoid (f;

DAD); Uterosis (f; SKJ); VD (f; DAD; JFM); Virus (1; FNF); Vomiting (f; SKJ); Water Retention (1; FNF); Worm (f; IED); Wound (f; DAD); Yeast (1; JFM; MPI).

Coconut for cancer:

- AntiHIV: lignin
- Antiaggregant: ferulic-acid; ligustrazine; menthol; salicylates
- Anticancer: alpha-terpineol; ferulic-acid; lignin; limonene; p-coumaric-acid; p-hydroxy-benzoic-acid; squalene; syringaldehyde; vanillic-acid; vanillin
- Anticarcinogenic: ferulic-acid
- Antihepatotoxic: ferulic-acid; p-coumaric-acid
- Antiinflammatory: alpha-amyrin; cycloartenol; ferulic-acid; gentisic-acid; menthol; n-hentriacontane; salicylates; syringaldehyde; vanillic-acid
- Antimutagenic: ferulic-acid; limonene; n-nonacosane; p-hydroxy-benzoic-acid; vanillin
- Antineoplastic: ferulic-acid
- Antinitrosaminic: ferulic-acid; lignin; p-coumaric-acid
- Antioxidant: ferulic-acid; gamma-tocopherol; lignin; p-coumaric-acid; p-hydroxy-benzoic-acid; squalene; syringaldehyde; vanillic-acid; vanillin
- Antiperoxidant: p-coumaric-acid
- Antistress: gaba
- Antitumor: alpha-amyrin; ferulic-acid; lignin; limonene; p-coumaric-acid; squalene; vanillic-acid; vanillin
- Antiviral: ferulic-acid; gentisic-acid; lignin; limonene; vanillin
- Anxiolytic: gaba
- Chemopreventive: limonene; squalene
- Cytotoxic: alpha-amyrin; p-coumaric-acid
- Hepatoprotective: ferulic-acid
- Immunostimulant: ferulic-acid; squalene
- Lipoxygenase-Inhibitor: p-coumaric-acid; squalene
- Ornithine-Decarboxylase-Inhibitor: ferulic-acid; limonene
- Prostaglandigenic: ferulic-acid; p-coumaric-acid; p-hydroxy-benzoic-acid
- Sunscreen: ferulic-acid; squalene

Coconut for infection:

- Analgesic: ferulic-acid; gentisic-acid; menthol
- Anesthetic: menthol
- Antibacterial: alpha-terpineol; cycloartenol; ferulic-acid; gentisic-acid; lignin; limonene; menthol; p-coumaric-acid; p-hydroxy-benzoic-acid; squalene; vanillic-acid
- Antiedemic: alpha-amyrin; beta-amyrin; syringaldehyde
- Antiinflammatory: alpha-amyrin; cycloartenol; ferulic-acid; gentisic-acid; menthol; n-hentriacontane; salicylates; syringaldehyde; vanillic-acid
- Antiseptic: alpha-terpineol; limonene; menthol
- Antiviral: ferulic-acid; gentisic-acid; lignin; limonene; vanillin
- Bacteristat: malic-acid; n-hexacosane
- Fungicide: capric-acid; caprylic-acid; ferulic-acid; octanoic-acid; p-coumaric-acid; phyl-loquinone; vanillin
- Fungistat: limonene; p-hydroxy-benzoic-acid
- Immunostimulant: ferulic-acid; squalene
- Lipoxygenase-Inhibitor: p-coumaric-acid; squalene

C

Other Uses (Coconut) — One of the 10 most useful trees in the world, providing food for millions of people, especially in the tropics. If the tree can be spared, the cabbage-like heart makes a tasty treat, a "millionaire's salad." Terminal buds are eaten raw (FAC). Inflorescences (flower clusters) are eaten as a vegetable and are a source of sugar, vinegar, and palm wine (FAC). Flowers provide good honey for bees. A clump of unopened flowers may be bruised and "weep" sweet juice, up to a gallon per day. The brown liquid is boiled down to syrup. Left standing, it ferments quickly into "toddy." After a few weeks, it becomes a vinegar. "Arrack" is the distilled fermented "toddy." Pith of stems is made into bread, added to soups, fried or pickled (FAC). Sprouting seeds may be eaten like celery. One cv "Nawasi" is said to have an edible husk. Coconut sport or makapuno is a delicacy that comes from a cv with the coconut water replaced by a thick curd. Coconut apple, a spongy mass that forms inside a germinating seed, is another delicacy. Nutmeat of immature coconuts is like a custard in flavor and consistency (NUT). Arndt (1999) says, "the versatile coconut palm offers its trunk for structures and utensils, its fronds for thatch, its fiber for ropes and mats, and its seed and sap for a variety of nutritious foods. For many people, this is truly the tree of life." Coconut water is used as a braising liquid to flavor and tenderize meats, and occasionally to enrich the liquid in a stew. To make coconut chips, slice a piece of coconut meat into very thin strips, bake in a single layer on a cookie sheet in a low oven (about 200°F) ca. $\frac{1}{2}$ hour or until they are deep golden brown. Sprinkle with salt while still hot. Shredded or grated coconut is used in cakes, pies, candies, and in curries and sweets. Mixed with orange segments, shredded coconut makes ambrosia, a Christmas treat. Try substituting grated coconut for bread crumbs in your meatball recipe. These are even better if you add coriander, toasted and ground; coriander marries most happily with the coconut flavor. Coconut milk is obtained by pouring boiling water over grated flesh and letting it soak until cool. Use equal volumes of grated nut meat and water. Strain, pressing down on the coconut. Caribbean cooks use coconut with bananas, limes, mangoes, papaya, and pineapples. Antiguans mix grated coconut, grated sweet potato, a bit of flour and seasonings (sugar, nutmeg, and vanilla). The mixture is wrapped in a banana leaf or a sea-grape leaf and boiled. This dumpling-like "doucana" is traditionally served with salt fish. "Crème de coco" often indicates a mix of coconut milk, coconut oil, and sugar, for piña coladas and other tropical cocktails (AAR). Cooked with rice to make Panama's famous "arroz con coco"; also cooked with taro leaves or game and used in coffee as cream. Ontjom, tempeh bongrek, nata de coco, and dageh kelapa are fermented products made from coconut. Nata de coco, or coconut gel, is a common ingredient of halo-halo, a bottled dessert that also contains aduki beans and sugar syrup (FAC). When nuts are open and dried, meat becomes copra, which is processed for oil, used to make soaps, shampoos, shaving creams, toothpaste, lotions, lubricants, hydraulic fluid, paints, synthetic rubber, plastics, margarine, and in ice cream. Leaves are sometimes used to wrap foods for cooking (FAC). Hindus make vegetarian butter, "ghee," from coconut oil. Coconut roots provide a dye; frayed out, it makes toothbrushes; scorched, it is used as coffee substitute (NUT). The husk has a mass of packed fibers, "coir," used for mattresses, upholstery, and life preservers. Coconut fiber, resistant to sea water, is used for cables and rigging on ships, for making mats, rugs, bags, brooms, brushes, and olive oil filters, and for fires and mosquito smudges. Charcoal is used for cooking fires.

For more information on activities, dosages, and contraindications, see the *CRC Handbook of Medicinal Herbs*, ed. 2, Duke et al., 2002.

Cultivation (Coconut) — Nuts are planted right away in the nursery or stored in a cool, dry shed. Soil should be sandy or light loamy, free from water logging, close to source of water, devoid of heavy shade. Beds should be dug and loosened to a depth of 30 cm. Nuts spaced in beds ca. 20 × 30 cm; a hectare of nursery accommodating 100,000 seed-nuts. Nuts planted horizontally produce better seedlings than those planted vertically. About 15 weeks after the nut is planted, the shoot appears. At about 30 weeks, when three seed-leaves have developed, seedlings should be planted out. Best spacing 9–10 m on the square, planting 70–150 trees/ha. Young plantation should be fenced to protect plants from cattle, goats, and other wild animals. During the first 3 years, seedlings should be watered during drought. Cover crops (*Centrosema pubescens, Calopogonium mucunoides,* or *Pueraria phaseoloides*) are used and turned under. Catch-crops may be used. There is no evidence that salt is beneficial, as sometimes claimed. Trees begin to bear in 5–6 years, more likely 7–9 years, reach full bearing in 12–13 years. Fruit set to maturity is 8–10 months; 12 months from setting of female flowers. Under good climatic conditions, a fully productive palm produces 12–16 bunches of coconuts per year, each bunch with 8–10 nuts, or 60–100 nuts/tree. Nuts must be harvested fully ripe for making copra or desiccated coconut. For coir, they are picked about one month short of maturity so that husks will be green. Coconuts are usually picked by human climbers or cut by knives attached to end of long bamboo poles, this being the cheapest method. With a pole, a man can pick from 250 palms in a day; by climbing, only 25. In some areas, nuts are allowed to fall naturally and collected regularly. Nuts are husked in field, a good husker handling 2000 nuts/day. Then, nut is split (up to 10,000 nuts/working day). For copra, an average of 6000 nuts are required for 1 ton; 1000 nuts yield 500 lbs or copra, which yields 250 lbs of oil. Average copra yield is 3–4 tons/ha. The U.S. annually imports 190 million lb of coconut oil and more than 650 million lb of copra; some sources state 300,000 tons copra and over 200,000 tons coconut oil annually (NUT). Copra may be cured by sun-drying, or by kiln-drying, or by a combination of both. Sun-drying requires 6–8 consecutive days of good bright sunshine to dry meat without its spoiling. Drying reduces moisture content from 50% to below 7%. Copra is stored in a well-ventilated, dry area. Extraction of oil from copra is one of the oldest seed-crushing industries of the world. Coconut cake is usually retained to feed domestic livestock. When it contains much oil, it is not fed to milk cows but is used as fertilizer. Efficient pressing will yield from 100 kg of copra, approximately 62.5 kg of coconut oil, and 35 kg coconut cake, which contains 7–10% oil. Desiccated coconut is just the white meat; the brown part is peeled off. Dried in driers similar to those for tea. Good desiccated coconut should be white in color, crisp, with a fresh nutty flavor, and should contain less than 20% moisture and 68–72% oil, the extracted oil containing less than 0.1% of free for oil, yielding about 55%. The resulting "poonac" is used for feeding draught cattle. Coconut flour is made from desiccated coconut with oil removed, and the residue dried and ground. However, it does not keep well. Coir fiber obtained from slightly green coconut husks by retting in slightly saline water that is changed frequently (requires up to 10 months); then, husks are rinsed with water and fiber separated by beating with wooden mallets. After drying, the fiber is cleaned and graded. Greater part of coir produced in India is spun into yarn, a cottage industry, and then used for rugs and ropes. In Sri Lanka, most coir consists of mechanically separated mattress and bristle fiber. To produce this, husks are soaked or retted for 1–4 weeks and then crushed between iron rollers before fibers are separated. Bristle fibers are 20–30 cm long; anything shorter is sold as superior mattress fiber. In some areas, dry milling of husks, without retting, is carried on and produces only mattress fiber. For coir, 1000 husks yield about 80 kg per year, giving about 25 kg of bristle fiber and 55 kg of mattress fiber. The separated pith, called bast or dust, is used as fertilizer, since the potash is not leached out. Coconuts may be stored at temperature of 0–1.5°C with relative humidity of 75% or less for 1–2 months. In storage, they are subject to loss in weight, drying up of nut milk, and mold. They may be held for 2 weeks at room temperature without serious loss.

Chemistry (Coconut) — In the *CRC Handbook of Phytochemical Constituents*, coconut towers above the other entries for lauric acid, attaining 36%, on a calculated dry weight basis, followed by uchuba (Virola) at 11.5, betel nut at 9.0, datepalm at 5.4, calendula at 1.8, macadam at 1.1, cantaloupe seed at 0.9, cashew at 0.8, ginger at 0.4, water melon seed at 0.3, and mace and thyme at 0.2%, on a rounded and calculated ZMB. Here are a few of the more notable chemicals found in coconut. For a complete listing of the phytochemicals and their activities, see the CRC phytochemical compendium, Duke and duCellier, 1993 (DAD) and the USDA database http://www.ars-grin.gov/duke/.

Alpha-Tocopherol — Anticancer; Anticonvulsant Synergen; Antimutagenic; Antioxidant (5 × quercetin); Antiradicular (5 × quercetin); Antitumor; RDA = 3–12 mg/day.

Lauric-acid — Antibacterial; Antiviral.

Monolaurin—Antiadenomic ED50 = 23 µg/ml; Antiplaque; Antitumor (Kidney) ED50 = 4 µg/ml; Antitumor (Pancreas) ED50 = 2 µg/ml; Antitumor (Prostate) ED50 = 23 µg/ml; Artemicide LC50 = 79 µg/ml; Carioatatic X6963883.

Tocopherol — ADI = 2 mg/kg; Analgesic 100 IU 3 × day; Antiaging; Antiaggregant; Antialzheimeran 2000 IU; Antianginal 1067 mg/man/day; Antiatherosclerotic; Antibronchitic; Anticancer; Anticariogenic; Anticataract; Antichorea; Anticoronary 100–200 IU/day; Antidecubitic; Antidermatitic; Antidiabetic 600–1200 mg/day; Antidysmenorrheic; Antiepitheleomic; Antifibrositic; Antiglycosation; Antiherpetic; Antiinflammatory; Antiischaemic; Antileukemic 100–250 µ*M*; Antileukotrienic; Antilithic 600 mg/day; Antilupus; Antimastalgic; AntiMD; AntiMS; Antimyoclonic; Antineuritic; Antinitrosaminic; Antiophthalmic; Antiosteoarthritic; Antioxidant IC50 = 30 µg/ml, IC95 = 650 µ*M*; Antiparkinsonian?; AntiPMS 300 IU 2 × day; Antiproliferant IC50 = 150 µg/ml; Antiradicular; Antiretinopathic?; Antisenility; Antisickling; Antispasmodic 300 mg/man/day; Antisterility; Antistroke; Antisunburn; Anti-Syndrome-X; Antithalassemic; Antithrombic 600 IU/day; Anti-Thromboxane-B2; Antitoxemic; Antitumor 7 µ*M* ckn; Antitumor (Breast) IC50 = 125 µg/ml, 100–250 µ*M*; Antitumor (Colorectal) 500–10,000 µ*M*; Antitumor (Prostate)100–250 µ*M*; Antiulcerogenic 67 mg/man/3 ×/day/orl; Apoptotic 100–250 µ*M*; Cerebroprotective; Circulostimulant; Hepatoprotective; 5-HETE-Inhibitor; Hypocholesterolemic 100–450 IU/man/day; Hypoglycemic 600 IU/man/day; Immunostimulant 60–800 IU; Insulin-Sparing 1000 IU; Lipoxygenase-Inhibitor; ODC-Inhibitor 400 mg/kg; p21-Inducer 500–10,000 µ*M*; Phospholipase-A2-Inhibitor; Protein-Kinase-C-Inhibitor 10–50 µ*M*, IC50 = 450 µ*M*; Vasodilator; RDA = 2–10 mg/day; PTD = 800 mg/day.

Commiphora myrrha (Nees) Engl. (Burseraceae)
African Myrrh, Herabol Myrrh, Myrrh, Somali Myrrh

Synonyms — *Balsamodendrum myrrha* Nees, *Commiphora molmol* (Engl.) Engl., *C. myrrha* var. *molmol* Engl.

    "I have perfumed my bed with myrrh…" Proverbs 7:17.

Medicinal Uses (Myrrh) — Myrrh was all but panacea to the ancients; in Mesopotamia and the Greco-Roman world, it was used for abscesses, bladder stones, chilblains, dropsy, earache, eye problems, fever, hemorrhoids, and rhinosis. Bitter and pungent, the myrrh was esteemed by orientals as an astringent tonic internally and as a cleansing agent externally. In Algeria, myrrh is used to dress suppurations. According to Grieve's *A Modern Herbal*, myrrh is an emmenagogue, a tonic in dyspepsia, an expectorant, a mucosal stimulant, a stomachic carminative, exciting appetite and gastric juices, and an astringent wash. It is occasionally combined with aloes and iron. To this day, in the U.S., it is used as a mouthwash in gingivosis, for spongy gums, ulcerated throat, and aphthous stomatitis. The tincture is also applied externally to foul and indolent sores. It is useful in bronchorrhea and leucorrhea, and has served as a vermifuge (BIB).

White et al. (2000) say that myrrh stimulates the thyroid, although scientists still don't know the mechanism. They recommend tinctures rather than tea, since myrrh does offer up its phytochemicals to water as well as to alcohol. They suggest $^1/_8$ to $^1/_4$ tsp tinctures thrice daily (WAF). Zhu et al. (2001), reporting on six sesquiterpenoids, including two new furanosesquiterpenoids, note that the myrrh extracts or the furanosesquiterpenoids have demonstrated anesthetic, antibacterial, antifungal, and anithyperglycemic activities (JNP64:1460). So the Biblical myrrh is still revealing new secrets two millennia after Christ.

Indications (Myrrh) — Abrasion (1; CAN); Adnexosis (f; MAD); Alopecia (f; MAD); Amenorrhea (f; BGB; FEL; MAD; PH2); Aphtha (1; CAN); Asthma (1; APA; FEL; FNF); Atherosclerosis (f; MAD); Athlete's Foot (1; SKY); Bacteria (1; JNP64:1460); Bedsore (f; APA); Bladder Stone (f; BIB); Boil (f; PNC); Bronchosis (1; APA; BGB; FEL; FNF); Bruise (f; BOW); Cancer (f; APA; PH2); Cancer, abdomen (f; PH2); Cancer, colon (f; PH2); Candida (1; BGB; FNF); Canker Sore (1; APA; SKY); Carbuncle (f; PH2); Caries (f; FEL); Catarrh (f; BGB; CAN; FEL); Chilblain (f; BIB); Chlorosis (f; BIB); Circulosis (f; BOW); Cold (1; BGB; CAN; FNF; SKY); Congestion (1; APA; BGB); Cough (f; PH2); Cramp (1; FNF); Decubosis (f; BGB; BOW); Dermatosis (1; APA; FNF; MAD; PH2); Diabetes (1; JNP64:1460); Diarrhea (f; MAD; JNP64:1460); Dropsy (f; BIB); Dysentery (f; MAD); Dysmenorrhea (1; BGB; FNF; PH2); Dyspepsia (f; APA; FEL); Dysuria (f; MAD); Earache (f; BIB); Enterosis (f; PH2); Erysipelas (f; MAD); Fever (f; BIB; MAD); Freckle (f; MAD); Fungus (1; FNF; JNP64:1460); Furunculosis (1; CAN; PH2); Gangrene (f; FEL); Gas (f; APA; MAD); Gastrosis (f; FEL; PH2; PNC; JNP64:1460); Gingivosis (1; APA; FEL; FNF; PNC; SKY); Gleet (f; FEL); Gonorrhea (f; FEL); Halitosis (f; FEL); Hemorrhoid (f; APA; BGB; BIB); Hepatosis (f; MAD); Hoarseness (f; APA); Hypothyroidism (1; WAF); Infection (1; FNF; PH2; JNP64:1460); Infertility (f; MAD); Inflammation (1; BGB; FNF; PH2); Laryngosis (f; FEL); Leprosy (f; APA); Leukorrhea (f; FEL; MAD); Menopause (1; BGB); Menorrhagia (f; MAD); Mononucleosis (f; BOW); Mucososis (1; APA; FEL; PH2); Mycosis (1; JNP64:1460); Nervousness (1; FNF); Odontosis (f; MAD); Ophthalmia (f; BIB); Osteosis (f; BGB); Otosis (f; BOW); Pain (1; JNP64:1460); Pharyngosis (2; APA; FEL; KOM; MAD; PH2; PNC); Pulmonosis (f; MAD); Respirosis (f; BGB); Rheumatism (f; BGB); Rhinosis (f; APA; BIB); Salpingosis (f; MAD); Side Ache (f; MAD); Sinusosis (1; APA); Sore (1; APA; FEL; FNF; PNC); Sore Throat (2; BGB; FEL; KOM; MAD; SKY); Stomatosis (2; APA; KOM; MAD; PH2; PIP); Swelling (f; APA); Tonsilosis (1; APA; BGB; FEL; PNC); Tuberculosis (f; MAD); Ulcer (f; APA; PH2; X11113992); Uterosis (f; MAD); Uvulosis (f; FEL); VD (f; FEL); Water Retention (f; MAD); Worm (f; FEL; MAD); Wound (f; APA; BGB); Wrinkle (f; MAD); Yeast (1; BGB; FNF).

Myrrh for gingivosis:

- Antibacterial: acetic-acid; alpha-pinene; cinnamaldehyde; cuminaldehyde; dipentene; eugenol; limonene
- Antiinflammatory: alpha-pinene; cinnamaldehyde; eugenol
- Antioxidant: eugenol
- Antiseptic: eugenol; formic-acid; limonene; m-cresol
- Astringent: formic-acid
- COX-2-Inhibitor: eugenol

- Candidicide: cinnamaldehyde; eugenol
- Candidistat: limonene
- Cyclooxygenase-Inhibitor: cinnamaldehyde
- Fungicide: acetic-acid; cinnamaldehyde; cuminaldehyde; eugenol

Myrrh for infection:

- Analgesic: eugenol
- Anesthetic: cinnamaldehyde; eugenol
- Antibacterial: acetic-acid; alpha-pinene; cinnamaldehyde; cuminaldehyde; dipentene; eugenol; limonene
- Antiedemic: eugenol
- Antiinflammatory: alpha-pinene; cinnamaldehyde; eugenol
- Antiseptic: eugenol; formic-acid; limonene; m-cresol
- Antiviral: alpha-pinene; cinnamaldehyde; dipentene; limonene
- Astringent: formic-acid
- COX-2-Inhibitor: eugenol
- Circulostimulant: cinnamaldehyde
- Cyclooxygenase-Inhibitor: cinnamaldehyde
- Fungicide: acetic-acid; cinnamaldehyde; cuminaldehyde; eugenol
- Fungistat: formic-acid; limonene
- Lipoxygenase-Inhibitor: cinnamaldehyde

Other Uses (Myrrh) — Long before questions of intellectual property and germplasm rights were issues, Queen Hatshepsut of Egypt, circa 3500 B.P,, sent a treasure hunting expedition to what is today Somalia, land of myrrh and qat. The Queen wanted more myrrh. Reliefs in the temple she built at Deir el Bahari suggest that they brought back plenty of resin and even potted plants. So, for 3500 years, germplasm has been moving from country to country. The Queen, you see, rubbed myrrh on her legs as perfume. During the feast of Isis, when the Egyptians burned oxen as offerings, they stuffed the carcass with frankincense and myrrh to mask the smell of burning flesh (Thieret, 1996). Ancient Egyptians burned it in their temples and used it for embalming. Much of the myrrh of commerce today comes from this or closely related species of Arabia, Ethiopia, and Somalia. Highly regarded in the Orient as an aromatic substance, perfume, and medicine. There is even a term "myrrophore" applied to the women who bore spices to the sepulcher of Jesus with aloes, cassia, and cinnamon; it was an ingredient of the holy oil and a domestic perfume. Myrrh is a Jewish holy oil and gets honorable mention in the first books of Christian, Jewish, and Muslim holy texts (BGB). Some authorities maintain that the Biblical myrrh was in reality a mixture of myrrh and ladanum. Myrrh was burned as incense for the sun god at Heliopolis at noon. Myrrh was introduced into Chinese and Tibetan pharmacy as early as the 7th century C.E. Persian Kings wore myrrh in their crowns. Around 1342, the khan of China sent myrrh among other gifts to Pope Benedict at Avignon, but the shipment was plundered, never reaching its destination (Thieret, 1996). Myrrh and frankincense were even burned during the reign of King George III. One old legend says that Myrrha, daughter of the King of Cyprus, became unnaturally obsessed with her father, who exiled her to the Arabian desert where the gods transformed her into the myrrh tree, "in which guise she remains, weeping tears perfumed of repentance." The pungent gum-resin is used commercially to flavor baked goods, beverages, candy, chewing gum, gelatin, puddings, and soups (FAC). It is also an ingredient in Swedish bitters (FAC). The gum makes a good mucilage, and the insoluble residue from the tincture can be used as a glue (BIB). Import statistics are hard to come by. Scholarly historian Thieret (1996) suggests total yearly production of myrrh is perhaps 500 tons, and of frankincense 1000 tons. Recently, U.S. imports ran 5–20 tons. The U.K. imports ca. 30 tons frankincense each year, one perfume manufacturer alone consuming 5 tons annually (Thieret, 1996).

For more information on activities, dosages, and contraindications, see the *CRC Handbook of Medicinal Herbs*, ed. 2, Duke et al., 2002.

Cultivation (Myrrh) — Though much is harvested from the wild (Arabia, Somalia, Yemen), some is cultivated in Ethiopia, Kenya, and Tanzania. Needs well drained sunny areas, with minimum temperatures 10–15°C (50–60°F). Propagated by seeds sown in spring or hardwood cuttings at the end of the growing season (Bown, 2001). Whether harvested from the wild or cultivars, cuts in the bark ooze forth tears of myrrh (BGB). Natives cut the bark, causing the exudation of a yellowish oleoresin, which hardens on exposure to the air and turns reddish-brown (GEO).

Chemistry (Myrrh) — Here are a few of the more notable chemicals found in myrrh. For a complete listing of the phytochemicals and their activities, see the CRC phytochemical compendium, Duke and duCellier, 1993 (DAD) and the USDA database http://www.ars-grin.gov/duke/.

Cinnamaldehyde — See also *Cinnamomum verum*.

Curzerone — Antitumor.

Furano-Eudesma-1, 3–Diene — Analgesic; 50 mg/kg orl mus; Myorelaxant 50 mg/kg orl mus.

Lindstrene — Hepatoprotective.

Costus speciosus (J. König) Sm.
(Costaceae. Also placed in Zingiberaceae)
CANE REED, CREPE GINGER, WILD GINGER

Synonym — *Banksea speciosa* J. König

Medicinal Uses (Cane Reed) — Asians consider the rhizomes anthelminthic, depurative, laxative, and tonic (KAB, WOI). Ayurvedics use it in "kapha" and "vata," and in Western terms, for anemia, bronchitis, fever, hiccups, inflammation, and lumbago (KAB). It is reportedly used in India for colds, pneumonia, and rheumatism (IHB). The Malayan name, setawar, means "remover of virus." Malays chew it with betel for cough (IHB). The decoction is used to bathe patients with high fever. It is even splattered on elephants suffering fever. They also poultice bruised leaves onto the heads of feverish patients. Malays use it for humans inhabited by evil spirits. Javanese apply the rhizome in syphilis. Jain and Defilipps (1991) suggest that the rhizome is used as an antispasmodic, antipyretic, CNS-depressant, diuretic, and tonic (SKJ). Leaves are used for scabies and stomach ailments. Stem is used for blisters and burns. Bark used for cholera. Gruenwald et al. (2000) say that, in Asian Indian medicine, the roots and/or rhizomes are used for very different circumstances, making me suspect they may have confused kust ("*Costus*" Latin) with kuth or costus (colloquial) root (Saussurea). But how will we ever know? They suggest that the cane reed is used for insufficient uterine contractility, post-partum bleeding, retention of the placenta, threatening abortion (PH2). They also say that overdoses might lead to European cholera; maybe that means dysentery or diarrhea (JAD).

Steroid-like compounds isolated from the plant have antiarthritic, antifertility, and antiinflammatory capacity (SKJ). The saponin fraction has antiexudative, choleretic, estrogenic, and antispasmodic effects and potentiates or prolongs anesthesia (PH2). I am pleasantly surprised to find that PH2 lists curcumin, the active ingredient that I don't often find outside the ginger family, which is admittedly closely related to the costus or ginger lilies. If curcumin is confirmed, then the cane ginger shares in COX-2-Inhibitory activity, proportionate to the levels of curcumin.

Medicinal Plants of India gives some interesting data. Alkaloids from the plant show *in vitro* and *in vivo* anticholinesterase activities, perhaps explaining the depurative and ophthalmic activity (MPI). *C. speciosus* was found to be 2.5 times more ecbolic than *Gloriosa superba*, enough to make me advise pregnant women to avoid both. Saponins from the herb caused proliferation of uterine and vaginal tissues, similar to those produced by stilbosterol (MPI). These saponins also had antiarthritic and antiinflammatory activities. Habsah et al. (2000) screening dichloromethane and methanol extracts of 13 Zingiberaceae species (*Alpinia, Costus,* and *Zingiber*) for antimicrobial and antioxidant activities found all to be antibacterial. Only the methanol extract of *Costus discolor* showed potent antifungal activity against *Aspergillus ochraceous*. All extracts were antioxidant and comparable with alpha-tocopherol (X10996279).

Indications (Cane Reed) — Abortion (f; PH2); Adenoma (1; FNF; FT68:483); Alzheimer's (1; FNF); Anasarca (f; SKJ); Anemia (f; KAB); Arthrosis (1; FNF; MPI; SKJ); Asthma (1; SKJ; HG37:18); Atherosclerosis (1; FT68:483); Bite (f; PH2; SKJ); Bleeding (f; PH2); Blister (f; SKJ); Bronchosis (f; KAB); Burn (f; SKJ); Cancer (1; FNF; JLH); Cancer, colon (1; FNF); Cataract (1; FNF; FT68:483); Catarrh (f; DEP; MPI); Childbirth (f; PH2); Cholecystosis (1; PAM); Cholera (f; SKJ); Cold (1; FNF; IHB; SKJ); Colitis (1; HG37:18); Conjunctivosis (f; BOW); Constipation (f; BOW; SKJ); Cough (f; DEP; IHB; MPI; PH2; SKJ); Crohn's Disease (1; HG37:18); Dermatosis (f; DEP; IHB; MPI; PH2); Dog Bite (f; SKJ); Dropsy (f; SKJ); Dysentery (f; IHB); Dyspepsia (f; KAB; MPI); Eczema (1; IWU); Edema (1; FNF; PCF:338); Fever (f; DEP; IHB; MPI; PH2; SKJ); Fungus (1; FNF); Gastrosis (f; SKJ); Gravel (f; SKJ); Headache (f; SKJ); Hematuria (f; SKJ); Hepatosis (1; FNF); Hiccup (f; KAB); High Cholesterol (1; FNF); Impotence (f; BOW); Infection (1; FNF); Inflammation (1; FNF; KAB; MPI); Ischemia (1; PFH44:87); Leishmaniasis (1; X10865470); Leprosy (f; IHB); Leukemia (1; FT68:483); Lumbago (f; KAB); Lymphoma (1; FNF); Malaria (f; SKJ); Melanoma (1; FNF); Mycosis (1; FNF); Ophthalmia (f; IHB; MPI); Osteosis (f; KAB); Otosis (f; BOW); Pain (f; KAB); Phthisis (f; SKJ); Pneumonia (f; IHB); Psoriasis (1; FNF); Pulmonosis (1; FNF); Rabies (f; DAA); Rheumatism (1; FNF; IHB; KAB; SKJ); Scabies (f; SKJ); Smallpox (f; BOW; IHB); Snakebite (f; MPI; PH2; SKJ); Stomatosis (f; JLH); Stone (1; FNF); Swelling (1; FNF); Syphilis (f; IHB); Thirst (f; SKJ); Thrombosis (1; FNF); Tuberculosis (f; SKJ); Worm (f; MPI).

Cane Reed for arthrosis:

- 12-Lipoxygenase-Inhibitor: curcumin
- Antiarthritic: curcumin
- Antiedemic: curcumin
- Antiinflammatory: curcumin; curcuminoids; diosgenin
- Antiprostaglandin: curcumin; curcuminoids
- Antispasmodic: curcumin
- COX-2-Inhibitor: curcumin
- Cyclooxygenase-Inhibitor: curcumin

Cane Reed for infection:

- Antibacterial: curcumin
- Antiedemic: curcumin
- Antiinflammatory: curcumin; curcuminoids; diosgenin
- Antiviral: curcumin
- COX-2-Inhibitor: curcumin
- Cyclooxygenase-Inhibitor: curcumin
- Fungicide: curcumin
- Immunostimulant: curcumin

Other Uses (Cane Reed) — Used as a food plant in S.E. Asia, where it's called "Tebu" and "Setawar." Tender young shoots are boiled or steamed as a vegetable, often in coconut milk; the fruits and rhizomes are also eaten (FAC). In India, the nonaromatic, astringent, fibrous, mucilaginous rhizome is eaten, often cooked in syrup or preserves. The rhizome is said to taste like a cucumber (IHB). Often cultivated as an ornamental. It is used in fishing and magic ceremonies (IHB). Drug sometimes used to adulterate Gloriosa.

For more information on activities, dosages, and contraindications, see the *CRC Handbook of Medicinal Herbs*, ed. 2, Duke et al., 2002.

Cultivation (Cane Reed) — Cultivated and harvested like ginger relatives and other ginger lilies. Apparently does best in moist, rich, neutral to acid, well-drained soils in shade, tropical climates with high humidity and minimum temperature 13°C (55°F). Seeds sown as soon as ripe at temperatures close to 20°C (68°F). Root divisions in fall (Bown, 2001). Harvested as needed.

Chemistry (Cane Reed) — With some 3% curcuminoids (FNF), cane reed may be a weak cousin to some of the other zingiberaceous species, like turmeric (which ranges from 3–8% curcuminoids). Here are a few of the more notable chemicals found in cane reed. For a complete listing of the phytochemicals and their activities, see the CRC phytochemical compendium, Duke and duCellier, 1993 (DAD) and the USDA database http://www.ars-grin.gov/duke/.

Curcumin — ADI = 100 µg/kg; Antiaflatoxic IC50 = 81 µ*M*; Antiadenomacarcinogenic 50–200 mg/kg ipr rat; Antiaggregant; Antiangiogenic; Antiarachidonate; Antiarthritic; Antiasthmatic; Antiatherosclerotic 0.4–20 mg/kg/day; Antibacterial; Antibronchitic; Anticancer; Anticataract 75 mg/kg orl rat; Anticholecystosic; Anticolitic; Anticollagenic; AntiCrohn's; AntiEBV IC50 = 5.4 µ*M*; Antieczemic; Antiedemic ED50 = 100.2 mg/kg orl mus (cf 78 for cortisone) ED50 = 48 mg/kg orl rat (cf 45 for cortisone, 48 for phenylbutazone); Antieicosanoid; Antihepatosis; Anti-HIV IC50 = 40 µ*M*; Antiinflammatory 1200 mg/man/day, 1 µ*M*; Antiintegrase 40–150 µ*M*; Antiischemic; Antileishmannic IC50 = 7.8 µg/ml; Antileukemic; Antileukotriene 170 mg/kg orl rat; Antilipoperoxidant; Antilithic 0.5% diet; Antilymphomic 0.4 mg/ml, 4 µg/ml; Antimelanomic 200 nM/kg orl mus; Antimetastatic 200 nM/kg orl mus; Antimutagenic; Antinitrosaminic; Antioxidant IC50 = 500 µ*M*; Antipapillomic; Antiperoxidant; Antiproliferant IC50 = 13 µ*M*; Antiprostaglandin 8.8 µ*M*; Antipsoriatic; Antispasmodic; Antithrombic; Antithromboxane; Antitumor (Breast); Antitumor (Colon); Antitumor (Duodenum); Antitumor (Liver) 10 µ*M*; Antitumor (Mammary) 50–200 mg/kg ipr rat; Antitumor (Skin); Antitumor-Promoter IC91 = 10 µ*M*; Antiulcer orl rat; Antiviral IC50 = 5.4 µ*M*; Apoptotic 30–90 µ*M*, 150–2000 ppm (diet) orl rat; Cardiodepressant; Chelator IC50 = 500 µ*M*; Chelator (Iron); Cholagogue; Choleretic; COX-2-Inhibitor 10–20 µ*M*; Cyclooxygenase-Inhibitor; Cytochrome-p450-Inhibitor; Cytotoxic 0.4–4 mg/ml, IC50 = 1 µg/ml, IC50 = 25 µ*M*; Deodorant; Detoxicant; DNA-Protectant; Dye; Fibrinolytic; Fungicide; Glutathionigenic 80 mg/kg igs rat; Hepatoprotective 30 ppm, 30 mg/kg/day; 5-HETE-Inhibitor IC50 = 3–10 µ*M*; 8-HETE-Inhibitor IC40 = 3 µ*M*; Hypocholesterolemic 0.15% diet 7 wks; Hypolipidemic 0.15% diet 7 wks; Hypotensive; Immunostimulant 40 mg/kg/5 wk orl rat; Liptase-Promoter; 5-Lipoxygenase-Inhibitor; 12-Lipoxygenase-Inhibitor; Litholytic 0.5% diet; Maltase-Promoter; Metal-Chelator; MMP-9-Inhibitor 10 µ*M*; Nematicide; Neuroprotective 80 mg/kg igs rat; NO-Scavenger; ODC-Inhibitor ED = ~150 mg/kg; P-450-Inhibitor IC50 = 2–14 µg/ml; PGE2-Inhibitor IC42 = 3 µ*M*; Phototoxic; Plasmodicide; Protease-Inhibitor IC50 = 11–250 µ*M*; Protein-Kinase-C-Inhibitor IC69 = 15 µ*M*, IC50 = 15 µ*M*; Protein-Kinase-Inhibitor; PTK-Inhibitor 5–100 µ*M*; Pulmonoprotective 200 mg/kg/7 d; Quinone-Reductase-Inducer 3.4 µ*M*; Radioprotective; 5-Alpha-Reductase-Inhibitor; Sucrase-Promoter; Topoisomerase-I-Inhibitor; Topoisomerase-II-Inhibitor; Ulcerogenic orl rat; TD = >5000 mg/kg orl rat; LDlo = >2000 mg/kg orl mus; LDlo = >1800 mg/kg orl rat.

Curcuminoids — Antiinflammatory; Antimitotic; Antiprostaglandin; Antitumor; Cholagogue; Cytotoxic; Leukotriene-Inhibition; Ulcerogenic?.

Diosgenin — Antifatigue; Antiinflammatory; Antistress; Estrogenic 20–40 mg/kg/day/15 day scu mus; Hepatoprotective; Hypocholesterolemic; Mastogenic 20–40 mg/kg/day/15 day scu mus.

Crocus sativus L. (Iridaceae)
SAFFRON

Medicinal Uses (Saffron) — Saffron is not included in American and British pharmacopoeias, but some Indian medical formulae still include it. Saffron is used to promote eruption of measles, and in small doses is considered antihysteric, antispasmodic, aphrodisiac, carminative, diaphoretic, ecbolic, emmenagogue, expectorant, sedative, stimulant, and stomachic. It is an oft-cited folk remedy for various types of cancer. It is sometimes used to promote menstruation. Early on, it was taken as a preventive against the plague and other epidemics. In Biblical times, saffron was important in some ancient herbals, its extracts used as an antispasmodic, emmenagogue, and general stimulant and tonic. Early Eclectics valued it for amenorrhea, chlorosis, dysmenorrhea, hysteria, menorrhagia, and suppression of lochial discharge. In India, saffron is regarded for bladder, kidney, and liver ailments, also for cholera. Mixed with "ghee," it is used for diabetes. Saffron oil is applied externally to uterine sores (BIB, CRC). Lebanese add a dozen pistils to a large cup hot water for children coming down with chickenpox, measles, or mumps. Germans take it in milk for measles (MAD). Algerians and Gypsies use the saffron infusion as a collyrium. Eight to ten filaments (stigmata) per cup of tea is suggested as a narcotic for asthma, hysteria, or whooping cough (BIB, LIL, MAD, PH2, RIN). Small saffron doses stimulate the flow of gastric juices, large doses stimulate the smooth muscle of the uterus (PH2). Overdoses are reportedly narcotic, and saffron corms are toxic to young animals. Apoplexy and extravagant gaiety are possible after effects.

Kazuho and Saito (2000) review pharmacological studies showing saffron extracts have antitumor effects, radical scavenger properties, or hypolipemic effects. Among the constituents of saffron extract, crocetin is mainly responsible for these pharmacological activities. They stress recent behavioral and electrophysiological studies, demonstrating that saffron extract affects learning and memory in experimental animals. Saffron extract improved ethanol-induced impairments of learning behaviors in mice and prevented ethanol-induced inhibition of hippocampal long-term potentiation, a form of activity-dependent synaptic plasticity that may underlie learning and memory. Memory-sparing effect of saffron extract is attributed to crocin (crocetin di-gentiobiose ester), but not crocetin. Saffron extract or its active constituents, crocetin and crocin, could be useful as a treatment for neurodegenerative disorders accompanying memory impairment. So maybe if you must imbibe, a saffron liqueur might be the liqueur of choice. The Saffron occurs in several liqueurs. In what I call "Brandy Swifter," a cup of saffron tea, heavily charged with brandy, has been used for measles. Others suggest three spoonfuls or more of Saffron Cordial for those who "have taken too liberal a cup over night" (LIL).

Crocin is choleretic, but many herbs, if not all, contain choleretics. Choleretics stimulate the production of bile, the bitter substance that emulsifies fats in the duodenum, stimulates peristalsis, thereby encouraging movement of food along the GI tract. While useful in moderation in healthy people, choleretics might be contraindicated for those with gallbladder or liver diseases.

Premkumar et al. (2001) found that saffron inhibited genotoxicity, at least in mice, hinting that it might spare some side effects of chemotherapy. Saffron modulated *in vivo* genotoxicity of cisplatin, cyclophosphamide, mitomycin C, and urethane. Swiss albino mice were pretreated five days with three doses (20, 40, and 80 mg/kg body weight) aqueous saffron extract. Treatment with the genotoxins alone significantly inhibited GST activity. Saffron pretreatment attenuated the inhibitory effects of the genotoxins on GST activity (X11665650).

Martinez-Tome et al. (2001) comparing antioxidant properties of Mediterranean spices, compared annatto, cumin, oregano, sweet and hot paprika, rosemary, and saffron, at 5% concentration with the common food additives (butylated hydroxyanisole [BHA], butylated hydroxytoluene [BHT], and propyl gallate) at 100 µg/g. For inhibiting lipid peroxidation: rosemary > oregano > propyl gallate > annatto > BHA > sweet paprika > cumin > hot paprika > saffron > BHT. So 5% saffron was slightly better than 100 µg/g BHT as an antioxidant (Martinez-Tome et al., 2001).

Kubo and Kinst-Hori (1999) reported that the common flavonol, kaempferol, isolated from the fresh flower petals of saffron, inhibited tyrosinase (ID50 = 6700 mg/ml (0.23 mM). Not very potent, methinks. More important perhaps is the COX-2-Inhibitory action of kaempferol.

Indications (Saffron) — Adenosis (f; JLH); Aegilops (f; JLH); Amenorrhea (1; CRC; MAD; PH2); Asthma (f; MAD); Bacteria (1; FNF); Bladder Ailment (f; CRC); Bleeding (f; DAA; MAD); Blood Disorder (f; CRC); Bronchosis (f; PH2); Burn (f; JLH); Cacoethes (f; JLH); Cancer (1; APA; FNF; PR14:149); Cancer, abdomen (1; APA; CRC); Cancer, bladder (1; APA; CRC); Cancer, breast (1; APA; CRC; JLH); Cancer, colon (1; APA; JLH); Cancer, diaphragm (1; APA; JLH); Cancer, ear (1; APA; CRC); Cancer, eye (1; APA; JLH); Cancer, kidney (1; APA; CRC); Cancer, larynx (1; APA; JLH); Cancer, liver (1; APA; CRC); Cancer, mouth (1; APA; CRC); Cancer, neck (1; APA; CRC); Cancer, spleen (1; APA; CRC); Cancer, stomach (1; APA; CRC; JLH); Cancer, testicle (1 APA; JLH); Cancer, throat (1; APA; JLH); Cancer, tonsil (1; APA; CRC); Cancer, uterus (1; APA; CRC; JLH); Cardiopathy (f; APA); Catarrh (f; CRC; SKJ); Childbirth (f; DAA; PH2); Cholera (f; CRC); Chorea (f; HHB; MAD); Cold (f; CRC); Condyloma (f; DAA); Conjunctivosis (f; MAD); Cough (f; DAA; MAD); Cramp (f; DAA; HHB); Cystosis (f; JLH); Depression (f; CRC; DAA; PNC); Dermatosis (f; CRC); Diabetes (f; CRC); Dysmenorrhea (f; DAA; HHB; MAD PNC); Edema (1; APA; FNF); Encephalosis (1; APA); Enterosis (f; JLH); Epistaxis (f; MAD); Fear (f; CRC; DAA); Fever (f; CRC; PH2); Fibroid (f; JLH); Fungus (1; FNF); Gas (f; MAD); Gastrosis (f; JLH); Gout (f; MAD); Hangover (f; LIL); Headache (f; PH2); Hemoptysis (f; DAA; MAD); Hepatosis (1; CRC; FNF; JLH; SKJ); High Blood Pressure (1; APA; FNF); High Cholesterol (1; APA; FNF); HIV (1; FNF); Hysteria (f; CRC; DAA; MAD); Induration (f; JLH); Infection (1; FNF); Inflammation (1; FNF; JLH); Lachrimosis (f; JLH); Laryngosis (f; JLH); Leukemia (f; JLH); Lochiostasis (f; PH2); Lymphoma (1; APA; JLH); Measles (f; CRC; DAA; MAD); Melancholy (f; CRC; HHB); Menorrhagia (f; HHB; PH2); Menoxenia (f; CRC); Nephrosis (f; JLH); Neurosis (1; CRC; FNF); Obesity (1; FNF; PR14:149); Ophthalmia (f; JLH); Orchosis (f; JLH); Pain (1; DAA; FNF); Parotosis (f; JLH); Pertussis (f; BIB; DAA; MAD); Phymata (f; JLH); Plague (f; MAD); Puerperium (f; CRC); Sclerosis (f; CRC); Shock (f; CRC; DAA); Snakebite (f; SKJ); Sore Throat (f; PH2); Spasm (f; CRC); Splenosis (f; CRC; JLH);

Swelling (1; APA); Tonsilosis (f; JLH); Twitching (f; MAD); Uterosis (f; CRC; DAA; JLH); VD (f; CRC; DAA); Vertigo (f; MAD); Vomiting (f; PH2); Wart (f; CRC).

Saffron for cancer:

- AntiHIV: myricetin; oleanolic-acid; quercetin
- Antiaggregant: kaempferol; quercetin; salicylates
- Anticancer: alpha-pinene; beta-myrcene; camphor; delphinidin; geraniol; kaempferol; limonene; linalool; lycopene; myricetin; oleanolic-acid; quercetin
- Antifibrosarcomic: quercetin
- Antihepatotoxic: oleanolic-acid; quercetin
- Antiinflammatory: alpha-pinene; beta-pinene; borneol; kaempferol; myricetin; oleanolic-acid; quercetin; salicylates
- Antileukemic: astragalin; kaempferol; quercetin
- Antileukotriene: quercetin
- Antilipoperoxidant: quercetin
- Antimelanomic: geraniol; quercetin
- Antimutagenic: crocetin; kaempferol; limonene; myricetin; quercetin
- Antinitrosaminic: quercetin
- Antioxidant: crocetin; cyanidin; delphinidin; gamma-terpinene; kaempferol; lycopene; malvidin; myricetin; oleanolic-acid; quercetin
- Antiperoxidant: quercetin
- Antiproliferant: quercetin
- Antisarcomic: oleanolic-acid
- Antitumor: crocetin; geraniol; kaempferol; limonene; lycopene; oleanolic-acid; quercetin
- Antiviral: alpha-pinene; kaempferol; limonene; linalool; myricetin; oleanolic-acid; p-cymene; quercetin
- Apoptotic: kaempferol; myricetin; quercetin
- Beta-Glucuronidase-Inhibitor: oleanolic-acid
- COX-2-Inhibitor: kaempferol; oleanolic-acid; quercetin
- Chemopreventive: limonene
- Cyclooxygenase-Inhibitor: kaempferol; oleanolic-acid; quercetin
- Cytochrome-p450-Inducer: 1,8-cineole
- Cytotoxic: quercetin
- Hepatoprotective: borneol; oleanolic-acid; quercetin; zeaxanthin
- Hepatotonic: 1,8-cineole
- Immunostimulant: astragalin
- Lipoxygenase-Inhibitor: kaempferol; myricetin; quercetin
- Mast-Cell-Stabilizer: quercetin
- Ornithine-Decarboxylase-Inhibitor: limonene; quercetin
- p450-Inducer: 1,8-cineole; quercetin
- PTK-Inhibitor: quercetin
- Previtamin-A: gamma-carotene
- Protein-Kinase-C-Inhibitor: quercetin
- Topoisomerase-II-Inhibitor: kaempferol; myricetin; quercetin
- Tyrosine-Kinase-Inhibitor: myricetin; quercetin

Saffron for infection:

- Analgesic: borneol; camphor; p-cymene; quercetin; vitamin-b-1
- Anesthetic: 1,8-cineole; camphor; linalool

- Antibacterial: 1,8-cineole; alpha-pinene; beta-phenylethanol; geraniol; kaempferol; limonene; linalool; myricetin; oleanolic-acid; p-cymene; pinene; quercetin; terpinen-4-ol
- Antiedemic: oleanolic-acid
- Antiinflammatory: alpha-pinene; beta-pinene; borneol; kaempferol; myricetin; oleanolic-acid; quercetin; salicylates
- Antiseptic: 1,8-cineole; beta-pinene; camphor; geraniol; kaempferol; limonene; linalool; myricetin; oleanolic-acid; pinene; terpinen-4-ol
- Antiviral: alpha-pinene; kaempferol; limonene; linalool; myricetin; oleanolic-acid; p-cymene; quercetin
- Bacteristat: quercetin
- COX-2-Inhibitor: kaempferol; oleanolic-acid; quercetin
- Cyclooxygenase-Inhibitor: kaempferol; oleanolic-acid; quercetin
- Fungicide: 1,8-cineole; camphor; geraniol; linalool; p-cymene; pinene; quercetin; terpinen-4-ol; terpinolene
- Fungistat: limonene
- Immunostimulant: astragalin
- Lipoxygenase-Inhibitor: kaempferol; myricetin; quercetin

Other Uses (Saffron) — In Biblical times, saffron was important to people of the East as a condiment and sweet perfume, the stigmas being particularly valued for their food-coloring property. Dioscorides mentions its use as a perfume. Pliny records that the benches of the public theaters were strewn with saffron, and the costly petals were placed in small fountains, to diffuse the scent into public halls. Today, saffron the dye, dried stigmata of the flowers, is more important as a culinary spice than as a medicine. In Europe, it is used as a flavoring and coloring ingredient, and druggists add it to medicines. In India, they are used to add yellow shades to curry. Saffron is cultivated for the dye obtained from the stigmas of the flowers; about 100,000 flowers yield 1 kg saffron. Dye used chiefly as a coloring agent and spice in cookery (especially Spanish), for biscuits, bouillabaisse, butter, cakes, cheese, creams, curries, eggs, liqueurs, preserves, puddings, rice dishes (e.g., biryani, paella, risotto), soups, stews, especially chicken dishes, and in confectionery to give color, flavor, and aroma. It is used as a spice with fish, shellfish, or poultry creating such famous dishes as Spanish paella, French bouillabaisse, and Pennsylvania Dutch chicken pot pies (AAR). For biryanis and elegant Persian rice dishes, some rice is spooned into the golden-warm saffron water, gelded, and returned to the ungelded rice. Fruits also blend well with saffron; saffron creams and ice creams are real treats. Never saute saffron threads with anything, especially onions, at the beginning of a recipe (AAR). Stigmata used as a tea substitute. Roots are eaten roasted (FAC). Used in cosmetics for eyebrows and nail polishes and as incense. It has also been used as a mild deodorant. Dissolved in water, it is used as an ink and is applied to foreheads on religious and ceremonial occasions. The gold of cookery is now as expensive as the gold of jewelry (BIB, CRC). Being very expensive, saffron is almost always adulterated with *Calendula, Carthamus, Curcuma,* or *Tagetes.* It should not be stored real hot or real cold. Do not refrigerate or freeze, but keep in a cool dry place (AAR). Properly stored, it can last three years (BIB, CRC, LIL, TAD).

For more information on activities, dosages, and contraindications, see the *CRC Handbook of Medicinal Herbs*, ed. 2, Duke et al., 2002.

Cultivation (Saffron) — For the gardener, bulbs (corms) are planted in late spring, 4–6 in deep in well-prepared, well-drained soil, in full sun or only partially shaded sites. Thriving in a dry Mediterranean climate, saffron can be grown by cautious herbalists in colder, wetter climates. Hardy to zone 6 (LIL). Usually planted in raised beds, occasionally with drains on four sides, especially in waterlogged areas, often monocultured. If skillfully intercropped, weeding may be kept to a minimum. Some recommend 6-in spacing between the corms. Flowering in fall, the flowers or stigmas are hand picked and quickly and carefully dried with low heat. One needs avoid high humidity and/or hard frost at flowering time, a period of only about three weeks. Yields run pretty close to 10 lb/a dried saffron but may exceed 25. After about three crops in Mediterranean countries, the bulbs are taken up and replanted elsewhere. It takes more than a ton of bulbs to plant an acre. For production agriculture, some 3500 kg corms will plant a hectare (ca. 3500 lb/acre). Planting begins in March by plowing to a depth of 25–30 cm, after removing all stones, roots, and rubbish. Second plowing and manuring are in April or early May, to a depth of 10–15 cm. Middle-sized tubers, sound and without external covering, planted in double rows at bottom of trenches, 15–20 cm apart in rows. Trenches should be 10–12 cm deep and 45 cm wide. Soil of new trench used to fill preceding trench, with a deep wide furrow being made every four to six rows. Thorough cultivation and freedom of weeds are essential. Saffron is usually grown in monoculture but may be overcropped with lettuce, cucumbers, or radishes during period when bulb is dormant, which is from late May until September, when the flowers come up. In September, soil is dug 6–7 cm deep between trenches, and in October, before flowering, soil is lightly hoed. After harvest, soil between trenches is dug to a depth of 12–15 cm. At end of April or beginning of May, leaves are cut and dried for fodder. After the fourth crop, the entire plot is dug up in May, tubers graded, the best ones saved for new plantings, and the remainder sold as livestock feed. Saffron is a good alternative for opium poppies; grows well under similar climate and soil conditions, requires much hand labor, and brings as high a price on world markets as opium. Harvest period does not coincide with rice harvest as opium tapping does (CFR, LIL).

Chemistry (Saffron) — Here are a few of the more notable chemicals found in saffron. For a complete listing of the phytochemicals and their activities, see the CRC phytochemical compendium, Duke and duCellier, 1993 (DAD) and the USDA database http://www.ars-grin.gov/duke/.

Crocetin — Anticancer (Skin) 3 µm; Antihypoxic; Antimutagenic; Antioxidant; Antitumor-Promoter 3 µm; Choleretic 100 mg/kg; Colorant; Hypocholesterolemic; Lipolytic; Neuroprotective.

Crocin — Choleretic 100 mg/kg; Colorant; Dye; Neuroprotective.

Kaempferol — See also *Alpinia officinarum.*

Cunila origanoides (L.) Britton (Lamiaceae)
AMERICAN DITTANY, DITTANY, FROST FLOWER, FROST MINT, MARYLAND DITTANY, MOUNTAIN DITTANY, STONE MINT

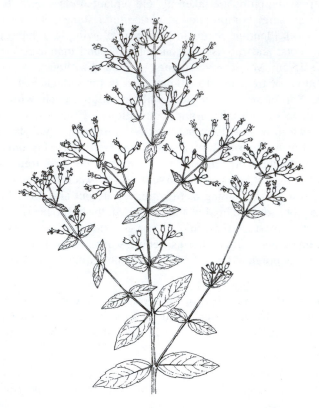

Synonyms — *Cunila mariana* L., *Satureja origanoides* L.

Medicinal Uses (Frost Mint) — Reportedly, dittany is called "feverwort" in Arkansas and is used for headache and snakebite. It is also used dry or fresh for tea. Wild dittany was reportedly used by Native Americans and early American settlers for colds and fever. I find the dittany attractive, summer or winter, and truly believe that a January tea of the aromatic dead stalks of *Cunila* and *Monarda* helped break up a bad, lingering cough. I suspect expectorant and antiseptic properties reside even in the dead stems persisting in the winter. The smell of thymol and carvacrol are still there, with all their medicinal virtues. I like the quote *Organic Gardening* attributes to famed pharmacognocist Norman Farnsworth in their first issue of the last decade of this millennium (January 1990, p. 54), "Thymol has been found to loosen phlegm in the respiratory tract.... It also has been shown to act as an antitussive which will relieve coughing." I think it will be just as promising for backache. If I had a backache and a lot of dittany, I'd drink dittany tea and maybe even add a little to my bath water. The oil is said to be a stimulant aromatic. Because of its high content of thymol, it is probably a good antiseptic as well. But don't overdo the thymol, it can irritate mucous membranes. Even GRAS herbs should be used in moderation. It seems that thymol and carvacrol often run in tandem. I suspect within a species, if one is high, the other is compensatorily low. Thymol is also said to be antispasmodic.

Indications (Frost Mint) — Acne (1; FNF); Alzheimer's (1; FNF); Arthrosis (1; FNF); Atherosclerosis (1; FNF); Backache (1; FNF); Bacteria (1; FNF); Bronchosis (1; FNF); Candida (1; FNF); Caries (1; FNF); Childbirth (f; BOW; DEM); Cold (1; FAD; FNF); Congestion (1; FNF);

Cough (1; FNF); Cramp (1; FNF); Depression (1; FNF); Dermatosis (1; FNF); Fever (f; BOW; DEM; FAD; HHB); Flu (1; FNF); Fungus (1; FNF); Headache (1; BOW; DEM; FAD; FNF); Halitosis (1; FNF); Herpes (1; FNF); Infection (1; FNF); Inflammation (1; FNF); Melancholy (1; FNF); Mycosis (1; FNF); Neurosis (1; FNF); Pain (1; FNF); Periodontosis (1; FNF); Plaque (1; FNF); Rheumatism (1; FNF); Snakebite (f; FAD; HHB); Staphylococcus (1; FNF); Strepto-coccus (1; FNF); Trichinosis (1; FNF); Trichomonas (1; FNF); UTI (1; FNF); Virus (1; FNF); Worm (1; FNF); Yeast (1; FNF).

Frost Mint for backache:

- Analgesic: myrcene; p-cymene
- Anesthetic: 1,8-cineole; carvacrol; myrcene; thymol
- Antiinflammatory: alpha-pinene; carvacrol; caryophyllene; thymol
- Antirheumatalgic: p-cymene
- Antirheumatic: thymol
- Antispasmodic: carvacrol; caryophyllene; limonene; myrcene; thymol
- Counterirritant: 1,8-cineole; thymol
- Myorelaxant: thymol
- Sedative: 1,8-cineole; alpha-pinene; limonene
- Tranquilizer: alpha-pinene

Frost Mint for cold:

- Analgesic: myrcene; p-cymene
- Anesthetic: 1,8-cineole; carvacrol; myrcene; thymol
- Antiallergic: 1,8-cineole
- Antibacterial: 1,8-cineole; alpha-pinene; carvacrol; caryophyllene; limonene; myrcene; p-cymene; thymol
- Antibronchitic: 1,8-cineole; thymol
- Antiflu: alpha-pinene; limonene; p-cymene
- Antiinflammatory: alpha-pinene; beta-pinene; carvacrol; caryophyllene; thymol
- Antioxidant: carvacrol; gamma-terpinene; myrcene; thymol
- Antipharyngitic: 1,8-cineole
- Antiseptic: 1,8-cineole; beta-pinene; carvacrol; limonene; thymol
- Antitussive: 1,8-cineole; carvacrol; thymol
- Antiviral: alpha-pinene; limonene; p-cymene
- Cyclooxygenase-Inhibitor: carvacrol; thymol
- Expectorant: 1,8-cineole; alpha-pinene; carvacrol; limonene; thymol

Other Uses (Frost Mint) — Dittany could be substituted for any of the other high-carvacrol/thymol plants (*Monarda*, *Origanum*, *Satureja*, *Thymus*), one for the other, as pizza herbs—at least in my kitchen. If I had pizza with cheese and tomato, and no spices, I'd add a little dittany in lieu of oregano. Grieve's Herbal speaks of "oil of dittany, which is stated to contain about 40 per cent of phenols, probably thymol." A former associate, Jeff Strachan, noted that when he brought potted plants into the greenhouse, most of the whiteflies (*Prialeuroydes vaporianum*) migrated to his dittany for a few days. Then, after populations built up on the dittany, the whiteflies returned to the other host plants, in even greater numbers. Some of the aromatic compounds in the dittany might lure whiteflies to some fatal trap. Dr. John Neil, USDA entomologist, looked into this at the USDA but was not able to confirm this trap concept. Too bad!

For more information on activities, dosages, and contraindications, see the *CRC Handbook of Medicinal Herbs, ed. 2*, Duke et al., 2002.

Cultivation (Frost Mint) — In early autumn, this old botanist's fancy turns to frost flowers. The first weekend in October, I head for what I call "frost flower fen," where there is an abundance of the plants, and I dig a new stash for the winter. Not for my pizza pies, but so I'll have flowers every month of the year. For almost a decade now, I have had flowers 12 months of the year, Thanksgiving, Christmas Day, New Year's Day, etc., at least when the temperatures got well below freezing the night before. I have grown the plant from seed and transplanted in spring and fall, and during thaws in midwinter. It seems to tolerate dry, well-drained slopes in semi-shady open hardwood forests in the Piedmont and low Mountains. Tucker and Debaggio (2000) note that it is hardy to zone 6, preferring light shade, moist but well-drained rich soil with leaf mold. Normally propagated by divisions or seeds in early spring.

Chemistry (Frost Mint) — The aromatic chemicals in frost flower share the essences of European oreganos, savory, and thyme, like our American dittany and horsebalm (*Monarda*), all good antispasmodic herbs, loaded with carvacrol and thymol. Here are a few of the more notable chemicals found in frost mint. For a complete listing of the phytochemicals and their activities, see the CRC phytochemical compendium, Duke and duCellier, 1993 (DAD) and the USDA database http://www.ars-grin.gov/duke/.

Carvacrol — Anesthetic; Anthelminthic; Antiatherosclerotic IC50 = 5.53 μM; Antibacterial MIC = 39–625, MIC 170–290; Anticholinesterase?; Antidiuretic; Antiinflammatory IC50 = 4 μM; Antimelanomic IC50 = 120 μM/l; Antioxidant; Antioxidant (LDL) IC50 = 5.53 μM; Antiplaque MIC = 39–625 μg/ml; Antiprostaglandin; Antiradicular (600 × thymol); Antiseptic (1.5 × phenol); Antispasmodic; Antistaphylococcic; Antistreptococcic; Antitussive; Candidicide MIC <0.1 μg/ml, 100–150 μg/ml; Carminative; Cyclooxygenase-Inhibitor (= indomethacin); Enterorelaxant; Expectorant; Fungicide; Insectifuge; Irritant; Nematicide MLC = 1 mg/ml; Tracheorelaxant; Trichomonicide LD100 = 150 μg/ml; Vermifuge; LD50 = 810 orl rat; LDlo = 100 orl rbt.

P-Cymene — Analgesic; Antiacetylcholinesterase IC40 = 1.2 mM; Antibacterial; Antiflu; Antirheumatalgic; Antiviral; Fungicide; Herbicide IC50 = 50 μM; Insectifuge; Irritant; Laxative; Sedative; Trichomonicide LD100 = 50 μg/ml; LD50 = 4750 mg/kg orl rat.

Thymol — Analgesic; Anesthetic; Ankylostomacide; Antiacne; Antiaggregant IC50 = 0.75; Antiarthritic; Antiatherosclerotic IC50 = 4 μM; Antibacterial MIC = 50–400 μg/ml; Antibronchitic; Anticariogenicic MIC = 200–400 μg/ml; Anticholinesterase; Antihalitosic; Antiherpetic; Antiinflammatory (= indomethacin); Antilepric; Antimelanomic IC50 = 120 μM/l; Antineuritic; Antioxidant 100 ppm; Antioxidant (LDL) IC50 = 4 μM; Antiperiodontic MIC = 50–200 μg/ml; Antiplaque MIC = 39–625 μg/ml; Antiradicular EC = 60; Antirheumatic; Antisalmonellic; Antiseptic 20 × phenol; Antispasmodic; Antistaphylococcic; Antistreptococcic MIC = 200–400 μg/ml; Ataxigenic; Antispasmodic; Antitrichinosic 5%; Antitussive; Candidicide 100 μg/ml (>Nystatin); Carminative; Counterirritant; Cyclooxygenase-Inhibitor (= indomethacin); Dentifrice; Deodorant; Dermatitigenic; Enterorelaxant; Enterotoxic; Expectorant; Fungicide; Gastroirritant; Gram(+)-icide MIC = 200–400 μg/ml; Gram(–)-icide MIC = 50–200 μg/ml; Insectifuge; Irritant; Larvicide 10 ppm, 58 ppm; Molluscicide; Myorelaxant; Nematicide MLC = 0.1 mg/ml; Sedative; Sprout-Inhibitor; Tracheorelaxant; Trichomonicide LD100 = 25 μg/ml; Urinary-Antiseptic; Vermicide; Vermifuge; LD50 = 980 mg/kg orl rat; LD50 = 1800 (orl mus).

Curcuma longa L. (Zingiberaceae)
INDIAN SAFFRON, TURMERIC

Synonym — *Curcuma domestica* Valeton

Medicinal Uses (Turmeric) — Regarded as carminative, choleretic, digestive, stomachic, a cure for liver troubles, taken internally for treating ulcers, or externally as an ointment to heal skin sores. Boiled with milk and sugar, it is used as a cold remedy. Reflecting the doctrine of signatures, orientals use the rhizome for jaundice. Mowrey (1988) says turmeric's hepatoprotective activity compares well with that of milk thistle and licorice. Chinese consider the root alterative, antiparasitic, antiperiodic, cholagogue, depurative, hemostat, stomachic, and tonic, and they apply it externally for inflammations, indolent ulcers, and purulent ophthalmia. They use turmeric for abdominal pain, bruises, chest pains, colic, coma, dysmenorrhea, epistaxis, fever, gas, hematuria, hematemesis, hepatitis, sores, and toothache, and poultice in onto itch, ringworm, and sores. They also employ turmeric for arsenic poisoning, hemoptysis, mania, post-partum hemorrhage, and primary syphilis. Ayurvedics use turmeric as alterative, antiperiodic, depurative, stomachic, and tonic, perhaps reflecting the fact that the EO is antiseptic, antacid, aperitif, stomachic, and tonic in small doses, acting as antispasmodic in larger doses. Ayurvedics also prescribe for boils, biliousness, bruises, dyspepsia, dysuria, elephantiasis, inflammations, leucoderma, scabies, smallpox, snakebite, and swellings. In Madagascar, the rhizome is used as an aperient, astringent, carminative, cordial, detergent, diuretic, emmenagogue, maturant, stimulant, and tonic. Yunani prescribe for affections of the liver and jaundice, urinary discharges, scabies, and bruises. Indians even apply the root to leech bites. Simonds (1999) says that, as a popular Indian culinary curry ingredient and cosmetic, turmeric increases the mucus content in gastric juices and is prescribed for stomach disorders. Indian women mix ground turmeric with water to make a paste that they rub on to clear blemishes (Simonds, 1999). Fumes of burning turmeric directed into the nostrils cause a copious mucous discharge and relieve the congestion. Turmeric is given in diarrhea, so difficult to correct in atonic subjects. Turmeric is used in dropsy and malaria. The root, parched, and powdered, is given in bronchitis; the fumes are used during hysteric fits. A paste made of fresh rhizome is applied on the head in cases of vertigo. Turmeric and alum is blown into the ear in chronic otorrhea. The flowers, pasted, are used in ringworm, other parasitic skin diseases, and gonorrhea. Cambodians consider the leaves antipyretic. Powdered turmeric is antioxidant (BGB, BIB, DAD).

I think of turmeric first as a medicine, queen of the natural COX-2-Inhibitors, and secondly as a spice. "Turmeric is a safer, more natural, and less expensive COX-Inhibitor than pharmaceutical COX-Inhibitor drugs" (BGB, quoting me). Like the so-called miracle aspirin of 1999 (Celebrex and Vioxx) and the centenarian aspirin, turmeric relieves pain gently, as its curcumin is a trimillennial COX-2

Inhibitor. Joseph et al. (2002) note that there's a flip side to the poor availability of turmeric in the stomach. The upside is that a lot of curcumin remains in the GI tract, where it appears to fight colon polyps and possibly colon cancer (JNU). Unlike Celebrex and Vioxx, which inhibit activities of COX-2-enzymes, curcumin prevents their production in the first place (Dr. Andrew Danneberg as quoted in JNU, 2002). Christian Jobin, professor of medicine at UNC, is reportedly checking out curcumin against other abdominal diseases, like colitis and IBD (JNU).

The EO of turmeric has greater antiinflammatory effects than hydrocortisone in test tube arthritis and edema (MPI). Some of the antiinflammatory fractions can reduce histamine significantly, at least in rat epidermis. Mills and Bone (2000) give a brilliant summary, citing *in vivo, in vitro,* and some clinical evidence regarding turmeric's antihepatotoxic, anticancer, antiinflammatory, antioxidant, antiseptic, antitussive, antiviral, cardiovascular, digestive, and hypolipidemic activities (MAB). Turmeric increased glutathione-S-transferase activity >78% in esophagus, liver and stomach, enough to be considered chemopreventive (Khan and Balick, 2001). In a study of clinical applications of ayurvedic herbs, Khan and Balick (2001) note human studies on turmeric for cancer, gastric ulcers, osteoarthritis, and scabies. Turmeric decreases serum lipid peroxides and urinary mutagens (Khan and Balick, 2001). As a COX-2-Inhibitor, curcumin is such a good antirheumatic that whole books have been written about it (Majeed et al., 1995, Newmark and Schulick, in ed.). Controlled, double-blind research confirms that curcumin taken internally helps arthritis, postoperative inflammation, and edema (swelling or fluid retention). Rheumatoid arthritis patients were given 1200 mg curcumin per day in a double-blind study. All 49 reported improvement in morning stiffness, pain, and physical endurance (Deohdhar et al., 1980). In a cross-over study of 45 osteoarthritis patients, a combination therapy of curcumin, ashwagandha, boswellia, and zinc produced significant reduction in pain and disability. Forty-five surgical patients between the ages of 15 and 68 were given 400 mg curcumin, 100 mg phenylbutazone, or lactose placebo 3 times daily for 5 days following their operations. Curcumin significantly reduced edema, inflammation, and pain with no side effects (Kulkarni et al., 1991). Curcumin's ability to reduce edema has been tested repeatedly on rats and mice. Chemicals that cause temporary edema are injected into the surface of their foot pads. Some of the rodents are then treated with curcumin, and the swelling diminishes faster than in those not treated (Duke, 1997, Majeed et al., 1995).

The COX-2-Inhibitory activity of curcumin has led some to study turmeric for Alzheimer's. *Science News* picked up on this, too (Travis, 2001), noting that India has one of the lowest Alzheimer's rates in the world (and one of the highest intakes of curry and curcumin, where turmeric is native). In mice, dietary curcumin reduced inflammation and free radical damage in the mice brains (Travis, 2001). Lim et al. (2001) note that inflammation in Alzheimer's patients is characterized by increased cytokines and activated microglia. Epidemiology suggests reduced AD risk with long-term use of NSAIDs. Curcumin is a potent polyphenolic antioxidant. Lim et al. tested low (160 ppm) and high (5000 ppm) doses curcumin for inflammation, oxidative damage, and plaque pathology in rats. Both significantly lowered oxidized proteins and interleukin-1beta, a proinflammatory cytokine elevated in the mice brains. With low but not high dose, insoluble beta-amyloid (Abeta), soluble Abeta, and plaque burden were significantly decreased (43–50%). However, levels of amyloid precursor (APP) in the membrane fraction were not reduced. Microgliosis was also suppressed in neuronal layers, but not adjacent to plaques. In view of efficacy and apparent low toxicity, turmeric shows promise for preventing Alzheimer's (X11606625).

Curcumin may be every bit as promising as the synthetic COX-2-Inhibitors, alleviating arthritis, gout, inflammation, and rheumatism, and possibly preventing Alzheimer's and cancer. Boik (2001) certainly praises curcumin, if not the mother turmeric, as a cancer preventive. Nearly 20 *in vitro* studies show that curcumin inhibits proliferation of several cancer cell lines, at levels of some 5.2–27 μM. Some found it prevented angiogenesis and decreased invasiveness. Boik reports three animal studies of curcumin, wherein it prolonged life span of cancerous rodents, inhibited tumor growth, or inhibited metastasis. Curcumin is a stronger antioxidant than vitamin E at preventing lipid peroxidation *in vitro*. We have one phytochemical with more than two dozen different activities that could reduce the incidence of cancer, and we have at least half a dozen other anticancer chemicals doing many of the same things in slightly different levels. I presume that is what emboldens Boik to predict synergies among these food farmaceuticals. Boldly, he calculates some tentative human dosages as 0.36–3.2 g/day (as scaled from animal antitumor studies), 0.23–3.2 g/day (as scaled from animal antiinflammatory studies, leading to a target dose of 360–8700 mg/day curcumin). Then even more speculatively, he suggests that synergies with other phytochemicals may reduce that minimum antitumor dose to 24–580 mg curcumin/day, paralleling his minimum anticancer dosage of 100 mg for genistein (e.g., in beans) and apigenin (e.g., in celery), 170 mg for luteolin, and 250 mg quercetin through similar calculations. If you find this paragraph rough sledding, may I refer you to Boik's very interesting book (BO2).

In turmeric, we note another case of opposing medicinal compounds, curcumin increasing bile secretion, bis-desmethoxycurcumin decreasing it. Can the homeostatic body select the activity it needs? I think so. Rinzler (1990) adds notes that turmeric is choleretic, but many herbs, if not all, contain choleretics. Choleretics stimulate the production of bile, the bitter substance that emulsifies fats in the duodenum, stimulates peristalsis, thereby encouraging movement of food along the GI tract. Turmeric is also a cholagogue, which stimulates the bile duct and gallbladder to discharge bile, in the process aiding the excretion of cholesterol. While useful in moderation in healthy people, choleretics and cholagogues may be contraindicated in patients with gallbladder or liver diseases. In *Medicinal Plants of India* (MPI), curcumin is noted to stimulate the musculature of the gall bladder, unlike other bile stimulants in use. Curcumin seems to combine the choleretic and hydro-cholagogic activities with antisepsis, perhaps making it ideal for biliary and gallbladder problems suspected to have been caused by *Staphylococcus* (MPI). Further, sodium curcuminate is an active choleretic, inducing nearly 100% increase in bile production of anesthetized dogs. Sodium cur-cuminate acts as a hydrocholagogue. Increased bile salt excretions seems to speak for curcumin in digestive disorders. EO and some of its distillates are also choleretic but less so.

Asai and Miyazawa (2001) hint that dietary curcuminoids might prevent accumulation of lipids in the liver and epididymal adipose tissues. Reporting the antioxidative, anticarcinogenic, and hypocholesterolemic activities of curcumin and other curcuminoids, they also note that dietary curcuminoids have lipid-lowering potency *in vivo*, probably due to alterations in fatty acid metabolism.

Negi et al. (1999) note that after curcumin is produced industrially from turmeric oleoresin, the mother liquor contains ca. 40% oil. Fractions from the oil were tested for antibacterial activity against *Bacillus cereus, Bacillus coagulans, Bacillus subtilis, Escherichia coli, Pseudomonas aeruginosa*, and *Staphylococcus aureus*. Fraction II (5% ethyl acetate in hexane) was most active. ar-Turmerone, turmerone, and curlone were major constituents of such fractions (X10552805). Alcoholic extracts of the rhizome were active against *Entameba histolytica*. Sharma et al. (2000) showed that, though bactericidal on their own, extracts of black pepper, capsicum, and turmeric partially protected *Bacillus megaterium, Bacillus pumilus,* and *Escherichia* from radiation, prob-ably protecting their DNA. Chile was strongest. Curcumin and piperine were also concluded to be radioprotective.

In my Peruvian pharmacy ecotours, I have long maintained that I think as much of the inexpensive Peruvian antiAIDS cocktail (turmeric, cat's claw, jergon sacha) as I do of the $18,000 antiAIDS cocktail back in the States. I doubt that either will eradicate the virus. But I'll wager the naturals will have fewer side effects, especially food farmaceuticals like turmeric. Certainly turmeric has anti-HIV activity. Infection with HIV involves a complicated command system which results in activation or inactivation. One critical part of that system is the long terminal repeat (LTR). Drugs that interfere with LTP can delay infection and slow progression. Curcumin inhibits activation of LTR and decreases HIV replication (Majeed et al., 1995). In AIDS and *Complementary & Alternative Medicine*, Standish et al. (2002) note that, *in vitro*, turmeric inhibits HIV integrase and HIV-1 LTR-directed transcription at fractional micromolar concentrations. Turmeric modestly inhibits HIV-1 and -2 proteases. It also inhibits nuclear factor kappa B, which regulates viral transcription. Although not tested by the NCI, curcumin is popular among HIV-positive CAM users. In one positive study reported by Standish et al., 18 HIV patients with CD4 T-cell counts 5 to 615 found significant increases in CD4 T-cell counts with curcumin, cf controls. Less positively, curcumin at neither 2700 mg/day nor 4800 mg/day had any effect on CD4 T-cell count or viral load. I think curried hyssop tea, sweetened with licorice, and with floating hypericum flowers, might be an even better, though perhaps equally futile, food farmacy approach. See also licorice (Standish et al., 2002).

Cheng et al. (2001) demonstrated that curcumin is not toxic in oral human doses up to 8000 mg/day for 3 months. Their results also suggest a biologic effect of curcumin in the chemoprevention of cancer, especially bladder cancer, Bowen's disease, cervical cancer, leukoplakia, stomach cancer, and uterine cancer (uterine cervical intraepithelial neoplasm). Urinary excretion of curcumin was not detectable. Serum concentrations of curcumin usually peaked 1–2 hr after intake, gradually declining within 12 hr. The average peak serum concentrations after taking 4000 mg, 6000 mg, and 8000 mg of curcumin were ca. 0.5 μM, 0.6 μM, and 1.8 μM respectively (Cheng et al., 2001).

Indications (Turmeric) — Abscess (1; FNF; TRA); Adenoma (1; X7954412); Adenosis (1; DAD; JLH; X7954412);); Allergy (1; WAM); Alzheimer's (1; COX; FNF); Amenorrhea (1; BGB; PH2; WHO); Anorexia (2; BGB; BRU; PHR; PH2); Arthrosis (1; KAP; MAB; WAM; WHO); Asthma (1; MAB; WHO); Atherosclerosis (1; MAB; SKY); Athlete's Foot (1; FNF); Bacillus (1; X10552805); Bacteria (1; FNF; X10552805); Bite (f; BIB; PH2); Bleeding (f; PED; PH2); Boil (1; DAD; WHO); Bowen's Disease (1; X11712783); Bronchosis (f; BIB; PH2); Bruise (f; DAV; PED; PH2; WHO); Bursitis (1; SKY); Cancer (1; FNF; MAB); Cancer, abdomen (1; COX; FNF; JLH); Cancer, bladder (1; X11712783); Cancer, breast (1; COX; FNF; MAB); Cancer, cervix (1; X11712783); Cancer, colon (1; COX; FNF; JLH; JNU); Cancer, duodenum (1; X7954412); Cancer, esophagus (1; JAC7:405); Cancer, joint (1; JLH; MAB); Cancer, liver (1; JAC7:405); Cancer, mouth (1; COX; FNF; JLH); Cancer, nose (1; COX; FNF; JLH); Cancer, sinew (1; COX; FNF; JLH); Cancer, skin (1; X7954412); Cancer, stomach (1; JAC7:405); Cancer, uterus (1; X11712783); Cardiopathy (1; AKT; MAB); Cataract (1; MAB); Catarrh (f; UPW); Chest Ache (f; PH2); Child-birth (f; DAD); Cholecystosis (2; APA; PHR); Circulosis (f; BOW); Cold (f; KAP; PH2); Colic (f; APA; PED; PH2); Coma (f; DAD); Congestion (f; APA; BIB); Conjunctivosis (f; KAB; MAB; PH2; SUW); Constipation (f; PH2); Coryza (f; KAB); Cramp (1; AKT; BIB; DAD); Cystosis (f; PH2); Dermatosis (1; AKT; MAB; PH2; SUW; WHO; WOI); Diabetes (f; BOW); Diarrhea (1; APA; WHO); Dropsy (f; DAD); Duodenosis (1; X7954412); Dusgeusia (f; KAB); Dysmenorrhea (1; AKT; APA; PED; WHO); Dyspepsia (2; KOM; MAB; PH2; WHO); Dysuria (f; DAD); Eczema (1; BGB; KAP; MAB); Edema (1; KAP; PH2); Elephantiasis (f; DAD); Enterosis (1; AKT; DAD; PH2; WHO); Epilepsy (f; WHO); Epistaxis (f; DAD; PH2); Esophagosis (1; JAC7:405); Fever (1; APA; BIB; COX); Fibrosis (1; BGB; MAB); Fungus (1; BIB; FNF; PH2); Gallstone (1; APA; MAB); Gas (1; APA; PH2); Gastrosis (f; PH2); Gonorrhea (f; BIB; KAB); Gray Hair (f; HAD); Fungus (1; FNF; LIB); Headache (f; PH2); Hematemesis (f; DAD; PH2); Hematuria (f; DAD); Hemorrhoid (f; MAB); Hepatosis (2; FNF; MAB; PED; PHR; PH2; TRA); High Blood Pressure (1; KAP); High Cholesterol (1; AKT; APA; MAB; TRA); High Triglycerides (1; MAB; TRA); Hyperlipidemia (1; MAB); Hysteria (f; DAD); IBS (1; PED); Immunodepression (1; FNF); Infection (2; FNF; MAB; MPI; PH2); Inflammation (1; FNF; PHR; PH2; WAM; WHO); Itch (f; APA; KAP; PH2); Jaundice (1; MAB; TRA); Laryngosis (1; BIB; COX); Leprosy (f; PH2); Leukemia (1; AKT); Leishmannia (1; X10865470); Leukoderma (f; DAD); Leukoplakia (1; X11712783); Lymphoma (1; BIB; COX; FNF); Malaria (f; KAP; PH2); Mania (f; DAD); Morning Sickness (1; MAB); Mucososis (f; PH2); Mycosis (1; FNF; PH2; X8824742); Nematode (1; X8221978); Nephrosis (1; AKT; PH2); Nervousness (1; FNF); Ophthalmia (1; AKT; DAD; PH2); Osteoarthrosis (1; MAB); Ozena (f; KAB); Pain (1; BIB; COX; FNF; WHO); Parasite (f; BIB; DAD; KAP; LIB); Polyp (1; COX; JLH; JNU); Psoriasis (1; FNF; MAB); Puerperium (f; MAB); Radiation (1; AKT); Restenosis (1; MAB); Rheumatism (1; BIB; COX; SKY); Rhinosis (1; COX; JLH); Ringworm (f; APA; BIB;

KAP; PH2); Scabies (2; BGB); Smallpox (f; DAD); Sore (f; PH2); Sore Throat (f; PH2); Sprain (1; MAB; SUW); Staphylococcus (1; MPI; UPW); Stone (1; HHB; MAB); Stroke (f; BOW; PH2); Swelling (1; AKT; COX; PH2); Syphilis (f; DAD); Trauma (f; AKT); Ulcer (1; BIB; COX; FNF; PED; WHO); Uveosis (2; AKT); VD (f; BIB; DAD); Vertigo (f; BIB; DAD); Vomiting (f; PH2); Wart (f; JLH); Water Retention (1; FNF); Whitlow (f; JLH); Worm (1; X8221978); Wound (1; APA; BGB; PH2; SUW; WAM); Yeast (1; PED).

C

Turmeric for cancer:

- AntiEBV: curcumin
- AntiHIV: caffeic-acid; curcumin
- AntiX-Radiation: curdione
- Antiadenomacarcinogenic: curcumin
- Antiaflatoxin: curcumin
- Antiaggregant: caffeic-acid; curcumin; eugenol; salicylates
- Antiangiogenic: bis-desmethoxycurcumin; curcumin; demethoxycurcumin
- Antiarachidonate: curcumin; eugenol
- Anticancer: alpha-pinene; alpha-terpineol; caffeic-acid; camphor; cinnamic-acid; curcumenol; curcumin; curcuminoids; eugenol; limonene; linalool; p-coumaric-acid; terpineol; vanillic-acid
- Anticarcinogenic: caffeic-acid
- Antihepatotoxic: caffeic-acid; p-coumaric-acid; protocatechuic-acid
- Antiinflammatory: alpha-curcumene; alpha-pinene; ar-turmerone; azulene; beta-pinene; bis-(4-hydroxy-cinnamoyl)-methane; bis-desmethoxycurcumin; borneol; caffeic-acid; caryophyllene; cinnamic-acid; curcumin; curcuminoids; demethoxycurcumin; eugenol; feruloyl-4-hydroxycinnamoyl-methane; germacrone; protocatechuic-acid; salicylates; tetrahydrocurcumin; triethylcurcumin; vanillic-acid
- Antileukemic: 2-hydroxy-methyl-anthraquinone; curcumin
- Antileukotriene: caffeic-acid; curcumin; curcuminoids
- Antilymphomic: curcumin
- Antimelanomic: curcumin
- Antimetastatic: curcumin
- Antimutagenic: caffeic-acid; cinnamic-acid; curcumin; eugenol; limonene; protocatechuic-acid
- Antinitrosaminic: caffeic-acid; curcumin; p-coumaric-acid
- Antioxidant: caffeic-acid; camphene; curcumin; eugenol; gamma-terpinene; p-coumaric-acid; protocatechuic-acid; tetrahydrocurcumin; turmerin; turmeronol-a; turmeronol-b; vanillic-acid
- Antiperoxidant: caffeic-acid; curcumin; p-coumaric-acid; protocatechuic-acid
- Antiproliferant: terpineol
- Antiprostaglandin: caffeic-acid; curcumin; curcuminoids; eugenol
- Antisarcomic: curcumol; curdione
- Antistress: germacrone
- Antithromboxane: curcumin; eugenol
- Antitumor: alpha-curcumene; ar-turmerone; caffeic-acid; caryophyllene; curcumin; curdione; eugenol; limonene; p-coumaric-acid; vanillic-acid
- Antiviral: alpha-pinene; beta-bisabolene; caffeic-acid; curcumin; limonene; linalool; p-cymene; protocatechuic-acid
- Apoptotic: curcumin
- COX-2-Inhibitor: curcumin; eugenol; ar-turmerone; beta-turmerone
- Cytochrome-p450-Inducer: 1,8-cineole

- Cytoprotective: caffeic-acid
- Cytotoxic: 2-hydroxy-methyl-anthraquinone; caffeic-acid; curcumin; curcuminoids; di-p-coumaroyl-methane; diferuloyl-methane; eugenol; feruloyl-p-coumaroyl-methane; p-coumaric-acid
- Fibrinolytic: curcumin
- Hepatoprotective: borneol; caffeic-acid; curcumin; di-p-coumaroyl-methane; eugenol; p-coumaroyl-feruloyl-methane
- Hepatotonic: 1,8-cineole; turmerone
- Immunostimulant: caffeic-acid; curcumin; protocatechuic-acid; ukonan-a
- Lipoxygenase-Inhibitor: caffeic-acid; cinnamic-acid; p-coumaric-acid
- Ornithine-Decarboxylase-Inhibitor: caffeic-acid; curcumin; limonene
- Prostaglandigenic: caffeic-acid; p-coumaric-acid; protocatechuic-acid
- Protease-Inhibitor: curcumin
- Protein-Kinase-Inhibitor: curcumin
- Sunscreen: caffeic-acid

Turmeric for dyspepsia:

- Analgesic: borneol; caffeic-acid; camphor; eugenol; germacrone; p-cymene
- Anesthetic: 1,8-cineole; camphor; cinnamic-acid; eugenol; guaiacol; linalool
- Antiemetic: camphor
- Antiinflammatory: alpha-curcumene; alpha-pinene; ar-turmerone; azulene; beta-pinene; bis-(4-hydroxy-cinnamoyl)-methane; bis-desmethoxycurcumin; borneol; caffeic-acid; caryophyllene; cinnamic-acid; curcumin; curcuminoids; demethoxycurcumin; eugenol; feruloyl-4-hydroxycinnamoyl-methane; germacrone; protocatechuic-acid; salicylates; tetrahydrocurcumin; triethylcurcumin; vanillic-acid
- Antioxidant: caffeic-acid; camphene; curcumin; eugenol; gamma-terpinene; p-coumaric-acid; protocatechuic-acid; tetrahydrocurcumin; turmerin; turmeronol-a; turmeronol-b; vanillic-acid
- Antistress: germacrone
- Antiulcer: azulene; beta-bisabolene; beta-sesquiphellandrene; caffeic-acid; curcumin; eugenol; germacrone; zingiberene
- Carminative: camphor; eugenol; zingiberene
- Secretagogue: 1,8-cineole; protocatechuic-acid
- Sedative: 1,8-cineole; alpha-pinene; alpha-terpineol; borneol; caffeic-acid; caryophyllene; eugenol; isoborneol; limonene; linalool; p-cymene
- Tranquilizer: alpha-pinene

Turmeric for infection:

- Analgesic: borneol; caffeic-acid; camphor; eugenol; germacrone; p-cymene
- Anesthetic: 1,8-cineole; camphor; cinnamic-acid; eugenol; guaiacol; linalool
- Antibacterial: 1,8-cineole; alpha-pinene; alpha-terpineol; azulene; caffeic-acid; caryophyllene; cinnamic-acid; curcumin; eugenol; guaiacol; limonene; linalool; o-coumaric-acid; p-coumaric-acid; p-cymene; protocatechuic-acid; terpineol; vanillic-acid
- Antiedemic: caffeic-acid; caryophyllene; curcumin; eugenol; germacrone
- Antiinflammatory: alpha-curcumene; alpha-pinene; ar-turmerone; azulene; beta-pinene; bis-(4-hydroxy-cinnamoyl)-methane; bis-desmethoxycurcumin; borneol; caffeic-acid; caryophyllene; cinnamic-acid; curcumin; curcuminoids; demethoxycurcumin; eugenol; feruloyl-4-hydroxycinnamoyl-methane; germacrone; protocatechuic-acid; salicylates; tetrahydrocurcumin; triethylcurcumin; vanillic-acid

C

- Antiseptic: 1,8-cineole; alpha-terpineol; azulene; beta-pinene; caffeic-acid; camphor; eugenol; guaiacol; limonene; linalool; terpineol
- Antiviral: alpha-pinene; beta-bisabolene; caffeic-acid; curcumin; limonene; linalool; p-cymene; protocatechuic-acid
- COX-2-Inhibitor: curcumin; eugenol; ar-turmerone; beta-turmerone
- Cyclooxygenase-Inhibitor: curcumin
- Fungicide: 1,8-cineole; alpha-phellandrene; caffeic-acid; camphor; caprylic-acid; caryophyllene; cinnamic-acid; curcumin; eugenol; linalool; o-coumaric-acid; p-coumaric-acid; p-cymene; p-methoxy-cinnamic-acid; protocatechuic-acid; terpinolene
- Fungistat: limonene
- Immunostimulant: caffeic-acid; curcumin; protocatechuic-acid; ukonan-a
- Lipoxygenase-Inhibitor: caffeic-acid; cinnamic-acid; p-coumaric-acid

Other Uses (Turmeric) — Cheaper and stronger, and probably healthier gram for gram than the world's most expensive saffron, turmeric is sometimes called Indian Saffron. Bown (2001) properly advises that, because of strong flavor, one might not wish to substitute turmeric for annatto or saffron. It can be used in such things as piccalilli (somewhere between a chutney and a pickle), popular in the U.K., especially at Christmas when eaten with ham and cold turkey. The word piccalilli appears to date from the 18th century, probably a blend of pickle and chilli, though recipes now use ginger and mustard rather than chilli as the hot component. Piccalilli is bright yellow in color due to the high turmeric content (Bown, 2001). Dried rhizomes are used as spice, whole or ground, to flavor meat and egg dishes, and to flavor and/or color pickles, relishes, prepared mustard, butter and cheese; an indispensable constituent of curry powder. Indian chefs agree that uncooked turmeric is too earthy to serve. They often sizzle their powdered turmeric in hot oil (AAR). Mine almost tastes like ginseng, even when cooked. Moroccans combine saffron and turmeric in their "harira soup," consumed after completion of the Ramadan fast. Mixed with coconut oil, turmeric imparts a special flavor to seafoods and soups (AAR). In Java, young rhizomes and shoots are also eaten in lablab (Ochse, 1931). In India they are eaten raw. Leaves, wrapped around fish, impart their own flavor (FAC). In West Sumatra, leaves are essential in "Rendang," a traditional buffalo dish (FAC). Turmeric provides a natural dye to color cloth, leather, silk, palm fiber, wool, and cotton. Sudanese use it as a cosmetic, smearing a turmeric ointment around their eyes (UPW). Anecdotally, I have learned of one dowager who accidentally dyed her gray hair blonde while killing lice with a neem/turmeric formula I had recited from Mills and Bone (2000). A paste with turmeric and neem cured 97% of 814 scabies patients within 3–15 days of treatment. "No toxic or adverse reactions were observed" (MAB). As a chemical indicator turmeric changes color in alkaline and acid substrates. Turmeric paper, prepared by soaking unglazed white paper in the tincture and then dried, is used as a test for alkaloids and boric acid. Turmeric rhizomes yield 2–6% orange-yellow EO (curcumin, upon oxidation becomes vanillin), used in flavoring spice products and in perfumery (BIB, DAD).

Here are some food farmacy formulae that may be of interest:

Gobo Gumbo — Burdock, heavy with turmeric as possible, heavily spiced with black pepper for its piperine (can increase absorption of curcumin up to twentyfold), plus a dash of grapefruit juice. Arthritis sufferers might wish to consider adding some or all of the following antiarthritic foods and spices: black cumin; cayenne (the hotter the pepper, the less the pain; rich also in salicylates), evening primrose (crushed seed or bran flakes), fenugreek (distasteful to some), garlic, ginger, job's tear (Coix, a weed in the tropics), licorice (distasteful to some), nettleleaf (self-flagellating as well), onion (leave the skin on), oregano, purslane, rosemary, sage, savory, and/or thyme. Several other spices are also well endowed with COX-2-Inhibitors.

High Colonic Tea — In addition to getting my 25–35 g fiber a day, still firmly believing that fiber prevents colorectal cancer, I have fashioned a new tea embracing some of the better

COX-2-Inhibiting spices, with which I further improve the flavor of my COX-2-Inhibiting Green Tea. To the usual cup of tea, I add a small dash each of up to seven of the herbs high in COX-2-Inhibitors, chamomile, clove, holy basil, ginger, lavender, marjoram, oregano, rosemary, sage, and turmeric.

For more information on activities, dosages, and contraindications, see the *CRC Handbook of Medicinal Herbs*, ed. 2, Duke et al., 2002.

Cultivation (Turmeric) — Propagated vegetatively by fingers or rhizomes with one or two buds. Older planting stock roots are said to yield earlier. Turmeric, rarely grown in pure stands, is usually rotated with either ragi (*Eleusine coracana* Gaertn.), paddy (*Oryza sativa* L.), or sugar-cane (*Saccharum officinarum* L.), or may be mixed with castorbeans, corn, eggplants, french beans, ragi, sun hemp, or tomatoes. West Africa, turmeric is cultivated as a dye plant along with ginger in the forest (DAD). Soil should be plowed to 30 cm deep and liberally manured with potash and organic manures. Ochse especially recommend stable dung. Manure should be plowed into the soil three or four times. Ridges 22.5–25 cm high and 45–55 cm broad are made with furrows between the rows for irrigation. Sets or fingers of the previous crop are planted during April to August, 7.5 cm deep and 30–45 cm apart in rows, at rate of 560–725 kg/ha. Plants spaced 15 cm by 15 cm yield significantly more rhizomes than plants placed 30 cm by 30 cm or more. Periodic weeding or hoeing may be necessary. Purseglove et al. (1981) suggest 2,4-D as a preemergent herbicide, hinting that simazine might also be effective. Animal manures have doubled yields. In India, 25 MT farmyard manure/ha has been recommended with 50 kg each N, P_2O_5, and K_2O. Rhizomes should be ready to harvest in 9–10 months, when lower leaves turn yellow. They are carefully dug with picks (to avoid bruising), then scalded with hot water to gelatinize the starch and to prevent sprouting and molding, then sun dried. The dried rhizome is rubbed on a rough surface or trampled to remove the outer skin and give an attractive color and polish. Sprinkling water during polishing diffuses the color throughout the rhizome. In India, rain fed yields of 7–9 MT/ha are reported, 17–22 irrigated.

Chemistry (Turmeric) — *The Wealth of India* indicates 50 IU vitamin A/100 g. See Purseglove et al. (1981) for striking differences in the chemicals found in the EOs of *C. aromatica, C. domestica, C. xanthorrhiza,* and *C. zedoaria.* Here are a few of the more notable chemicals found in turmeric. For a complete listing of the phytochemicals and their activities, see the CRC phytochemical compendium, Duke and duCellier, 1993 (DAD) and the USDA database http://www.ars-grin.gov/duke/.

AR-Turmerone — Antihemorrhagic; Antiinflammatory; Antilymphocytic; AntiNKC; Antiophidic; Antiproliferative; Antitumor; Insectifuge.

Curcumin — See also *Costus speciosus.*

Curcumol — See also *Curcuma zedoaria.*

Curdione — See also *Curcuma zedoaria.*

Turmerin — Antimutagenic; Antioxidant; DNA-Protectant.

Turmerone — Choleretic; Hepatotonic; Insectifuge.

Zingiberene — See also *Curcuma zedoaria.*

Curcuma xanthorrhiza Roxb. (Zingiberaceae)
Temu Lawak

Very closely related to zedoary. Important in Indonesian Jamu.

Medicinal Uses (Temu Lawak) — In Singapore as "Ubat jamu" and "Ubat maaju," it is used for many things, including indigestion. In Perak, it is used in infusion for dyspepsia and rheumatism. It is used to bring on the period in amenorrhea and as a puerpereal tonic. It is also pasted over the body following childbirth. In southeast Asia, it has a deserved reputation as a choleretic and cholagogue (BRU). Because of its reputation for the liver in the Dutch Indies, it has gained a reputation in Holland for gallstones (IHB).

Yasni et al. (1994) identified the major component (approx. 65%) of the EO as alpha-curcumene. Rats fed EO or hexane-soluble fraction had lower hepatic fatty acid synthase activity. Alpha-curcumene is one of the active principles exerting triglyceride-lowering activity in temu lawak (X8157222). Mata et al. (2001) found xanthorrhizol modestly active against gram(+) and gram(−) bacteria, including *Bacillus subtilis*, *Enterococcus faecum*, *Klebsiella pneumoniae*, and *Staphylococcus aureus*, including methicillin-resistant *Staphylococcus* and vancomycin-resistant *Enterococcus*. It was scarcely active against *Candida albicans* (MIC = 128 µg/ml). At 128 and 64 µg/ml, it inhibited DNA and RNA synthesis and protein synthesis. Hwang et al. (2000) compared the antibacterial activity of xanthorrhizol with chlorhexidine against oral microorganisms in comparison. Xanthorrhizol has shown bactericidal activity against *Streptococcus*, making it potentially useful at preventing caries (X10844172). Campos et al. (2000) described the calcium-channel-blocking, endothelium-independent relaxation, and vasorelaxing activities of xanthorrhizol at doses of 1–100 µg/ml. Xanthorrhizol is assuming new importance as a COX-2-Inhibitor.

Patel and Srinivasan (1985) noted that dietary curcumin significantly increased lipase, maltase, and sucrase activities. Lin et al. (1995) found that extracts reduced serum transaminase levels induced by hepatotoxins. Extracts reduced liver damage 24 hr after ipr administration of hepatotoxins. The authors conclude that temu lawak could be useful in treating liver injuries and has promise as a broad spectrum hepatoprotective agent (X8571920).

Indications (Temu Lawak) — Alzheimer's (1; COX; FNF; JAD); Amenorrhea (f; IHB); Anorexia (2; PHR; PH2); Arthrosis (1; COX; FNF; JAD); Bacillus (1; JNP64:911); Bacteria (1; X10844172); Cancer (1; COX; FNF); Candida (1; JNP64:911); Cardiopathy (1; JNP64:911); Caries (1; X10844172); Childbirth (f; IHB); Cholecystosis (2; PHR; PH2); Dyspepsia (2; FNF; KOM; PH2); Enterococcus (1; JNP64:911); Enterosis (f; BRU); Escherichia (1; JNP64:911); Fullness (f; PH2); Fungus (1; FNF; JNP64:911); Gallstone (f; IHB); Gas (1; PH2); Gastrosis (1; BRU); Hepatosis (2; HHB; IHB; PHR; PH2); High Triglycerides (1; X8157222); Infection (1; X10844172); Inflammation (1; COX; FNF); Jaundice (f; HHB); Klebsiella (1; JNP64:911); Mycosis (1; JNP64:911); Pain (1; COX; FNF) Rheumatism (1; COX; FNF; IHB); Stone (f; IHB; HHB); Staphylococcus (1; JNP64:911); Streptococcus (1; PM66:196).

Temu Lawak for cardiopathy:

- Antiaggregant: curcumin
- Antiatherosclerotic: curcumin
- Antiedemic: curcumin
- Antiischemic: curcumin
- Antioxidant: curcumin
- COX-2-Inhibitor: curcumin; turmerone; xanthorrhizol
- Cyclooxygenase-Inhibitor: curcumin
- Hypocholesterolemic: curcumin
- Hypotensive: curcumin
- Sedative: borneol

Temu Lawak for hepatosis:

- AntiEBV: curcumin
- Antiedemic: curcumin
- Antihepatosis: curcumin
- Antiinflammatory: borneol; curcumin; curcuminoids
- Antileukotriene: curcumin; curcuminoids
- Antioxidant: curcumin
- Antiperoxidant: curcumin
- Antiprostaglandin: curcumin; curcuminoids
- Antiviral: ar-curcumene; curcumin
- COX-2-Inhibitor: curcumin; turmerone; xanthorrhizol
- Cholagogue: curcumin; curcuminoids
- Choleretic: curcumin; turmerone
- Cyclooxygenase-Inhibitor: curcumin
- Detoxicant: curcumin
- Hepatoprotective: borneol; curcumin
- Hepatotonic: turmerone
- Immunostimulant: curcumin

Temu Lawak for infection:

- Analgesic: borneol
- Antibacterial: curcumin
- Antiedemic: curcumin
- Antiinflammatory: borneol; curcumin; curcuminoids
- Antiviral: ar-curcumene; curcumin

- COX-2-Inhibitor: curcumin; turmerone; xanthorrhizol
- Cyclooxygenase-Inhibitor: curcumin
- Fungicide: alpha-phellandrene; curcumin
- Immunostimulant: curcumin

Other Uses (Temu Lawak) — Another spicy ginger relative, temu lawak is also a source of starch, used in making oriental porridges or puddings. Roots like those of turmeric and zedoary are yellow to orange, especially the older roots. The starch is rendered grating the rhizome and kneading the gratings in water above a sieve, repeating for several days. A beverage is made by boiling the rhizomes in water and sweetening. Hearts of the stem and tips of the rhizomes are eaten raw. Cooked inflorescences are served with rice (FAC). The plant is also used in local dyes (IHB).

For more information on activities, dosages, and contraindications, see the *CRC Handbook of Medicinal Herbs*, ed. 2, Duke et al., 2002.

Cultivation (Temu Lawak) — Botanically closely related to turmeric, temu lawak is cultivated in Indonesia, where it, like many members of the ginger family, is important in Jamu, the traditional Indonesian medicine. Probably treated and cultivated like other members of the ginger family. The rhizome is cut after harvest (BRU).

Chemistry (Temu Lawak) — Bruneton (1999) notes that the plant is rich in sesquiterpenes, like bisacumol, bisacurol, bisacurone, ar-curcumene, turmerones, (R)-(+)-xanthorrhizol, and zingiberene. Curcuminoids (1–2%) include curcumin and is a monodemethoxylated derivative as well as di-, hexa-, and octahydrogenated derivatives (BRU). Here are a few of the more notable chemicals found in temu lawak. For a complete listing of the phytochemicals and their activities, see the CRC phytochemical compendium, Duke and duCellier, 1993 (DAD) and the USDA database http://www.ars-grin.gov/dukc/.

Alpha-Curcumene — Antiinflammatory; Antitumor; Hypotriglyceridemic.

Ar-Curcumene — Antirhinoviral IC50 = 1750; Antiulcer IC45 = 100 mg/kg; Antiviral IC50 = 1750; Stomachic.

Curcumin — See also *Costus speciosus.*

Curcuminoids — See also *Costus speciosus.*

Turmerone — See also *Curcuma longa.*

Xanthorrhizol — Antibacterial; Antiseptic; Artemicide; Bacillicide MIC = 16 µg/ml; Calcium-Channel-Blocker; Candidicide MIC = 69 µg/ml; Cytotoxic EC50 = 4.9 µg/ml; Fungicide MIC = 69 µg/ml; Gram(+)-icide MIC = 16–32 µg/ml; Gram(–)-icide MIC = 16–32 µg/ml; Myorelaxant; Uterorelaxant; Vasorelaxant.

Zingiberene — Antirhinoviral; Antiulcer IC54 = 100 mg/kg; Carminative; Insecticide.

Curcuma zedoaria (Christm.) Roscoe (Zingiberaceae)
KUA, ZEDOARY

C

Synonym — *Amomum zedoaria* Christm.

Medicinal Uses (Zedoary) — Considered antipyretic, aphrodisiac, aromatic, carminative, demulcent, expectorant, stomachic, stimulant, and tonic. Fresh rhizomes have diuretic properties and are used in checking leucorrhea and gonorrheal discharges and for purifying the blood. Rhizomes are simply chewed to alleviate cough (DEP). A decoction of the rhizome administered along with long pepper, cinnamon, and honey is said to be beneficial for colds, fevers, bronchitis, and coughs. Mixed with black pepper, licorice, and sugar, it is used for bronchitis and cough (DEP). Rhizomes are an important ingredient of preserves given as a tonic to women after childbirth. Externally, the rhizome is applied as a paste mixed with alum, to sprains and bruises. Asian Indians apply the root to dermatitis, sprains, ulcer, and wounds (UPW). Juice of leaves is given for dropsy (DEP, WOI). Zedoary is a food, so there's little cause for concern with rational use. Women who experience a heavy menstrual flow should avoid taking large dosages. Extremely large amounts of curcuminoids, more than you are liable to ingest, might cause ulcers or cancer, or might reduce the number of red and white corpuscles in the blood (Duke, 2000).

Containing up to 0.1% curcumin, zedoary, like turmeric and ginger, may be viewed as a COX-2-Inhibitor, of potential use in alzheimer's, arthritis, and cancer. Yoshioka et al. (1998) demonstrated the analgesic, antiedemic, antiinflammatory, antioxidant, and radical-scavenging potency of dehydrocurdione, the major component of zedoary. Curcumol and curdione are regarded in the Orient as effective anticancer compounds, especially for cervical cancer and lymphosarcoma. Polysaccharide fractions decrease tumor sizes in mice and prevent chromosomal mutation. One fraction, at 6.25 mg/kg/day inhibited solid tumor growth 50% (Kim et al., 2000, X10987135). Used with success in Chinese studies of cervical cancer and in improving the efficacy of chemotherapy and radiation (Bown, 2001).

Considering CAM HIV treatments, Standish et al. (2002) discuss ACA (also called Perthon), a westernized version of an Asian combo of eleven herbs including some spices discussed in this book, cinnamon, cloves, myrrh, Sumatra benzoin, and zedoary. The combo received an investigational new drug (IND) status from the FDA in 1994 based on preliminary Australian-collected data. Fifteen HIV-positives treated for up to 2 years significantly improved re body weight and CD4+/CD8+ ratios (Standish et al., 2002).

Indications (Zedoary) — Adenosis (f; DAA; KAB); Alzheimer's (1; COX; FNF; JAD); Anorexia (f; KAB; PH2); Arthrosis (1; COX; FNF; JAD); Asthma (f; KAB; PH2); Bronchosis (f; DEP; KAB; KAP; PH2); Bruise (f; DEP; KAB; KAP; SUW); Cancer (1; DAA; FNF); Cancer, cervix (1; FNF; JAD; PH2); Cancer, colon (1; COX; FNF; JLH; PH2); Cancer, liver (1; COX; FNF; JLH; PH2); Cancer, uterus (1; DAA; FNF); Cardiopathy (1; X10973625); Cervicosis (1; BOW); Childbirth (f; DEP; KAB); Cold (1; DEP; FNF; SUW; WOI); Colic (f; HDR); Convulsion (f; KAB); Cough (f; KAP); Cramp (1; PH2); Debility (f; PH2); Dermatosis (f; DEP); Dropsy (f; KAB; UPW); Dyspepsia (f; BOW; PH2); Enterosis (f; PH2); Epilepsy (f; KAB); Fever (f; DEP; KAB); Furuncle (f; KAB); Gas (1; FNF; KAP; WOI); Gastrosis (f; PH2); Gonorrhea (f; KAB; KAP); Halitosis (f; KAB); Hematoma (f; DAA); Hemorrhoid (f; KAB); Hepatosis (f; JLH); Inflammation (1; COX; FNF; KAB); Jaundice (f; SKJ); Leukoderma (f; PH2); Leukorrhea (f; KAB; KAP); Lymphadenosis (f; KAB); Lymphagosis (f; KAP); Lymphosarcoma (1; DAA; FNF); Malaise (f; KAB); Nausea (f; BOW); Neurosis (f; PH2); Pain (1; COX; DEP; FNF; SUW); Rheumatism (1; COX; FNF); Sore Throat (f; DEP; KAP); Splenosis (f; KAB); Sprain (f; DEP; KAB; KAP); Toothache (f; KAB); Tuberculosis (f; PH2); Vertigo (f; KAB); Wound (f; KAB; PH2).

Zedoary for Alzheimer's:

- Antiacetylcholinesterase: 1,8-cineole
- Antiaggregant: curcumin
- Antiatherosclerotic: curcumin
- Anticholinesterase: 1,8-cineole
- Antiinflammatory: alpha-pinene; bis-desmethoxycurcumin; curcumin
- Antiischemic: curcumin
- Antileukotriene: curcumin
- Antioxidant: curcumin
- Antiperoxidant: curcumin
- Antiprostaglandin: curcumin
- CNS-Stimulant: 1,8-cineole
- COX-2-Inhibitor: curcumin; beta-turmerone; ar-turmerone
- Chelator: curcumin
- Cyclooxygenase-Inhibitor: curcumin
- Metal-Chelator: curcumin

Zedoary for rheumatism:

- 12-Lipoxygenase-Inhibitor: curcumin
- Anesthetic: 1,8-cineole
- Antiarthritic: curcumin
- Antiedemic: curcumin
- Antiinflammatory: alpha-pinene; bis-desmethoxycurcumin; curcumin
- Antiprostaglandin: curcumin
- Antispasmodic: 1,8-cineole; curcumin
- COX-2-Inhibitor: curcumin; beta-turmerone; ar-turmerone
- Counterirritant: 1,8-cineole
- Cyclooxygenase-Inhibitor: curcumin
- Myorelaxant: 1,8-cineole

Other Uses (Zedoary) — Primarily cultivated for the starchy tubers that provide the Shoti Starch of commerce. It is used as a substitute for arrowroot and barley and is highly regarded as a diet food for infants and convalescents. Nearly all the starch is amylose and gives a highly viscous paste with water. A red powder, "Abir," is prepared from powdered rhizomes by treatment with a decoction of sappan wood. Abir is used in Hindu religious rituals. "Ghisi abir" mixes zedoary with artemisia, cardamom, cerasus, cloves, and deodar. Young rhizomes are diced and added to salads. Steam distillation of the rhizomes yields 1–1.5 io of a light yellow volatile EO. The cores of young shoots are eaten (FAC). Fresh leaves, scented like lemongrass, may be used as a vegetable or used for seasoning fish. Zedoary is used in the manufacture of liqueurs, various essences, and bitters, and in cosmetics and perfumes (FAC, UPW). Dried rhizomes spice various bitters, e.g., Swedish bitters, and liqueurs, like Italy's "Ramazzotti." Singers chew the rhizome to clear their throats (FAC, UPW).

For more information on activities, dosages, and contraindications, see the *CRC Handbook of Medicinal Herbs, ed. 2,* Duke et al., 2002.

Cultivation (Zedoary) — Propagated exclusively from divisions of the rhizomes. These are cut into pieces bearing buds and planted in raked soil at the beginning of the monsoon season. Bown (2001) suggests well-drained soil with ample humidity and tropical climate with min temp 13°C (55°F). They are usually cultivated in shaded deciduous forests or along shaded irrigation channels. A 2-year period is required for the full development of the rhizomes.

Chemistry (Zedoary) — Here are a few of the more notable chemicals found in zedoary. For a complete listing of the phytochemicals and their activities, see the CRC phytochemical compendium, Duke and duCellier, 1993 (DAD) and the USDA database http://www.ars-grin.gov/duke/.

Curcumol — Antisarcomic; Antitumor (Cervix).

Curdione — Antileukopenic; Antisarcomic; Antitumor; Antitumor (Cervix); Anti-X-radiation.

Dehydrocurdione — Analgesic 40–200 mg/kg; Antiarthritic 120 mg/kg/day/12 days; Antiedemic 200 mg/kg; Antiinflammatory; Antioxidant; Antipyretic; Antiradicular 100–5000 μM; Calcium-Channel-Blocker 0.1–1 mM.

Zingiberene — Antirhinoviral; Antiulcer IC54 = 100 mg/kg; Carminative; Insecticide.

Cymbopogon citratus (DC.) Staph (Poaceae)
LEMONGRASS, WEST INDIAN LEMONGRASS

Synonym — *Andropogon citratus* DC.

C

Medicinal Uses (Lemongrass) — Reported to be analgesic, antiseptic, antispasmodic, carminative, depurative, diaphoretic, digestive, diuretic, emmenagogue, expectorant, pectoral, stimulant, and tonic. The EO is rubefacient and used externally for chronic rheumatic ailments, lumbago, and sprains. It is also carminative, diaphoretic, spasmodic, stimulant, and tonic, used internally for catarrh, febrile conditions, and gas. With black pepper, it is given in congestive and neuralgic forms of dysmenorrhea, diarrhea, and dropsical condition caused by malaria and vomiting. Tea made from the leaves is a stomachic tonic; with buttermilk, it is used to treat ringworm. Leaves also used in vapor bath. Orientals add the leaves to baths to stop body odor, reduce swelling, improve circulation, and to treat bladder troubles, cuts, leprosy, skin eruptions, and wounds. Orientals also use it for asthma, convulsions, hemoptysis, oppression, puerperium, sprains, and toothache. Latin Americans chew the rhizome until frayed, then use it as a toothbrush. It is boiled whole, with dirt attached to the root, as an abortifacient.

Leung and Foster (1995) reports antioxidant characteristics of the EO, which sensationalists might promote as increasing longevity. Alpha-citral (geranial) and beta-citral (neral) individually elicit antibacterial action on gram(–) and gram(+) organisms. Myrcene did not show observable antibacterial activity on its own. But myrcene, also anesthetic, enhanced activities when mixed with either of the other two main components. Here's another hint that the whole herb is better than its individual components (LIL). Some folk uses are not holding up to scientific scrutiny. One unpublished document in the USDA files is entitled "Lemongrass, the Medicine that Wasn't." I think more highly of it (DAD, LIL).

According to Anon (1998), lemongrass oil, like that of orange and peppermint, will "kill most strains of fungal and bacterial infections." In a study of 52 plant oils and extracts for activity against *Acinetobacter baumanii, Aeromonas veronii* biogroup *sobria, Candida albicans, Enterococcus faecalis, Escherichia coli* bacteria, *Klebsiella pneumoniae, Pseudomonas aeruginosa, Salmonella enterica* subsp. *enterica* serotype *typhimurium, Serratia marcescens,* and *Staphylococcus aureus,* Hammer et al. (1999) noted that lemongrass, oregano, and bay inhibited all organisms at concentrations of ≤ 2.0% (X10438227). Lemongrass, eucalyptus, peppermint, and orange oils, were effective against 22 bacterial strains. Aegle and palmarosa oils inhibited 21 bacteria; patchouli and ageratum oils inhibited 20 bacteria (including Gram (+) and Gram (–). Twelve fungi were inhibited by citronella, geranium, lemongrass, orange, palmarosa, and patchouli oils. The MIC of eucalyptus, lemongrass, palmarosa, and peppermint oils ranged from 0.16 to > 20 µl/ml for 18 bacteria and 0.25–10 µl/ml for 12 fungi (X8893526).

Vapors of several EOs are active against dermatophytes, like *Tinea* and *Trichophyton*. Of those seven studied by Inouye et al. (2001), cinnamon was most potent, followed by lemongrass, thyme, and perilla (chiso) oils, which killed the conidia and inhibited germination and hyphal elongation at 1–4 µg/ml air (X11413931).

Aphid (*Aphis gossypii*) populations were significantly reduced with lemongrass extract prepared by grinding 10 g green leaf in 1 liter of water (DAD, LIL).

Indications (Lemongrass) — Acid Indigestion (f; DAV); Allergy (1; FNF); Athlete's Foot (1; BOW); Backache (f; AAB); Bacteria (1; AAB; FNF; PH2); Bronchosis (1; FNF; PH2; TRA); Bruise (f; MPG); Cancer (1; JNU); Candida (1; X10438227); Catarrh (f; MPG); Cholera (f; MPI; SKJ); Cold (1; AAB; FNF; TRA); Congestion (1; FNF); Cough (1; APA; FNF; MPG; TRA); Cramp (1; AAB; APA; FNF; MPG); Dermatosis (f; APA); Diabetes (f; HHB); Diarrhea (f; APA); Dysmenorrhea (f; DAV); Dyspepsia (f; APA; IED; MPG); Dysuria (f; JFM); Enterosis (f; APA; DAV; JFM; MPG; PH2); Escherichia (1; X10438227); Exhaustion (f; KOM); Fever (1; APA; DAV; IED; PH2; TRA); Flu (1; APA; FNF; MPG; TRA); Fungus (1; AAB; FNF); Gas (f; TRA); Gastrosis (f; AAB; APA; DAV; MPG; PH2); Gingivosis (f; JFM); Headache (f; AAB; SKJ); High Cholesterol (2; FNF; MPG); High Blood Pressure (1; APA; MPG; PH2; TRA); Gas (f; APA; PH2); Headache (f; WBB); Infection (1; DAA; FNF; JBU); Inflammation (1; FNF); Insomnia (f; APA); Klebsiella (1; X10438227); Leprosy (1; PH2; WBB); Lumbago (f; PH2); Malaria (f; JFM; SKJ); Mycosis (1;

X10438227); Myosis (1; AAB; KOM); Nervousness (1; FNF); Neuralgia (f; KOM; MPG; PH2); Pain (1; AAB; APA; FNF; JBU; PH2); Parasite (f; PH2); Pediculosis (f; BOW); Pneumonia (f; JFM); Pulmonosis (f; MPG); Pyorrhea (f; JFM); Rheumatism (f; APA; PH2); Ringworm (f; APA); Salmonella (1; X10438227); Scabies (1; BOW); Stomachache (f; DAA; DAV; MPG; TRA); Toothache (f; WBB); Tuberculosis (f; JFM); UTI (f; MPG); Wound (f; MPG); Yeast (1; AAB; FNF).

Lemongrass for cold/flu:

- Analgesic: myrcene; quercetin
- Anesthetic: 1,8-cineole; linalool; linalyl-acetate; myrcene
- Antiallergic: 1,8-cineole; citral; linalool; quercetin
- Antibacterial: 1,8-cineole; alpha-pinene; alpha-terpineol; caryophyllene; citral; citronellal; citronellol; dipentene; geraniol; limonene; linalool; luteolin; myrcene; neral; nerol; quercetin; rutin
- Antibronchitic: 1,8-cineole
- Antiflu: alpha-pinene; limonene; quercetin
- Antihistaminic: citral; linalool; luteolin; quercetin; rutin
- Antiinflammatory: alpha-pinene; caryophyllene; luteolin; quercetin; rutin
- Antioxidant: linalyl-acetate; luteolin; myrcene; quercetin; rutin
- Antipharyngitic: 1,8-cineole; quercetin
- Antiseptic: 1,8-cineole; alpha-terpineol; citral; citronellal; citronellol; furfural; geraniol; limonene; linalool; nerol
- Antitussive: 1,8-cineole; hcn; luteolin
- Antiviral: alpha-pinene; dipentene; limonene; linalool; luteolin; quercetin; rutin
- Bronchorelaxant: citral; linalool
- COX-2-Inhibitor: quercetin
- Cyclooxygenase-Inhibitor: quercetin
- Expectorant: 1,8-cineole; alpha-pinene; citral; dipentene; geraniol; limonene; linalool

Lemongrass for pain:

- Analgesic: myrcene; quercetin
- Anesthetic: 1,8-cineole; linalool; linalyl-acetate; myrcene
- Antiinflammatory: alpha-pinene; caryophyllene; luteolin; quercetin; rutin
- Antileukotriene: quercetin
- Antispasmodic: 1,8-cineole; caryophyllene; farnesol; geraniol; limonene; linalool; linalyl-acetate; luteolin; myrcene; quercetin; rutin
- COX-2-Inhibitor: quercetin
- Cyclooxygenase-Inhibitor: quercetin
- Lipoxygenase-Inhibitor: luteolin; quercetin; rutin
- Myorelaxant: 1,8-cineole; luteolin; rutin
- Sedative: 1,8-cineole; alpha-pinene; alpha-terpineol; caryophyllene; citral; citronellal; citronellol; dipentene; farnesol; geraniol; geranyl-acetate; isovaleric-acid; limonene; linalool; linalyl-acetate; nerol

Other Uses (Lemongrass) — Very important in Asian cookery, lemongrass leaf blades are used for teas, but the fibrous stems are used in cuisine, flavoring beef, fish, poultry, seafood, and typical Thai and Vietnamese soups. Skinned chicken, boiled with two stalks, is great cold or in cold salads. The juicier lower end of the stalk is smashed with other spices to season Indonesian spiced coconut milk with crab, Thai curries, and Vietnamese marinated beef (AAR). Asians blend it skillfully with chiles or with coconut milk. Basal portions of the leafy shoots are chopped to flavor curries, fish, soups, and sauces. Center of tender stems is used in curry powders. Outer leaves are tied in loops and cooked with food as a flavorant; they are removed before serving. Japanese add to spiced sherbet. Hearts of the young shoots are eaten with rice. Lemongrass is substituted for yogurt in the preparation of "nistisemos trahanas" (fasting trahanas), a fermented milk and cereal food. This type of trahanas is used during religious holidays in Greece and Turkey, when it would be sacrilegious to consume animal milks (FAC). Oil distilled from leaves is used as a popular ingredient of food and drinks. According to Leung (1995), lemongrass oil (GRAS §182.20) is used in most major food categories, including alcoholic and nonalcoholic beverages, frozen dairy deserts, candy, baked goods, gelatins and puddings, meat and meat products, and fats and oils, at about 33 ppm and 36 ppm for highest average maximum use levels for candy and baked goods, respectively. It is also used in creams, lotions, and perfumes with maximum use level at 0.7% oil in the perfumes. Citral, principal constituent of the EO, varies with age from 78–85%. Oil is a good source of synthetic violet odor (ionones), used to perfume soaps, bath salts, and cosmetics. It is one major source of vitamin A. Oil is used in the U.S. for furniture polish. The rhizome is used to flavor tobacco. Lemongrass residue with salt is readily eaten by cattle and gives no flavor residue in milk from cows fed the pulp.

One of America's best selling herbal teas (1978), "Red Zinger," contains, in addition to lemongrass, hibiscus flowers, rose hips, orange peel, and peppermint leaves. Wild cherry bark is being phased out of this popular tea, because it is difficult to obtain. I've made a "Red Zapper" liqueur by steeping my "Red Ringer" tea base in vodka. I have made my "Red Ringer" with lemonbalm and peppermint for the snap, bergamot petals or wild cherries for the red, and rosehips for the vitamins.

For more information on activities, dosages, and contraindications, see the *CRC Handbook of Medicinal Herbs*, ed. 2, Duke et al., 2002.

Cultivation (Lemongrass) — In the temperate zone, a large potted frost-sensitive lemongrass in a sun-parlor or sunny window can provide for a family. It can be outplanted during the frost free season, subdividing and bringing under glass for winter (LIL). Tucker and DeBaggio (2000) say the frost sensitive plant does best in full sun, well drained sandy, moist but not constantly wet, soil pH 4.3–8.4 (average 6.0) soil temperatures 64°–100°F (18°–38°C), relative humidity 40–100. Tops are killed at 28°F (–2°C) (TAD). Since seed rarely if ever form, propagation is mainly by division of clumps. Division, 25–50 from a mature clump, are separated by tearing them off base of plant, retaining a few roots on each. Before planting, tops of divisions are cut back to about 7.5 cm. Sometimes divisions are torn off mature plants left in place. A complete fertilizer, containing ammonium sulfate, calcium superphosphate, and potassium sulfate, at rate of 100 kg/ha, and calcium hydrate, is applied at bottom of furrows just before planting. Application of ammonium nitrate, urea, and IAA enhances vegetative growth (fresh weight of leaves more than 60%) with concomitant increase of volatile oils and their main constituents (MPI). Plants or stools are set in early spring in rows 1 m apart, spaced about 45 cm apart in rows. Planting is done just after a rain, or at a time when sufficient moisture is available not to require artificial watering. Irrigation best be frequent as lemongrass is shallow rooted and does not tolerate much water stress. Planting may be mechanical. Potash is required in Florida soil. As soon as plants become well established in field, side application of fertilizer is added and well worked in at first cultivation. Cultivation should be frequent throughout spring to conserve moisture and throughout summer to control weeds, as a few ill-smelling weeds in crop at harvest time may greatly damage odor of oil. After first year, only slight cultivation is necessary, as a well-established lemongrass stand tends to retard weed growth. On some soils, potash increases weight of grass, but N plus K produces highest yields, up to 41 tons/ha. No additional fertilizer is required for ratoon crop. Under irrigated conditions, lemongrass can remove 186 kg/ha N, 26 P and 384 K. In India, hillsides are burned in January to burn down the old useless straw; 6 months later, the fresh crop is ready to cut (DAD). First cutting should be made 4–5 months after planting, when plants are 75–100 cm tall, clumps 20–25 cm in diameter. In fall of first year, a second cutting is made. After the first year, spring growth is more rapid, and you can take three harvests a year. Mechanically harvest by mowing with blade adjusted to cut plants 20 cm above ground. Cut material raked up with horserake run crosswise of rows. Closer cutting is not profitable because of low oil content in lower portion of plant. Harvest by hand with a machete. Equipment for distillation of lemongrass oil is the same as for other volatile oils. Plants are passed through a fodder cutter, cut into about 5 cm lengths, and loaded about 45 kg/2 cm m of space in retort. In retort of 10 cu m, a charge of 1.5 tons can be distilled in 2–2.5 hours by steam. Distillation under 20 lb pressure increases yield of oil, but oil is dark and of low citral content. After oil is distilled and freed of water, it is dried by shaking with anhydrous calcium chloride and filtered. In the West Indies, sea or salt water distillation increases yield of oil and produces a better quality oil than fresh water. Yields of grass vary from 18–30 MT/ha (fresh weight) for first cutting, 6.2–10 MT/ha for ratoon cutting. Fresh yields reported as high as 140 MT/ha (DAD). Oil yields may run 10–40 lb/a (LIL). Maximum oil yields are 419 liters/ha (45 gallons/acre) when cut at 60-day intervals (TAD).

Chemistry (Lemongrass) — Here are a few of the more notable chemicals found in lemongrass. For a complete listing of the phytochemicals and their activities, see the CRC phytochemical compendium, Duke and duCellier, 1993 (DAD) and the USDA database http://www.ars-grin.gov/duke/.

Citral—ADI = 500 µg/kg; Antiallergic; Antianaphylactic; Antibacterial; Anticancer; Antihistaminic; Antiseptic 5.2 × phenol; Antishock; Barbiturate-Synergist; Bronchorelaxant; Expectorant; Fungicide MIC 625 µg/ml; Glaucomagenic; Herbicide IC50 = 115 μM; Nematicide IC52 = 100 µg/ml, MLC = 100–260 µg/ml; p450(2B1)-Inhibitor IC50 = 1.19 μM; Prostatitigenic 185 mg/kg/day/3 ms; Sedative ED 1–32 mg/kg; Teratogenic; Trichomonicide LD100 = 150 µg/ml; Tyrosinase-Inhibitor ID50 = 1.5 mM; LD50 = 4960 orl rat; ADI 0.5 gm/kg.

Citronellal — Antibacterial; Antiseptic 3.8 × phenol; Antistaphylococcic; Antistreptococcic; Candidacide; Embryotoxic; Fungicide; Insectifuge; Irritant; Motor-Depressant; Mutagenic?; Nematicide MIC 1 mg/ml; p450(2B1)-Inhibitor IC50 = 1.56 μM; Sedative ED = 1 mg/kg; Teratogenic; LD50 = >5000 mg/kg orl rat.

Citronellol — ADI = 500 µg/kg; Antibacterial; Antiseptic; Antistaphylococcic; Antistreptococcic; Candidicide; Fungicide; Herbicide IC50 = 160 μM; Irritant; Nematicide MLC = 100 µg/ml; Sedative; Trichomonicide LD100 = 300 µg/ml; LD50 = 4000 ims mus.

Geraniol — Anthelminthic; Antibacterial MIC = 64 µg/ml, MIC = 400 µg/ml, MBC = 800 µg/ml; Anticancer; Anticariogenic MIC = 400 µg/ml; Antimelanomic IC50 = 150 μM/l; Antimycobacterial MIC = 64 µg/ml PL; Antisalmonella MIC = 400 µg/ml; Antiseptic MIC = 64 µg/ml, 400–800 µg/ml, 7 × phenol; Antispasmodic; Antitubercular MIC = 64 µg/ml; Antitumor (Pancreas) IC50 = 265 μM; Ascaricide; Candidicide; CNS-Stimulant; Embryotoxic; Emetic (3 × ipecac); Expectorant; Fungicide IC93 = 2 mM; Herbicide IC100–2000 μM; Insectifuge 50 ppm; Insectiphile; Nematicide IC86 = 100 µg/ml; MLC = 1000 µg/ml; Sedative; Trichomonicide LD100 = 300 µg/ml; LD50 = 3600 mg/kg orl rat.

Geranyl-Acetate — Insectiphile; Sedative; ADI = 500 µg/kg.

Myrcene — See also *Alpinia galanga.*

D

Dipteryx odorata (Aubl.) Willd. (Fabaceae)
Dutch Tonka Bean, Tonga Bean, Tonka Bean, Cumaru, Tonquin

Synonyms — *Coumarouna odorata* Aubl., *Coumarouna punctata* Blake

Medicinal Uses (Tonka Bean) — The plant is used as an anticoagulant, antidyspeptic, antipyretic, antitussive, cardiotonic, diaphoretic, fumigant, narcotic, stimulant, and stomachic (DAV). The fluid extract has been recommended in whooping cough. In China, seed extracts are used rectally for schistosomiasis. Guyanese use the astringent gum for sore throat. Brazilians apply the seed oil for buccal ulcers and earache. Black Caribs are said to use the fruits as an aphrodisiac (CRC). Brazilians make a cough pill by balling up the crushed seed. In Peru, seeds soaked in rum (alias "agiadiente") are used by locals for snakebite, contusions, and rheumatism. Wayãpi Indians use the bark decoction as antipyretic baths; Palikur as fortifying baths for infants.

The spice plant is used in folk remedies for cacoethes, cancers, indurations, and tumors, especially of the diaphragm, abdomen, liver, spleen, stomach, and uterus (Hartwell, 1982). I think of the coumarin as more checking than causing cancer, at least in reasonable doses. L-Dopa (from faba and velvet beans) may possibly activate malignant melanomas, which, however, can be checked by 100 mg/day coumarin, found in the tonka bean. Marles et al. (1987) and Tisserand (1995) have dispelled the myth of the carcinogenicity and toxicity of coumarin, not to be confused with the furanocoumarins. Marles et al. (1987) suggest that the potential for coumarin toxicity to humans is quite low, if the subject has normal liver function. "Coumarin does not appear to have anticoagulant, carcinogenic, mutagenic, teratogenic, or allergenic properties" (Marles et al., 1987). Tisserand (1995) concluded rather adamantly that "coumarin cannot be regarded as hepatotoxic in humans." In clinical trials, only 0.37% of patients developed abnormal liver function (reversible). Most (of >2000) patients got 100 mg/day coumarin for 1 month, followed by 50 mg/day for 2 yr. Only 8 patients developed elevated liver enzyme levels, which returned to normal following coumarin curtailment (Tisserand, 1995). Coumarin is, however, a highly active, interesting compound. Reported to anesthetize and depress the heart and respiratory rates. In a review of more than 300 coumarins, including the tonka coumarin (1,2-benzopyrone), Hoult and Paya (1996) note coumarin's long-established efficacy in slow-onset long-term reduction of lymphedema in man (confirmed in recent double-blind trials for elephantiasis and postmastectomy swelling). The mechanism may possibly involve macrophage-induced proteolysis of edema protein. Coumarin has low absolute bioavailability in man (< 5%), due to extensive first-pass hepatic conversion to 7-hydroxycoumarin followed by glucuronidation. It may, therefore, be a pro-drug.

Indications (Tonka Bean) — Bacteria (1; FNF); Brucellosis (1; FNF); Bruise (f; DAV); Cachexia (f; APA; CRC); Cancer, kidney (1; FNF); Cancer, prostate (1; FNF); Canker (1; CRC; FNF); Cardiopathy (f; DAW); Cough (f; DAV); Cramp (f; APA; CRC); Diabetes (1; FNF); Dyspepsia (f; DAW); Earache (1; CRC; FNF); Edema (1; FNF); Elephantiasis (2; X8853310); Escherichia (1; FNF); Fever (1; DAV; FNF); Fungus (1; FNF); Hyperglycemia (1; FNF); Infection (1; FNF); Inflammation (1; FNF); Insomnia (1; FNF); Lymphedema (2; X8853310); Lymphoma (1; FNF); Melanoma (1; FNF); Metastasis (1; FNF); Mononucleosis (1; FNF); Mycoplasm (1; FNF);

Mycosis (1; FNF); Nausea (f; APA; CRC; DAD); Nephrosis (1; FNF); Pain (1; FNF); Pertussis (f; APA; CRC; FEL; PHR; PH2); Prostatosis (1; FNF); Psittacosis (1; FNF); Rheumatism (1; DAV; FNF); Schistosomiasis (f; CRC; DAD); Snakebite (f; DAV); Sore (1; CRC; FNF); Sore Throat (f; CRC); Spasm (f; CRC); Stomatosis (1; CRC; FNF); Toxoplasmosis (1; MAB); Tuberculosis (1; APA); Ulcer (1; CRC; FNF).

D

Tonka Bean for cancer:

- Antiaggregant: coumarin; ferulic-acid
- Antiandrogenic: coumarin
- Anticancer: coumarin; ferulic-acid; p-hydroxy-benzoic-acid; salicylic-acid; umbelliferone
- Anticarcinogenic: betulin; ferulic-acid
- Antihepatotoxic: ferulic-acid
- Antiinflammatory: betulin; coumarin; ferulic-acid; gentisic-acid; lupeol; salicylic-acid; umbelliferone
- Antimelanomic: coumarin
- Antimetastatic: coumarin
- Antimutagenic: coumarin; ferulic-acid; p-hydroxy-benzoic-acid; umbelliferone
- Antineoplastic: ferulic-acid
- Antinitrosaminic: ferulic-acid
- Antioxidant: ferulic-acid; lupeol; p-hydroxy-benzoic-acid; salicylic-acid
- Antiperoxidant: lupeol
- Antiprostaglandin: umbelliferone
- Antitumor: betulin; coumarin; ferulic-acid; lupeol; retusin; salicylic-acid
- Antiviral: betulin; ferulic-acid; gentisic-acid; lupeol
- COX-2-Inhibitor: salicylic-acid
- Chemopreventive: coumarin
- Cyclooxygenase-Inhibitor: salicylic-acid
- Cytotoxic: betulin; lupeol; retusin
- Hepatoprotective: ferulic-acid
- Immunostimulant: coumarin; ferulic-acid
- Lipoxygenase-Inhibitor: umbelliferone
- Lymphocytogenic: coumarin
- Lymphokinetic: coumarin
- Ornithine-Decarboxylase-Inhibitor: ferulic-acid
- Prostaglandigenic: ferulic-acid; p-hydroxy-benzoic-acid
- Sunscreen: ferulic-acid; umbelliferone

Tonka Bean for infection:

- Analgesic: coumarin; ferulic-acid; gentisic-acid; salicylic-acid
- Antibacterial: ferulic-acid; gentisic-acid; o-coumaric-acid; p-hydroxy-benzoic-acid; salicylic-acid; umbelliferone
- Antiedemic: coumarin; lupeol
- Antiinflammatory: betulin; coumarin; ferulic-acid; gentisic-acid; lupeol; salicylic-acid; umbelliferone
- Antilymphedemic: coumarin
- Antiseptic: salicylic-acid; umbelliferone
- Antiviral: betulin; ferulic-acid; gentisic-acid; lupeol
- Bacteristat: coumarin
- COX-2-Inhibitor: salicylic-acid

- Cyclooxygenase-Inhibitor: salicylic-acid
- Fungicide: coumarin; ferulic-acid; o-coumaric-acid; salicylic-acid; umbelliferone
- Fungistat: p-hydroxy-benzoic-acid
- Immunostimulant: coumarin; ferulic-acid
- Lipoxygenase-Inhibitor: umbelliferone

D

Other Uses (Tonka Bean) — *D. odorata, D. oppositifolia* (syn. of *Taralea oppositifolia* Aubl.), and *D. pteropus* are cultivated for the seed, which yield coumarin, used to give a pleasant fragrance to tobacco, a delicate scent to toilet soaps, and a piquant taste to liqueurs (EB2:337). Extract is also used in foodstuffs, e.g., baked goods, cakes, candies, cocoa, ice cream, preserves, and as a substitute for vanilla; as a fixing agent in manufacturing coloring materials; in snuffs; and in the perfume industry. Black Caribs of Nicaragua celebrate "maypole" festivities when the Central American species *D. oleifera* is ripe. The Caribs make a paste of the seeds, mix it with coconut water or milk, and make a rich nut-flavored beverage, more satisfying than a malted milk (CRC, FAC). The most important use of coumarin in the U.S. is for flavoring tobacco. Coumarin has been more maligned than deserved (unlike furanocoumarins). More than a century ago, the Eclectics said Tonka depends undoubtedly upon coumarin for its virtues; it's narcotic effects from coumarin, which is also a cardiac stimulant, can paralyze the heart. Dr. Laurence Johnston attributes the effects of cigarette smoking to this principle, since substances used in preparing cigarettes are plants which contain coumarin, notably *Liatris odoratissima* (FEL).

Around Explorama Lodge, Iquitos, Peru, the timber of this huge buttressed tree, or a closely related congener, is used for bridges, dormers, posts, etc. The buttresses are used to make jungle telegraphs, the flat drums made from such buttresses. The edible solitary coumariniferous seed is imbedded in a clam-like shell, and several of these are strung on a waistelet (bracelet-like waist band) which clatters like castanets when one dances.

The timber is said to be resistant to marine borers, perhaps because it contains 0.01% silicon dioxide. And from the new CD, *Forestry Compendium* (CAB International, 2000), one can find several forestry descriptors: round wood, transmission poles, posts, stakes, building poles, sawn or hewn building timbers for heavy construction, beams for light construction, carpentry/joinery, flooring, shingles, railway sleepers, woodware, industrial and domestic woodware, tool handles, sports equipment, wood carvings, turnery, furniture, boats, vehicle bodies, pulp, short fibre pulp. For more information, the CAB International e-mail address is cabi@cabi.org.

For more information on activities, dosages, and contraindications, see the *CRC Handbook of Medicinal Herbs*, ed. 2, Duke et al., 2002.

Cultivation (Tonka Bean) — Usually harvested from the wild, but sometimes cultivated in plantations or planted as windbreaks for cacao. Propagation usually by seed, but the trees can be propagated by budding, cuttings, and marcottage. Seeds lose viability soon after ripening. Germination takes place in 4–6 weeks for whole seeds, and in 1–2 weeks for endocarpless seeds. Seed should be sown in place as they do not transplant well. No particular cultivation is required in native regions. Bown (2001) recommends well-drained gravelly or sandy soil, with ample rainfall and humidity. Bulk of crop is still produced from wild trees (CRC). The ripe seeds after removal from their shells (shells made into castanets) are dried in the shade and then immersed in 65% alcohol for a half day. The alcohol is then poured off and the beans allowed dry 5–6 days in the shade. In this, they accumulate an exterior "frost" of the coumarin crystals (Anon, 1948).

Chemistry (Tonka Bean) — Showing how little we know about tropical forest products, 138 volatile constituents were found, 131 of which have not previously been described as tonka-bean constituents. One compound, undecylfuran, identified for the first time in nature, could be used as a marker for tonka bean, often used as an adulterant or fraudulent substitute for vanilla. Tonka beans are a rich source of coumarin, to 3.5% ZMB; deer tongue, to 1.6%; woodruff, to 1.3%; peru balsam seeds, 0.4%; jujube leaves, 0.3%; and sweet clover, to 0.2% coumarin, on a dry weight basis (ZMB). Deer's tongue, at 1.6% coumarin, is second only to the tonka bean as a cheap source of the controversial coumarin. Too many alarmists warn about the toxicity of the sweet-smelling coumarin, used to make the poor man's vanilla. Lest you be frightened off, let me remind you that coumarin is the odor that gives new-mown hay its pleasing aroma. Coumarin is the lactone of cis-o-hydroxycinnamic acid. Related compounds occur as bound coumarin in sweet clover (*Melilotus* spp) and other aromatic species like sweet vernal grass, vanilla grass, and woodruff. The derivative dicoumarol is a serious anticoagulant. Here are a few of the more notable chemicals found in tonka bean. For a complete listing of the phytochemicals and their activities, see the CRC phytochemical compendium, Duke and duCellier, 1993 (DAD) and the USDA database http://www.ars-grin.gov/duke/.

Coumarin — Allelochemic IC100 = 2 mM; Analgesic; Anesthetic; Antiaggregant; Antiandrogenic; Antibrucellosic; Anticancer 5–25 μg/ml; Antidiuretic; Antiedemic; Antiescherichic; Antiinflammatory; Antilymphedemic; Antimelanomic 50 mg/day; Antimetastatic 50 mg/man/day; Antimitotic; Antimononucleotic; Antimutagenic; Antimycoplasmotic; Antipsittacotic; Antipsoriac; Antitoxoplasmotic; Antitumor 50 mg/day; Antitumor (Kidney) 400–7000 mg/day; Antitumor (Prostate) 400–7000 mg/day; Bacteristat; Bruchiphobe; Carcinogenic 200 mg/kg orl mus; Cardiodepressant; Cardiotonic; DME-Inhibitor IC50 = 57.5 μM; Chemopreventive; Emetic; Estrogenic; Fungicide; Hemorrhagic; Hepatotoxic 0.8–1.71 mM/kg orl rat, 2500 ppm diet, 100 mg/kg dog; Hypnotic; Hypoglycemic 250–1000 mg/kg orl; Immunostimulant; Juvabional; Larvistat; Lymphocytogenic 100 mg/day; Lymphokinetic; Narcotic; Ovicide; Phagocytogenic; Piscicide; Respiradepressant; Rodenticide; Sedative; LD50 = 202 orl gpg; LD50 = 293 orl rat; LD50 = 680 orl rat; LD50 = 720 orl rat.

Linoleic-Acid — 5-Alpha-Reductase-Inhibitor; Antianaphylactic; Antiarthritic; Antiatherosclerotic; Anticancer; Anticoronary; Antieczemic; Antifibrinolytic; Antigranular; Antihistaminic; Antiinflammatory IC50 = 31 μM; Antileukotriene-D4 IC50 = 31 μM; Antimenorrhagic; Anti-MS; Antiprostatitic; Carcinogenic; Hepatoprotective; Hypocholesterolemic; Immunomodulatory; Insectifuge; Metastatic; Nematicide.

Oleic-Acid — 5-Alpha-Reductase-Inhibitor; Anticancer; Antiinflammatory IC50 = 21 μM; Antileukotriene-D4 IC50 = 21 μM; Choleretic 5 ml/man; Dermatitigenic; Hypocholesterolemic; Insectifuge; Irritant; Percutaneostimulant; LD50 = 230 ivn mus; LDlo = 50 ivn cat.

E

Elettaria cardamomum (L.) Maton (Zingiberaceae)
CARDAMON, MALABAR OR MYSORE CARDAMON

Synonym — *Amomum cardamomum* L.

Medicinal Uses (Cardamon) — Ranked as the world's third most expensive spice (saffron number one, vanilla number two), cardamom is almost as good a medicine as it is a spice, but there are cheaper alternatives. As McCormick (1981) notes, "the value of spices to Europeans in the late Middle Ages can hardly be imagined today. A handful of cardamom was worth as much as a poor man's yearly wages. Many a slave was bought and sold for a few handfuls of peppercorn" (McCormick, 1981). Reported to be antidotal, aperitif, balsamic, carminative, diuretic, stimulant, and stomachic. Finely powdered seed are snuffed for headache. Cardamoms, fried and mixed with mastic and milk, are used for bladder problems (DEP). For nausea and vomiting, they are mixed into a pomegranate sherbet. The seeds are popularly believed to be aphrodisiac (DAD, DEP).

Elgayyar et al. (2001) compared antiseptic activity of several EOs against selected pathogenic and saprophytic microbes (anise, angelica, basil, carrot, celery, cardamom, coriander, dill weed, fennel, oregano, parsley, and rosemary). Oregano was strongest, inhibiting all test strains (*Listeria monocytogenes, Staphylococcus aureus, Escherichia coli, Yersinia enterocolitica, Pseudomonas aeruginosa, Lactobacillus plantarum, Aspergillus niger, Geotrichum,* and *Rhodotorula*) at an MLC ca. 8 ppm. Inhibition was complete for oregano, completely nil with carrot oil.

So far, cardamom is the richest source of 1,8-cineole in my database. That can be good, that can be bad. Cineole can enhance dermal absorption of other drugs, temporarily speeding up transdermal absorption of topical drugs. But in the long run, it may nullify this advantage, by inducing detoxification enzymes. Yes, like hypericin and hundreds of other phytochemicals eaten dietarily every day, 1,8-cineole induced p450 detoxication. As Blumenthal et al. (1998), in their translation of Commission E, remind us that cineole induces liver detoxification enzymes, thereby reducing the longevity and/or effectiveness of many natural and prescription drugs. Under myrtle, which contains a theoretical maximum (as calculated in FNF) of 2250 ppm cineole, Gruenwald et al. (1998) caution that more than 10 g myrtle oil can threaten life, "due to the high cineole content" (myrtle contains 135–2250 ppm cineole according to my calculations, meaning 10 g myrtle would contain a maximum 22.5 mg cineole). Several herbs may attain higher levels of cineole: bay, beebalm, betel pepper, biblical mint, boldo, cajeput, cardamom, eucalyptus, ginger, greater galangal, horsebalm, hyssop, lavender, nutmeg, rosemary, sage, spearmint, star anise, sweet annie, thyme, turmeric. So, reductionistically, assuming no synergies or antagonisms or additivities, a ridiculous assumption, one would assume that any goods (and evils) accruing to the cineole in myrtle should apply even more so to those listed above, which, theoretically at least, may attain higher levels of cineole—some (e.g., cardamom) attaining levels more than twentyfold higher. Symptoms of this alleged cineole intoxication may include circulatory disorders, collapse, lowered blood pressure, and respiratory failure. So, rather than placing all this under the obscure spice, myrtle, why not put it under the GRAS cardamom, which can contain up to 5.6% cineole (theoretical max in my database), compared to a mere 0.225% in myrtle?

Indications (Cardamon) — Ague (f; DAD); Alcoholism (f; DAD); Allergy (1; FNF); Anorexia (2; PHR; PH2); Asthma (f; APA; KAP; SKJ); Bacteria (1; FNF); Biliousness (f; KAP); Bleeding (f; DAD); Bronchosis (2; FNF; KAP; PHR; PH2; SKJ); Cacoethes (f; JLH); Cancer (1; FNF; JLH); Cancer, abdomen (1; FNF; JLH); Cancer, colon (1; FNF; JLH); Cancer, diaphragm (1; FNF; JLH); Cancer, liver (1; FNF; JLH); Cancer, spleen (1; FNF; JLH); Cancer, stomach (1; FNF; JLH); Cancer, uterus (1; FNF; JLH; KAB); Catarrh (f; DAD); Cholecystosis (2; PHR; PH2); Cholera (f; DEP); Cold (2; FNF; PHR; PH2); Colic, liver (1; APA); Congestion (1; FNF); Constipation (1; FNF); Cough (f; PH2); Cramp (1; APA); Cystosis (f; DEP; KAB; KAP); Debility (f; DAD); Dermatosis (f; KAB); Diaphragmosis (f; JLH); Diarrhea (f; PH2); Dysmenorrhea (f; DAD); Dyspepsia (2; APA; DAD; KAP; KOM; PH2); Dysuria (f; APA; KAB); Earache (f; KAB); Enterosis (f; JLH); Enuresis (f; BOW; DAD); Fatigue (1; APA); Fever (2; FNF; PHR; PH2); Flu (1; FNF); Fungus (1; FNF); Gas (1; APA; KAP; PH2; RIN); Gastrosis (1; JLH; PH2); Halitosis (1; APA; DAD); Headache (f; DEP); Hemorrhoid (f; KAB; KAP); Hepatosis (2; FNF; JLH; PHR; PH2); Hyperacidity (f; DAD); Impotence (f; APA); Induration (f; JLH); Infection (2; FNF; PHR; PH2); Inflammation (1; FNF; KAB); Intoxication (f; DAD); Insomnia (1; FNF); Lethargy (1; FNF); Malaria (f; DAD); Morning Sickness (f; PH2); Mycosis (1; FNF); Nausea (f; DEP; PH2); Nervousness (1; FNF); Nephrosis (f; KAB); Pain (1; DAD; FNF); Pharyngosis (2; KAB; PHR; PH2); Proctosis (f; KAB); Pulmonosis (f; DAD); Roemheld Syndrome (f; PH2); Scabies (f; KAB); Snakebite (f; KAB); Splenosis (f; JLH); Spermatorrhea (f; DAD); Stomachache (1; APA; PH2); Stomatosis (2; PHR; PH2); Strangury (f; KAP); Toothache (f; KAB); Trichomonas (1; FNF); Tuberculosis (f; DAD; SKJ); Urethrosis (f; PH2); Urogenitosis (f; DAD); Uterosis (f; JLH); Virus (1; FNF); Vomiting (f; DEP; PH2).

Cardamon for cold/flu:

- Analgesic: ascaridole; borneol; caffeic-acid; camphor; menthone; myrcene; p-cymene
- Anesthetic: 1,8-cineole; camphor; linalool; linalyl-acetate; myrcene
- Antiallergic: 1,8-cineole; linalool; menthone; terpinen-4-ol
- Antibacterial: 1,8-cineole; acetic-acid; alpha-pinene; alpha-terpineol; caffeic-acid; caryophyllene; citronellal; citronellol; geraniol; limonene; linalool; myrcene; nerol; nerolidol; p-coumaric-acid; p-cymene; sinapic-acid; terpinen-4-ol; vanillic-acid
- Antibronchitic: 1,8-cineole; borneol
- Antiflu: alpha-pinene; caffeic-acid; limonene; neryl-acetate; p-cymene
- Antihistaminic: caffeic-acid; linalool; menthone
- Antiinflammatory: alpha-pinene; beta-pinene; borneol; caffeic-acid; caryophyllene; eugenyl-acetate; salicylates; vanillic-acid

- Antioxidant: caffeic-acid; camphene; cyanidin; gamma-terpinene; gamma-tocopherol; linalyl-acetate; myrcene; p-coumaric-acid; sinapic-acid; vanillic-acid
- Antipharyngitic: 1,8-cineole
- Antipyretic: borneol; salicylates
- Antiseptic: 1,8-cineole; alpha-terpineol; beta-pinene; caffeic-acid; camphor; carvone; citronellal; citronellol; geraniol; limonene; linalool; menthone; nerol; terpinen-4-ol
- Antitussive: 1,8-cineole; terpinen-4-ol
- Antiviral: alpha-pinene; caffeic-acid; limonene; linalool; neryl-acetate; p-cymene
- Bronchorelaxant: linalool
- Decongestant: camphor
- Expectorant: 1,8-cineole; acetic-acid; alpha-pinene; beta-phellandrene; camphene; camphor; geraniol; limonene; linalool
- Immunostimulant: caffeic-acid

Cardamon for hepatosis:

- Antiedemic: caffeic-acid; caryophyllene
- Antihepatotoxic: caffeic-acid; p-coumaric-acid; sinapic-acid
- Antiherpetic: caffeic-acid
- Antiinflammatory: alpha-pinene; beta-pinene; borneol; caffeic-acid; caryophyllene; eugenyl-acetate; salicylates; vanillic-acid
- Antileukotriene: caffeic-acid
- Antioxidant: caffeic-acid; camphene; cyanidin; gamma-terpinene; gamma-tocopherol; linalyl-acetate; myrcene; p-coumaric-acid; sinapic-acid; vanillic-acid
- Antiperoxidant: caffeic-acid; p-coumaric-acid
- Antiprostaglandin: caffeic-acid; eugenyl-acetate
- Antiradicular: caffeic-acid; vanillic-acid
- Antiviral: alpha-pinene; caffeic-acid; limonene; linalool; neryl-acetate; p-cymene
- Cholagogue: caffeic-acid
- Choleretic: 1,8-cineole; caffeic-acid; p-coumaric-acid; vanillic-acid
- Cytoprotective: caffeic-acid
- Hepatoprotective: borneol; caffeic-acid
- Hepatotonic: 1,8-cineole
- Hepatotropic: caffeic-acid
- Immunostimulant: caffeic-acid
- Lipoxygenase-Inhibitor: caffeic-acid; p-coumaric-acid

Other Uses (Cardamon) — Called the "Queen of Spices" in India (where black pepper is king). Whole cardamom pods are tossed into curries and biryanis. Ground cardamom seasons tandoori chicken and enhances sweets such as "gajar ka halva" (carrot and milk pudding), rice pudding, and "gulab jamun" (fried balls of thick reduced milk, in rose-scented sugar syrup). In Ethiopia, cardamoms are important in the hot pepper mix, a condiment called "mit'mit'a" (FAC). Cardamom imparts the distinctive flavors to Danish pastries, Swedish coffee cakes, Norwegian Christmas cakes, Finnish puula, and Icelandic pönnukökur. Scandinavian cooks use it to season cookies, waffles, crispbreads, and many other baked goods. Cardamom is also used to flavor "glögg," the hot spiced Scandinavian wine popular at Christmas. A half-teaspoon can improve gingerbread, spice cakes, and chocolate cake, apple or peach pie, fruit salads, jams, and custard or ice cream (AAR). Decorticated cardamon seeds are used for seasoning sausages and hamburgers, pastries and confections. Seeds used to flavor bitters, cakes, candies, coffee, cordials, curries, drinks, gingerbreads, liqueurs, and pickles (CFR, FAC). Arabs and East Indians chew the seed as a candy. The seeds impart a warm, slightly pungent, highly aromatic taste and are used in breath sweeteners. Occasionally, the seed are chewed with betel-leaf (DEP). Cardamom is important in some curries. In traditional "Massaman" (Muslims in S. Thailand) cardamom seed, roasted in their shell, are garnished onto curries. Cardamon is used in coffee (from a dash to 2 tsp of pods and seeds per cup of coffee), especially by Arabs, giving it a double CNS-stimulant whammy. An EO, obtained from green cardamon pods and seeds by steam distillation, serves for flavoring cakes, in confectionery, sausages, pickles, table sauces, curry preparations, and in certain bitters and liqueurs. As with many ginger relatives, cardamom leaves are used to wrap foods over the fire or stove. The young shoots, like those of many ginger relatives, are eaten raw, roasted, or steamed. If, as Rinzler (1990) notes (for high cineole content), bay leaves seem to repel fleas, moths, and roaches, cardamom should do it better. A fruit in a canister of flour just might keep out the bugs, and a fruit by the pipes coming up from the basement might keep the cockroaches underneath. Cardamom is one of the richest sources of cineole. But Rinzler was way ahead of me. She doesn't recommend them in your flour cannister unless you wish your flour and resultant baked goods to smell of cardamom (DAD, FAC).

For more information on activities, dosages, and contraindications, see the *CRC Handbook of Medicinal Herbs*, ed. 2, Duke et al., 2002.

Cultivation (Cardamon) — May be raised from seed, which require 2–3 months to germinate. Sites must be sheltered from direct sun and strong winds. Clearings in forest often provide such sites. It is more usually started from selected rhizomes or bulbs 1.5 to 2 years old and with at least 2 growing stems; mature plants can be divided after fruiting. Flowers are said to be self-sterile, so it is necessary to plant a mixture of clones. In India, the fruits are harvested from August through December, as they ripen; dividing and planting is done later. Plants are set 3–4 m apart. Manual labor is generally used to keep weeds down during the first 2 years. Plants should be periodically cleaned of dried leaves and damaged or decaying stems. Raised alone in monoculture or with pepper and/or coffee. If a swampy site runs through a coffee plantation, it is often planted with cardamon (DAD, DEP). Plants mature in 2–3 years, ripening more slowly at higher elevations. Most growers have three fields of plants: plants just set, those in second year growth, and those fruiting. Fruiting plants are dug, divided, and replanted. Ripe fruits must be severed carefully with scissors so as not to injure the flowers and unripe fruit. Fruits are picked green (yellow ones split and shatter = shed their seeds). Yields all year round in the humid tropics, but chiefly during dry period. Crop is gathered every 2–3 weeks. In India and Sri Lanka, fruit is prepared for market by: (1) sulfur bleaching — sulfur fumigation, alternated with soaking and drying, carried out in four stages with a final sun drying; whole process takes from 10–12 days to complete, and bleached cardamons are creamy-white; (2) green curing — cardamons are dried on trays in a heated chamber or over an open charcoal fire in a closed chamber. These cardamons are green. Stalks and calyxes are removed from dried cardamons by cutting or grating, then sorted and graded according to size. Yields of 110–330 kg/ha of dried cardamons have been obtained, but yields of 40–80 kg/ha seem to represent more customary yields. Scientists have increased yields to 625 kg/ha by clonal propagation (DAD).

Chemicals (Cardamon) — All things in moderation, health food or poison. And don't be alarmed by the LD50 (2480 mg/kg orally in rats), which indicates that cineole is less than 1/10th as toxic acutely as caffeine (192 mg/kg orally in rat). Cardamom is also my richest source of p-cymene, and surely the cymene also contributes to the medicinal rationale of cardamom.

1,8-Cineole — Acaricide; Allelopathic; Anesthetic; Antiacetylcholinesterase IC50 = 41 µg/ml; Antiallergic; Antibacterial 50 ppm; Antibronchitic; Anticariogenic; Anticatarrh; Anticholinesterase; Antifatigue; Antihalitosic; Antiinflammatory; Antilaryngitic; Antipharyngitic; Antirheumatic; Antirhinitic; Antiseptic; Antisinusitic; Antispasmodic; Antistaphylococcic 50 ppm; Antitussive; Antiulcer; Candidicide; Carcinogenic; Choleretic; CNS-stimulant; Convulsant; Counterirritant; Decongestant; Degranulant 0.3 µl/ml; Dentifrice; Edemagenic; Expectorant; Fungicide; Gastroprotective;

Gram(+)-icide; Gram(−)-icide; Hepatotonic; Herbicide IC50 = 78 μ*M*; Hypotensive; Insectifuge; Irritant; Myorelaxant; Nematicide; Negative Chronotropic 87 nl/ml; Negative Inotropic 87 nl/ml; p450-Inducer; Neurotoxic; Perfume; Rubefacient; Secretagogue; Sedative; Spasmogenic; Surfactant; Testosterone-Hydroxylase-Inducer; Trichomonicide LD100 = 1000 μg/ml; Vermifuge; LD50 = 2480 orl rat; LD50 = 3480 mg/kg; LD50 = >5000 der rbt.

Palmitic-Acid — 5-Alpha-Reductase-Inhibitor; Antifibrinolytic; Hemolytic; Hypercholesterolemic; Lubricant; Nematicide; Soap; LD50 = 57 ivn mus.

P-Cymene — Analgesic; Antiacetylcholinesterase IC40 = 1.2 mM; Antibacterial; Antiflu; Antirheumatalgic; Antiviral; Fungicide; Herbicide IC50 = 50 μ*M*; Insectifuge; Irritant; Laxative; Sedative; Trichomonicide LD100 = 50 μg/ml; LD50 = 4750 mg/kg orl rat.

Stearic-Acid — 5-Alpha-reductase-inhibitor; Cosmetic; Hypocholesterolemic; Lubricant; Propecic; Suppository; LD50 = 22 ivn rat.

Terpinen-4-OL — Antiacetylcholinesterase IC21–24 = 1.2 mM; Antiallergic; Antiasthmatic; Antibacterial; Antioxidant; Antiseptic; Antispasmodic; Antitussive; Antiulcer; Bacteristat; Diuretic 0.1 ml/rat; Fungicide; Herbicide IC50 = 200 mM, IC50 = 22 μ*M*; Insectifuge; Irritant; Nematicide MLC = 1 mg/ml; Renoirritant; Spermicide ED100 = 0.015; Vulnerary; LD50 = 0.78 ml/kg ims mus; LD50 = 0.25 ml/kg ipr mus; LD50 = 1.85 ml/kg orl mus; LD50 = 0.75 ml/kg scu mus.

Eryngium foetidum L. (Apiaceae)
Cilantro, Culantro, False Coriander, Shadow Beni, Stinkweed

Medicinal Uses (Culantro) — Considered antimalarial, antispasmodic, carminative, and pectoral (DAW). Roots contain saponin and have an offensive odor. They are used as a stomachic. A decoction of root is valued in Venezuela as a stimulant, antipyretic, a powerful abortive, and sedative. In Cuba, it is a valuable emmenagogue. It is used throughout tropical America as a remedy for seizures and high blood pressure. Its infusion with salt is taken for colic. Leaf infusion used for stomachaches. The seeds and roots are used for earache (Mustalish and Baxter, 2001). Around our camps in Peru, "sacha culantro" is boiled for cramps and stomachache. Around Pucallpa, Peru, culantro with meat broth is taken for bronchitis and fever. Chamis Indians braise the dried fruits and have the children inhale smoke to treat diarrhea. Green fruits are crushed and mixed with food to treat insomnia. Créoles drink the decoction for colds and flu, and rub crushed leaves over the body to reduce high fever.

Indications (Culantro) — Anemia (f; MPG); Anorexia (f; JFM); Arthrosis (f; DAV; MPG); Asthma (f; IED; MPG); Biliousness (f; JFM); Bronchosis (f; DAV); Cardiopathy (f; IED; MPG); Catarrh (f; IED); Chill (f; BOW); Cold (f; DAV; JFM); Colic (f; DAV); Constipation (f; JFM); Convulsion (f; JFM); Cough (f; DAV; JFM); Debility (f; MPG); Diabetes (1; JFM; MPG; JAC7:405); Diarrhea (f; DAV); Dyspepsia (1; BOW; DAV); Earache (f; MPG; TRA); Edema (1; PR13:75); Epilepsy (f; BOW); Fever (1; DAV; JFM; TRA); Fits (f; BOW; JFM); Flu (f; DAV; JFM); Gas (1; DAV; JFM); High Blood Pressure (f; DAV; IED; MPG); High Cholesterol (f; MPG); Inflammation (1; PR13:75); Insomnia (f; DAV); Malaria (f; DAV); Nausea (f; DAV; TRA); Obesity (f; MPG); Parasite (f; IED); Pneumonia (f; DAV); Rheumatism (f; DAV; JFM); Snakebite (f; HHB; JFM); Stomachache (f; DAV; MPG); Swelling (1; PR13:75); Syncope (f; JFM); Tumor (f; DAV; JLH); Water Retention (f; HHB); Worm (f; IED; JFM); Yellow Fever (f; JFM).

Culantro for dyspepsia:

- Analgesic: p-cymene
- Antiinflammatory: alpha-pinene
- Sedative: alpha-pinene; p-cymene
- Tranquilizer: alpha-pinene

Other Uses (Culantro) — How well I remember this as a vital constituent of the chicken caldos, called "sancocho" in Panama. Elsewhere, it's almost as much a medicine as a spice. The Spanish name "recao de monte" indicates that it is a wild spice. It is a weed. The leaves constitute a powerful spice, a love-it or hate-it spice. Leaves are used in tropical America and elsewhere, as condiment in stews, pastries, soups, and meat dishes, to impart an agreeable flavor (though some say it smells like bedbugs). One spice mixture, called "sofrito"(chiles, cilantro, and culantro), is sold in the West Indian markets of New York and other large cities (FAC). Javanese add the tenderest leaves to rice as a lablab. Roots are also used, almost as spice vegetables, in meat dishes and soups. The leaves of culantro (*Eryngium*) retain their aroma and flavor better than leaves of cilantro (*Coriandrum*) on drying. It contains 0.02–0.04% of a volatile oil (DAV, FAC, TAD).

For more information on activities, dosages, and contraindications, see the *CRC Handbook of Medicinal Herbs*, ed. 2, Duke et al., 2002.

Cultivation (Culantro) — Ochse (1931) says the plant multiplies only by its seed, produced in great numbers. Seed lose their viability quickly and so must be sown as soon as they are ripe. It is easily grown in moist U.S. garden soils if started indoors and outplanted after the last frost. Minimum temp. = 15–18°C (59–64°F) (Bown, 2001). After spring-planted *Coriandrum* has quit producing leaves due to the heat, *Eryngium* continues to produce, rewarding the gardener's efforts (TAD). Leaves are best picked before flowering but may be used fresh or dried. Two-year-old roots are also utilized (Bown, 2001). Slugs and mealy bugs tend to like culantro. Storage at 50°F (10°C) can extend shelf life of the leaves (ambient shelf life 4 days) up to 2 weeks. Blanching at 205°F (96°C) before drying preserves the green color (TAD).

Chemistry (Culantro) — As so often happens, the phytochemicals don't occur in isolation; e.g., the phytosterols in a single extract are many and possibly synergistic: alpha-cholesterol, brassicasterol, campesterol, stigmasterol (as the main component, 95%), clerosterol, beta-sitosterol, delta 5-avenasterol, delta (5)24-stigmastadienol, and delta 7-avenasterol (Garcia et al., 1999). Here are a few of the more notable chemicals found in culantro. For a complete listing of the phytochemicals and their activities, see the CRC phytochemical compendium, Duke and duCellier, 1993 (DAD) and the USDA database http://www.ars-grin.gov/duke/.

Beta-Sitosterol — ADI = 9–30 g/day/man; Androgenic; Anorectic; Antiadenomic; Antiandrogenic; Antibacterial; Anticancer; Antiestrogenic; Antiedemic IC54 = 320 mg/kg orl; Antifeedant; Antifertility; Antigonadotropic; Antiinflammatory; Antileukemic; Antilymphomic; Antimutagenic 250 µg/ml; Antiophidic 2.3 mg mus; Antiprogestational; Antiprostaglandin 30 mg/day/12 wk; Antiprostatadenomic; Antiprostatitic 10–20 mg/3×/day/orl man; Antitumor (Breast); Antitumor (Cervix); Antitumor (Lung); Antiviral; Artemicide LC50 = 110 ppm; Candidicide; Estrogenic; Gonadotropic; Hepatoprotective; Hypocholesterolemic 2–6 g/man/day/orl; Hypoglycemic; Hypolipidemic 2–6 g/day; Hypolipoproteinaemic; Spermicide; Ubiquict; Ulcerogenic 500 mg/kg ipr rat; LD50 = 3000 mg/kg ipr mus; LDlo = >10,000 inj rat.

Delta-5-Avenasterol — Antioxidant.

Delta-7-Avenasterol — Antioxidant.

Stigmasterol — Anesthetic; Anticancer; Antihepatotoxic; Antiinflammatory; Antiophidic 2.3 mg/ipr mus; Antiviral; Artemicide LC50 = 110 ppm; Estrogenic; Hypocholesterolemic; Ovulant; Sedative.

E

F

Ferula assa-foetida L. (Apiaceae)
ASAFETIDA

Medicinal Uses (Asafetida) — If you think medicines stink and spices smell good, you may regard asafetida as more medicine than food. Like those unrelated alliums, this is loaded with sulfureous medicinal compounds, some of which are called mercaptans. Some superstitious types hang asafetida (as was done with garlic) around the neck to ward off colds and infectious diseases (RIN). Reported to be analgesic, antispasmodic, aperient, aphrodisiac, carminative, diuretic, emmenagogue, expectorant, laxative, nervine, sedative, stimulant, and vermifuge. Asafetida seems to have hypotensive activity and antiaggregant activity, slowing blood clotting (like the similar smelling garlic). "Herbal Highs" recommends 1/2 teaspoon in warm water as a tranquilizer. Malays take asafetida for abdominal trouble, broken bones, and rheumatism. Javanese use it for stomachache and worms. Asafetida is used as an enema for intestinal flatulence. Homeopathically used for gas, osteosis, and stomach cramps (CRC).

Sulfur compounds in the oil may protect against fat-induced hyperlipidemia. Two double-blind studies report asafetida useful for IBS (just below 5% significance level in one, near 1% in the other) (CAN). That might seem odd when one considers that Desai and Kalro (1985) demonstrated under their experimental conditions that powdered black pepper does not damage the gastric mucosa, but that *Ferula* does, based on the rate of exfoliation of human gastric mucosa surface epithelial cells.

Saleem et al. (2001) show that asafetida inhibits early events of carcinogenesis, but they may have been studying a closely related species. They report antioxidant and anticarcinogenic potential of asafoetida (*Ferula narthex* to them) in mice. Pretreatment of animals with asafoetida can protect against free radical mediated carcinogenesis. Unnikrishnan and Kuttan (1990) claim that oral extracts of black pepper, asafetida, pippali, and garlic could increase the life span in mice by 64.7%, 52.9%, 47%, and 41.1%, respectively. Such results suggest the use of spices as anticancer agents and antitumor promoters.

According to Sambaiah and Srinivasan (1989), asafetida stimulated liver microsomal cytochrome p450 dependent aryl hydroxylase. Asafetida increased glutathione-S-transferase activity >78% in esophagus, liver, and stomach, enough to be considered chemopreventive (JAC7:405). Aruna and Sivaramakrishnan (1990) and Patel and Srinivasan (1985) suggest that asafetida decreased levels of phosphatase and sucrase activities.

The gum may induce contact dermatitis. Generally not regarded as toxic. Ingestion of 15 g produced no untoward effects, but related *Ferula sumbul* has produced narcosis at 15 g. Approved by FDA for use in food (§182.20, CRC).

Indications (Asafetida) — Amenorrhea (f; CRC); Angina (f; KAB); Arthrosis (1; BOW); Ascites (f; KAB); Asthma (1; APA; CRC; FNF; WOI); Bacteria (1; FNF); Bite (f; KAB); Bronchosis (1; APA; CAN; WOI); Callus (f; JLH); Cancer (1; APA; FNF; PH2); Cancer, abdomen (1; APA); Cancer, colon (f; KAB); Cancer, gum (f; JLH); Cancer, liver (f; JLH); Cardiopathy (f; KAB); Caries (f; KAB); Cholera (f; CRC; SKJ; WOI); Colic (f; APA; CAN; CRC); Cold (f; TAD); Colitis (f; APA; PHR; PH2); Conjunctivosis (f; KAB); Constipation (f; PH2); Convulsion (f; BOW; CRC); Corn (f; JLH); Cough (f; BOW; PNC); Cramp (1; CAN; CRC; FNF; SKJ); Croup (f; CRC); Deafness (f; KAB); Dermatosis (f; KAB); Diarrhea (f; PH2); Dysentery (f; BOW); Dyspepsia (1; APA; CAN; FNF; PH2); Dyspnea (f; KAB); Enterosis (f; APA; CRC; PH2); Epilepsy (f; APA; CRC; PH2; WOI); Felon (f; JLH); Fracture (f; CRC); Frigidity (f; APA); Fungus (f; APA); Gas (1; APA; CAN; CRC; HHB; PNC; WOI); Gastrosis (f; PHR; PH2); Gingivosis (f; JLH); Hemiplegia (f; CRC); Hepatosis (f; JLH; PH2); High Blood Pressure (f; DAA); Hyperlipidemia (f; CAN); Hypoacidity (f; PH2); Hypoglycemia (f; APA); Hysteria (f; APA; CAN; WOI); IBS (2; CAN); Impotence (f; APA); Indigestion (1; APA; CAN); Induration (f; JLH); Infection (f; PHR; PH2); Inflammation (1; FNF; KAB); Insanity (f; CRC); Insomnia (f; PHR); Jaundice (f; KAB); Laryngismus (f; CAN); Mucososis (1; APA; CAN); Mycosis (f; KAB); Nervousness (1; FNF); Neurasthenia (f; CRC; DAA); Neurosis (1; APA); Obesity (1; CRC; FNF); Ophthalmia (f; KAB); Osteosis (f; CRC); Pain (1; FNF); Paralysis (f; KAB); Parasite (f; PH2); Pertussis (f; CAN; CRC; PH2; WOI); Pneumonia (f; KAB; SKJ); Polyp (f; CRC); Rheumatism (f; CRC; KAB); Rinderpest (f; CRC); Ringworm (f; KAB); Sarcoma (f; CRC); Snakebite (f; KAB); Sore Throat (f; KAB); Spasm (f; CRC); Splenosis (f; CRC; PH2); Stomachache (1; APA; CRC); Thrombosis (1; CAN; FNF; PNC); Tumor (1; APA; FNF); Wart (f; JLH); Whitlow (f; JLH); Worm (f; CRC).

Asafetida for cancer:

- AntiHIV: diallyl-disulfide; luteolin
- Antiaggregant: ferulic-acid
- Anticancer: alpha-pinene; alpha-terpineol; diallyl-disulfide; ferulic-acid; isopimpinellin; luteolin; umbelliferone; vanillin
- Anticarcinogenic: ferulic-acid; luteolin
- Antihepatotoxic: ferulic-acid; glucuronic-acid
- Antihyaluronidase: luteolin
- Antiinflammatory: alpha-pinene; azulene; beta-pinene; ferulic-acid; isopimpinellin; luteolin; umbelliferone
- Antileukemic: luteolin
- Antimutagenic: diallyl-sulfide; ferulic-acid; luteolin; umbelliferone; vanillin
- Antineoplastic: ferulic-acid
- Antinitrosaminic: ferulic-acid
- Antioxidant: ferulic-acid; luteolin; vanillin
- Antiproliferative: diallyl-disulfide
- Antiprostaglandin: umbelliferone
- Antitumor: diallyl-disulfide; diallyl-sulfide; ferulic-acid; luteolin; vanillin
- Antiviral: alpha-pinene; diallyl-disulfide; ferulic-acid; luteolin; vanillin
- Apoptotic: luteolin
- Beta-Glucuronidase-Inhibitor: luteolin
- Cytotoxic: luteolin
- Hepatoprotective: ferulic-acid; luteolin
- Immunostimulant: diallyl-disulfide; ferulic-acid
- Lipoxygenase-Inhibitor: luteolin; umbelliferone
- Ornithine-Decarboxylase-Inhibitor: ferulic-acid
- PTK-Inhibitor: luteolin
- Prostaglandigenic: ferulic-acid
- Protein-Kinase-C-Inhibitor: luteolin
- Sunscreen: ferulic-acid; umbelliferone

Asafetida for IBD:

- Analgesic: ferulic-acid
- Antibacterial: alpha-pinene; alpha-terpineol; azulene; diallyl-disulfide; diallyl-sulfide; ferulic-acid; luteolin; umbelliferone

- Antiinflammatory: alpha-pinene; azulene; beta-pinene; ferulic-acid; isopimpinellin; luteolin; umbelliferone
- Antioxidant: ferulic-acid; luteolin; vanillin
- Antiseptic: alpha-terpineol; azulene; beta-pinene; diallyl-sulfide; umbelliferone
- Antispasmodic: azulene; ferulic-acid; luteolin; umbelliferone; valeric-acid
- Antiulcer: azulene
- Diuretic: isopimpinellin; luteolin
- Lipoxygenase-Inhibitor: luteolin; umbelliferone
- Myorelaxant: luteolin; valeric-acid
- Sedative: alpha-pinene; alpha-terpineol; valeric-acid
- Tranquilizer: alpha-pinene; valeric-acid

F

Other Uses (Asafetida) — With a taste stronger than onion or even garlic, asafetida is still used as a spice in the Middle East. Iranians rub asafetida on warmed plates on which meat is to be served. One-fourth teaspoon powder may suffice in a dish for four. You could substitute 1 tbsp freshly grated white onion for $\frac{1}{4}$ tsp asafetida powder. Young shoots are consumed as a cooked green vegetable. The heads, rather resembling cabbage, are eaten raw as a delicacy. Roots are also roasted and eaten in the Southwest (FAC). Also called "hing" and "perunkayam" in Asia (AAR). In Kashmir, asafetida is eaten with vegetables and pulses. It is all but essential in Kashmiri lamb with yogurt sauce. It is said to be good with legumes, like lentils and beans, and cruciferous vegetables, like cabbage and cauliflower; also good with fish and seafood soups. It is mixed with ground meat to make "kofta" (meatballs) (AAR). India likes crispy asafetida-flavored "papadams," flat, fried crackers made with lentil flour. Asafetida is important in "chaat masala." Hindus, particularly in Kashmir, may avoid garlic and onions (inflaming the baser passions), turning instead to asafetida. But some Indian recipes call for all three, asafetida, garlic, and/or onions. Indian housewives add a lump to stored spices, hoping to deter insects. Roots are the source of a gum resin used as a flavoring in bean and lentil soup, curried fish, vegetarian dishes, sauces, drinks, pickles, cakes, etc. It is a standard ingredient of Worcestershire sauce and is widely employed in spice blends and condiments. Also popular in natural foods cuisine as a substitute for garlic (FAC). Alcoholic tinctures of the gum-resin, or the oil and/or fluid extract, are reportedly used, at very low levels, in baked goods, beverages, candies, frozen deserts, gelatins, meat and meat products, relishes, sauces, and spices. Its main use, however, is as a fixative or fragrance component in perfumery. The volatile oil has not attained commercial importance, because the flavoring and pharmaceutical industries utilize instead the tincture of asafetida (CRC). Asafetida is reportedly used in veterinary practice to repel cats and dogs (CRC). Rinzler (1990) mentions something I want to try in my Green Farmacy Garden; she suggests a 2% asafetida solution for the garden to repel deer and rabbits (1 oz powdered asafetida well shaken in 1.5 quarts water) (maybe it'll repel my repugnant groundhog, too) (AAR, CRC, FAC, RIN).

For more information on activities, dosages, and contraindications, see the *CRC Handbook of Medicinal Herbs*, ed. 2, Duke et al., 2002.

Cultivation (Asafetida) — Asafetida grows in Asian high plains region 2000–4000 ft above sea level, arid and bare in winter but covered with a thick growth of *Ferula foetida* and other *Ferula* species in summer. Hardy from zone 7–9 (Bown, 2001). Asafetida is gathered in June. The gum is caused to form by bleeding the root of the plant. Preparatory to cutting, soil is scraped away from around the root to about 6 in. The top is severed at the crown and several lacerations made around the head of the root. Fresh cuts are made every 3 or 4 days until sap ceases to run (a week or a month), depending on the vigor and size. After cutting, the root stump is covered with a dome of twigs, herbs, stones, or other trash to protect it from the sun so that it will not wither. As gum exudes it hardens into tears, or lumps. This is collected and spread in the sun to harden. Quantities ranging from a few ounces to several pounds per root have been reported (Source: USDA fact sheet).

Chemistry (Asafetida) — Contains 40–64% resinous material composed of ferulic acid, umbelliferone, asaresinotannols, farnesiferols A, B, and C, etc., about 25% gum composed of glucose, galactose, l-arabinose, rhamnose, and glucuronic acid, and volatile oil (3–17%) consisting of disulfides as its major components, notably 2-butyl propenyl disulfide (E- and Z-isomers), with monoterpenes (alpha- and beta-pinene, etc.), free ferulic acid, valeric acid, and traces of vanillin (LAF). The disagreeable odor of the oil is reported to be due mainly to the disulphide $C_{11}H_{20}S_2$. Analysis of bazaar samples from Mysore gave the following values: ash, 4.4–44.3% and alcohol soluble matter, 20.8–28.0%; samples obtained from Teheran contained: ash, 6.3–8.9% and alcohol soluble matter, 28.3–40.9% (WOI).

Here are a few of the more notable chemicals found in asafetida. For a complete listing of the phytochemicals and their activities, see the CRC phytochemical compendium, Duke and duCellier, 1993 (DAD) and the USDA database http://www.ars-grin.gov/duke/.

Ferulic-Acid — Allelopathic; Analgesic; Antiaggregant; Antiallergic; Antiarrythmic; Antibacterial; Anticancer; Anticarcinogen; Antidysmenorrheic; Antiestrogenic; Antihepatotoxic; Antiherpetic; Antiinflammatory; Antimitotic; Antimutagenic; Antineoplastic (Stomach); Antinitrosaminic; Antioxidant 3000 μM, IC51 = 200 ppm, $^1/_3$ quercetin; Antiserotonin; Antispasmodic; Antithrombic; Antitumor; Antitumor (Colon); Antitumor (Forestomach); Antitumor (Liver); Antitumor (Skin); Antitumor-Promoter IC46 = 10 μM; Antiviral; Arteriodilator; Candidicide; Cardiac; Cholagogue; Choleretic; Fungicide; Hepatoprotective; Herbicide; Hydrocholeretic; Hypolipidemic; Immunostimulant; Insectifuge; Metal-Chelator; Ornithine-Decarboxylase-Inhibitor; Phagocytotic; Preservative; Prostaglandigenic; Sunscreen; Uterosedative 30–100 mg/kg ivn rat; LDlo = 1200 par mus; LD50 = 416 ivn mus; LD50 = 837–895 ivn mus.

Foetidin — Hypoglycemic

Umbelliferone — Allelochemic IC94 = 2 mM; Antibacterial; Anticancer 5–25 μg/ml; Antihistaminic; Antiinflammatory; Antimitotic 5–25 μg/ml; Antimutagenic; Antiprostaglandin; Antiseptic; Antispasmodic; Antistaphylococcic; Candidicide; Choleretic; Fungicide; Lipoxygenase-Inhibitor; Photoactive; Sunscreen; Xanthine-Oxidase-Inhibitor.

G

Gaultheria procumbens L. (Ericaceae)
Box Berry, Checker Berry, Creeping Wintergreen, Mountain Tea, Teaberry, Wintergreen Teaberry

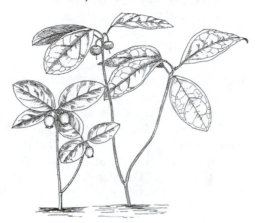

Medicinal Uses (Wintergreen) — I find the aroma of the methyl salicylate, the active main ingredient in wintergreen, very pleasant. I frequently use a boswellin cream with wintergreen when my knee acts up. In Maine, we make wintergreen tea, drinking it and applying it topically for chronic or temporary pain. There's more than analgesic methyl-salicylate, there are at least three more analgesics: caffeic-acid, ferulic-acid, and gentisic-acid. And it complements the red pepper with its analgesic capsaicin, and the peppermint with its menthol. You'll find several OTC pain relievers using these herbs alone, or any one of their constituents, or in various combination. Methyl salicylate has been employed in baths, liniments, and ointments, for pain relief, e.g., in gout, lumbago, rheumatism, and sciatica. Children who chew the roots for six weeks each spring reportedly suffer less tooth decay. With leaves shaped like South America's coca leaves, these wintergreen leaves and/or fruits were used by North American Indians to keep their breath when portaging heavy loads. Algonquin guides chewed the leaves to improve their breathing (and I expect their breath as well) during hunting. Amerindians smoked and chewed the dried leaves. Quebec Indians rolled the leaves around aching teeth. The EO is used as an analgesic, antiseptic, and counterirritant in rheumatism, lumbago, and sciatica. Like so many other aromatic EOs, its oil can be fatal if ingested in large quantities. The whole plant is used as an antiseptic, carminative, diuretic, emmenagogue, lactagogue, nervine, rubefacient, stimulant, antirheumatic, and as a flavoring in medicine. The tea is used as a gargle for sore throat and stomatosis, as a douche for leucorrhea, and as a collyrium for conjunctivitis. Small doses stimulate the stomach, large doses cause vomiting. It may be used for diarrhea, and as an infant's carminative. Leaves are used in the treatment of asthma.

In an interesting case of synergy, Jin et al. (1999) show that aloesin and arbutin inhibit tyrosinase synergistically, acting via different mechanism.

And death from stomach inflammations have resulted from frequent and large doses of the oil. The highest average maximum use level is ca. 0.04% in candy. Not listed under FDA §172.510, 182.10, or 182.20. Wintergreen has lectinic, including mitogenic properties. Salicylism usually marked by tinnitus, nausea, and vomiting and may result from excessive dosage of salicylic acid and/or its salts.

Indications (Wintergreen) — Amenorrhea (f; CEB); Arthrosis (1; DEM; FNF; PHR; PH2); Asthma (f; CEB; PHR; PH2); Bacteria (1; FNF); Cancer (1; CEB; FNF; LEL); Candida (1; FNF); Caries (1; CEB; CRC; FNF); Catarrh (f; CRC); Cold (1; DEM; FAD; FNF); Colic (1; APA; MAD); Congestion (f; MAD); Conjunctivosis (f; CRC); Cough (1; X1798722); Cramp (1; FNF); Cystosis (1; FNF); Debility (f; CEB); Dermatosis (f; CRC); Diabetes (1; CRC; FNF); Diaphragmosis (1; MAD; PHR); Diarrhea (f; CEB; CRC); Dropsy (f; CRC); Dysentery (f; DEM); Dysmenorrhea (1; CRC; FNF; MAD; PHR); Dyspepsia (1; APA; DEM); Dyspnea (f; CEB; CRC); Edema (1; APA; FNF); Epididymosis (1; CRC; MAD; PHR); Fever (1; APA; CRC; FAD; MAD); Flu (1; DEM; FNF); Fungus (1; FNF); Gas (1; APA); Gastrosis (f; CRC; MAD); Gingivosis (1; DEM; FNF); Gonorrhea (f; CRC); Gout (1; CRC; FNF; MAD); Headache (1; APA; DEM; FAD; FNF); Heart (f; MIC); Immunodepression (1; FNF); Infection (1; FNF; MAD); Inflammation (1; APA; FNF); Ischiosis (f; MAD); Leukorrhea (f; CRC); Lumbago (1; CRC; DEM; FAD); Myosis (1; APA; FAD); Nephrosis (f; DEM; FAD); Neuralgia (1; FAD; PH2); Ophthalmia (f; CEB); Orchosis (1; CRC; MAD; PHR); Ovariosis (1; PHR); Pain (1; APA; MAD); Pertussis (f; MAD); Pleurisy (1; PHR); Pleurosis (f; MAD); Pleurodynia (1; CRC; PHR); Rheumatism (1; APA; CRC; FAD; FNF); Sciatica (1; CRC; FAD; PH2); Scrofula (f; CRC); Sniffle (f; MAD); Sore Throat (1; APA; CRC); Sprain (1; BOW; FAD); Stomachache (1; DEM; FAD); Stomatosis (1; CRC; FNF); Streptococcus (1; FNF); Stroke (f; MIC); Swelling (1; X1875280); Tapeworm (f; DEM); Toothache (1; CRC; FNF); Typhus (f; MAD); Uterosis (f; MAD); UTI (1; FNF); VD (f; CRC; DEM); Water Retention (1; APA; CEB; FNF); Worm (f; DEM); Yeast (1; FNF).

Wintergreen for cold/flu:

- Analgesic: caffeic-acid; ferulic-acid; gallic-acid; gentisic-acid; methyl-salicylate; ursolic-acid
- Antiallergic: ferulic-acid
- Antibacterial: arbutin; caffeic-acid; ferulic-acid; gallic-acid; gentisic-acid; p-coumaric-acid; p-hydroxy-benzoic-acid; tannic-acid; vanillic-acid
- Antibronchitic: gallic-acid
- Antiflu: caffeic-acid; gallic-acid; lupeol
- Antihistaminic: caffeic-acid; ursolic-acid
- Antiinflammatory: alpha-amyrin; caffeic-acid; ferulic-acid; gallic-acid; gaultherin; gentisic-acid; lupeol; methyl-salicylate; ursolic-acid; vanillic-acid
- Antioxidant: caffeic-acid; ferulic-acid; gallic-acid; lupeol; p-coumaric-acid; p-hydroxy-benzoic-acid; tannic-acid; ursolic-acid; vanillic-acid
- Antipharyngitic: tannic-acid
- Antipyretic: methyl-salicylate
- Antiseptic: arbutin; caffeic-acid; ericolin; gallic-acid; methyl-salicylate; tannic-acid
- Antitussive: arbutin
- Antiviral: caffeic-acid; ferulic-acid; gallic-acid; gentisic-acid; lupeol; tannic-acid; ursolic-acid
- COX-2-Inhibitor: ursolic-acid
- Cyclooxygenase-Inhibitor: gallic-acid; ursolic-acid
- Immunostimulant: caffeic-acid; ferulic-acid; gallic-acid; tannic-acid
- Phagocytotic: ferulic-acid

Wintergreen for rheumatism:

- Analgesic: caffeic-acid; ferulic-acid; gallic-acid; gentisic-acid; methyl-salicylate; ursolic-acid
- Antiarthritic: ursolic-acid
- Antiedemic: alpha-amyrin; beta-amyrin; caffeic-acid; lupeol; ursolic-acid
- Antiinflammatory: alpha-amyrin; caffeic-acid; ferulic-acid; gallic-acid; gaultherin; gentisic-acid; lupeol; methyl-salicylate; ursolic-acid; vanillic-acid
- Antiprostaglandin: caffeic-acid
- Antirheumatic: gentisic-acid; lupeol; methyl-salicylate
- Antispasmodic: caffeic-acid; ferulic-acid; p-coumaric-acid
- COX-2-Inhibitor: ursolic-acid
- Counterirritant: methyl-salicylate
- Cyclooxygenase-Inhibitor: gallic-acid; ursolic-acid
- Elastase-Inhibitor: ursolic-acid
- Lipoxygenase-Inhibitor: caffeic-acid; p-coumaric-acid; ursolic-acid
- Myorelaxant: gallic-acid

Other Uses (Wintergreen) — Oil of Wintergreen is used as a flavoring agent in beers, beverages, candies, chewing gums (e.g., the now rare Teaberry Gum), soft drinks, and dental preparations, often combined with menthol and eucalyptus. One root beer remedy called for 4 drachms wintergreen oil, 2 sassafras oil, 1 clove oil, and ca. 120 g alcohol. The red to pinkish spicy fruits are eaten raw and used in jams, jellies, pies, and syrup. Amerindians ate the berries, even in the snow. Leaves used to make an herbal tea (Mountain Tea), as a condiment, and a nibble. Stronger teas, candies, and wines, are made from the fermented bright-red leaves. Amerindians smoked and chewed the dried leaves. I have steeped the leaves and berries in vodka for my "Teaberry Trip," even in midwinter. In summer, I like to add wild ginger and beebalm. Old timers steeped the leaves in brandy as a tonic liqueur. Weed (1985) recommends the vinegar tincture of wintergreen (CRC, FAC, LIL).

According to the Annals on Endocrinology, and Dominic and Pandey (1979), female mice do not return to estrus following exposure to males perfumed with oil of wintergreen (or a commercial perfume). Male urine may be the source of the primer pheromone involved in estrus induction, ineffective because of the masking by wintergreen oil. Unable to perceive the male pheromone, due to wintergreen oil, females remain in anestrus following exposure to perfumed males.

For more information on activities, dosages, and contraindications, see the *CRC Handbook of Medicinal Herbs*, ed. 2, Duke et al., 2002.

Cultivation (Wintergreen) — It's hard transplanting this common herb from the edges of the peat bogs and from acid sterile soils, probably because of an obligate mycorrhiza. Bown (2001) suggest propagation by seed, autumn sown on the soil surface, or by semi-ripe cuttings in summer, or by separating rooted suckers in spring. Hardy zones 3–7.

Chemistry (Wintergreen) — The active ingredient is methyl salicylate, now made synthetically. Commercial oil of wintergreen, or oil of checkerberry, is obtained from distillation of the twigs of black birch. The volatile oil contains 98–99% methyl salicylate. Arbutin, ericolin, gallic acid, gaultherine, gaultherilene, gaultheric acid, mucilage, tannin, wax, an ester, triacontane, and a secondary alcohol are also reported. Other acids reported include O-pyrocatechusic-, gentisinic-, salicylic-, p-hydroxybenzoic-, protocatechuic-, vanillic-, syringic-, p-coumaric-, caffeic-, and ferulic-acids. Here are a few of the more notable chemicals found in wintergreen. For a complete listing of the phytochemicals and their activities, see the CRC phytochemical compendium, Duke and duCellier, 1993 (DAD) and the USDA database http://www.ars-grin.gov/duke/.

Aloesin — Laxative; Sunscreen

Arbutin — Allelochemic IC51 = 1.1 mM; Antibacterial MIC = 4000–8000 ppm; Antiedemic 50 mg/kg; Antimelanogenic IC50 = 40 μ*M*; Antiseptic 60–200 mg/man; Antistreptococcic MIC = 4000–8000 ppm; Antitussive; Artemicide; Candidicide; Diuretic 60–200 mg/man; Insulin-Sparer; Mycoplasmistat; Tyrosinase-Inhibitor IC50 = 40 μ*M*; Urinary-Antiseptic.

Gaultherin — Antiinflammatory; Diuretic.

Methyl-Salicylate — ADI = 500 μg/mg; Analgesic; Anticancer; Antiinflammatory; Antipyretic; Antiradicular; Antirheumatalgic; Antiseptic; Carminative; Counterirritant; Dentifrice; Insectifuge; LDlo = 170 orl hmn; LD50 = 887 orl rat; LD50 = 1110 orl mus; LD50 = 4 ml orl chd; LD50 = 30 ml orl man.

G

Glycyrrhiza glabra L. (Fabaceae)
LICORICE

Medicinal Uses (Licorice) — It is considered alexeritic, alterative, antipyretic, demulcent, deodorant (lf), depurative, diuretic, emollient, expectorant, estrogenic, laxative, pectoral, and sudorific. With a long history of use for indigestion and inflamed stomach, licorice provides two derivatives that reduce or cure ulcers. When I was being heavily medicated for slipped disk, especially with ulcerogenic NSAIDs, I took a lot of licorice to prevent the ulcer the NSAIDs might cause. Its mucilage makes a natural demulcent, not only for the stomach but for the throat and other mucous membranes. Modern studies hint that sweets trigger endorphin production, and that may be why we have so many sweet OTC cough drops. The licorice treatments help too. One could boil frayed roots in water to relieve a cough or sore throat. Not too much though! Ingestion of 280 mg/kg licorice per day for four weeks triggered cardiac problems, GI problems, and hypertension (RIN). In India, it is chewed with betel. Chinese use the frayed roots as a preferred "toothbrush," which cleans plaque better with a scouring motion than most types of toothbrush, while not irritating the gums like the brushing action currently used here.

Heartwise, four isoflavans and two chalcones proved to be very potent antioxidants against LDL oxidation, possibly preventing formation of atherosclerotic lesions (glabradin and isoflavan were both most abundant and active), according to Israel researchers (TAD). UK researchers conclude, as do I, that the antioxidant activity of the whole root is more a synergistic result of the whole mix rather than isolated flavonoid. May I predict that allopathic pharmaceutical manufacturers will seize on one of the stronger, modifying and making it more unnatural but much more patentable (JAD).

Isoflavonoids are antiseptic against bacteria, candida, mycobacterium, and staphylococcus (CAN), and antiviral against Epstein-Barr, herpes, Newcastle, vaccinia, and vesicular stomatitis virus with no activity toward polio (CAN). Glycyrrhizin and/or GA inhibit *in vitro* viruses like chickenpox, herpes, HIV-1, HHV-6, HHB-7 et al. Glycyrrhizin reduces morbidity and mortality of mice infected with lethal doses of flu virus. And long-time administration of licorice to hepatitis C patients prevented liver cancer. Glycyrrhizin is also active against test tube leukemia (TAD). Ma et al. (2001) note that apoptosis, induced by isoliquiritigenin, may be helpful in stomach cancer. Isoliquiritigenin may be a principal antitumor constituent of licorice.

Yarnell and Abascal (2000) described licorice as one of the most thoroughly studied botanical immunomodulators for HIV patients, both *Glycyrrhiza glabra* and *Glycyrrhiza uralensis*. Glycyrrhizin, the major active ingredient seems to serve both as immunomodulator and an antiviral, an ideal combination for HIV infection. Intravenous glycyrrhizin also improves liver function in uncontrolled clinical trials in HIV patients. But there are other useful constituents besides glycyrrhizin. Hence, whole plant extracts should be studied. Whole licorice extracts should be compared head-to-head with glycyrrhizin to determine relative efficacy and safety (Yarnell and Abascal, 2000). Standish et al. (2002) note that glycyrrhizin *in vitro* inhibits viral attachment or fusion. "It may also inhibit protein kinase C (PKC), an activator of NF-kB. When 400–1600 mg glycyrrhizin was administered ivn to three HIV-positive hemophiliacs, their viral load was substantially lower after a month. Another small ivn study in hemophiliacs showed lymphocyte improvement in all nine patients, CD4+/CD8+ ratios increased in six, and CD4 lymphocyte levels increased in eight. In a third observational study (4 and 7 years, daily doses 150–225 mg/day), patients who received glycyrrhizin when CD4 T-cell counts were above 200 maintained those counts; ditto for those starting with counts above 500. Two researchers independently reported immune enhancements (CD4 T-cell counts, CD4+/CD8+ ratios, lymphocytes, NKC) in 22 hemophiliac patients over periods of 3–11 years (Standish et al., 2002).

And glycyrrhizin blocks estrogen effects binding to estrogen receptors, hence the antiestrogenic activities reported. Estrogenic activity has also been attributed to the isoflavones, but these too may bind to estrogen receptors (JAD, CAN). Maybe this is one of those amphoteric herbs. "Liquorice exhibits an alternative action on estrogen metabolism, causing inhibition if oestrogen concentrations are high and potentiation when concentrations are low" (CAN). I've heard the same things about clover's phytoestrogenic isoflavones, some of which are shared with licorice. Strandberg et al. (2001) studied birth outcome in relation to licorice consumption during pregnancy. Heavy glycyrrhizin exposure during pregnancy did not significantly affect birth weight or maternal blood pressure but was significantly associated with lower gestational age.

Glycyrrhizin not only has its own antiarthritic, antiedemic, and antiinflammatory activities, it potentiates the antiarthritic activities of hydrocortisone, at least in rats (MPI). Glycyrrhizin even potentiates cocoa. Licorice seems also to potentiate prednisolone in five pemphigus patients kept free of bullae with prednisolone. Licorice seems to potentiate by inhibiting metabolic degradations of prednisolone (MPI).

In a study of clinical applications of ayurvedic herbs, Khan and Balick (2001) note human studies on licorice for acne vulgaris, chronic duodenal ulcers, chronic hepatitis, and diabetic hyperkalemia. Oral DGGL (380 mg, 3 × day) equaled antacids or cimetidine in 169 patients with chronic duodenal ulcers. Interestingly, GA inhibits growth of the ulcer bacteria, *Helicobacter pylori* (TAD). Oral dose of GA as antitussive orally as codeine (LEG, LIL, MAB, TAD).

Indications (Licorice) — Abscess (f; DAA); Addison's Disease (1; DAA; FAY; PED; WHO); Adenosis (f; JLH); Adrenal Insufficiency (1; CAN; PNC; WHO); Allergy (f; BOW); Alzheimer's

G

(1; COX; FNF); Ameba (1; FAD); Anemia (f; DAA); Anorexia (f; DAA; WHO); Anxiety (1; BGB); Appendicitis (f; PH2; VVG; WBB; WHO); Arthrosis (1; COX; MAB; WHO); Asthenia (f; DAA); Asthma (1; BGB; DEP; FAD; FAY; KAB; SKY); Atherosclerosis (1; AKT); Bacteria (1; DAA; FNF); Biliousness (f; KAB); BO (f; KAB); Boil (f; DAA; MAB); BPH (1; FNF); Bronchosis (2; DEP; FAD; FAY; FNF; KAB; PHR; PH2; SKY; WHO); Bug Bite (f; VVG); Burn (f; DAA); Cancer (1; COX; DAA; FNF; HOX); Cancer, abdomen (1; FNF; JLH); Cancer, bladder (1; FNF; JLH); Cancer, breast (1; FNF; JLH); Cancer, colon (1; FNF; JLH); Cancer, gland (1; FNF; JLH); Cancer, kidney (1; FNF; JLH); Cancer, liver (1; FNF; JLH); Cancer, neck (1; FNF; JLH); Cancer, spleen (1; FNF; JLH); Cancer, stomach (1; FNF; JLH; PM67:754); Cancer, throat (1; FNF; JLH); Cancer, uterus (1; FNF; JLH); Cancer, uvula (1; FNF; JLH); Candida (1; APA; FNF); Canker Sore (f; SKY); Carbuncle (f; FAY; PH2); Caries (1; WHO); Cardiopathy (f; WHO); Cataract (1; CAN); Catarrh (2; DEP; KOM; PH2; PIP; WHO); CFS (1; MAB; SKY); Chickenpox (1; TAD); Cholecystosis (1; FAD); Cirrhosis (f; AKT); Cold (1; APA; CRC); Colic (f; CAN; KAB); Condyloma (f; JLH); Congestion (1; APA; FNF); Conjunctivosis (1; MAB; MPI; PH2); Constipation (f; APA; MAB; PH2; WAM); Cough (2; APA; DAA; DEP; FAD; FAY; FNF; KAB; PHR; PH2; PED; SUW; VVG); Cramp (1; FAY; FNF; MAB; VVG); Cystosis (f; CRC; MAD); Cytomegalovirus (1; PH2); Depression (f; MAB); Dermatosis (f; PH2); Diabetes (1; MAB); Diarrhea (f; DAA); Diphtheria (f; WHO); Duodenosis (f; PH2); Dysmenorrhea (f; APA); Dyspepsia (1; CRC; SKY; WHO); Dyspnea (f; DAA); Dysuria (f; MAD); Earache (f; APA); Eczema (1; SKY; WAM); Encephalosis (1; MAB); Enterosis (f; KAB; MPI); Epigastrosis (1; BGB; VVG); Epilepsy (f; KAB; WHO; PH2); Fatigue (f; KAB); Fibromyalgia (f; SKY); Flu (1; MAB; PH2; TAD); Fungus (1; FNF); Gastrosis (2; CAN; DAA; FAD; FAY; PHR; PH2); Hay Fever (1; WAM); Headache (f; PH2); Heartburn (f; SKY); Hemicrania (f; KAB); Hemophilia (1; BGB); Hemoptysis (f; KAB); Hemorrhoid (f; DAA; WHO; VVG); Hepatosis (2; APA; FAY; FNF; PHR; PH2; PNC); Hepatosis C (1; MAB); Herpes (1; AKT; APA; MAB; WAM); Hiccup (f; KAB); HIV (1; FNF; MAB; TAD); Hoarseness (f; DEM; DEP; FAY; HHB); Hot Flash (1; AKT); Hyperphagia (1; MAB); Hyperthyroid (f; DAA); Hysteria (f; FAY); IBD (1; WAM); Immunodepression (1; FNF); Induration (f; JLH); Infection (1; FNF); Inflammation (1; DAA; DEP; FNF; MPI; WBB); Itch (f; VVG); Kidney Stone (f; WHO); Laryngosis (f; DAA); Leukemia (1; FNF); Lichen Planus (1; MAB); Low Blood Pressure (1; MBB; PH2); Lupus (SLE) (1; APA); Malaria (1; DAA; MAB); Malaise (f; FAY); Melanoma (1; FNF; TAD); Mucososis (1; FAD; MAB); Nausea (f; DAA; KAB); Nephrosis (1; CAN; MAD); Neuropathy (1; CAN); Ophthalmia (f; KAB); Otosis (f; KAB); Pain (1; DAA; FNF; KAB; KAP; MBB); Pemphigus (1; MPI); Pharyngosis (1; BGB; DAA); PMS (1; WAM); Pneumonia (f; MAD); Polycystic Ovary Syndrome (1; BGB; MAB); Polyp (f; JLH); Psoriasis (1; WAM); Pterygium (f; JLH); Respirosis (2; APA; DEP; KOM; PIP); Retinosis (1; CAN); Rheumatism (1; FAY; WHO); Rhinosis (f; JLH); Senility (f; DAA); Shingles (1; BOW; MAB); Snakebite (f; KAB; WHO); Sore (f; DAA; KAB); Sore Throat (1; APA; DAA; KAB; PH2; SUW; WAM; WHO); Splenosis (1; DAA; FAY; MAD; PH2); Staphylococcus (1; FAY); Sting (f; SUW); Stomatosis (f; MAB); Strangury (f; MAD); Sunburn (f; VVG); Swelling (f; DAA); Tetanus (f; WHO); Thirst (f; CRC; DAA; DEP); Thrombosis (f; PH2); Trichomonas (1; FAY); Tuberculosis (1; DAA; FAY; KAB; MAB; MAD; VVG; WBB; WHO); Ulcer (2; AKT; DAA; FAY; FNF; KOM; PHR; PH2; PIP; WAM); Urethrosis (f; WBB); Urogenitosis (f; DEP; HHB; SUW); UTI (1; MAB); Vaginosis (1; APA); Vertigo (f; BGB; WHO); Viral Hepatosis (f; PHR); Virus (1; FNF; PH2); Voice (f; KAB); Wound (f; KAB; PH2); Yeast (1; APA; PH2).

Licorice for cancer:

- Adaptogen: paeonol
- Alpha-Reductase-Inhibitor: genistein
- AntiEBV: glycyrrhetinic-acid
- AntiHIV: apigenin; betulinic-acid; glycycoumarin; glycyrrhisoflavone; glycyrrhizin; iso-licoflavonol; licochalcone-a; licopyranocoumarin; lignin; naringenin; quercetin

- Antiaggregant: apigenin; bergapten; estragole; eugenol; ferulic-acid; genistein; isoliquiritigenin; kaempferol; ligustrazine; naringenin; paeonol; quercetin; tetramethyl-pyrazine; thymol
- Antiangiogenic: apigenin; genistein
- Antiarachidonate: eugenol
- Anticancer: alpha-terpineol; anethole; apigenin; benzaldehyde; bergapten; camphor; estragole; eugenol; ferulic-acid; formononetin; galangin; genistein; geraniol; glabrene; glabridin; glabrol; glycyrrhetic-acid; glycyrrhetinic-acid; glycyrrhizic-acid; glycyrrhizin; indole; isoliquiritigenin; isoquercitrin; kaempferol; licocoumarone; licoflavanone; lignin; linalool; liquiritigenin; maltol; methyl-salicylate; naringenin; o-cresol; p-hydroxy-benzoic-acid; p-methoxy-phenol; phenol; quercetin; salicylic-acid; sinapic-acid; soyasaponin; umbelliferone; vitexin; xanthotoxin
- Anticarcinomic: betulinic-acid; ferulic-acid; hederasaponin-c
- Antiestrogenic: apigenin; estriol; ferulic-acid; genistein; glycyrrhetic-acid; glycyrrhetinic-acid; glycyrrhizic-acid; glycyrrhizin; quercetin
- Antifibrosarcomic: quercetin
- Antihepatotoxic: ferulic-acid; glucuronic-acid; glycyrrhetic-acid; glycyrrhetinic-acid; glycyrrhizin; naringenin; quercetin; sinapic-acid
- Antihyaluronidase: apigenin
- Antiinflammatory: apigenin; bergapten; betulinic-acid; carvacrol; eugenol; ferulic-acid; galangin; genistein; glycyrrhetic-acid; glycyrrhetinic-acid; glycyrrhizic-acid; glycyrrhizin; isoliquiritin; kaempferol; licochalcone-a; liquiritic-acid; liquiritigenin; liquiritin; lupeol; mannitol; methyl-salicylate; naringenin; neoisoliquiritin; neoliquiritin; paeonol; quercetin; salicylic-acid; thymol; umbelliferone; vitexin; xanthotoxin
- Antileukemic: apigenin; astragalin; genistein; kaempferol; liquiritigenin; naringenin; pinocembrin; quercetin
- Antileukotriene: genistein; licochalcone-a; quercetin
- Antilipoperoxidant: quercetin
- Antilymphomic: genistein; xanthotoxin
- Antimelanomic: apigenin; betulinic-acid; carvacrol; genistein; geraniol; glycyrrhetic-acid; glycyrrhetinic-acid; quercetin; thymol
- Antimetastatic: apigenin; tetramethyl-pyrazine
- Antimicrobial: genistein
- Antimutagenic: apigenin; eugenol; ferulic-acid; galangin; genistein; glabrene; glycyrrhetic-acid; glycyrrhetinic-acid; glycyrrhizin; kaempferol; mannitol; n-nonacosane; naringenin; o-cresol; p-hydroxy-benzoic-acid; paeonol; quercetin; saponins; umbelliferone; xanthotoxin
- Antineoplastic: ferulic-acid
- Antinephrotic: anethole
- Antineuroblastomic: genistein
- Antinitrosaminic: ferulic-acid; lignin; quercetin
- Antioxidant: apigenin; carvacrol; eugenol; ferulic-acid; genistein; glycyrrhetic-acid; glycyrrhetinic-acid; glycyrrhizin; isoquercitrin; kaempferol; lignin; lupeol; maltol; mannitol; naringenin; p-hydroxy-benzoic-acid; phenol; quercetin; salicylic-acid; sinapic-acid; thymol; vitexin
- Antiperoxidant: galangin; lupeol; quercetin
- Antiperoxidative: naringenin
- Antiproliferant: apigenin; quercetin
- Antiproliferative: genistein
- Antiprostaglandin: carvacrol; eugenol; glycyrrhizin; umbelliferone
- Antistress: paeonol
- Antithromboxane: eugenol

- Antitumor: apigenin; benzaldehyde; bergapten; betulinic-acid; eugenol; ferulic-acid; geraniol; glycyrrhetic-acid; glycyrrhetinic-acid; isoquercitrin; kaempferol; lignin; lupeol; naringenin; quercetin; salicylic-acid; xanthotoxin
- Antiviral: apigenin; betulinic-acid; ferulic-acid; galangin; genistein; glabranin; glycycoumarin; glycyrrhetic-acid; glycyrrhetinic-acid; glycyrrhisoflavone; glycyrrhizic-acid; glycyrrhizin; isolicoflavonol; kaempferol; licochalcone-a; licopyranocoumarin; lignin; linalool; lupeol; naringenin; p-cymene; phenol; quercetin
- Anxiolytic: apigenin
- Apoptotic: apigenin; genistein; kaempferol; quercetin
- Beta-Glucuronidase-Inhibitor: apigenin
- COX-2-Inhibitor: apigenin; eugenol; kaempferol; quercetin; salicylic-acid
- Cyclooxygenase-Inhibitor: apigenin; carvacrol; galangin; kaempferol; quercetin; salicylic-acid; thymol
- Cytotoxic: apigenin; betulinic-acid; eugenol; genistein; lupeol; pinocembrin; quercetin; xanthotoxin
- DNA-Binder: estragole
- Estrogen-Agonist: genistein
- Hepatoprotective: betaine; eugenol; ferulic-acid; glycyrrhetic-acid; glycyrrhetinic-acid; glycyrrhizic-acid; glycyrrhizin; herniarin; quercetin; soyasaponin
- Immunostimulant: anethole; astragalin; benzaldehyde; ferulic-acid; glycyrrhetic-acid; glycyrrhetinic-acid; glycyrrhizic-acid; glycyrrhizin
- Interferonogenic: glycyrrhetic-acid; glycyrrhetinic-acid; glycyrrhizin
- Leucocytogenic: anethole
- Lipoxygenase-Inhibitor: galangin; kaempferol; quercetin; umbelliferone
- Mast-Cell-Stabilizer: quercetin
- Mitogenic: glycyrrhizin
- Ornithine-Decarboxylase-Inhibitor: apigenin; ferulic-acid; genistein; quercetin
- p450-Inducer: quercetin
- PKC-Inhibitor: apigenin
- PTK-Inhibitor: apigenin; genistein; quercetin
- Prostaglandigenic: ferulic-acid; p-hydroxy-benzoic-acid
- Protein-Kinase-C-Inhibitor: apigenin; quercetin
- Reverse-Transcriptase-Inhibitor: glycyrrhizin
- Sunscreen: apigenin; ferulic-acid; umbelliferone
- Topoisomerase-II-Inhibitor: apigenin; genistein; isoquercitrin; kaempferol; quercetin
- Topoisomerase-II-Poison: genistein
- Tyrosine-Kinase-Inhibitor: genistein; quercetin

Licorice for ulcer:

- Analgesic: camphor; eugenol; ferulic-acid; glycyrrhizin; methyl-salicylate; p-cymene; paeonol; phenol; quercetin; salicylic-acid; thymol
- Anesthetic: benzaldehyde; benzoic-acid; benzyl-alcohol; camphor; carvacrol; eugenol; guaiacol; linalool; phenol; pinocembrin; thymol
- Antibacterial: acetic-acid; acetophenone; alpha-terpineol; anethole; apigenin; benzaldehyde; benzoic-acid; carvacrol; eugenol; ferulic-acid; geraniol; glabridin; glabrol; glycyrrhetic-acid; glycyrrhetinic-acid; glycyrrhizic-acid; glycyrrhizin; guaiacol; herniarin; hispaglabridin-a; hispaglabridin-b; indole; isoquercitrin; kaempferol; lignin; linalool; naringenin; p-cymene; p-hydroxy-benzoic-acid; paeonol; phenethyl-alcohol; phenol; pinocembrin; quercetin; salicylic-acid; sinapic-acid; terpinen-4-ol; tetramethyl-pyrazine; thujone; thymol; umbelliferone; xanthotoxin

- Antiinflammatory: apigenin; bergapten; betulinic-acid; carvacrol; eugenol; ferulic-acid; galangin; genistein; glycyrrhetic-acid; glycyrrhetinic-acid; glycyrrhizic-acid; glycyrrhizin; isoliquiritin; kaempferol; licochalcone-a; liquiritic-acid; liquiritigenin; liquiritin; lupeol; mannitol; methyl-salicylate; naringenin; neoisoliquiritin; neoliquiritin; paeonol; quercetin; salicylic-acid; thymol; umbelliferone; vitexin; xanthotoxin
- Antioxidant: apigenin; carvacrol; eugenol; ferulic-acid; genistein; glycyrrhetic-acid; glycyrrhetinic-acid; glycyrrhizin; isoquercitrin; kaempferol; lignin; lupeol; maltol; mannitol; naringenin; p-hydroxy-benzoic-acid; phenol; quercetin; salicylic-acid; sinapic-acid; thymol; vitexin
- Antiprostaglandin: carvacrol; eugenol; glycyrrhizin; umbelliferone
- Antiseptic: alpha-terpineol; anethole; benzaldehyde; benzoic-acid; benzyl-alcohol; camphor; carvacrol; cresol; eugenol; furfural; geraniol; glabranin; glabrene; glabridin; glabrol; glycyrrhizin; guaiacol; hexanol; hispaglabridin-a; hispaglabridin-b; kaempferol; linalool; methyl-salicylate; o-cresol; oxalic-acid; phenethyl-alcohol; phenol; pinocembrin; salicylic-acid; terpinen-4-ol; thujone; thymol; umbelliferone
- Antispasmodic: anethole; apigenin; benzaldehyde; bergapten; camphor; carvacrol; eugenol; ferulic-acid; genistein; geraniol; herniarin; isoliquiritigenin; kaempferol; linalool; liquiritigenin; mannitol; naringenin; quercetin; tetramethyl-pyrazine; thujone; thymol; umbelliferone; xanthotoxin
- Antiulcer: eugenol; glycyrrhetic-acid; glycyrrhetinic-acid; glycyrrhizic-acid; glycyrrhizin; isoliquiritigenin; kaempferol; licochalcone-a; liquiritigenin; liquiritin; naringenin
- Antiviral: apigenin; betulinic-acid; ferulic-acid; galangin; genistein; glabranin; glycycoumarin; glycyrrhetic-acid; glycyrrhetinic-acid; glycyrrhisoflavone; glycyrrhizic-acid; glycyrrhizin; isolicoflavonol; kaempferol; licochalcone-a; licopyranocoumarin; lignin; linalool; lupeol; naringenin; p-cymene; phenol; quercetin
- Bacteristat: malic-acid; quercetin
- COX-2-Inhibitor: apigenin; eugenol; kaempferol; quercetin; salicylic-acid
- Cyclooxygenase-Inhibitor: apigenin; carvacrol; galangin; kaempferol; quercetin; salicylic-acid; thymol
- Fungicide: acetic-acid; acetophenone; anethole; benzoic-acid; camphor; carvacrol; eugenol; ferulic-acid; formononetin; furfural; genistein; geraniol; hederasaponin-c; herniarin; isoliquiritin; isomucronulatol; licoisoflavone-a; linalool; liquiritigenin; liquiritin; naringenin; octanoic-acid; p-cymene; paeonol; phenol; pinocembrin; propionic-acid; prunetin; quercetin; salicylic-acid; sinapic-acid; terpinen-4-ol; thymol; umbelliferone; xanthotoxin
- Lipoxygenase-Inhibitor: galangin; kaempferol; quercetin; umbelliferone
- MDR-Inhibitor: genistein
- Vulnerary: benzoic-acid; terpinen-4-ol

Other Uses (Licorice) — Grown primarily for its dried rhizome and roots, it is a condiment and used to flavor candies and tobaccos. Roots are the source of licorice powder and extract, widely employed in baked goods, confection, ice cream, soft drinks, etc. Dutch confectioners produce sweet and salty versions of licorice drops. Some candy makers augment the anethole flavor with oil of anise. Most "licorice" candy contains little or no licorice but is flavored with anise oil. Glycyrrhizin is some 50–100 times sweeter than sugar. Glycyrrhizin potentiates the flavor of cocoa, replacing 25% cocoa in manufactured products. A tea made from the powdered rootstock is consumed directly or added to other herbal teas to sweeten them naturally. I used them to stir herbal teas, sweetening them gently in the process. I also use them to make liqueurs. One old cold remedy, which I call "licorice liqueur," was a British beer brewed of licorice, fennel, anise seed, elecampane, and sassafras. I steep them all in vodka to make the licorice liqueur. Maybe it cures the cold; surely it intoxicates. Others put 4 tbsp licorice root in a fifth of vodka and let it stand a couple of days in a dark place, then strain and add sugar syrup (1 cup sugar dissolved in 1/2 cup water). In nineteenth-century Paris, a drink was made from licorice and lemon called "coco" (AAR). Adding lemon juice to a licorice infusion produces a rather full-bodied lemonade (AAR). Legend suggests that Scythian warriors could survive 12 days without drink if supplied with licorice and mare's milk cheese (TAD). According to Rose, quantities of licorice were stored in King Tut's tomb. Licorice sticks and brewer's licorice are added to beer for increased head retention and to give thickness, flavor, and blackness to porter and stout (FAC). Singers chew the root to strengthen the throat. In India, licorice root is chewed with betel. Licorice is used in Chinese cooking in soups and meats slow-braised until tender in multispiced soy sauce. Ground licorice root sometimes appears in Chinese five-spice powder (AAR). Grain coffees can be flavored with the extract (FAC). The leaves, called "nakhalsa," are used as a substitute for tea in Mongolia (FAC). Most licorice production is used by the tobacco industry and for the preparation of licorice paste, licorice extract, powdered root, and mafeo syrup. Spent licorice serves in fire-extinguishing agents, to insulate fiberboards, and as a compost for growing mushrooms; also, in feed for cattle, horses, and chickens. (AAR, CRC, FAC, LEG, LIL).

For more information on activities, dosages, and contraindications, see the *CRC Handbook of Medicinal Herbs*, ed. 2, Duke et al., 2002.

Cultivation (Licorice) — As Tucker and DeBaggio (2000) put it, licorice is so easily grown that it has the potential to become a weed. Bown (2001) lists it for zones 7–9. Seed should be scarified or soaked overnight and sown in spring. Perennial plants can be divided in spring or autumn. Stolon cuttings better made in spring (Bown, 2001). Licorice seems to thrive in full sun, on deep sandy soils, moist but not waterlogged, pH 5.5–8.2 (average 7.1). It is usually planted in spring from crown divisions spaced about 18 in apart in the rows. Plants should be weeded or intercropped with catch crops such as cabbage, carrot, or potato. It takes 3–4 years to produce marketable roots. Flowers are often pinched off to encourage root production. Apparently, glycyrrhizin is maximal at flowering time. And the thicker the root, the more glycyrrhizin. Harvest is labor intensive as the roots are deep and as long as 25 ft (8 m). Roots are washed after digging and then dried. Root yields often run 2000–4000 lb/a (= ~2000–4000 kg/ha), but yields of 22,000 lb/a dried root have been attained, from 50,000 lb/a in Russia. Cultivated stands produce two to three times more root than weedy stands in Eurasia (LEG, TAD).

Chemistry (Licorice) — Here are just a few of the chemicals, some almost restricted to licorice. For a complete listing of the phytochemicals and their activities, see the CRC phytochemical compendium, Duke and duCellier, 1993 (DAD) and the USDA database http://www.ars-grin.gov/duke/.

Glycyrrhetinic-Acid — Beta-Reductase-Inhibitor; Antiaddisonian; Antiallergic 600 mg/kg ipr; Anti-anaphylactic 600 mg/kg ipr; Antiarthritic; Antiasthmatic; Antibacterial; Anticancer; Anticirrhotic; Anticomplement 0.1 μM; Antidiuretic; AntiEBV; Antiedemic; Antierythemic; Antiestrogenic; Anti-hepatotoxic 50–500 μg/ml; Antiherpetic; Antiinflammatory; Antimelanomic; Antimutagenic; Anti-oxidant; Antiprostatic; Antirheumatic; Antistomatitic; Antitumor-Promoter; Antitussive; Antiulcer; Antivaccinia; Antiviral; Hepatoprotective; Hypertensive; Immunostimulant; Interferongenic; Minertalcorticoid; Ornithine-Decarboxylase-Inhibitor.

Glycyrrhinic-Acid — Antihistaminic; Antiophidic; Antitetanic; Antitoxic.

Glycyrrhisoflavanone — Xanthine-Oxidase-Inhibitor IC50 = >100 μM.

Glycyrrhisoflavone — Antiherpetic EC50 = >10 μg/ml; AntiHIV; Antiviral EC50 = >10 μg/ml, 20 μg/ml; MAOI IC50 = 60–140 μM, IC50 = 95; Xanthine-Oxidase-Inhibitor IC50 = 53 μM.

Glycyrrhizan-UC — Immunomodulator; RES-Activator.

Glycyrrhizic-Acid — Antiaddisonian; Antibacterial; Anticancer; Anticirrhotic; Antiestrogenic; Anti-herpetic; Antiinflammatory; Antirheumatic; Antistomatitic; Antitussive; Antiulcer; Antiviral; 11B-HSD-Inhibitor; Hepatoprotective; Hypertensive; Immunostimulant.

Glycyrrhizin — Adrenocorticotropic; Amphiestrogenic; Analgesic; Antiaddisonian; Antiaggregant; Antiallergic; Antianaphylactic; Antiarthritic; Antiasthmatic; Antibacterial; Anticancer; Anticapillary-Fragility; Anticariogenic; Anticataract; Anticirrhotic; Antidiptheric; Antidote; Antiedemic; Antiestrogenic; Antiflu; Antigingivitic; Antihepatosis; Antihepatotoxic 1000 µg/ml; Antiherpetic; AntiHIV 0.6 mM; Antiinflammatory; Antimutagenic; Antioxidant; Antiplaque; Antiprostaglandin; Antiradicular; Antirheumatic; Antiseborrheic; Antiseptic; Antistomatitic; Antitetanic; Antithrombic; Antitussive; Antiulcer; Antivaccinic; Antiviral 8 mM; Candidicide; Choleretic; Detoxicant; Ergogenic 40–80 mg/may/day/shortterm; Estrogenic; Expectorant; Hepatoprotective; Hypertensive; Immunostimulant; Interferonogenic; MAOI IC50 = 160 µM; Mineralcorticoid; Mitogenic; Pancreaprotective; Pseudoaldosteronistic; Reverse-Transcriptase-Inhibitor; Sweetener (50 × sucrose).

Isolicoflavanol — AntiHIV 20 µg/ml; Antiviral 20 µg/ml; Aromatase-Inhibitor IC50 = 0.1 µM.

Isoliensinine — Antitumor; Cytotoxic 16 ppm.

Isoliquiritigenin — Aldose-Reductase-Inhibitor IC50 = 0.32 µM; Antiaggregant; Anticancer; Antidepressant; Antidiabetic; Antiperoxidase; Antispasmodic; Antiulcer; Apoptotic; Cyclooxygenase-Inhibitor; Lipoxygenase-Inhibitor; MAOI EC50 = 17 µM, IC50 = >200; Pigment.

Isoliquiritin — Aldose-Reductase-Inhibitor IC50 = 0.72 µM; Antiangiogenic; Antiinflammatory; Fungicide; MAOI; Phytoalexin.

Licocoumarone — Anticancer; MAOI IC50 = 60–140 µM, IC50 = 60 µM; Xanthine-Oxidase-Inhibitor IC50 = 13 µM.

Licoflavanone — Anticancer.

Licofuranone — MAOI IC50 = 60–140 µM, IC50 = 87 µM.

Licoisoflavone-A — Antifeedant 1.2 ppm; Fungicide ED50 = <50; Phytoalexin.

Licoisoflavone-B — Antifeedant ED50 = <1 ppm.

Licopyranocoumarin — Antiherpetic EC50 = >10 µg/ml; AntiHIV 20 µg/ml; Antiviral EC50 = >10 µg/ml; MAOI IC50 = 60–140 µM, IC50 = 140 µM; Xanthine-Oxidase-Inhibitor IC50 = >100 µM.

Licoricin — Antiaggregant; Antiinflammatory.

Licorione — Antiulcer.

Liquiritigenin — Anticancer; Antidepressant; Antiinflammatory; Antileukemic IC50 = 0.290 µg/ml; Antispasmodic; Antiulcer; CNS-Active; Fungicide ED50 = >100; Hemoglobin-Inducer; MAOI; Phytoalexin.

Liquiritigenin-Chalcone — Antispasmodic; Antiulcer.

Liquiritin — Antiinflammatory; Antiulcer; Fungicide; Xanthine-Oxidase-Inhibitor.

Liquiritone — Antiinflammatory.

Liquiritoside — Analgesic; Anticonvulsant.

H

Hibiscus sabdariffa L. (Malvaceae)
INDIAN SORREL, JAMAICA SORREL, KHARKADI, RED SORREL, ROSELLE

Medicinal Uses (Roselle) — Writing this at 6:00 A.M., January 1, 2001, starting the new millennium, I couldn't help but be delighted at my new hangover remedy, possibly aphrodisiac, that just might prevent cancer, lower blood pressure and blood sugar, and could be a cosmetic for face peels. I call it "Kharkarindo," reflecting its major ingredients, kharkadi (dried-flowers minus-ovary) and tamarindo, also covered in this spice book. Yes, both kharkadi and tamarindo are loaded with tasty hydroxycitric "fruit" acids. Calling it "sour tea," Haji Faraji and Haji Tarkhani (1999) studied ca. 50 patients with moderate essential hypertension. Patients with secondary hypertension or consuming more than two drugs were excluded. Statistical findings showed 11.2% lowering of the systolic blood pressure and a 10.7% decrease of diastolic pressure in the experimental group 12 days after beginning the sour tea treatment, as compared with the first day, proving the "public belief" in hypotensive activity of sour tea (X10404421). Adegunloye et al. (1996) demonstrated an antihypertensive effect of aqueous roselle extracts, which caused a dose-dependent decrease in mean arterial pressure of rats.

Reported to be antipyretic, antiseptic, aphrodisiac, astringent, cholagogue, demulcent, digestive, diuretic, emollient, laxative, resolvent, sedative, stomachic, and tonic. The flowers contain gossypetin, anthocyanin, and glucoside hibiscin, which may have diuretic and choleretic effects, decreasing the viscosity of the blood, reducing blood pressure, and stimulating intestinal peristalsis. A drink, made by placing the calyx in water, is even a folk remedy for cancer. Medicinally, leaves are emollient and are much used in Guinea as a antipyretic, diuretic, and sedative. Fruits are antiscorbutic. Leaves, seeds, and ripe calyces are diuretic and antiscorbutic, and the succulent calyx, boiled in water, is used as a drink in bilious attacks. In Burma, the seed are used for debility, the leaves as emollient. Angolans use the mucilaginous leaves as an emollient and as a soothing cough remedy. Central Africans poultice the leaves on abscesses. Taiwanese regard the seed as diuretic, laxative, and tonic. Philippines use the bitter root as an aperitif and tonic. Alcoholics might consider one item: simulated ingestion of the plant extract decreased the rate of absorption of alcohol, lessening the intensity of alcohol effects in chickens (DAD).

Sachdewa et al. (2001) demonstrated the hypoglycemic activity of the leaf extract in glucose and streptozotocin induced hyperglycemic rats. Administration of the extract once a day for seven days, at an oral dose equivalent to 250 mg/kg, significantly improved glucose tolerance in rats. The peak blood glucose level was obtained at 30 min of glucose load (2 g kg(–1)), thereafter a decreasing trend was recorded up to 120 min. At 250 mg/kg, the efficacy of the extract was 51.5% of tolbutamide at 100 mg/kg. The data suggest that the hypoglycemic activity is comparable to tolbutamide and not to glibenclamide treatment.

Chewonarin et al. (1999) found the 80% ethanol extract of roselle was antimutagenic and chemopreventive in several colon cancer models. Tseng et al. (2000) reported that hibiscus protocatechuic acid induced apoptosis in human leukemia cells. This phenolic compound from the dried flower also had antioxidant and antitumor promotion effects in their studies (Tseng et al., 2000).

If we can believe Kasture et al. (2000), the ethanolic extracts of the flowers have anticonvulsant, anxiogenic, CNS-depressant, and serotoninergic activities. The extracts protected animals from induced convulsions in mice and raised brain contents of gamma-aminobutyric acid (GABA) and serotonin. They were found to be anxiogenic and to depress the central nervous system. Wang et al. (2000) showed that the floral anthocyanins protected against hepatic toxicity in rats. They quenched free radicals at 100 to 200 µg/ml. Oral pretreatment of the anthocyanins (100 and 200 mg/kg) for 5 days significantly lowered serum levels of hepatic enzyme markers (alanine and aspartate aminotransferase) and reduced oxidative liver damage, inflammatory liver lesions, necrosis. Hansawasdi et al. (2000, 2001) reported on three alpha-amylase inhibitors from roselle tea extract which strongly inhibits porcine pancreatic alpha-amylase (citric acid, hibiscus acid and its 6-methyl ester).

Indications (Roselle) — Abscess (1; CRC; WBB); Anorexia (f; APA; CRC; KOM; PH2); Aphtha (1; FNF); Atherosclerosis (1; CRC); Bacteria (1; HHB); Biliousness (f; CRC); Calculus (1; FNF); Cancer (f; JLH); Cancer, colon (1; X10478827); Carbuncle (f; PH2); Cardiopathy (f; APA); Catarrh (f; KOM); Chill (f; APA; PED); Circulosis (f; PH2); Cold (1; FNF; PHR; PH2); Conjunctivosis (f; PH2); Constipation (f; APA; KOM); Convulsion (1; X10904147); Cough (1; CRC; WBB); Cramp (1; APA; HHB; PED); Debility (f; CRC); Diabetes (1; X11495291); Dyspepsia (1; CRC; FNF; PHR); Dysuria (f; CRC); Enterosis (1; CRC); Fever (f; CRC; HHB); Fungus (1; FNF); Gastrosis (f; PH2); Hangover (f; CRC); Heart (f; CRC); Herpes zoster (f; PH2); High Blood Pressure (1; CRC; X10457797); Infection (1; APA; CRC; FNF); Inflammation (f; PH2); Intoxication (f; CRC; WBB); Kidney Stone (1; APA; FNF); Neuropathy (1; APA); Neurosis (f; CRC); Ophthalmia (f; PH2); Respirosis (f; APA; PED; PH2); Scurvy (f; CRC); Seborrhea (1; FNF); Strangury (f; CRC); Swelling (f; PH2); Tuberculosis (1; HHB); Virus (f; PH2). (Commission E list it as unapproved; KOM.)

Roselle for cardiopathy:

- ACE-Inhibitor: gossypetin
- Antiarrhythmic: protocatechuic-acid
- Antiischemic: protocatechuic-acid
- Antioxidant: delphinidin; delphinidin-3-glucoside; gossypetin; protocatechuic-acid
- Diuretic: glycolic-acid
- Sedative: benzaldehyde; benzyl-alcohol

Roselle for cramp:

- Anesthetic: benzaldehyde; benzyl-alcohol
- Antiinflammatory: protocatechuic-acid
- Antispasmodic: benzaldehyde; protocatechuic-acid
- Lipoxygenase-Inhibitor: gossypetin
- Sedative: benzaldehyde; benzyl-alcohol

Other Uses (Roselle) — Source of a red beverage known as "jamaica" in Mexico (said to contain citric acid and salts, serving as a diuretic). Calyx is what gives the red to my favorite commercial tea, Red Zinger (lemongrass, hibiscus flowers, rose hips, orange peel, and peppermint leaves). Calyx is called "karkade" in Switzerland, a name not too different from the Arabic kharkadi. Karkade is used in jams, jellies, sauces, and wines. In the West Indies and elsewhere in the tropics, the fleshy calyces are used fresh for making roselle wine, jelly, syrup, gelatin, refreshing beverages, pudding and cakes, and to color and flavor rum. The dried roselle is used for tea, jelly, marmalade, ices, ice cream, sherbets, butter, pies, sauces, tarts, and other desserts. Tender leaves and stalks are eaten as salad and as a pot-herb and are used for seasoning curries. Fruits are edible. Seeds, containing an edible oil, have been used as an aphrodisiac coffee substitute. Seed has properties similar to those of cotton seed oil and is used as a substitute for crude castor oil. Seed can be eaten or made into an oily sauce (Bown, 2001). Sudanese ferment the seeds to make a meat substitute called "furundu." Roselle is also cultivated for the bast fiber obtained from the stems. The fiber strands, up to 1.5 m long, are used for cordage and as a substitute for jute in the manufacture of burlap. Residues can be used as a cheap fuel (DAD).

For more information on activities, dosages, and contraindications, see the *CRC Handbook of Medicinal Herbs*, ed. 2, Duke et al., 2002.

Cultivation (Roselle) — Bown (2001) suggests well-drained but moist, circumneutral to alkaline soils, in full sun in warm climates (zone 9–11). Soil should be tilled deep, about 20 cm, and thoroughly. Seed, 10–22 kg/ha depending upon the soil, is drilled about 15 cm by 15 cm at beginning of rainy season, mid-April in India, planting to a depth of about 0.5 cm. Seed at 5.5–7.5 kg/ha for a population of 60,000 plants/ha. Broadcasting results in uneven stand and hence lack of uniformity in fiber. Seeds can be started in flats and transplanted. When grown for its fiber, it is planted closely to produce long stems with little foliage. Weed carefully during first month. A dressing of NPK before sowing or planting may promote early growth on soils that are marginally fertile. Roselle responds favorably to nitrogen, and 45 kg/ha is a safe level in India, applied in the form of compost or mineral fertilizer with a small quantity of phosphate. In Java, green-manure (*Mimosa invisa*) is plowed under before it matures seeds. Javanese recommend 80 kg N/ha, 36–54 kg P_2O_5/ha, and 75–100 kg K_2O/ha. Rotations are sometimes used, since the root-knot nematode is a pest. A sequence of a legume green-manure crop, then roselle, and then corn is suggested. For home gardens of roselle, seeds are sown directly in rows about May 15. After germination, seedlings are thinned to stand 1 m apart. For larger plantings, seeds are sown in protected seedbeds and the seedlings transplanted to 1.3–2.6 m apart in rows 2–3.3 m apart. Applications of manure or commercial fertilizers are beneficial. The red calyces ripen about 3 weeks after the onset of flowering (some 100–160 days after transplanting). The fruit consists of the large reddish calyces surrounding the small seed pods (capsules). Capsules are easily separated but need not be removed before cooking. Calyx production ranges from ca. 1.5 kg (California), to 2 kg (Puerto Rico), and to 7.5 kg/plant in South Florida. Production yields are 8–10 MT/ha calyces, 10 MT leaves, 200 kg/ha seed. Dual-purpose plantings can yield 17,000 kg of herbage in three cuttings and later 6300 kg of calyces. Average fiber production is 1700 kg/ha, with as much as 3500 kg/ha reported (Malaya) (DAD).

Chemistry (Roselle) — Wrobel et al. (2000) compared aluminum, chromium, copper, iron, manganese, and nickel contents in teas of black tea, green tea, mate, and roselle. Roselle contained (272 ± 19 µg/g), mate (369 ± 22 µg/g) as compared to black tea (759 ± 31 µg/g) or green tea (919 ± 29 µg/g). The authors suggested that mate could be a good dietary source of manganese (total content 2223 ± 110 µg/g, 48.1% leached to the infusion). The roselle tea could supply more iron (111 ± 5 µg/g total, 40.5% leached) and copper (5.9 ± 0.3 µg/g total, 93.4% leached). The lower the tannins level, the better the mineral extraction into the tea (X11314985). Here are a few of the more notable chemicals found in roselle. For a complete listing of the phytochemicals and their activities, see the CRC phytochemical compendium, Duke and duCellier, 1993 (DAD) and the USDA database http://www.ars-grin.gov/duke/.

Anthocyanins — Antiarthritic 120 mg/man orl; Anticollegenase; Antielastase; Antimutagenic; Antimyopic 600 mg/man orl; Antinyctalopic 600 mg/man orl; Antioxidant; Antitumor 120 mg/man orl; Antiviral; Hemostat; Goitrogenic; NO-Scavenger; Vasoactive; Vasodilator.

Citric-Acid — Alpha-Amylase-Inhibitor; Antiaggregant; Antiaphthic 20,000 ppm; Anticalculic; Antimutagenic; Antioxidant Synergist; Antipyretic; Antiseborrheic; Antiseptic; Antitumor; Hemostat; Irritant; Laxative?; Litholytic; Odontolytic; LD50 = 975 ipr rat; LD50 = 6730 mg/kg orl rat.

Gossypetin — ACE-Inhibitor IC60–90 = 333 µg/ml; Antibacterial; Antimutagenic; Antioxidant; Pigment.

Hibiscin-Chloride — Antiseptic.

Hibiscus-Acid — Alpha-Amylase-Inhibitor.

Malic-Acid — Antibacterial; Antimycobacterial; Antioxidant Synergist; Antiseborrheic; Antiseptic; Antitubercular; Antitumor; Bruchiphobe; Hemopoietic; Laxative?; Sialagogue; LDlo = 1600 orl rat.

I

Illicium verum Hook. f. (Illiciaceae)
CHINESE STAR ANISE

Medicinal Uses (Star Anise) — Reported to be analgesic, antirheumatic, antiseptic, carminative, diuretic, expectorant, lactagogue, pediculicide, piscicide, stimulant, stomachic, and vermifuge, star anise is also a folk remedy for back ailments, bladder ailments, croup, diarrhea, fever, nervousness, vomiting, and whooping cough. Used in cough medicines and cough drops, perhaps due to expectorants cineole and terpineol, which increase liquid secretions from mucous membranes, facilitating productive coughs. A medicinal tea is made from the leaves in China (DAD). Anise oil is a good carminative, settling the stomach (RIN).

Teissedre and Waterhouse (2000) look at the antioxidant activities of EOs of spices and culinary herbs. They first assessed EOs as antibacterial and antifungal, second as flavorants and preservatives when added to foods, and third as used in cosmetology, for their aromatic and antioxidant properties. In their survey of 23 EOs, star anise was the most potent LDL antioxidant (IC83 = 2 μM) with Spanish red thyme (*Thymus zygis*). Maybe a little star anise in red wine would please the imbiber if not the enologists (JAF4:3801).

Chinese star anise should not be confused with Japanese star anise, *I. lanceolatum* A. C. Smith, which is said to be highly poisonous. Toxicities of anethole, isosafrole, and safrole were discussed by Buchanan (1978) (GRAS §182.10 and 182.20) (DAD).

Indications (Star Anise) — Anemia (1; APA); Anorexia (2; PHR; PH2); Arthrosis (f; PH2); Bacteria (1; APA; FNF); Bronchosis (2; APA; FNF; PHR; PH2); Candida (1; FNF); Catarrh (2; KOM; PHR; PH2); Cholecytosis (f; CRC); Colic (1; APA; CRC); Congestion (1; APA; FNF); Constipation (f; CRC); Cough (2; APA; CRC; PHR; PH2); Cramp (1; APA; DEP; FNF; PH2); Dysentery (f; CRC; DEP; PH2); Dyspepsia (2; APA; CRC; FNF; KOM; PH2); Enterosis (1; PH2); Extrophy (f; CRC); Favus (f; CRC); Frigidity (f; APA); Fungus (1; FNF; LAF); Gas (1; APA; DEP; PH2); Gastrosis (1; APA; PHR; PH2); Halitosis (f; APA; CRC); Hemopareisis (f; PH2); Hernia (f; CRC); Infection (1; APA; CRC; FNF); Inflammation (1; FNF); Insomnia (f; CRC); Lumbago (f; CRC); Morning Sickness (f; APA); Mycosis (1; FNF; LAF); Nausea (f; APA); Nervousness (1; FNF); Otosis (f; CRC); Pain (f; CRC; PH2); Paralysis (f; PH2); Parturition (f; APA); Respirosis (2; KOM; LAF; PHR; PH2); Rheumatism (f; CRC; PH2); Scabies (1; APA; CRC); Spasm (2; CRC; FNF; LAF); Stomach Distress (1; APA); Toothache (f; CRC).

Star Anise for bronchosis:

- Antibacterial: 1,8-cineole; alpha-pinene; alpha-terpineol; anethole; caryophyllene; delta-3-carene; delta-cadinene; hydroquinone; limonene; linalool; myrcene; nerolidol; p-cymene; rutin; terpinen-4-ol
- Antibronchitic: 1,8-cineole
- Antihistaminic: linalool; proanthocyanidins; rutin

- Antiinflammatory: alpha-pinene; beta-pinene; caryophyllene; delta-3-carene; quercetin-3-o-galactoside; rutin
- Antioxidant: camphene; gamma-terpinene; hydroquinone; myrcene; proanthocyanidins; rutin
- Antipharyngitic: 1,8-cineole
- Antispasmodic: 1,8-cineole; anethole; caryophyllene; limonene; linalool; myrcene; rutin
- Antitussive: 1,8-cineole; terpinen-4-ol
- Antiviral: alpha-pinene; beta-bisabolene; kaempferol-3-o-glucoside; limonene; linalool; p-cymene; proanthocyanidins; rutin
- Bronchorelaxant: linalool
- Candidicide: 1,8-cineole; beta-pinene; caryophyllene
- Candidistat: limonene; linalool
- Expectorant: 1,8-cineole; alpha-pinene; anethole; beta-phellandrene; camphene; limonene; linalool
- Immunostimulant: anethole

Star Anise for dyspepsia:

- Analgesic: anisatin; myrcene; p-cymene
- Anesthetic: 1,8-cineole; linalool; myrcene
- Antiinflammatory: alpha-pinene; beta-pinene; caryophyllene; delta-3-carene; quercetin-3-o-galactoside; rutin
- Antioxidant: camphene; gamma-terpinene; hydroquinone; myrcene; proanthocyanidins; rutin
- Antiulcer: beta-bisabolene
- Carminative: anethole; carvone
- Digestive: anethole
- Gastrostimulant: anethole
- Secretagogue: 1,8-cineole; anethole
- Sedative: 1,8-cineole; alpha-pinene; alpha-terpineol; anisatin; carvone; caryophyllene; limonene; linalool; p-cymene
- Tranquilizer: alpha-pinene

Star Anise for fungus:

- Analgesic: anisatin; myrcene; p-cymene
- Anesthetic: 1,8-cineole; linalool; myrcene
- Antibacterial: 1,8-cineole; alpha-pinene; alpha-terpineol; anethole; caryophyllene; delta-3-carene; delta-cadinene; hydroquinone; limonene; linalool; myrcene; nerolidol; p-cymene; rutin; terpinen-4-ol
- Antidermatitic: rutin
- Antiinflammatory: alpha-pinene; beta-pinene; caryophyllene; delta-3-carene; quercetin-3-o-galactoside; rutin
- Antiseptic: 1,8-cineole; alpha-terpineol; anethole; beta-pinene; carvone; hydroquinone; limonene; linalool; proanthocyanidins; terpinen-4-ol
- Candidicide: 1,8-cineole; beta-pinene; caryophyllene
- Candidistat: limonene; linalool
- Fungicide: 1,8-cineole; alpha-phellandrene; anethole; beta-phellandrene; caryophyllene; linalool; myrcene; p-anisaldehyde; p-cymene; terpinen-4-ol; terpinolene
- Fungistat: limonene
- Immunostimulant: anethole
- Mycoplasmistat: hydroquinone

Other Uses (Star Anise) — Dried fruit has a pleasant, aromatic, anise-like aroma and taste. Dried fruits contribute a licorice flavor to cakes, cookies, coffee, curries, pickles, sweetmeats, tea, and Chinese five-spice powder. Used whole, not ground, as a flavoring agent in confections, candy, chewing gum, and tobacco. Orientals chew the seeds after meals to promote digestion and sweeten the breath. The distilled oil is used in candy, ice cream, soft drinks, and liqueurs. Oil is used in animal feeds, in scenting soaps, toothpaste, creams, detergents, perfumes, etc., and to improve the flavor of some medicines. Highest use levels are ca. 570 ppm in alcoholic beverages and 680 ppm in candies (numbers derived from *Pimpinella)*. Anise oil and star anise candies are used interchangeably in the U.S., both being officially recognized as anise oil. Japanese use the ground bark as incense. A 10–15% aqueous extract is used as an agricultural insecticide in China. The fine-grained wood (oven-dried density 0.58) contains 30% parenchyma and 43% fiber, and though suitable for pulping is not recommended for forest plantations because of its slow growth (DAD).

For more information on activities, dosages, and contraindications, see the *CRC Handbook of Medicinal Herbs*, ed. 2, Duke et al., 2002.

Cultivation (Star Anise) — Propagated by seeds and by semi-ripe cuttings in summer. Seems to do better in damp but well-drained acid to circumneutral soils in subtropical climates, e.g., zone 8 (Bown, 2001). Only fresh seed will germinate, those planted within 3 days of harvest germinating readily. In China, seed are sown in nursery beds in October and November. Seedlings are transplanted when one year old (fourth leaf stage) to nursery beds, set about 25 cm apart. Plants are allowed to grow for three years and then planted out about 5–7 m apart. Plants should be weeded to facilitate fruit picking and to reduce fire hazards. Soil should be plowed and mulched in fall to provide sufficient moisture for the dry season. At the beginning of each summer, each tree should receive about 6.8 kg of stable manure and 45 kg of ammonium sulfate. Trees yield 6–10 years from planting and continue to bear more than 100 years. Children usually climb the trees and handpick the fruits. A mature tree, 25 years old, may yield 23–27 kg of dried fruits, containing up to 3% EO. About 100 kg of fresh green fruit yield 25–30 kg of dried fruit (CFR).

Chemistry (Star Anise) — Contains ca. 5% volatile oil (ca. 2.5% in seed, 10% in follicle) with trans-anethole as its major ingredient. Presence of safrole disputed. Here are a few of the more notable chemicals found in star anise. For a complete listing of the phytochemicals and their activities, see the CRC phytochemical compendium, Duke and duCellier, 1993 (DAD) and the USDA database http://www.ars-grin.gov/duke/.

Anethole — See also *Osmorhiza* spp.

Anisatin — Analgesic 0.03 mg/kg; Convulsant 3 mg/kg orl mus; GABA-Antagonist; Neurotropic 0.03–3 mg/kg orl mus; Poison; Sedative 0.03 mg/kg; Toxic; LD50 = 1.46 mus.

P-Anisaldehyde — Antimutagenic; Cosmetic; Fungicide; Insecticide; Irritant; Nematicide MLC = 1 mg/ml; Sedative; LD50 = 1510 mg/kg orl rat.

1,8-Cineole — See also *Elettaria cardamomum*.

Trans-Anethole — See also *Osmorhiza* spp.

J

Juniperus communis L. (Cupressaceae)
JUNIPER

Medicinal Uses (Juniper) — Reported to be carminative, cephalic, deobstruent, depurative, diaphoretic, digestive, diuretic, emmenagogue, and stimulant. Berries, wood, and oil are used in folk remedies. The berries are sometimes chewed to alleviate halitosis. The Herbal PDR reports antidiabetic, antiexudative, and hypotensive activities in animals, antiviral effects *in vitro*. But MPI recounts less than mediocre results. Checking *J. communis* var. *saxatilis*, they reported only abortifacient effects, and no anthelminthic, antibacterial, antifertility, antifungal, antiviral, diuretic, and no CNS, CVS, and smooth muscle effect (MPI). More positively, MPI reports the berries active against the parasites that cause mange in sheep. Ether extracts of the berries inhibited *Trichophyton* both *in vitro* and *in vivo* (MPI). And if juniper really does prevent uric acid build-up, and red wine is really bad for gout, perhaps gout sufferers might try juniper tea (or maybe even a martini) in lieu of red wine! But remember that alcohol, especially red wine, is generally considered bad for gout. In a study of clinical applications of ayurvedic herbs, Khan and Balick (2001) note studies on juniper reducing levels of hyperglycemia. They also cite antifertility and antiimplantation activities.

Indications (Juniper) — Ache (1; FAD; FNF); Amenorrhea (f; MAD); Anasarca (f; DEP); Anorexia (2; BGB; KAB; PH2); Arthrosis (1; APA; CAN; CRC; FNF); Ascites (1; FEL); Asthma (f; DEM); Atherosclerosis (f; CRC; PH2); Backache (f; DEM); Bite (f; CRC; MAD); Blenorrhea (f; CRC); BPH (1; PED); Bright's Disease (f; DEP); Bronchosis (f; APA; CRC; FAD; KAB); Burn (f; MIC); Calculus (f; CRC); Cancer (1; CRC; FAD; FNF); Cancer, kidney (1; FNF; JLH); Cancer, leg (1; FNF; JLH); Cancer, liver (1; FNF; JLH); Cancer, spleen (1; FNF; JLH); Cardiopathy (1; APA; FNF); Catarrh (f; MAD); Chest (f; DEM); Childbirth (f; CEB; DEM); Chlorosis (f; MAD); Cholecystosis (f; CRC); Cold (f; APA; CEB; FAD); Colic (f; CAN; CRC); Condyloma (1; FNF); Congestion (f; APA); Constipation (f; KAB); Cough (f; DEM; FAD; MAD); Cramp (1; APA; FNF); Cystosis (1; APA;

CAN; CEB; FAD; FEL); Dermatosis (f; CRC; FEL; SUW); Diabetes (1; APA; FNF; MAD; PHR); Diarrhea (f; DEM); Dropsy (f; CEB; CRC; FEL; KAB; MAD); Dysentery (f; CRC); Dysmenorrhea (f; APA; MAD; PH2); Dyspepsia (1; APA; BGB; KAB; KOM; PH2); Dyspnea (f; CRC; DEM); Dysuria (f; CEB; MIC); Edema (1; FNF); Encephalosis (f; KAB); Enterosis (f; CEB; CRC; FAD; KAB); Enuresis (f; MAD); Epilepsy (f; CEB); Eructation (f; PHR); Fever (f; DEM; MAD); Fistula (f; MAD); Flu (f; DEM; MIC); Fungus (1; KAP; MPI); Gallstone (f; MAD); Gas (1; APA; BGB; CAN; CEB; FAD; MAD); Gastrosis (f; CRC; MIC); Gleet (f; CRC; FEL; KAP); Gonorrhea (f; CRC; FEL; KAP); Gout (1; APA; FNF; PH2); Gravel (f; CRC; MAD); Halitosis (1; PH2); Headache (f; CEB); Heart (f; DEM); Heartburn (1; APA; DEM; PHR); Hemicrania (f; KAB); Hepatosis (1; CEB; FNF; JLH; KAB); Herpes (1; CAN; MAD); High Blood Pressure (1; DEM; FNF; PHR); High Cholesterol (1; FNF); HIV (1; FNF); Hydrocele (f; KAB); Hyperglycemia (1; FNF; JAC7:405); Hysteria (f; CRC); Immunodepression (1; FNF); Induration (f; CRC; JLH); Infection (1; APA; FNF); Inflammation (1; FNF; PH2); Jaundice (f; MAD); Kidney Stone (f; MAD); Leukemia (1; FNF); Leukorrhea (f; CRC; DEP; FEL; KAP); Lumbago (f; CRC); Malaria (1; ABS; FNF; MAD); Mange (1; MPI); Melanoma (1; FNF); Miscarriage (f; CEB); Mycosis (1; KAP; MPI); Myosis (f; CAN; DEM); Nephrosis (f; BGB; CRC; FEL; MIC); Neuralgia (f; APA); Neurasthenia (f; APA); Neurosis (f; APA); Odontosis (f; CEB); Ophthalmia (f; DEM); Otosis (f; KAB); Pain (1; FNF; JBU; KAB; PH2); Palsy (f; CEB); Polyp (f; CRC; JLH); Psoriasis (f; PED); Pulmonosis (f; CRC; MAD); Pyelosis (f; CRC; FEL); Rheumatism (1; CAN; CRC; FAD; FNF; KAP; MAD; PH2); Rhinosis (f; CRC); Scabies (f; MAD); Scrofula (f; CRC); Snakebite (f; CRC); Sore (f; CEB; FAD; MIC); Sore Throat (f; CEB; DEM); Splenosis (f; CEB; JLH; KAB); Sprain (f; MIC); Stone (2; PHR); Snakebite (f; FAD); Stomachache (f; APA; DEM; FAD); Strangury (f; KAB); Swelling (f; CRC; KAP; MAD); Tenesmus (f; CRC); Tonsilosis (f; DEM); Toothache (f; CEB; KAB); Tuberculosis (f; CEB; CRC); Tumor (1; CRC; FNF); Urogenitosis (f; CRC); Ulcer (f; CEB; DEM); Urethrosis (f; CEB); UTI (2; FAD; FNF; PHR; SKY); Uterosis (f; CEB; MAD); Vaginosis (f; KAB); VD (f; CRC); Virus (1; FNF; PH2); Wart (1; CRC; FNF); Water Retention (1; FNF; MAD); Worm (f; APA); Wound (f; DEM; MIC).

Juniper for cancer:

- AntiEBV: (–)-epicatechin; bilobetin; chlorogenic-acid; hinokiflavone
- AntiHIV: (+)-catechin; (–)-epicatechin; amentoflavone; apigenin; caffeic-acid; chlorogenic-acid; gallic-acid; hinokiflavone; quercetin
- Antiadenomic: farnesol
- Antiaggregant: (+)-catechin; (–)-epicatechin; 3-alpha-hydroxymanool; apigenin; caffeic-acid; ferruginol; ferulic-acid; menthol; quercetin
- Antiangiogenic: apigenin
- Anticancer: (+)-catechin; (–)-epicatechin; alpha-pinene; alpha-terpineol; apigenin; aromadendrene; beta-myrcene; caffeic-acid; camphor; cedrene; chlorogenic-acid; cinnamic-acid; ferulic-acid; gallic-acid; isoquercitrin; limonene; linalool; p-coumaric-acid; quercetin; rutin; terpineol; umbelliferone; vanillic-acid
- Anticarcinomic: betulin; caffeic-acid; chlorogenic-acid; cis-aconitic-acid; desoxypodophyllotoxin; ferulic-acid; gallic-acid
- Antiestrogenic: apigenin; ferulic-acid; quercetin
- Antifibrosarcomic: quercetin
- Antihepatocarcinogenic: fumaric-acid
- Antihepatotoxic: caffeic-acid; chlorogenic-acid; ferulic-acid; gallic-acid; glucuronic-acid; p-coumaric-acid; protocatechuic-acid; quercetin; rutin
- Antihyaluronidase: apigenin
- Antiinflammatory: (+)-catechin; (–)-epicatechin; alpha-amyrin; alpha-pinene; amentoflavone; apigenin; beta-pinene; betulin; borneol; caffeic-acid; caryophyllene; caryophyllene-oxide; chlorogenic-acid; cinnamic-acid; cis-communic-acid; cuparene; delta-3-carene;

ferulic-acid; gallic-acid; lupeol; menthol; n-hentriacontane; protocatechuic-acid; quercetin; rutin; sciadopitysin; umbelliferone; vanillic-acid

- Antileukemic: (–)-epicatechin; amentoflavone; apigenin; deoxypodophyllotoxin; farnesol; quercetin
- Antileukotriene: caffeic-acid; chlorogenic-acid; quercetin
- Antilipoperoxidant: (–)-epicatechin; quercetin
- Antimelanomic: apigenin; farnesol; quercetin
- Antimetastatic: apigenin
- Antimutagenic: (+)-catechin; (+)-gallocatechin; (–)-epicatechin; apigenin; caffeic-acid; chlorogenic-acid; cinnamic-acid; ferulic-acid; gallic-acid; l-ascorbic-acid; limonene; myrcene; n-nonacosane; protocatechuic-acid; quercetin; rutin; umbelliferone
- Antineoplastic: ferulic-acid
- Antinitrosaminic: caffeic-acid; chlorogenic-acid; ferulic-acid; gallic-acid; p-coumaric-acid; quercetin
- Antioxidant: (+)-catechin; (–)-epicatechin; amentoflavone; apigenin; caffeic-acid; camphene; chlorogenic-acid; ferulic-acid; fumaric-acid; gallic-acid; gamma-terpinene; isoquercitrin; leucoanthocyanin; linalyl-acetate; lupeol; myrcene; p-coumaric-acid; protocatechuic-acid; quercetin; rutin; vanillic-acid
- Antiperoxidant: (+)-catechin; (–)-epicatechin; amentoflavone; caffeic-acid; chlorogenic-acid; gallic-acid; lupeol; p-coumaric-acid; protocatechuic-acid; quercetin
- Antiproliferant: apigenin; quercetin; terpineol
- Antiprostaglandin: (+)-catechin; caffeic-acid; umbelliferone
- Antiretroviral: isoquercetin
- Antitumor: alpha-amyrin; alpha-humulene; apigenin; betulin; caffeic-acid; caryophyllene; caryophyllene-oxide; chlorogenic-acid; deoxypodophyllotoxin; ferulic-acid; fumaric-acid; gallic-acid; isoquercitrin; limonene; lupeol; nepetin; p-coumaric-acid; quercetin; rutin; vanillic-acid
- Antiviral: (–)-epicatechin; alpha-pinene; amentoflavone; apigenin; betulin; bilobetin; bornyl-acetate; caffeic-acid; chlorogenic-acid; deoxypodophyllotoxin; desoxypodophyllotoxin; ferulic-acid; gallic-acid; hinokiflavone; isoquercetin; limonene; linalool; lupeol; neryl-acetate; p-cymene; protocatechuic-acid; quercetin; rutin
- Anxiolytic: apigenin
- Apoptotic: apigenin; farnesol; gallic-acid; quercetin
- Beta-Glucuronidase-Inhibitor: apigenin
- COX-2-Inhibitor: (+)-catechin; apigenin; quercetin
- Chemopreventive: limonene
- Cyclooxygenase-Inhibitor: (+)-catechin; apigenin; gallic-acid; quercetin
- Cytochrome-p450-Inducer: delta-cadinene
- Cytoprotective: caffeic-acid
- Cytotoxic: (+)-catechin; (–)-epicatechin; alpha-amyrin; apigenin; betulin; caffeic-acid; deoxypodophyllotoxin; hinokiflavone; lupeol; p-coumaric-acid; quercetin
- Hepatoprotective: (+)-catechin; borneol; caffeic-acid; chlorogenic-acid; desoxypodophyllotoxin; ferulic-acid; quercetin
- Hepatotonic: glycolic-acid
- Immunostimulant: (+)-catechin; (–)-epicatechin; caffeic-acid; chlorogenic-acid; ferulic-acid; gallic-acid; protocatechuic-acid
- Interferonogenic: chlorogenic-acid
- Lipoxygenase-Inhibitor: (–)-epicatechin; caffeic-acid; chlorogenic-acid; cinnamic-acid; p-coumaric-acid; quercetin; rutin; umbelliferone
- Mast-Cell-Stabilizer: quercetin

- Ornithine-Decarboxylase-Inhibitor: apigenin; caffeic-acid; chlorogenic-acid; ferulic-acid; limonene; quercetin
- p450-Inducer: delta-cadinene; quercetin
- PKC-Inhibitor: apigenin
- PTK-Inhibitor: apigenin; quercetin
- Prostaglandigenic: caffeic-acid; ferulic-acid; p-coumaric-acid; protocatechuic-acid
- Protein-Kinase-C-Inhibitor: apigenin; quercetin
- Reverse-Transcriptase-Inhibitor: (−)-epicatechin
- Sunscreen: apigenin; caffeic-acid; chlorogenic-acid; ferulic-acid; rutin; umbelliferone
- Topoisomerase-II-Inhibitor: apigenin; deoxypodophyllotoxin; isoquercitrin; quercetin; rutin
- Tyrosine-Kinase-Inhibitor: quercetin

Juniper for rheumatism:

- Analgesic: borneol; caffeic-acid; camphor; chlorogenic-acid; ferulic-acid; gallic-acid; menthol; myrcene; p-cymene; quercetin
- Anesthetic: benzoic-acid; camphor; cinnamic-acid; linalool; linalyl-acetate; menthol; myrcene
- Antidermatitic: apigenin; fumaric-acid; quercetin; rutin
- Antiedemic: alpha-amyrin; amentoflavone; beta-amyrin; caffeic-acid; caryophyllene; caryophyllene-oxide; lupeol; rutin; sciadopitysin
- Antiinflammatory: (+)-catechin; (−)-epicatechin; alpha-amyrin; alpha-pinene; amentoflavone; apigenin; beta-pinene; betulin; borneol; caffeic-acid; caryophyllene; caryophyllene-oxide; chlorogenic-acid; cinnamic-acid; cis-communic-acid; cuparene; delta-3-carene; ferulic-acid; gallic-acid; lupeol; menthol; n-hentriacontane; protocatechuic-acid; quercetin; rutin; sciadopitysin; umbelliferone; vanillic-acid
- Antiprostaglandin: (+)-catechin; caffeic-acid; umbelliferone
- Antirheumatalgic: p-cymene
- Antirheumatic: lupeol; menthol
- Antispasmodic: apigenin; borneol; bornyl-acetate; caffeic-acid; camphor; caryophyllene; cinnamic-acid; farnesol; ferulic-acid; limonene; linalool; linalyl-acetate; menthol; myrcene; p-coumaric-acid; protocatechuic-acid; quercetin; rutin; umbelliferone
- COX-2-Inhibitor: (+)-catechin; apigenin; quercetin
- Counterirritant: camphor; formic-acid; menthol
- Cyclooxygenase-Inhibitor: (+)-catechin; apigenin; gallic-acid; quercetin
- Lipoxygenase-Inhibitor: (−)-epicatechin; caffeic-acid; chlorogenic-acid; cinnamic-acid; p-coumaric-acid; quercetin; rutin; umbelliferone
- Myorelaxant: apigenin; borneol; bornyl-acetate; gallic-acid; menthol; rutin

Other Uses (Juniper) — The dried berries and the oil distilled from them are utilized commercially to flavor gin, liqueurs such as Ginepro, and cordials. The berries are also used in alcoholic bitters. Berries of *J. communis* contain 0.2–3.42% volatile oil, the principal flavoring agent in gin. Those who prefer the taste of gin (but the price of vodka) might upgrade cheap vodka by steeping a berry or two. In France, a kind of beer called "genevrette" is made from fermented juniper berries and barley (FAC). Extracts and oils are used in most food categories, including alcoholic and nonalcoholic beverages (cola, root beer), baked goods, brines, candy, chewing gum, frozen dairy desserts, gelatins, ice cream, meat and meat products, pates, puddings, sauerkraut, and stuffings. Highest average maximum use level reported for the oils is 0.006% in alcoholic beverages and 0.01% for the extract in alcoholic and nonalcoholic beverages. The oil is a fragrance component in soaps, detergents, creams, lotions, and perfumes (maximum use level 0.80/c). Used as a spice in pickled fish, kraut, and gravies. In Sweden, they are made into a conserve. They are often used in cooking to cut the odor of cabbage and turnips (FAC). One can use juniper berries to give bean, meat, and soup dishes a gin-like flavor. One tsp berries = $^1/_2$ cup gin, from a flavoring point of view (RIN). Aromatic berries are used as a pepper substitute. Roasted berries are used as a poor-man's coffee. Tea made from the berries has a spicy, gin-like flavor. Westphalian ham is smoked with both juniper twigs and berries. Swedes make a "wholesome" beer from cedar. In hot climates, the incised tree yields a gum or varnish. Deer and moose graze the plant. Sheep readily eat the fruit. (DAD).

For more information on activities, dosages, and contraindications, see the *CRC Handbook of Medicinal Herbs*, ed. 2, Duke et al., 2002.

Cultivation (Juniper) — Hardy to zone 5 (Bown, 2001 says zones 2–8), this juniper fares best in full sun, in moist, well-drained, but not constantly waterlogged soil. Junipers can withstand moderate drought. Seeds, not always coming true, need to be stratified (alternating moist freezing and thawing). Home gardeners may just clean their seed and sow them, with fingers crossed, in the garden to await the spring and its eternal hope. If you want a shrub or tree like the parent, you are better off with cuttings. Cuttings, especially of named cultivars, are taken in late fall and overwintered in a cold frame or cool greenhouse. Many cultivars of juniper are male, and some of the female cultivars refuse to bear cones (dry berries), making it rough on us cheapscates who want to convert a cheap vodka into a respectable gin.

Chemistry (Juniper) — Here are a few of the more notable chemicals found in juniper. For a complete listing of the phytochemicals and their activities, see the CRC phytochemical compendium, Duke and duCellier, 1993 (DAD) and the USDA database http://www.ars-grin.gov/duke/.

Alpha-Pinene — See also *Amomum compactum*.

Camphor — See also *Amomum compactum*.

Cedrol — Irritant; Termiticide; LD50 = >5000 mg/kg orl rat.

Junene—Diuretic.

Menthol — ADI = 200 µg/kg; Analgesic; Anesthetic 2000 ppm; Antiacetylcholinesterase IC50 = 2.0 mM; Antiaggregant IC50 = 750; Antiallergic; Antiasthmatic; Antibacterial; Antibronchitic; Antidandruff; Antihalitosic; Antihistaminic; Antiinflammatory; Antiitch; Antineuralgic; Antiodontalgic; Antipyretic; Antirheumatic; Antiseptic 4 × phenol; Antisinusitic; Antispasmodic ED50 = 0.01 mg/ml; Antivaginitic; Antivulvitic; Bradycardic 65 mg/3 × day/woman; Bronchomucolytic; Bronchomucotropic; Bronchorrheic; Calcium-Antagonist; Carminative; Choleretic; Ciliotoxic; CNS-Depressant; CNS-Stimulant; Congestant; Convulsant; Counterirritant; Decongestant 11 mg/man; Dermatitigenic; Diaphoretic; Enterorelaxant; Expectorant; Gastrosedative; Irritant; Myorelaxant; Nematicide MLC = 1 mg/ml; Neurodepressant; Neuropathogenic 40–100 mg/day/rat; Nociceptive; Rubefacient; Vibriocide; LDlo = 2000–>9000 mg/man; LD50 = 700–3180 orl rat; LD50 3300 mg/kg orl rat.

Myrcene — See also *Alpinia galanga*.

K

Kaempferia galanga L. (Zingiberaceae)
GALANGA

See also lesser and greater galangal (*Alpinia galanga* and *A. officinarum*).

Medicinal Uses (Galanga) — Reported to be carminative, diuretic, expectorant, pectoral, pediculicide, stimulant, stomachic, and tonic. The rhizome, called "gisol," has been used to treat sore throat. Philipinos use the plant for headache and parturition. They also mix the rhizome with oils as a cicatrizant, applying it to boils and furuncles. Bown (2001) recounts a mix of four ginger relatives (*Alpinia, Curcuma, Kaempferia*, and *Zingiber*) called "awas empas," a Jamu remedy for headache, stiff joints, and UTIs.

Many zingiberaceous plants exhibit antitumor activities, over and beyond COX-2-Inhibitory and antimutagenic activities. Vimala et al. (1999) reported seven zingiberaceous rhizomes which inhibited EBV activation (induced by TPA): *Curcuma domestica, C. xanthorrhiza, Kaempferia galanga, Zingiber cassumunar, Z. officinale,* and *Z. zerumbet.* Lack of serious cytotoxicity led the authors to conclude that naturally occurring non-toxic compounds inhibited the EBV activation.

Pitasawat et al. (1998) screened ten carminative species and found larvicidal activity against *Culex quinquefasciatus* (exposing early fourth instar larvae to ethanolic extracts). They found significant larvicidal effects with *Kaempferia galanga, Illicium verum,* and *Spilanthes acmella,* which had LC50 values of 50.54, 54.11, and 61.43 ppm respectively. Chu et al. (1998) found that galangal extracts were amebicidal for three species of *Acanthamoeba.* The extracts induced encystment.

Indications (Galanga) — Amebiasis (1; X9766904); Boil (f; CRC; DAA); Bruise (f; DAA); Cancer (1; JLH; X10389986); Childbirth (f; CRC; SKJ); Chill (f; CRC; DAA; WOI); Cholera (f; DAA); Cough (f; CRC; DAA; KAB); Dandruff (f; CRC; WOI); Dyspepsia (f; CRC; WOI); EBV (1; X10389986); Enterosis (f; DAA); Fever (f; CRC; WOI); Furuncle (f; CRC); Headache (f; CRC; WOI); Inflammation (f; CRC; WOI); Lameness (f; DAA); Lice (f; DAA); Lumbago (f; DAA); Malaria (f; CRC; DAA; WOI); Myosis (f; DAA); Ophthalmia (f; CRC; WOI); Pain (f; DAA); Parasite (1; X9766904); Rheumatism (f; CRC; DAA); Rhinosis (f; KAB); Scabies (f; CRC); Sore

Throat (f; CRC; DAA; WOI); Sting (f; WOI); Swelling (f; CRC; DAA; WOI); Toothache (f; CRC; DAA); Tumor (f; CRC; WOI); Virus (1; X10389986).

Other Uses (Galanga) — An attractive spice plant used in various culinary applications, it is called "kencur" in Indonesia and "krachai" ("grachai," "kachai") in Thailand. The aromatic EOs of the roots are used widely in perfumery, as a condiment, and as folk medicine. In Java, the rhizomes of "kentjoor," as it is called, are used in seasoning many dishes, especially rice dishes. The rhizomes are also pickled, or used to make "beras," a sweet, spicy beverage. They are chewed with the betelnuts. A Javan beverage called "beras kentjoor" is made from the roots (Ochse, 1931). Dried rhizomes are used as a substitute for turmeric in curry powder (FAC). In Java and elsewhere in Asia, the galanaga is used, almost interchangeably, with the greater galangal. The galanga is a bit more pungent. One tsp powdered galanga is roughly equal, culinarily, and probably medicinally, to a quarter-inch (diameter) slice of the fresh rhizome or a one-eighth thick dried slice (length not mentioned). Leaves and rhizomes may be used in curries, eaten raw or steamed, or cooked with chili (FAC). The leaves of a narrow-leafed variety are also consumed as food. But leaves of all varieties may be used in lalabs. Asians employ the rhizomes and leaves as a perfume in cosmetics, hair washes, and powders. They are also used to protect the clothing against insects. More rarely it is said to be used as an hallucinogen.

For more information on activities, dosages, and contraindications, see the *CRC Handbook of Medicinal Herbs*, ed. 2, Duke et al., 2002.

Cultivation (Galanga) — Seems to be cultivated in humid tropical Asia (zones 9–11), like several other members of the ginger family (Bown, 2001). Propagated by dividing plants in spring or sowing ripe seed at ca. 20°C (68°F). Pieces of the rhizome are planted in fertile, moderately drained soils, often at 40–60 cm a part, both ways, with interplants grown in between. In very rich soils they may be grown as a solitary crop, spaced closer at 15–25 cm apart. They are dug when the plant loses its leaves during the monsoon (to prevent its decay in the excessively humid soils).

Chemistry (Galanga) — The EO contains n-pentadecane, ethyl-p-methoxycinnamate, ethyl cinnamate, carene, camphene, borneol, and p-methoxystyrene. Narcotic hallucinogen. Here are a few of the more notable chemicals found in galanga. For a complete listing of the phytochemicals and their activities, see the CRC phytochemical compendium, Duke and duCellier, 1993 (DAD) and the USDA database http://www.ars-grin.gov/duke/.

Borneol — Allelochemic; Analgesic; Antiacetytlcholine; Antibacterial MIC = 125–250 µg/ml; Antibronchitic; Antiescherichic MIC = 125 µg/ml; Antifeedant; Antiinflammatory; Antiotitic; Antipyretic; Antisalmonella MIC = 125 µg/ml; Antispasmodic ED50 = 0.008 mg/ml; Antistaphylococcic MIC = 250 µg/ml; Candidicide MIC = 250 µg/ml; Choleretic; CNS-Stimulant; CNS-Toxic; Fungicide; Hepatoprotectant; Herbicide IC50 = 470 mM; IC50 = 470 µ*M*; Inhalant; Insectifuge; Irritant; Myorelaxant; Negative Chronotropic 29 µg/ml; Negative Inotropic 29 µg/ml; Nematicide MLC = 1 mg/ml; Sedative; Tranquilizer; LDlo = 2000 orl rbt.

Camphene — Antilithic; Antioxidant; Expectorant; Hypocholesterolemic; Insectifuge; Spasmogenic.

Carene — Antibacterial; Antiseptic; Fungicide; Irritant; LD50 = 4800 mg/kg orl rat.

L

Laurus nobilis L. (Lauraceae)
BAYLEAF LAUREL, GRECIAN LAUREL, LAUREL, SWEET BAY

Medicinal Uses (Bayleaf) — Regarded as aperitif, carminative, diuretic, emetic, emmenagogue, narcotic, nervine, stimulant, stomachic, and sudorific. Oleum lauri, the fixed oil derived from the fruits (as opposed to EO) is widely expressed and used in Turkey. It is used in massage for rheumatism and to kill body parasites. Vets use the oleum as an analgesic and fly repellent (SPI). Bay oil sometimes used as a liniment or analgesic for earache. This may well be the ointment of unguent derived from the plant said to remedy sclerosis of the spleen and liver and tumors of the uterus, spleen, parotid, testicles, liver, and stomach (JLH). Southern Anatolians make a yellowish soap from the oleum to treat hair loss and skin ailments (SPI). The fruit, prepared in various manners, is said to help uterine fibroids, tuberosities of the face, scirrhus and scleroma of the uterus, scirrhus of the liver, indurations of the joints, spleen, and liver, internal tumors, wens, and tumors of the eye. Leaves and fruits are said to possess aromatic, stimulant, and narcotic properties. In small doses, leaves are diaphoretic; in large doses, emetic. In Lebanon, an extract from the leaves and berries, used as a carminative, is tightly corked and steeped in brandy in the sun for several days. The residue, after subsequent distillation, is used as a liniment for rheumatism and sprains, the distillate as an emmenagogue. Lebanese mountaineers are said to use raw berries to induce abortion. Berries macerated in flour were poulticed onto dislocations.

As a natural COX-2 Inhibitor, parthenolide, probably more prevalent in feverfew than bay (they should be carefully compared), may alleviate arthritis, gout, inflammation, and rheumatism, and possibly prevent cancer. Boik (2001) suggests parthenolide, if not feverfew or bayleaf, as a cancer preventive. According to Boik, three *in vitro* studies suggest parthenolide inhibits proliferation of multiple cancer lines, usually at levels of 3–9 μM. Parthenolide reduced proliferation of human cervical cancer cells (IC50 = 3 μM) and nasopharyngeal cancer cells (IC50 = 9.3 μM). Boik (2001) identifies parthenolide as a PTK-Inhibitor which may secondarily help affect CAMs, induce apoptosis, inhibit angiogenesis, inhibit AP-1 activity, inhibit eicosanoid effects, inhibit histaminic effects, inhibit NF-κB activity, inhibit platelet aggregation, inhibit TNF, and inhibit VEGF. So we have one phytochemical, albeit at low levels in the Biblical laurel with more than a dozen different activities that could reduce the incidence of cancer. Boik (2001) calculates tentative human dosages as 96 mg/day (as scaled from animal antitumor studies), 57 mg/day (cytotoxic dose as determined from pharmacokinetic calculations) suggesting a target dose of 77 mg/day parthenolide. He suggests that 15-fold synergies with other phytochemicals may reduce that minimum antitumor dose closer to 5 mg parthenolide/day. Much lower doses (ca. 0.5–0.6 mg/day) are used classically for other conditions.

The EO has bactericidal and fungicidal properties, e.g., the dried leaves and its EOs inhibit *Bacillus cereus, Candida albicans, Clostridium botulinum, Escherichia coli, Mycobacterium smegmatis, Proteus vulgaris, Staphylococcus aureus, Vibrio parahaemolyticus*, etc. (SPI). Hammer et al. (1999), in a study of 52 plant oils and extracts for activity against *Acinetobacter baumanii, Aeromonas veronii* biogroup *sobria, Candida albicans, Enterococcus faecalis, Escherichia coli, Klebsiella pneumoniae, Pseudomonas aeruginosa, Salmonella enterica* subsp. *enterica* serotype *typhimurium, Serratia marcescens* and *Staphylococcus aureus*, noted that lemongrass, oregano, and bay inhibited all organisms at concentrations of ≤2.0%. Other organisms arrested or killed by bayleaf, its oils or its extracts: among bacteria, *Bacillus subtilis, Hafnea alnei, Micrococcus luteus*, and *Salmonella enteridis*, and among fungi, *Aspergillus niger, A. terreus, Fusarium moniliforme, Phytophthora capsici, Rhizoctonia solani*, and *Sclerotinia sclerotiorum*. Leaf extracts reduce aflatoxin production by *Aspergillus parasiticus* and botulitoxin from *Clostridium botulinum* (HH3). Peirce (1999), in the American Pharmaceutical Association's guide to natural medicines, states, "Test tube studies indicate that bay extracts kill a number of disease-causing bacteria, fungi, and viruses…. Some commercial German antiviral products contain bay extracts." And the bay might repel some germ-laden cockroaches. According to a 1981 study, bay contains cockroach-repelling compounds, including cineole (up to 50% in the volatile oil) (Peirce,1999). Rinzler (1990) adds that fleas (carriers of some plagues) and moths are also repelled.

There was a catchy title on the Internet, *Bayleaf and Bubonic Buzzwords* (admittedly on my website (http://www.fathernaturesfarmacy.com), hypothesizing that the parthenolide in bay might be useful in disarming the anthrax toxin, at least *in vitro*. Parthenolide is said to (1) Inhibit IL-1 secretion, (2) Inhibit phospholipase A2, (3) Inhibit TNF-a secretion, and (4) Inhibit tyrosine kinase. More importantly, it can be used in an immune boosting lentil soup. Ingredients per person and directions are:

- 1/2 cup dry lentils (wash and strain off any stones; soak 1 hr)
- 1 heaping tsp dry fenugreek seed (wash and strain off any stones; soak 1 hr) (may be bitter; my sample soup was much better the day after, losing most of the bitterness)
- 1/2 cup chopped onions, chives, leeks, ramps, scallions, mix and match as available
- Dash of curry (or powdered turmeric), oregano, or Biblical hyssop
- 1 whole bayleaf
- 1/2 clove garlic
- Salt and pepper and paprika to taste
- Simmer 2 hr or until tender

When it comes to stray anthrax, bubonic plague, chicken pox, fox pox, and smallpox, it might help, at least by complementing the Cipro. But remember, there have been no clinical trials comparing Cipro with lentil soup, or just plain garlic or bayleaf. I'd bet more on the garlic or bayleaf than the lentil soup, which dilutes the garlic and the bayleaf. But lentil soup, if pleasing, can boost your immune system, while providing hundreds of gentle antiseptic phytochemicals. I won't name all the side effects of Cipro. But it will soon lead to Cipro-resistant anthrax if we are not careful. That won't happen with the garlic or the bayleaf.

3,4-Dimethoxyallylbenzene produces sedation in mice at low doses; a reversible narcosis at higher doses. It prevented the death of mice treated with lethal convulsant doses of strychnine. It may have relatively specific nervous or myoneural effects, perhaps suggesting a clinical potential.

Indications (Bayleaf) — Alzheimer's (1; COX; FNF); Amenorrhea (f; CRC; SPI); Anorexia (1; APA; BOW); Arthrosis (1; APA; COX; FNF); Bacteria (1; CRC; FNF; HHB); Bruise (f; APA); Bugbite (f; APA); Cancer (1; CRC; FNF; JLH); Cancer, anus (1; JLH); Cancer, colon (1; COX; FNF); Cancer, eye (1; CRC; JLH); Cancer, face (1; CRC; JLH); Cancer, joint (1; CRC; JLH); Cancer, liver (1; CRC; JLH); Cancer, mouth (1; JLH); Cancer, parotid (1; CRC; JLH); Cancer, spleen (1; CRC; JLH); Cancer, stomach (1; CRC; JLH); Cancer, testicle (1; CRC; JLH); Cancer, uterus (f; CRC; JLH); Candida (1; HH3; SPI); Cheilosis (f; HH3); Colic (f; APA; CRC; SPI); Condyloma (f; CRC); Cough (f; CRC); Cramp (1; FNF); Dandruff (f; APA); Deafness (f; JFM); Debility (f; JFM); Dermatosis (f; APA; SPI); Dyspepsia (1; APA; FNF; JFM); Earache (f; CRC); Escherichia (1; X10438227); Fibroid (f; CRC; JLH); Fungus (1; CRC; FNF); Furuncle (f; HH3); Gas (1; APA; SPI); Gastrosis (f; CRC); Hepatosis (1; CRC; FNF); High Blood Pressure (1; APA; FNF); HIV (1; FNF); Hysteria (f; CRC; SPI); Impostume (f; CRC; JLH); Infection (1; CRC; FNF; SPI); Inflammation (1; FNF); Klebsiella (1; X10438227); Leukemia (1; FNF); Mange (f; JFM); Melanoma (1; FNF); Migraine (1; FNF; HAD); Mycosis (1; CRC; FNF; SPI); Orchosis (f; JLH); Pain (1; APA; FNF); Parasite (1; HHB; SPI); Pediculosis (f; HH3); Polyp (f; CRC); Proctosis (f; JLH); Rheumatism (f; CRC; HH3; PHR; PH2; SPI); Salmonella (1; HH3); Scabies (f; BOW); Sclerosis (f; CRC); Sore (f; APA; HH3; JFM); Spasm (1; CRC; FNF); Sprain (f; APA; CRC; WOI); Staphylococcus (1; HH3; SPI); Stomatosis (f; HH3); Tumor (1; FNF); Ulcer (f; JFM); Uterosis (f; JLH); Virus (1; FNF); Water Retention (1; FNF); Wen (f; CRC); Wound (1; APA); Yeast (1; X10438227).

Bayleaf for cancer:

- AntiEBV: (–)-epicatechin; beta-eudesmol
- AntiHIV: (+)-catechin; (–)-epicatechin; caffeic-acid; methanol; quercetin
- Antiaggregant: (+)-catechin; (–)-epicatechin; acetyl-eugenol; artecanin; caffeic-acid; elemicin; estragole; eugenol; eugenyl-acetate; kaempferol; parthenolide; quercetin; salicylates; thymol
- Antiarachidonate: eugenol
- Anticancer: (+)-catechin; (–)-epicatechin; alpha-pinene; alpha-terpineol; benzaldehyde; beta-myrcene; caffeic-acid; camphor; carvone; cinnamic-acid; citral; estragole; eugenol; eugenol-methyl-ether; geraniol; isoquercitrin; kaempferol; limonene; linalool; methyl-eugenol; p-coumaric-acid; parthenolide; quercetin; quercitrin; rutin; terpineol
- Anticarcinogenic: (–)-epigallocatechin; caffeic-acid
- Antiestrogenic: quercetin
- Antifibrosarcomic: quercetin
- Antihepatotoxic: caffeic-acid; p-coumaric-acid; quercetin; quercitrin; rutin
- Antihyaluronidase: proanthocyanidins
- Antiinflammatory: (+)-catechin; (–)-epicatechin; alpha-pinene; beta-pinene; boldine; borneol; caffeic-acid; carvacrol; caryophyllene; caryophyllene-oxide; cinnamic-acid;

delta-3-carene; eugenol; eugenyl-acetate; kaempferol; mannitol; parthenolide; quercetin; quercitrin; rutin; salicylates; santamarin; santamarine; thymol
- Antileukemic: (–)-epicatechin; astragalin; kaempferol; quercetin
- Antileukotriene: caffeic-acid; quercetin
- Antilipoperoxidant: (–)-epicatechin; quercetin
- Antimelanomic: carvacrol; geraniol; perillyl-alcohol; quercetin; thymol
- Antimutagenic: (+)-catechin; (+)-gallocatechin; (–)-epicatechin; (–)-epigallocatechin; beta-eudesmol; caffeic-acid; cinnamic-acid; costunolide; eugenol; kaempferol; limonene; mannitol; myrcene; quercetin; quercitrin; rutin
- Antinitrosaminic: caffeic-acid; p-coumaric-acid; quercetin
- Antioxidant: (+)-catechin; (–)-epicatechin; (–)-epigallocatechin; acetyl-eugenol; boldine; caffeic-acid; camphene; carvacrol; cyanidin; eugenol; gamma-terpinene; isoquercitrin; kaempferol; mannitol; methyl-eugenol; myrcene; p-coumaric-acid; proanthocyanidins; quercetin; quercitrin; rutin; thymol
- Antiperoxidant: (+)-catechin; (–)-epicatechin; caffeic-acid; p-coumaric-acid; quercetin
- Antiproliferant: perillyl-alcohol; quercetin; terpineol
- Antiprostaglandin: (+)-catechin; caffeic-acid; carvacrol; eugenol; eugenyl-acetate; parthenolide
- Antistress: elemicin
- Antithromboxane: eugenol
- Antitumor: benzaldehyde; beta-eudesmol; caffeic-acid; caryophyllene; caryophyllene-oxide; costunolide; eugenol; geraniol; isoquercitrin; kaempferol; limonene; p-coumaric-acid; parthenolide; proanthocyanidins; quercetin; quercitrin; rutin; santamarin; santamarine; tulipinolide
- Antiviral: (–)-epicatechin; alpha-pinene; beta-bisabolene; bornyl-acetate; caffeic-acid; kaempferol; limonene; linalool; neryl-acetate; p-cymene; proanthocyanidins; quercetin; quercitrin; rutin
- Apoptotic: kaempferol; perillyl-alcohol; quercetin
- Beta-Glucuronidase-Inhibitor: proanthocyanidins
- COX-2-Inhibitor: (+)-catechin; eugenol; kaempferol; parthenolide; quercetin
- Chemopreventive: limonene; perillyl-alcohol
- Cyclooxygenase-Inhibitor: (+)-catechin; carvacrol; kaempferol; parthenolide; quercetin; thymol
- Cytochrome-p450-Inducer: 1,8-cineole; delta-cadinene
- Cytoprotective: caffeic-acid
- Cytotoxic: (+)-catechin; (–)-epicatechin; caffeic-acid; eugenol; p-coumaric-acid; parthenolide; quercetin; santamarin; tulipinolide
- DNA-Binder: elemicin; estragole; eugenol-methyl-ether
- Hepatoprotective: (+)-catechin; beta-eudesmol; boldine; borneol; caffeic-acid; eugenol; proanthocyanidins; quercetin
- Hepatotonic: 1,8-cineole; quercitrin
- Immunostimulant: (+)-catechin; (–)-epicatechin; astragalin; benzaldehyde; caffeic-acid
- Lipoxygenase-Inhibitor: (–)-epicatechin; (–)-epigallocatechin; caffeic-acid; cinnamic-acid; kaempferol; p-coumaric-acid; quercetin; rutin
- Mast-Cell-Stabilizer: quercetin
- Ornithine-Decarboxylase-Inhibitor: caffeic-acid; limonene; proanthocyanidins; quercetin
- p450-Inducer: 1,8-cineole; delta-cadinene; quercetin
- PTK-Inhibitor: quercetin
- Prostaglandigenic: caffeic-acid; p-coumaric-acid
- Protein-Kinase-C-Inhibitor: quercetin

- Reverse-Transcriptase-Inhibitor: (–)-epicatechin
- Sunscreen: caffeic-acid; rutin
- Topoisomerase-II-Inhibitor: isoquercitrin; kaempferol; quercetin; rutin
- Tyrosine-Kinase-Inhibitor: quercetin

Bayleaf for infection:

- Analgesic: borneol; caffeic-acid; camphor; eugenol; myrcene; p-cymene; quercetin; reticuline; thymol
- Anesthetic: 1,8-cineole; benzaldehyde; camphor; carvacrol; cinnamic-acid; eugenol; linalool; methyl-eugenol; myrcene; thymol
- Antibacterial: (–)-epicatechin; (–)-epigallocatechin; 1,8-cineole; acetic-acid; alpha-pinene; alpha-terpineol; benzaldehyde; bornyl-acetate; caffeic-acid; carvacrol; caryophyllene; cinnamic-acid; citral; delta-3-carene; delta-cadinene; eugenol; geraniol; isoquercitrin; kaempferol; limonene; linalool; methyl-eugenol; myrcene; neral; nerol; p-coumaric-acid; p-cymene; parthenolide; perillyl-alcohol; quercetin; quercitrin; reticuline; rutin; terpinen-4-ol; terpineol; terpinyl-acetate; thymol
- Antiedemic: boldine; caffeic-acid; caryophyllene; caryophyllene-oxide; eugenol; proanthocyanidins; quercitrin; rutin
- Antiinflammatory: (+)-catechin; (–)-epicatechin; alpha-pinene; beta-pinene; boldine; borneol; caffeic-acid; carvacrol; caryophyllene; caryophyllene-oxide; cinnamic-acid; delta-3-carene; eugenol; eugenyl-acetate; kaempferol; mannitol; parthenolide; quercetin; quercitrin; rutin; salicylates; santamarin; santamarine; thymol
- Antiseptic: 1,8-cineole; actinodaphnine; alpha-terpineol; benzaldehyde; beta-pinene; caffeic-acid; camphor; carvacrol; carvone; citral; eugenol; formic-acid; geraniol; guaijaverin; hexanal; hexanol; kaempferol; limonene; linalool; methyl-eugenol; nerol; parthenolide; proanthocyanidins; terpinen-4-ol; terpineol; thymol
- Antiviral: (–)-epicatechin; alpha-pinene; beta-bisabolene; bornyl-acetate; caffeic-acid; kaempferol; limonene; linalool; neryl-acetate; p-cymene; proanthocyanidins; quercetin; quercitrin; rutin
- Astringent: formic-acid
- Bacteristat: quercetin
- COX-2-Inhibitor: (+)-catechin; eugenol; kaempferol; parthenolide; quercetin
- Cyclooxygenase-Inhibitor: (+)-catechin; carvacrol; kaempferol; parthenolide; quercetin; thymol
- Fungicide: 1,8-cineole; acetic-acid; alpha-phellandrene; beta-phellandrene; caffeic-acid; camphor; caprylic-acid; carvacrol; caryophyllene; caryophyllene-oxide; cinnamic-acid; citral; elemicin; eugenol; geraniol; linalool; methyl-eugenol; myrcene; p-coumaric-acid; p-cymene; parthenolide; perillyl-alcohol; propionic-acid; quercetin; reticuline; terpinen-4-ol; terpinolene; thymol
- Fungistat: formic-acid; limonene; methyl-eugenol
- Immunostimulant: (+)-catechin; (–)-epicatechin; astragalin; benzaldehyde; caffeic-acid
- Lipoxygenase-Inhibitor: (–)-epicatechin; (–)-epigallocatechin; caffeic-acid; cinnamic-acid; kaempferol; p-coumaric-acid; quercetin; rutin

Other Uses (Bayleaf) — In Biblical times, the bay was symbolic of wealth and wickedness. In the ancient Olympic games, the victorious contestant was awarded a chaplet of bay leaves, placed on his brow. The Roman gold coin of 342 B.C. has a laurel wreath modeled on its surface (BIB). The evergreen leaves, when broken, emit a sweet scent and furnish an extract used by the Orientals in making perfumed oil. Dried bay leaves are used to flavor meats, fish, poultry, shrimp and crab boils, soups, stews, stuffings, tomato sauces, and vegetables. They are also used as an ingredient in pickling spices and vinegars, even toothpastes (AAR, BIB, FAC, RIN). One or two medium-sized leaves will flavor almost any family dish. Rinzler (1990) suggests a leaf or two in the water to boil carrots, noodles, pasta, or potatoes, not to mention the spaghetti sauce itself. Bay leaves can create a flavorful bed for roasted or steamed meats and fish. Foods cooked en papillote are often seasoned with bay leaf, included in the wrappings. The leaves are indispensable in French cooking and basic bouquet garni (soups, stews, and stocks). Court boullion, for poaching fish, is also flavored with bay leaf, as are many marinades. Bay leaves are somewhat less bitter after drying than when fresh. Leaves can obstruct or even puncture the intestine, so they should be removed after cooking (AAR, RIN). In Corfu, the leaves are wrapped around "sikopsoma," a flattened cake of dried, spiced figs (FAC, SPI). Leaves are used as packing for fig and licorice, to deter weevils (Bown, 2001). Leaves have served as a tea substitute. And a leaf in a canister of flour just might keep out the bugs. If bay leaves will keep moths out of your woolens, maybe cardamom (richer in cineole) will do better. Twigs may be used as skewers for kebabs. Bark was used as a flavoring in ancient Rome and was one of the spices added to "mustacei," predecessor of the modern day mostaccioli biscuits. Spice cookies and dishes based on soft cheeses (ricotta) can be brightened by grated bay bark; one-eighth teaspoon suffices for a cup of ricotta or a cup of cookie dough (AAR). The berries are distilled to make a liqueur called "fioravanti." Dark berries are edible (*bacca laureus*) (AAR). Dried fruits and the leaf oil are also used for flavoring (FAC). The EO, distilled from the leaves, is used in perfumery and for flavoring food products such as baked goods, confectionary, meats, sausages, and canned soups. The oil can be measured more precisely and provides more uniform results. Oil and oleoresin of bay are used as soluble pickling spices, frequently in producing corned beef. The fat oleum lauri, from the edible fruits, has served in soap making and veterinary medicine. Juniper berries, or big black peppercorns, can be used as substitutes for bay leaves in some recipes (AAR). The wood, resembling walnut, can be used for cabinetry. Wood is used as an aromatic smoke flavoring (FAC).

For more information on activities, dosages, and contraindications, see the *CRC Handbook of Medicinal Herbs*, ed. 2, Duke et al., 2002.

Cultivation (Bayleaf) — Hardy from zone 10–8, marginally hardy to zone 7. Fares best in full sun, and moist but well-drained, friable garden loam, pH 4.5–8.2 (average 6.2). Warm humid summers are best (JAD, TAD). As many Mediterranean plants, it can tolerate a spell of drought. Sweet Bay is propagated by cuttings from half-ripened shoots, placed in a frame in July, forming roots in a few weeks; otherwise, they may require 6–9 months to root. Tucker and DeBaggio (2000) suggest rooting the suckers that arise at the base of the plant or with top shoots (severed small branches). They got their best results with tip cuttings of half-hardened wood soaked in water 4–6 weeks (changing the water every day), then moved to the rooting medium. All rooted within 3 weeks. Plants grown in large tubs or in conservatories may require one or more clippings per year, an occasional washing of the leaves, and a feeding once in a while to keep them healthy. In areas where the plants are grown commercially and suitable soils do not occur naturally, it is necessary to plant each tree in a large hole 75 cm across by 60 cm deep, filled with sand and compost. Plant easily pruned to desired shape. Leaves harvested manually and dried on trays in a moderately warm sheltered place, out of direct sun. Leaves may be flattened under flat plywood to prevent their curling. After two weeks or so drying, the dry leaves are containerized for shipment (CFR, TAD).

Chemistry (Bayleaf) — Yoshikawa et al. (2000) found seven alcohol-absorption-inhibitors in the leaves: costunolide, dehydrocostus lactone, zaluzanin D, reynosin, santamarine, 3alpha-acetoxy-eudesma-1,4(15),11(13)-trien-12,6alpha-+++olide, and 3-oxoeudesma-1,4,11(13)-trien-12,6alpha-olide), identified as the active principles from the leaves of *Laurus nobilis*. These sesquiterpenes appear to selectively inhibit ethanol absorption. Artemorin, costunolide, costuslactone, deacetly-laurenobiolide, laurenobiolide, reynosin, santamarin, and verlorin, are eight alpha-methylene-gamma-butyrolactones documented to be the chief cause of allergy (contact dermatitis) in *Laurus* (TAD). Here are a few of the more notable chemicals found in bayleaf. For a complete listing of the phytochemicals and their activities, see the CRC phytochemical compendium, Duke and duCel-lier, 1993 (DAD) and the USDA database http://www.ars-grin.gov/duke/.

1,8-Cineole — See also *Elettaria cardamomum*.

Costunolide — Antimutagenic; Antitumor; Chemopreventive 50–200 mg/kg/day; Dermatitigenic; Detoxicant 50–200 mg/kg/day; Ethanol-Absorption-Inhibitor; Fungicide EC50 = 6 µg/ml; GST-Inducer 50–200 mg/kg/day; NO-Inhibitor IC50 = 1.2–3.8 µ*M*; Schistosomicide; Trematodicide.

Lauric-Acid — Antibacterial; Antiviral; Candidicide; Hypercholesterolemic; LD50 = 131 ivn mus.

Neral — Antibacterial; Antispasmodic.

Nerol — Antibacterial; Antiseptic; Sedative; Trichomonicide ED100 = 300 µg/ml; LD50 = 3600 mg/kg orl rat.

Neryl-Acetate — Antiflu; Antiviral.

Oleic-Acid — See also *Dipteryx odorata*.

Parthenolide — Allelopathic; Antiaggregant IC50 = 50 µ*M*; Antiarthritic; Antibacterial MIC = 16–64 µg/ml; Anticarcinomic 0.45–1.1 µg/ml; Antieicosanoid µ*M*; Antifibrosarcomic 6–9 µ*M*; Antiinflammatory; Antilymphomic 6–9 µ*M*; Antimigraine 250 µg/man/day; Antimycobacterial MIC = 16–64 µg/ml; Antineuralgic; Antiproliferant 3–9 µ*M*; Antiprostaglandin; Antisecretory; Antiseptic MIC = 16–64 µg/ml; Antispasmodic; Antitubercular MIC = 16–64 µg/ml; Antitumor 3–9 µ*M*; Candidicide; COX-2-Inhibitor IC50 = 0.8 µ*M*; Cytotoxic 2.3 ppm; Dermatitigenic; Fungicide; Gram(+)-icide; 5-HT-Inhibitor; Hypercalcuric; 5-Lipoxygenase-Inhibitor IC50 = 20–200 µ*M*; NF-kB-Inhibitor; Phospholipase-Inhibitor; Prostaglandin-Synthetase-Inhibitor; PTK-Inhibitor 10–100 µ*M*, IC50 = 20 µ*M*.

L

Lindera benzoin (L.) Blume (Lauraceae)
BENJAMIN BUSH, SPICEBUSH, WILD ALLSPICE

Synonyms — *Benzoin aestivale* (L.) Nees, *Laurus aestivalis* L., *L. benzoin* L.

Medicinal Uses (Spicebush) — Cherokee Indians used spicebush for amenorrhea, blood disorders, cold, cough, croup, dysmenorrhea, dyspepsia, flu, gas, hives, nausea, phthisis and swellings. They also drank spicebush tea as a spring tonic and steeped the bark with wild cherry and dogwood in corn whiskey to break out measles (probably also good for cold, cough, and malaria) (HAD).

Cherokee inhaled the steam to clear sinuses and used the twig decoction in baths for arthritic pain, sometimes also drinking the tea (Winston, 2001). Creek Indians used the tea for pains of rheumatism, for purifying the blood, and making themselves sweat and throw up, a ritual cleansing. Wisely, they added willow to spicebush tea for drinking and using in the sweat lodges for rheumatism. The drug of choice today is still usually based on salicylates, which are copious in most willows. Iroquois used spicebush for colds, fevers, gonorrhea, measles, and syphilis. Mohegans chewed the leaves or took the tea for worms. Ojibwa took the tea for anemia and that "tired rundown feeling." Rappahannock used the tea for menstrual pain or delayed periods. Medicinally, the berries were used as a carminative for gas and colic. The oil from the fruits was applied to bruises, and muscles or joints, for chronic rheumatism. The tea made from the twigs was popular with the settlers (and available all year) for colds, colic, fevers, gas, and worms. The bark tea was used for various fevers, including typhoid, and to expel worms (FAD). Boiling a cupful chopped bark in one quart of maple sap that has been reduced to one quarter makes a pleasant tea, useful for fatigue.

There's not much folklore to anticipate that spicebush might be useful in yeast (candida), but maybe the Indians didn't have yeast. Apparently, the yeast is a normal component of the flora of all human beings. Maybe candida is mostly an iatrogenic ailment, induced by our medicines. Respected naturopaths Murray and Pizzorno (1991) say that, when antibiotic use first became widespread, it was noted immediately that yeast infections increased. White man's alcohol, anti-ulcer drugs, corticosteroids, increase in diabetes, oral contraceptives, tights instead of cotton undergarments, and too much sugar in the diet all may have contributed to the emergence of candidiasis as a major ailment, today afflicting half our womenfolk. The total incidence and relative frequency of vaginal candidiasis have increased more than two-fold since the late 1960s. Studying 54 plant species for antimicrobial effects, Heisey and Gorham (1992) found their extract of spicebush bark strongly inhibited yeast (*Candida albicans*) much better than any of the other 53 species. Walnut husks also showed some activity. Now, if garlic vinegar with spicebush and walnut hulls is as safe and effective as the commercial candidicides, it could save a lot of ladies (and gents) some $2000. Garlic is reportedly better than Nystatin. I suspect that my mix might be as safe and efficacious as any of the antiyeast drugs named above, but who will pay for the clinical trials to prove it?

Indications (Spicebush) — Ague (f; FEL); Amenorrhea (f; FAD); Anemia (f; FAD); Arthrosis (f; FAD; JAH2:45); Bacteria (1; FNF); Bruise (f; FAD; FEL); Cold (1; DEM; FAD; FNF); Colic (f; FAD; FEL); Congestion (1; FNF); Cough (f; FAD); Cramp (1; FAD; FNF); Croup (f; FAD); Dermatosis (1; FEL; FNF); Dysentery (f; BOW); Dysmenorrhea (f; FAD); Enterosis (f; BOW); Fatigue (f; EB48:333); Fever (f; FAD; FEL); Flu (1; FNF; JAH2:45); Fungus (1; FNF); Gas (f; FAD; FEL); Gonorrhea (f; DEM); Hives (f; HAD); Infection (1; FNF); Inflammation (1; FNF); Itch (f; FEL); Malaria (f; EB48:333); Measles (f; HAD); Myosis (f; FAD); Nausea (f; JAH2:45); Nervousness (f; FEL); Neuralgia (f; EB48:333); Pain (1; DEM; FNF); Parasite (f; BOW); Phthisis (f; HAD); Pulmonosis (f; EB48:333); Respirosis (f; EB48:333); Rheumatism (1; FAD; FEL; FNF); Sinusosis (f; JAH2:45); Spasm (1; FNF); Swelling (f; HAD); Syphilis (f; DEM); Tuberculosis (f; HAD); Typhoid (f; FAD; FEL); VD (f; DEM); Worm (f; BOW; DEM; FEL); Yeast (1; ABS; FNF).

Spicebush for cold/flu:

- Analgesic: borneol; myrcene; p-cymene
- Anesthetic: 1,8-cineole; myrcene
- Antiallergic: 1,8-cineole; terpinen-4-ol
- Antibacterial: 1,8-cineole; alpha-pinene; alpha-terpineol; delta-cadinene; limonene; myrcene; p-cymene; terpinen-4-ol
- Antibronchitic: 1,8-cineole; borneol
- Antiflu: alpha-pinene; limonene; p-cymene

- Antiinflammatory: alpha-curcumene; alpha-pinene; beta-pinene; borneol; caryophyllene-oxide
- Antioxidant: camphene; gamma-terpinene; myrcene
- Antipharyngitic: 1,8-cineole
- Antipyretic: borneol
- Antirhinoviral: beta-bisabolene; beta-sesquiphellandrene
- Antiseptic: 1,8-cineole; alpha-terpineol; beta-pinene; limonene; terpinen-4-ol
- Antitussive: 1,8-cineole; terpinen-4-ol
- Antiviral: alpha-pinene; beta-bisabolene; limonene; p-cymene
- Expectorant: 1,8-cineole; alpha-pinene; beta-phellandrene; beta-sesquiphellandrene; camphene; limonene

Spicebush for fungus:

- Analgesic: borneol; myrcene; p-cymene
- Anesthetic: 1,8-cineole; myrcene
- Antibacterial: 1,8-cineole; alpha-pinene; alpha-terpineol; delta-cadinene; limonene; myrcene; p-cymene; terpinen-4-ol
- Antiinflammatory: alpha-curcumene; alpha-pinene; beta-pinene; borneol; caryophyllene-oxide
- Antiseptic: 1,8-cineole; alpha-terpineol; beta-pinene; limonene; terpinen-4-ol
- Candidicide: 1,8-cineole; beta-pinene
- Candidistat: limonene
- Fungicide: 1,8-cineole; alpha-phellandrene; beta-phellandrene; caryophyllene-oxide; myrcene; p-cymene; terpinen-4-ol; terpinolene
- Fungistat: limonene

Other Uses (Spicebush) — An aromatic tea is made from young leaves, twigs, and fruits, which contain a fragrant EO. The twigs are best gathered when in flower, as the nectar adds considerably to the flavor (FAC). Pioneers of Ohio and African Americans use young twigs and leaves as a substitute for spice and tea. Cherokee and Chippewa Indians used the plant to make herb teas and used the leaves, and/or seeds or twigs, to flavor or mask the flavor of naturally strong meats, like groundhog and 'possum (DEM). Iroquois and Ojibwa used leaves and twigs to brew teas and season meat (EB48:333). Herbal vinegar made from the twigs and fruits was used to preserve beets. Fruits can be used as a spice in baking (Winston, 2001). Dried and powdered fruits can be used as a substitute for allspice. Dried leaves are great in potpourri (TAD). The new bark makes a pleasant antiseptic chewstick, and can be used, frayed, like a toothbrush (FAC, HAD, TAD).

For more information on activities, dosages, and contraindications, see the *CRC Handbook of Medicinal Herbs*, ed. 2, Duke et al., 2002.

Cultivation (Spicebush) — Tucker and DeBaggio (2000) give us some cultural detail, recommending this spicebush as an excellent background in shaded herb gardens. Hardy from zone 9–4 (BOW), spicebush grows naturally in shady situations but will survive full sun if adequately watered. It seems to fare best in subacid soils, rich in organic matter, and can be propagated by divisions in spring (TAD) and by seed in fall (BOW). Bown (2001) suggests moist acid soil and partial shade.

Chemistry (Spicebush) — Here are a few of the more notable chemicals found in spicebush. For a complete listing of the phytochemicals and their activities, see the CRC phytochemical compendium, Duke and duCellier, 1993 (DAD) and the USDA database http://www.ars-grin.gov/duke/.

Alpha-Phellandrene — Dermal; Emetic; Fungicide; Insectiphile; Irritant; Laxative.

Borneol — Allelochemic; Analgesic; Antiacetytlcholine; Antibacterial MIC = 125–250 µg/ml; Antibronchitic; Antiescherichic MIC = 125 µg/ml; Antifeedant; Antiinflammatory; Antiotitic; Antipyretic; Antisalmonella MIC = 125 µg/ml; Antispasmodic ED50 = 0.008 mg/ml; Antistaphylococcic MIC = 250 µg/ml; Candidicide MIC = 250 µg/ml; Choleretic; CNS-Stimulant; CNS-Toxic; Fungicide; Hepatoprotectant; Herbicide IC50 = 470 mM, IC50 = 470 µ*M*; Inhalant; Insectifuge; Irritant; Myorelaxant; Negative Chronotropic 29 µg/ml; Negative Inotropic 29 µg/ml; Nematicide MLC = 1 mg/ml; Sedative; Tranquilizer; LDlo = 2000 orl rbt.

Camphene — Antilithic; Antioxidant; Expectorant; Hypocholesterolemic; Insectifuge; Spasmogenic.

Laurotetanine — Curaroid; Cytotoxic.

Limonene — ACHe-Inhibitor; Allelochemic; Antiacetylcholinesterase IC22–26 = 1.2 mM; Antibacterial; Anticancer; Antifeedant; Antiflu; Antilithic; Antimutagenic; Antiseptic; Antispasmodic ED50 = 0.197 mg/ml; Antitumor; Antitumor (Breast); Antitumor (Pancreas); Antitumor (Prostate); Antiviral; Candidistat; Chemopreventive; Detoxicant; Enterocontractant; Expectorant; Fungistat; Fungiphilic; Herbicide IC50 = 45 µ*M*; Insectifuge; Insecticide; Irritant; Nematicide IC = 100 µg/ml; ODC-Inhibitor ~750 mg/kg (diet); p450-Inducer; Photosensitizer; Sedative ED = 1–32 mg/kg; Transdermal; LD50 = 4600 orl rat.

Myrcene — Analgesic; Anesthetic 10–20 mg/kg ipr mus, 20–40 mg/kg scu mus; Antibacterial; Anticonvulsant; Antimutagenic; Antinitrosaminic; Antioxidant; Antipyretic; Antispasmodic; Fungicide; Insectifuge; Irritant; p450(2B1)-Inhibitor IC50 = 0.14 µ*M*.

L

M

Moringa oleifera Lam. (Moringaceae)
Benzolive Tree, Drumstick Tree, Horseradish Tree, West Indian Ben

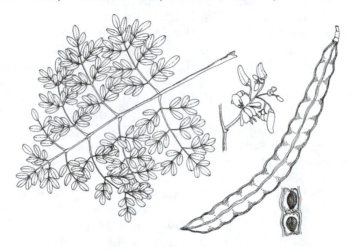

Synonyms — *Guilandina moringa* L., *Moringa moringa* (L.) Small, *M. pterygosperma* Gaertn.

Medicinal Uses (Horseradish Tree) — In rural Sudan, powdered seeds are used to purify drinking water by coagulation. In trials, the powder was toxic to guppies (*Poecilia reticulata*), protozoa (*Tetrahymena pyriformis*), and bacteria (*Escherichia coli*), and it inhibited acetylcholinesterase. It might serve as a fruit- and vegetable-preservative. In low concentrations, it protects mice against *staphylococcus* infections. Juice from the leaves and stem bark inhibits *Staphyloccoccus aureus* but not *Escherichia coli*. One study showed bark extracts active against *Bacillus subtilis, Dip. pneumoniae, Escherichia coli, Micrococcus pyogenes, Salmonella typhosa, Shigella dysenteri, Streptococcus pyogenes,* and *Vibrio comma.* Bark extract fungicidal to *Candida albicans, Helminthosporium sativum, Microsporum gypseum*, and *Trichiphyton mentagrophytes*. The 50% ethanolic extract of root bark (devoid of antibacterial activity) was antiviral to the vaccinia virus but not Ranikhet disease virus, but it did inhibit its replication. Ether leaf extracts were bacteriostatic to *Staphylococcus aureus* and *S. typhosa*. Alcohol extracts may contain an adrenergic neurone blocking agent (MPI).

The root alkaloid, spirochin, paralyzes the vagus nerve, hinders infection, and has antimycotic and analgesic activity. In doses of 15 g, the root bark is abortifacient. Alcoholic root bark extract is analgesic, antiedemic, and antiinflammatory at 500–1000 mg/kg orally in albino rats (MPI). Aqueous and ethanolic leaf extracts antibacterial, hypotensive, sedative, and respirastimulant.

Root-bark yields two alkaloids, moringine and moringinine. Moringinine acts as a cardiac stimulant, produces rise of blood pressure, acts on sympathetic nerve endings as well as smooth muscles all over the body, and depresses the sympathetic motor fibers of vessels in large doses only.

Extracts of Moringa increased glutathione-S-transferase activity >78% in esophagus, liver, and stomach, enough to be considered chemopreventive (Aruna and Sivaramakrishnan, 1990).

Indications (Horseradish Tree) — Abscess (f; KAB; PH2); Adenosis (f; KAP; NUT); Alopecia (f; NUT; SKJ); Ameba (1; TRA); Arthrosis (1; FNF; KAB; KAP; PH2; SUW); Ascites (f; HHB; NUT); Asthma (f; IED; KAP); Bacteria (1; FNF; KAP; MPI; WBB); Biliousness (f; KAB); Boil (f; KAP; NUT); Burn (f; JLH; NUT; TRA); Calculus (f; KAB); Cancer (1; FNF; JLH; JAC7:405); Cancer, abdomen (1; PH2; JAC7:405); Cancer, colon (1; JLH; JAC7:405); Cancer, esophagus (1; JAC7:405); Cancer, liver (1; JLH; JAC7:405); Cancer, nasopharynx (1; KAP; MPI); Cancer, spleen (f; JLH); Cancer, stomach (1; JAC7:405); Cardiopathy (f; PH2); Caries (f; SKJ; SUW); Catarrh (f; HHB; KAP; NUT); Cholera (1; SKJ; WBB); Circulosis (f; SUW); Cold (f; JFM); Colic (f; PH2); Constipation (f; PH2); Convulsion (f; NUT); Cough (f; JFM; KAP); Cramp (f; SUW); Cystosis (f; BOW); Dandruff (f; PH2); Debility (f; SUW); Dermatosis (f; JFM; PH2); Diabetes (f; PH2); Dropsy (f; IED; KAP; NUT); Dysentery (f; NUT); Dysmenorrhea (f; SKJ); Dyspepsia (f; KAP; PH2); Dysuria (f; NUT); Earache (f; IED); Edema (1; JFM; PH2; JAC7:405); Enterosis (f; JLH; PH2); Epilepsy (1; ABS; IED; PH2; SUW); Erysipelas (f; NUT); Escherichia (1; TRA; WOI); Esophagosis (1; JAC7:405); Fever (f; IED; JFM; PH2; SUW); Fracture (f; SKJ); Fungus (1; FNF; MPI; WBB); Gas (f; KAB; SUW); Gastrosis (f; PH2); Gingivosis (f; KAB); Gout (f; IED; KAP); Gravel (f; NUT; SKJ); Hallucination (f; KAB); Headache (f; JFM; PH2); Heart (f; KAB); Hematuria (f; NUT; SKJ); Hepatosis (f; HHB; JLH; SUW); Hiccup (f; KAB); Hoarseness (f; KAB); Hysteria (f; IED; KAB; SUW); Induration (f; JLH); Infection (1; FNF; KAP; WBB); Infertility (f; NUT); Inflammation (1; FNF; KAB; KAP; PH2; JAC7:405); Leprosy (f; KAB); Leukemia (f; KAP; MPI); Lumbago (f; KAB; PH2); Madness (f; NUT); Maggot (f; NUT); Malaria (f; JFM; KAP; PH2; SUW); Mycosis (1; HHB; MPI; NUT); Myosis (f; KAB); Nephrosis (f; JFM); Neuralgia (f; KAB; NUT); Odontosis (f; BOW); Oligolactea (f; BOW); Ophthalmia (f; KAB); Pain (f; JFM; KAP; SKJ; SUW); Palsy (f; KAB; SUW); Pancreatosis (f; WBB); Paralysis (f; KAB; PH2; SUW); Pharyngosis (f; KAB; KAP); Pneumonia (f; NUT; SKJ); Rheumatism (1; FNF; IED; JFM; KAP; PH2; SUW); Rhinosis (1; KAP); Salmonella (1; TRA; WOI); Scabies (f; NUT); Scirrhus (f; JLH); Scrofula (f; NUT); Septicemia (f; BOW); Shigella (1; TRA; WOI); Snakebite (f; IED; PH2); Sore (f; KAB; PH2); Sore Throat (f; KAB); Spasm (f; IED); Splenomegaly (f; PH2); Splenosis (f; JLH; HHB; PH2; SUW); Staphylococcus (1; MPI; WBB; WOI); Stomatosis (f; KAB); Stone (f; BOW); Streptococcus (1; WBB); Swelling (f; JFM; KAP); Syncope (f; KAB; SUW); Syphilis (f; NUT); Tetanus (f; KAB; SUW); Toothache (f; NUT); Tuberculosis (1; KAP); Tumor (1; FNF; NUT); Ulcer (f; BOW; IED); VD (f; NUT; SUW); Vertigo (f; NUT; PH2); Virus (1; FNF; KAP; MPI); Wart (f; JFM); Worm (f; JFM; PH2); Wound (f; IED; PH2); Yellow Fever (f; IED; NUT).

Horseradish Tree for cancer:

- AntiHIV: caffeic-acid; quercetin
- Antiaggregant: caffeic-acid; kaempferol; quercetin
- Anticancer: caffeic-acid; kaempferol; quercetin; vanillin
- Anticarcinogenic: caffeic-acid
- Antiestrogenic: quercetin
- Antifibrosarcomic: quercetin
- Antihepatotoxic: caffeic-acid; quercetin
- Antiinflammatory: caffeic-acid; kaempferol; quercetin
- Antileukemic: kaempferol; quercetin
- Antileukotriene: caffeic-acid; quercetin
- Antilipoperoxidant: quercetin
- Antimelanomic: quercetin
- Antimutagenic: caffeic-acid; kaempferol; quercetin; vanillin

- Antinitrosaminic: caffeic-acid; quercetin
- Antioxidant: caffeic-acid; delta-5-avenasterol; delta-7-avenasterol; gamma-tocopherol; kaempferol; quercetin; vanillin
- Antiperoxidant: caffeic-acid; quercetin
- Antiproliferant: quercetin
- Antiprostaglandin: caffeic-acid
- Antitumor: caffeic-acid; kaempferol; quercetin; vanillin
- Antiviral: caffeic-acid; kaempferol; quercetin; vanillin
- Apoptotic: kaempferol; quercetin
- COX-2-Inhibitor: kaempferol; quercetin
- Cyclooxygenase-Inhibitor: kaempferol; quercetin
- Cytoprotective: caffeic-acid
- Cytotoxic: caffeic-acid; quercetin
- Hepatoprotective: caffeic-acid; quercetin
- Immunostimulant: caffeic-acid
- Lipoxygenase-Inhibitor: caffeic-acid; kaempferol; quercetin
- Mast-Cell-Stabilizer: quercetin
- Ornithine-Decarboxylase-Inhibitor: caffeic-acid; quercetin
- p450-Inducer: quercetin
- PTK-Inhibitor: quercetin
- Prostaglandigenic: caffeic-acid
- Protein-Kinase-C-Inhibitor: quercetin
- Sunscreen: caffeic-acid
- Topoisomerase-II-Inhibitor: kaempferol; quercetin
- Tyrosine-Kinase-Inhibitor: quercetin

Horseradish Tree for infection:

- Analgesic: caffeic-acid; quercetin; spirochin
- Antibacterial: caffeic-acid; kaempferol; pterygospermin; quercetin
- Antiedemic: caffeic-acid
- Antiinflammatory: caffeic-acid; kaempferol; quercetin
- Antiseptic: caffeic-acid; kaempferol; oxalic-acid; pterygospermin; spirochin
- Antiviral: caffeic-acid; kaempferol; quercetin; vanillin
- Bacteristat: quercetin
- Candidicide: quercetin
- COX-2-Inhibitor: kaempferol; quercetin
- Cyclooxygenase-Inhibitor: kaempferol; quercetin
- Fungicide: caffeic-acid; pterygospermin; quercetin; spirochin; vanillin
- Immunostimulant: caffeic-acid
- Lipoxygenase-Inhibitor: caffeic-acid; kaempferol; quercetin

Other Uses (Horseradish Tree) — Described as "one of the most amazing trees God has created." Almost every part of the *Moringa* is said to be of value for food. Thickened pungent root used as substitute for horseradish. Mustard-flavored foliage eaten as greens, in salads, in vegetable curries, and for seasoning. Leaves pounded up and used for scrubbing utensils and for cleaning walls. Young, tender seedlings make an excellent cooked green vegetable (FAC). Flowers are said to make a satisfactory vegetable; interesting, particularly in subtropical places like Florida, where it is said to be the only tree species that flowers every day of the year. Flowers good for honey production. Young pods cooked as a vegetable, in soups and curries, or made into pickles. Immature seeds are eaten like peas while mature seeds, roasted or fried, are said to suggest peanuts (FAC). Seed is said to be eaten like a peanut in

Malaya. Seeds yield 38 to 40% of a nondrying oil, known as Ben Oil, used in arts and for lubricating watches and other delicate machinery. The oil contains high levels of unsaturated fatty acids, up to 75.39% oleic, making it competitive with olive oil and avocado as a poor man's source of MUFAs. The dominant saturated acids were behenic (to 6.73%) and palmitic (to 6.04%). The oil also rich in beta-sitosterol (to 50.07%), stigmasterol (to 17.27%), and campesterol (to 15.13%), and the various-tocopherols totaled more than 220 mg/kg of oil, respectively (JAF47:4495). Haitians obtain the oil by crushing browned seeds and boiling in water. Oil is clear, sweet and odorless, said never to become rancid. It is edible, used in salads and cooking, and in the manufacture of perfumes and hair dressings. Leaves and young branches are relished by livestock. A reddish gum produced by the bark, called Ben Gum, is used as a seasoning (FAC). Wood yields a blue dye. Commonly planted in Africa as a living fence (Hausa) tree. Trees planted on graves are believed to keep away hyenas and its branches are used as charms against witchcraft. Ochse notes an interesting agroforestry application: the thin crown throws a slight shade on kitchen gardens, which is "more useful than detrimental to the plants." In Taiwan, treelets are spaced 15 cm apart to make a living fence, the top of which is lopped off for the calcium- and iron-rich foliage. Bark can serve for tanning and also yields a coarse fiber. Trees are being studied as pulpwood sources in India. Analyses indicate that the tree is a suitable raw material for producing high alpha-cellulose pulps for use in cellophane and textiles (NUT).

For more information on activities, dosages, and contraindications, see the *CRC Handbook of Medicinal Herbs*, ed. 2, Duke et al., 2002.

Cultivation (Horseradish Tree) — In India, the plant is propagated by planting limb cuttings 1–2 m long, from June through August, preferably. The plant starts bearing pods 6–8 months after planting, but regular bearing commences after the second year. The tree bears for several years. Fruits are harvested as needed, or, perhaps in India, there may be two peak periods (March–April and September–October). A single tree may yield some 600–1000 pods a year. A single fruit may have 20 seeds, each weighing some 300 mg, suggesting a yield of 6 kg/tree and an oil yield of 2 kg/tree (NUT).

Chemistry (Horseradish Tree) — With the recent flurry of interest in "stanols," the health food industry might be interested in a closer look at Ben Oil, which has more stanols than olive oil. See table below. Ben Oil, as analyzed, is also well endowed with tocopherols, making it a double whammy for heart health, with a slight taste of horseradish. It's oleic-acid (MUFA) levels can be as high as 75.39%. The ben oil contained ca. 100 ppm alpha-tocopherol (cf 90 in olive oil); 35 ppm gamma-tocopherol (cf 10 in olive oil); 75 ppm delta-tocopherol (cf 2 in olive oil). These high levels of tocopherols, especially delta-tocopherol, the better antioxidant, offer some protection in processing and storage.

Ben Oil Analyzed

Sterol	% of sterols	Olive oil
Delta-5-Avenasterol	(8.84–12.79%)	16.77
Delta-7-Avenasterol	(0.94–1.11%)	0.29
Brassicasterol	(ND-0.06%)	TR
Campestanol	(ND-0.35%)	0.29
Campesterol	(14.03–15.13%)	3.20
Cholesterol	(0.12–0.13%)	0.15
Clerosterol	(0.84–22.52%)	0.54
Ergostadienol	(ND-0.39%)	ND
28-Isoavenasterol	(1.01–1.40%)	ND
24-Methylenecholesterol	(0.85–0.98%)	ND
Beta-Sitosterol	(49.19–50.07%)	64.3
Stigmastanol	(0.80–1.05%)	0.40
Delta-7,14-Stigmastanol	(0.44–0.52%)	TR
Stigmasterol	(16.78–17.27%)	0.60

Source: JAF47:4495

Here are a few of the more notable chemicals found in horseradish tree. For a complete listing of the phytochemicals and their activities, see the CRC phytochemical compendium, Duke and duCellier, 1993 (DAD) and the USDA database http://www.ars-grin.gov/duke/.

Alpha-Tocopherol — Anticancer; Anticonvulsant Synergen; Antimutagenic; Antioxidant (5 × quercetin); Antiradicular (5 × quercetin); Antitumor; RDA = 3–12 mg/day.

Gamma-Tocopherol — Antioxidant 10–15 μg/g.

Leucine — Antiencephalopathic.

Lysine — Antialkalotic; Antiherpetic 0.5–3 g/day; Essential; Hypoarginanemic 250 mg/kg; LD50 = 181 ivn mus.

Moringinine — Cardiotonic; Enterodepressant; Hypertensive; Myodepreessant; Sympathomimetic; Vasoconstrictor.

Oleic-Acid — 5-Alpha-Reductase-Inhibitor; Anemiagenic; Anticancer; Antiinflammatory IC50 = 21 μM; Antileukotriene-D4 IC50 = 21 μM; Choleretic 5 ml/man; Dermatitigenic; Hypocholesterolemic; Insectifuge; Irritant; Percutaneostimulant; LD50 = 230 ivn mus; LDlo = 50 ivn cat.

Spirochine — Analgesic; Antipyretic; Antiseptic; Cardiodepressant 350 mg/kg; Cardiotonic 35 mg/kg; CNS-Paralytic; Fungicide; Hypotensive; Myocardiotonic; Uterotonic.

Murraya koenigii (L.) Spreng. (Rutaceae)
CURRY LEAF

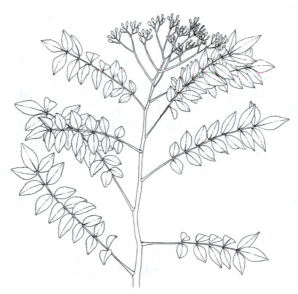

M

Synonyms — *Bergera koenigii* L., *Chalcas koenigii* (L.) Kurz.

Medicinal Uses (Curry Leaf) — Reported to be analgesic, carminative, stomachic, and tonic (DAW). Medicinally, leaves, roots, and bark are considered carminative, stomachic, and tonic, especially in India. Leaves used internally in dysentery and diarrhea and for checking vomiting; used externally, it is applied to bruises and skin eruptions (WOI). Juice of root taken to relieve pain associated with kidney ailments (KAB). Powdered root mixed with betel nut juice and honey as an antiperiodic (DEP).

Extracts antispasmodic and protisticidal (MPI). The 50% ethanolic extracts showed antiprotozoal activity against *Entameba histolytic* and antispasmodic activity on guinea pig ileum (no anticancer, antiviral, or CNS effects, nor hypolipemic activity in some Indian assays). Aqueous extracts (= tea) of leaves hypoglycemic in normal and alloxan diabetic dogs (MPI). Aqueous extracts inhibit ethanol-induced stomach ulcers (TAD). Leaf EO antibacterial against *B. subtilis* (at 2000 ppm), *C. pyogenes* (at 2000 ppm), *P. vulgaris, Pasteurella multicida*, and *Staphylococcus aureus* (at 2000 ppm) (MPI). Leaf EO fungicidal against *Aspergillus fumigatis, A. niger, Candida albicans* (at 2000 ppm), *C. tropicalis*, and *Microsporum gypseum* (MPI). Ethanolic leaf extract active against *Colletotrichum falcatum* and *Rhizoctonia solani* (MPI). Some of the carbazole alkaloids were fungicidal against *Microsporum* and *Trichophyton*. At 10 µg/ml, girinimbine inhibited their growth.

Tachibana et al. (2001), comparing five antioxidant carbazole alkaloids, found three were better antioxidants than tocopherol, two less so. Euchrestine, bismurrayafoline, mahanine, mahanimbicine, and mahanimbine all showed radical scavenging activity (JAF49:5594). Similarly, Ramsewak et al. (1999) isolated three bioactive carbazole alkaloids, mahanimbine, murrayanol, and mahanine, all of which were antiseptic, insecticidal, and inhibited topoisomerase I and II inhibition activities. Adebajo and Reisch (2000) reported several minor furocoumarins, possibly additive or synergistic, in the seeds: byakangelicol, byakangelicin, gosferol, isobyakangelicol, isogosferol, neobyakangelicol, phellopterin, and xanthotoxin. It seems to me the rule, rather than the exception, that biologically active compounds of a type (like an alkaloid, flavonoid, furanocoumarin, lignan, monterpene, an OPC, phytostanol, phytosterol, polyphenol, saponin, tannin, or a triterpenoid), don't usually occur alone in their class, but in a suite of closely related, often synergistic compounds in that class. Not only is that additivity or synergy lost when we seek out the single most active compound, we enable the enemy to develop resistance. Resistance is a response to the monochemical silver bullet, not the polychemical herbal shotgun.

M

Indications (Curry Leaf) — Ameba (1; MPI); Bacteria (1; FNF; MPI; TAD); Biliousness (f; DEP); Bite (f; DEP); Blood (f; KAB); Bruise (f; WOI); Cancer (1; FNF; X9366097); Colic (f; BOW); Constipation (f; BOW); Cramp (1; FNF; MPI); Dermatosis (f; DEP; WOI); Diabetes (1; MPI); Diarrhea (f; SKJ); Dysentery (f; DEP; SKJ); Fever (f; KAB); Fungus (1; FNF); Gastrosis (1; TAD); Hemorrhoid (f; KAB); Infection (1; FNF; TAD); Inflammation (1; FNF; KAB); Itch (f; KAB); Leukemia (1; ABS); Leukoderma (f; KAB); Malaria (f; DEP); Melanoma (1; FNF; X9366097); Nausea (f; SKJ); Nephrosis (f; SKJ); Nervousness (1; FNF); Pain (f; SKJ); Snakebite (f; KAB); Staphylococcus (1; MPI); Thirst (f; KAB); Tumor (1; FNF); Ulcer (1; FNF; TAD); Virus (1; FNF); Vomiting (f; DEP); Worm (f; KAB).

Curry Leaf for cramp:

- Analgesic: myrcene; p-cymene
- Anesthetic: benzaldehyde; carvacrol; linalool; myrcene
- Anticonvulsant: bergapten; gaba; linalool
- Antiinflammatory: alpha-pinene; bergapten; beta-pinene; carvacrol; caryophyllene; isopimpinellin
- Antispasmodic: benzaldehyde; bergapten; bornyl-acetate; carvacrol; caryophyllene; limonene; linalool; myrcene
- Carminative: carvacrol
- Cyclooxygenase-Inhibitor: carvacrol
- Myorelaxant: bornyl-acetate
- Sedative: alpha-pinene; alpha-terpineol; benzaldehyde; bornyl-acetate; caryophyllene; dipentene; gaba; limonene; linalool; p-cymene
- Tranquilizer: alpha-pinene; gaba

Curry Leaf for infection:

- Analgesic: myrcene; p-cymene
- Anesthetic: benzaldehyde; carvacrol; linalool; myrcene
- Antibacterial: alpha-pinene; alpha-terpineol; benzaldehyde; bornyl-acetate; carvacrol; caryophyllene; delta-cadinene; dipentene; limonene; linalool; myrcene; nerolidol; p-cymene; terpinen-4-ol
- Antiedemic: caryophyllene
- Antiinflammatory: alpha-pinene; bergapten; beta-pinene; carvacrol; caryophyllene; isopimpinellin
- Antiseptic: 2-nonanone; alpha-terpineol; aromadendrene; benzaldehyde; beta-pinene; carvacrol; hexanol; limonene; linalool; oxalic-acid; terpinen-4-ol
- Antiviral: alpha-pinene; beta-bisabolene; bornyl-acetate; dipentene; limonene; linalool; neryl-acetate; p-cymene
- Cyclooxygenase-Inhibitor: carvacrol
- Fungicide: alpha-phellandrene; beta-phellandrene; carvacrol; caryophyllene; isopimpinellin; linalool; myrcene; oxypeucedanin; p-cymene; terpinen-4-ol; terpinolene
- Fungistat: limonene
- Immunostimulant: benzaldehyde

Other Uses (Curry Leaf) — The pungently aromatic curry leaves, dried or fresh, may be found in Indian grocery shops. Much cultivated in India for its aromatic leaves and used as a spice in curries, chutneys, and stews (CFR). Tucker and Debaggio (2000), advise that, though not GRAS, the herb harmonizes well with curry mixtures for buttermilks, chutneys, fish dishes, meat dishes, pickles, scrambled eggs, soups (rasams), etc. Leaves retain their pungent flavor even after drying. Curry leaves, with stems removed, can be sautéed lightly in oil. This method should be chosen when making a dry curry and also for the savory spiced-rice dishes of south India (AAR). Leaves are fried in ghee or oil until crisp (FAC). It was found that fresh curry leaves (1%) prevented oxidation of ghee better than the commercial antioxidants BHA and BHT. Leaves are an ingredient of the Tamil Nadu spice blend called curry powder. After being powdered in a grinder or blender, the dried leaves can also be used in marinades or sprinkled on vegetables or yogurt. They are used almost like bay leaves, a couple of sprigs with a total of six to ten leaflets are needed for a dish serving four. Remove sprigs before serving. Curry leaves are sometimes called "sweet neem" (AAR). Fruit is edible and yields 0.76% of a yellow volatile oil with a neroli-like odor and a pepper-like taste, giving an agreeable sensation of coolness on the tongue (Reed, 1976). Wood (wt. 43–50 lb/cu ft) is grayish-white, even and close-grained, durable, used for agricultural implements. Plants often grown as an ornamental (CFR).

For more information on activities, dosages, and contraindications, see the *CRC Handbook of Medicinal Herbs*, ed. 2, Duke et al., 2002.

Cultivation (Curry Leaf) — Propagation by seeds, which germinate freely under partial shade. When cultivated, seedlings are spaced 3–7 m each way. Semiripe cuttings possible in late summer (Bown, 2001). Moist, rich, sunny, or partially shaded sites recommended (Bown, 2001). Hardy only in zones 9 and 10, plants tolerate full sun and evenly moist garden loams (TAD). Plants are outplanted in May in India. Plants may also be propagated by cuttings of ripened wood, which should be taken with leaves intact or by root suckers. These are inserted in moist sand in a shady site. Rooted cuttings are spaced as seedlings. Once established, shrubs require little attention. Native plants often form undergrowth in forest (CFR). Harvested as needed, leaves rarely enter into commercial statistics.

Chemistry (Curry Leaf) — Leaves contain 6.1% protein; 16% carbohydrate, calcium, phosphorus, and iron; and the vitamins A and C and nicotinic acid, but lack thiamine and riboflavin. Leaves are fair source of vitamin A and a rich source of calcium, but the high total oxalate content is a hindrance. The most important constituents of leaf EO are beta-caryophyllene, beta-gurjunene, beta-elemene, beta-phellandrene, and beta-thujene. It produces nearly 2000 times as many aroma volatiles as does *Pandanus latifolius*. Prakash and Natarajan (1976) tabulate the composition of young, medium, and older leaves:

	Tender	Medium	Mature
Moisture	70.1%	65.1%	63.2%
Total nitrogen	0.87%	1.03%	1.15%
Crude protein (Nx6.25)	5.48%	6.43%	6.92%
Fat	3.30%	4.74%	6.15%
Total sugars	14.86%	17.95%	18.92%
Starch	11.4%	14.2%	14.6%
Crude fibre	5.8%	6.2%	6.8%
Ash	12.54%	12.68%	13.06%
Volatile oil	0.82%	0.55%	0.48%

Here are a few of the more notable chemicals found in curryleaf. For a complete listing of the phytochemicals and their activities, see the CRC phytochemical compendium, Duke and duCellier, 1993 (DAD) and the USDA database http://www.ars-grin.gov/duke/.

Alpha-Phellandrene — Dermal; Emetic; Fungicide; Insectiphile; Irritant; Laxative.

Beta-Elemene — Antitumor (Cervix).

Beta-Phellandrene — Expectorant; Fungicide.

Mahanimbine — Antioxidant 33 µg/ml; Antiseptic; Insecticide; Mosquitocide; Topoisomerase-I-Inhibitor; Topoisomerase-II-Inhibitor.

Mahanine — Antioxidant (>tocopherol); Antiseptic; Insecticide; Mosquitocide; Topoisomerase-I-Inhibitor; Topoisomerase-II-Inhibitor.

Myristica fragrans Houtt. (Myristicaceae)
MACE, NUTMEG

Synonym — *Myristica officinalis* L.

Medicinal Uses (Nutmeg) — Considered aphrodisiac, astringent, carminative, narcotic, and stimulant. Mace has been used for putrid and intermittent fevers and mild indigestion. The expressed oil of nutmeg is used externally as a stimulant. They are used to allay both gas and nausea. Mixed with lard, grated nutmeg is applied to piles. Roasted nutmeg is used internally for leucorrhea. The EO is recommended for inflammation of the bladder and urinary tract. Chinese use powdered seeds for pediatric and geriatric fluxes, for cardosis, cold, cramps, and chronic rheumatism. Indonesians boil the powdered seed for anorexia, colic, diarrhea, dyspepsia, dysentery, and malaria. Seed oil is rubbed on the temples for headache or dropped in tea for dyspepsia and nausea. Indonesians use the leaf tea for gas. Malayans use the nutmeg for madness, malaria, puerperium, rheumatism, and sciatica. Arabians, as early as the seventh century B.P., recommended nutmeg for digestive disorders, kidney troubles, lymphatic ailments, etc. Even earlier, Indians used it for asthma, fever, heart disease, and tuberculosis. In India, nutmeg is prescribed for dysentery, gas, malaria, leprosy, rheumatism, sciatica, and stomachache. Arabs still use nutmeg as an aphrodisiac in love potions. Nutmeg butter is used in massage for arthritis, paralysis, rheumatism, sciatica, and sprains. It is also used as an external stimulant in hair lotions, ointments, and plaster (MPI). Yemenites recommend its use for the liver and spleen, for colds, fevers, and respiratory ailments.

Nutmeg may alleviate some symptoms of certain types of cancer, suggested in a case study presented in the *New England Journal of Medicine*. Dr. Ira Shafran and Daniel MaCrone of Ohio State University say, "further study may substantiate the speculation that inhibition of prostaglandin E_2 by nutmeg affords symptomatic improvement of hypercalcemia in medullary carcinoma of the thyroid and other prostaglandin-secreting neoplasms." They prescribed 4 to 6 tbsp of nutmeg per day to their patient, because nutmeg is known to improve diarrhea associated with medullary carcinoma of the thyroid. The patient also suffered from hypercalcemia that did not respond to standard calcium-reducing treatment. After 12 days of nutmeg therapy, the calcium levels were reduced by almost one-third. The medical team says that medullary carcinoma of the thyroid is known to produce "copious amounts of prostaglandin E_2... (and) inhibition of prostaglandin E_2 may be nutmeg's antidiarrheal mechanism of action" (DAD). Mace, at 10 mg per mouse, significantly decreased carcinoma of the uterine cervix. At 1% of diet, mace significantly decreases skin papillomas in mice. At 1–2% levels, for 10 days, mace significantly increased glutathione-S-transferase activities (JAC7:405).

Khan and Balick (2001) note human studies on Myristica for bladder and kidney stones. In rabbits, ethanolic fruits extracts at 500 mg/kg/day orally for 60 days, lowered LDL and total cholesterol and triglycerides (JE55:49). Nutmeg proved antiseptic to *Bacillus subtilis, Escherichia coli,* and *Saccharomyces cerevisiae,* confirming traditional uses of spices as food preservatives, disinfectants, and antiseptics (De et al., 1999).

Four or five grams of myristicin, a narcotic, produce toxic symptoms in man. Many women, in hopes of inducing abortion, have failed yet suffered the intoxication due to myristicin. Ingestion of 1–2 oz ground nutmeg produces a prolonged delirium, disorientation, and drunkenness. "Stirred into a glass of cold water, a penny matchbox full of nutmeg had the kick of three or four reefers," Malcolm X, as quoted by Schleiffer. Nutmeg taken as a psychotropic often causes reactions similar to those of other hallucinogenic drugs, quite unlike the classic account of *Myristica* poisoning. Myristicin alone does not give the reaction, but eating the whole seed does. Rinzler (1990) says boldly in bold print, "...high doses, defined as one to three whole seeds or 5 to 15 g (0.5 oz) grated nutmeg, can cause euphoria, a sensation of floating, flushed skin, vomiting, circulatory collapse, and visual or auditory hallucinations, within one to six hours after the nutmeg is consumed. Very large doses may be fatal" (RIN). Doses exceeding 1 tsp take effect within 2 to 5 hr, producing time-space distortions, feelings of unreality, and sometimes visual hallucinations accompanied by dizziness, headache, illness, and rapid heartbeat (anonymous). Reviewing research on myristicin, which occurs also in black pepper, carrot seed, celery seed, and parsley, it is noted that the psychoactive and hallucinogenic properties of mace, nutmeg, and purified myristicin have been studied. It has been hypothesized that myristicin and elemicin can be readily modified in the body to amphetamines.

The oral LD_{50} for nutmeg oil in rats, mice, and hamsters is 2600, 5620, and 6000 mg/kg, respectively. (For more data on the phytochemicals and their activities, see the CRC phytochemical compendia) (CRC, DAD).

Indications (Nutmeg) — Agoraphobia (f; HHB); Anorexia (f; CRC); Arthrosis (f; JLH) Asthma (f; CRC); Bacillus (1; X10548758); Bacteria (1; FNF; X10548758); Cancer (1; APA; CRC; FNF); Cancer, gum (f; CRC; JLH); Cancer, joint (f; CRC; JLH); Cancer, liver (f; CRC; JLH); Cancer, mouth (f; CRC; JLH); Cancer, spleen (f; CRC; JLH); Childbirth (f; BOW; CRC); Cholera (f; FEL; PH2); Cold (f; CRC; FEL); Colic (f; AHP; CRC; HHB); Cramp (1; BOW; CRC; FNF; PH2); Cystosis (f; CRC; MPI); Debility (f; PH2); Delirium (f; BOW); Diarrhea (1; AHP; APA; FNF; PH2); Dysentery (1; CRC; FNF; PH2); Dysmenorrhea (f; HHB); Dyspepsia (f; AHP; APA; CRC; PH2); Eczema (f; BOW); Enterosis (f; BOW); Escherichia (1; X10548758); Fever (f; CRC; FEL; PH2); Fungus (1; FNF; X10548758); Gas (f; AHP; APA; CRC; PH2); Gastrosis (f; CRC; PHR; PH2); Headache (f; CRC; PH2); Heart (f; CRC); Heartburn (f; HHB); Hemorrhoid (f; CRC; FEL); Hepatosis (1; CRC; FNF); High Cholesterol (1; APA; FNF); High Triglycerides (1; JE55:49); HIV (1; FNF); Hypercalcemia (1; CRC); Hypochondria (f; HHB); Hysteria (f; HHB); Impotence (f; PH2); Incontinence (f; BOW); Induration (f; CRC; JLH); Infection (1; FNF; X10548758); Inflammation (1; CRC; FNF; PH2); Insanity (f; CRC); Insomnia (f; APA; PH2); Lachrimosis (f; HHB); Leprosy (f; CRC); Leukemia (1; FNF); Leukorrhea (f; CRC; FEL); Lymphosis (f; CRC); Malaria (f; CRC; FEL; PH2); Mycosis (1; X10548758); Nausea (f; BOW; CRC); Nephrosis (f; APA; CRC); Neuralgia (f; PH2); Neurasthenia (f; HHB); Neurosis (f; PH2); Ophthalmia (f; PH2); Pain (1; APA; FNF); Paralysis (f; MPI); Pneumonia (f; FEL); Respirosis (f; CRC; PH2); Rheumatism (1; APA; CRC; FNF; MPI; PH2); Sciatica (f; CRC; MPI; PH2); Splenosis (f; CRC); Spermatorrhea (f; BOW); Sprain (f; MPI); Stomachache (f; CRC; FEL; MPI); Stomatosis (f; APA); Toothache (f; APA); Tuberculosis (f; CRC); Tumor (1; CRC; FNF; JLH); Urethrosis (f; MPI); UTI (f; CRC); Virus (1; FNF); Vomiting (f; PH2); Water Retention (1; FNF); Xerostomia (f; HHB); Yeast (1; X10548758).

M

Nutmeg for cancer:

- AntiEBV: (−)-epicatechin
- AntiHIV: (−)-epicatechin; caffeic-acid; oleanolic-acid; quercetin
- Antiaggregant: (−)-epicatechin; caffeic-acid; elemicin; eugenol; isoeugenol; kaempferol; myristicin; quercetin; safrole; salicylates
- Antiarachidonate: eugenol
- Anticancer: (−)-epicatechin; alpha-pinene; caffeic-acid; camphor; delphinidin; eugenol; eugenol-methyl-ether; geraniol; isoeugenol; kaempferol; limonene; linalool; myristicin; oleanolic-acid; p-coumaric-acid; quercetin; safrole; terpineol; vanillin
- Anticarcinogenic: caffeic-acid
- Antiestrogenic: quercetin
- Antifibrosarcomic: quercetin
- Antihepatotoxic: caffeic-acid; oleanolic-acid; p-coumaric-acid; quercetin
- Antiinflammatory: (−)-epicatechin; alpha-pinene; beta-pinene; borneol; caffeic-acid; caryophyllene; eugenol; gentisic-acid; kaempferol; myristicin; oleanolic-acid; quercetin; salicylates
- Antileukemic: (−)-epicatechin; kaempferol; quercetin
- Antileukotriene: caffeic-acid; quercetin
- Antilipoperoxidant: (−)-epicatechin; quercetin
- Antimelanomic: geraniol; quercetin
- Antimutagenic: (−)-epicatechin; caffeic-acid; eugenol; kaempferol; limonene; myrcene; quercetin; vanillin

- Antinitrosaminic: caffeic-acid; p-coumaric-acid; quercetin
- Antioxidant: (–)-epicatechin; caffeic-acid; camphene; cyanidin; delphinidin; eugenol; gamma-terpinene; isoeugenol; kaempferol; myrcene; myristicin; oleanolic-acid; p-coumaric-acid; quercetin; vanillin
- Antiperoxidant: (–)-epicatechin; caffeic-acid; p-coumaric-acid; quercetin
- Antiproliferant: quercetin; terpineol
- Antiprostaglandin: caffeic-acid; eugenol
- Antisarcomic: oleanolic-acid
- Antistress: elemicin; myristicin
- Antithromboxane: eugenol
- Antitumor: alpha-humulene; caffeic-acid; caryophyllene; eugenol; geraniol; kaempferol; limonene; oleanolic-acid; p-coumaric-acid; quercetin; vanillin
- Antiviral: (–)-epicatechin; alpha-pinene; beta-bisabolene; caffeic-acid; dipentene; gentisic-acid; kaempferol; limonene; linalool; oleanolic-acid; p-cymene; quercetin; vanillin
- Apoptotic: kaempferol; quercetin
- Beta-Glucuronidase-Inhibitor: oleanolic-acid
- COX-2-Inhibitor: eugenol; kaempferol; oleanolic-acid; quercetin
- Chemopreventive: limonene
- Cyclooxygenase-Inhibitor: kaempferol; oleanolic-acid; quercetin
- Cytochrome-p450-Inducer: 1,8-cineole; delta-cadinene; safrole
- Cytoprotective: caffeic-acid
- Cytotoxic: (–)-epicatechin; caffeic-acid; eugenol; isoeugenol; p-coumaric-acid; quercetin
- DNA-Binder: elemicin; eugenol-methyl-ether; safrole
- Hepatoprotective: borneol; caffeic-acid; eugenol; oleanolic-acid; quercetin
- Hepatotonic: 1,8-cineole
- Immunostimulant: (–)-epicatechin; caffeic-acid
- Lipoxygenase-Inhibitor: (–)-epicatechin; caffeic-acid; kaempferol; p-coumaric-acid; quercetin
- Mast-Cell-Stabilizer: quercetin
- Ornithine-Decarboxylase-Inhibitor: caffeic-acid; limonene; quercetin
- p450-Inducer: 1,8-cineole; delta-cadinene; quercetin
- PTK-Inhibitor: quercetin
- Prostaglandigenic: caffeic-acid; p-coumaric-acid
- Protein-Kinase-C-Inhibitor: quercetin
- Reverse-Transcriptase-Inhibitor: (–)-epicatechin
- Sunscreen: caffeic-acid
- Topoisomerase-II-Inhibitor: kaempferol; quercetin
- Tyrosine-Kinase-Inhibitor: quercetin

Nutmeg for infection:

- Analgesic: borneol; caffeic-acid; camphor; eugenol; gentisic-acid; myrcene; p-cymene; quercetin
- Anesthetic: 1,8-cineole; camphor; eugenol; linalool; myrcene; myristicin; safrole
- Antibacterial: (–)-epicatechin; 1,8-cineole; acetic-acid; alpha-pinene; alpha-terpineol; caffeic-acid; caryophyllene; citronellol; dehydroisoeugenol; delta-cadinene; dipentene; eugenol; gentisic-acid; geraniol; kaempferol; limonene; linalool; malabaricone-b; malabaricone-c; myrcene; nerol; oleanolic-acid; p-coumaric-acid; p-cymene; pinene; quercetin; safrole; sclareol; terpinen-4-ol; terpineol

- Antiedemic: caffeic-acid; caryophyllene; eugenol; oleanolic-acid
- Antiinflammatory: (–)-epicatechin; alpha-pinene; beta-pinene; borneol; caffeic-acid; caryophyllene; eugenol; gentisic-acid; kaempferol; myristicin; oleanolic-acid; quercetin; salicylates
- Antiseptic: 1,8-cineole; alpha-terpineol; beta-pinene; caffeic-acid; camphor; citronellol; eugenol; formic-acid; furfural; geraniol; kaempferol; licarin-a; limonene; linalool; nerol; oleanolic-acid; pinene; safrole; terpinen-4-ol; terpineol
- Antiviral: (–)-epicatechin; alpha-pinene; beta-bisabolene; caffeic-acid; dipentene; gentisic-acid; kaempferol; limonene; linalool; oleanolic-acid; p-cymene; quercetin; vanillin
- Astringent: formic-acid
- Bacteristat: isoeugenol; quercetin
- COX-2-Inhibitor: eugenol; kaempferol; oleanolic-acid; quercetin
- Cyclooxygenase-Inhibitor: kaempferol; oleanolic-acid; quercetin
- Fungicide: 1,8-cineole; acetic-acid; alpha-phellandrene; beta-phellandrene; caffeic-acid; camphor; caprylic-acid; caryophyllene; citronellol; dehydroisoeugenol; elemicin; eugenol; furfural; geraniol; isoelemicin; linalool; malabaricone-b; malabaricone-c; myrcene; myristicin; octanoic-acid; p-coumaric-acid; p-cymene; pinene; quercetin; sclareol; terpinen-4-ol; terpinolene; vanillin
- Fungistat: formic-acid; isoeugenol; limonene
- Immunostimulant: (–)-epicatechin; caffeic-acid
- Lipoxygenase-Inhibitor: (–)-epicatechin; caffeic-acid; kaempferol; p-coumaric-acid; quercetin

M

Other Uses (Nutmeg) — Both nutmeg and mace are used as spices in many American and exotic dishes. Seeds are the source of nutmeg, used to flavor cakes, custards, eggnog, pies, puddings, punches, possets, sauces, and vegetables, such as cabbage, cauliflower, and spinach. Mace, the dried aril, is favored to flavor baked goods, cakes, curries, fruits, salads, ketchups, pickles, soups, and sauces. Powdered mace, sprinkled on cooked cabbage, masks the sulfide odor. Nutmeg appears in several spice blends, curry powders, garam masala, jerk seasonings, mixed spices, mulling spices, and quatre épices. Both nutmeg and mace are indispensable ingredients of ras el hanout. Caribbean islands use some nutmeg for jerked meats, and in curry powder. Whole nutmegs can be quartered and boiled with cream, sugar syrup, broth, or some other element of a recipe to impart a good nutmeg flavor. Mace can be used in place of nutmeg, but use a pinch less mace. Fruits are sometimes gathered before maturing to make jelly. The flesh of the fruit is cut into slices and eaten as a delicacy with sambal (hot-pepper sauce), pickled, candied, preserved, or made into nutmeg-flavored jams, jellies, and syrups (FAC). Both nutmeg and mace yield an EO used for flavoring foods and liqueurs. Oil of Nutmeg is distilled for external medicine and perfumes. Oil, distilled from the leaves, has a spicy, aromatic, pleasant flavor and is used in making toilet and medicinal products, as a flavoring essence, and in chewing gum. Nutmeg butter, derived from broken seeds and poor grade mace, is used in medicinal ointments, suppositories, and perfumery. In Ayurvedic documents, nutmeg was called "madashaunda," a term meaning "narcotic fruits." Betel chewers in India often add nutmeg, and it is also added to chewing tobaccos and snuffs. On the island of Banda, a pap is made of the bark preserved with sugar and tasting like sour apples. The bark is pickled in brine in Java but tends to induce sleep. Juice from the pericarp is an efficacious mordant for fixing dyes (AAR, DAD, FAC).

Alcoholic extracts of nutmeg are antibacterial, and aqueous extracts kill cockroaches, while the volatile oils from the leaf are herbicidal. Myristicin enhances the toxicity of pyrethrum to house-flies. Perhaps the natural products might be less damaging environmentally than synthetics.

For more information on activities, dosages, and contraindications, see the *CRC Handbook of Medicinal Herbs*, ed. 2, Duke et al., 2002.

Cultivation (Nutmeg) — Propagated by seed. Selected fresh seed are planted in the shell, spaced about 10 × 30 cm, in nurseries where they germinate in about 6 weeks. When the seedlings are ca. 15 cm tall (ca. 6 months), they are outplanted, spaced about 8 m apart. Male and female trees occur in about a 50% ratio. Most male plants are cut out when recognized (when they flower at age of 5–8 years), leaving spacing about 12–14 m apart. In Grenada, marcots and approach-grafts, when outplanted, start bearing in 18 months. Full-fruiting and best yields obtained when trees are 12–20 years old; may continue to bear 40–75 or more years. Seedling trees may start bearing in 5–8 years, vegetative specimens earlier. Yields increase up to age 15 or so. Fruiting, scattered over the year, has two peaks, following the flowering period by about 6 months. Trees should be protected from extreme heat and strong winds. Without enough sun, however, EO does not develop. During dry spells, the trees must be watered frequently. Shade is provided by *Canarium*, *Gliricidia*, or *Musa*. Nutmeg may be mixed with arecanut, coconut, coffee, rubber, and tea. Little cultivation is required, as the oil in the leaves inhibits the growth of weeds and grass. Ripe fruits are gathered and may split or burst shortly after gathering from the tree or off the ground. Fruits that lie on the ground too long may discolor. The reddish aril, known as mace, is recovered first, flattened, and dried. There may be 4000 fruits per tree (in India—20,000 elsewhere are reported) the average running 1250 fruits per tree per year. In Grenada, trees average about 1000 nuts with 90–100 trees to the acre. Average nutmeg yields are about 720 lb dry (1500 lb green) with mace yields of about 150 lb green, ca. 35 lb dry to the acre. Dried nutmegs are subject to insect damage and should be limed and/or stored in sealed containers (BOW). (For more detail, see Duke and duCellier, 1993 and Purseglove, 1981).

Chemistry (Nutmeg) — Here are a few of the more notable chemicals found in nutmeg. For a complete listing of the phytochemicals and their activities, see the CRC phytochemical compendium, Duke and duCellier, 1993 (DAD) and the USDA database http://www.ars-grin.gov/duke/.

Elemicin — Antiaggregant IC50 = 360 μ*M*; Antidepressant; Antifeedant; Antihistaminic; Antiserotonic; Antistress ihl; DNA-Binder; Fungicide MIC = 8 μg; Hallucinogenic; Hypotensive ihl; Insecticide 100 ppm; Insectifuge; Larvicide 100 ppm; Neurotoxic; Schistosomicide.

Furfural — Antiseptic; Fungicide; Insecticide; Irritant; LD50 = 127 orl rat.

Myristic-Acid — Anticancer; Cosmetic; Hypercholesterolemic; Lubricant; Nematicide; LD50 = 43 ivn mus.

Myristicin — Amphetaminagenic; Anesthetic; Antiaggregant IC50 = 250 μ*M*; Anticancer 10 mg/mus/orl/day; Antidepressant; Antiinflammatory; Antioxidant 25–100 mg/kg/orl; Antispasmodic; Antistress ihl; Calcium-Antagonist IC50 = 88 μ*M*; Diuretic; Fungicide MIC = 20 μg; Hallucinogenic; Hepatotoxic; Hypotensive ihl; Hypnotic; Insecticide 25 ppm; Insecticide-Synergist; Larvicide 25 ppm; MAOI; Neurotoxic; Oxytocic; Paralytic; Psychoactive; Sedative 300 mg/kg ipr; Serotoninergic 1000 ppm orl; Tachycardic; Uterotonic; LDlo = 570 orl cat; LD50 = 200 mg/kg ivn mus; LD50 = 1000 mg/kg ipr rat.

Safrole — See also *Sassafras albidum*.

M

Myrtus communis L. (Myrtaceae)
MYRTLE

Medicinal Uses (Myrtle) — The leaf is used for condylomata, figs, whitlows, warts, figs, parotid tumors, cancer of the gums, ulcerated cancers, and polyps. Iranians make a hot poultice for boils from the plant. The oil, in plasters or unguents, is said to help indurations of the breast, condyloma of the genitals, and cancer. The berries and seed are said to cure tumors and uterine fibroids (JLH). An infusion or tincture of leaves is given for prolapsus and leucorrhoea, and for washing incisions and joints. It is also used to check night sweats of phthisis and for all types of pulmonary disorders. Unani direct smoke from the leaves onto hemorrhoids. Italians make a bolus of the leaves in turpentine for the same indication. Algerians recommend the leafy infusion for asthma. Unani use fruits for bronchitis, headache, and menorrhagia (KAB). They consider the fruits useful for the blood, brain, hair, and heart. North Africans use the dry flower buds for smallpox (BIB, PH2, WOI). Lebanese consider the plant binding and diuretic, believing it holds loose things in place — the bowels, the emotions, or the teeth. The EO and tincture have analgesic properties but not as strong as menthol and peppermint oil. Wine of myrtle corrects the bad odor and stimulates healing in offensive sores and ulcers, threatening gangrene (FEL).

Aqueous and ethanolic extracts of leaves, roots, and stems are active against Gram(–) and Gram(+) bacteria. The plant contains antibacterial phenols. One thermolabile principle was highly active against *Micrococcus pyogenes* var. *aureus*. The principle resembled streptomycin in its action on *Mycobacterium tuberculosis* (WOI). Aqueous berry extract is active against carrageenan-induced edema in rats paw [(=)comparable to oxyphenylbutazone] (MPI).

Large doses of myrtle may cause diarrhea, nausea, and vomiting, according to Gruenwald et al. (2000). More than 10 g myrtle oil can threaten life, due to high cineole content (myrtle contains 135–2250 ppm cineole according to my calculations, meaning 10 g myrtle would contain a maximum 22.5 mg cineole). Several herbs and spices may attain higher levels of cineole; see cardamom for the longer list. Myrtle phytochemicals are said to be quickly absorbed and to impart a violet aroma to the urine within 15 min (BOW).

Indications (Myrtle) — Acne (f; BOW); Adenosis (f; JLH); Allergy (1; FNF); Alopecia (f; DEP); Aphtha (f; BIB; DEP); Aposteme (f; JLH); Arthrosis (1; FNF; MPI); Bacteria (1; BIB;

FNF; WOI); Bleeding (f; BIB); BPH (f; PH2); Bronchosis (1; BIB; FEL; FNF; HHB; PH2); Cacoethes (f; BIB); Cancer (f; JLH); Cancer, breast (f; JLH); Cancer, colon (f; JLH); Cancer, gum (f; JLH); Cancer, liver (f; JLH); Cancer, spleen (f; JLH); Cancer, throat (f; JLH); Cancer, uterus (f; JLH); Candida (1; FNF); Catarrh (f; FEL); Cold (1; PH2); Condylomata (f; BIB); Congestion (1; FNF); Conjunctivosis (f; FEL); Cough (1; FNF; MAD); Cramp (1; FNF); Cystosis (1; BIB; FEL; FNF; PH2); Diarrhea (1; BIB; FNF; MAD; PH2); Dropsy (f; MAD); Dysentery (f; BIB); Dyspepsia (f; BIB); Eczema (f; BIB); Edema (1; FNF; MPI); Encephalosis (f; BIB; DEP); Enterosis (f; JLH); Epilepsy (f; BIB; WOI); Fatigue (f; PH2); Fever (f; BIB); Fibroid (f; JLH); Flu (1; FNF); Fungus (1; FNF); Gangrene (f; FEL); Gastrosis (f; BIB; MAD); Gingivosis (f; BOW; JLH); Gonorrhea (f; MAD); Gray Hair (f; BIB); Headache (f; BIB); Hemorrhoid (f; FEL; PH2); Hepatosis (f; BIB; JLH; WOI); Induration (f; JLH); Infection (1; FNF; PH2); Inflammation (1; FNF; MPI); Intertrigo (f; FEL); Leishmania (1; ABS; FT68:276); Leukorrhea (f; BIB; FEL; PH2); Mastosis (f; JLH); Menorrhagia (f; FEL); Nephrosis (f; FEL); Night Sweat (f; BIB); Otosis (f; PH2); Pain (1; FEL; FNF; MAD); Parotosis (f; JLH); Pertussis (1; PH2); Pharyngosis (f; FEL); Phthisis (f; BIB); Pleurodynia (f; MAD); Polyp (f; BIB; JLH); Proctosis (f; JLH); Prolapse (f; BIB); Pulmonosis (1; BIB; MAD); Pyelosis (f; BIB); Rheumatism (f; BIB); Sinusosis (f; PH2); Smallpox (1; BIB); Sore (f; BIB); Splenosis (f; JLH); Tonsilosis (f; JLH); Tuberculosis (1; MAD; PH2; WOI); Urogenitosis (f; BIB); Uterosis (f; BIB; JLH); UTI (f; BOW); Vaginosis (f; BOW); Virus (1; FNF); Wart (f; JLH); Water Retention (1; FNF); Whitlow (f; BIB); Worm (f; PH2); Wound (1; BIB; FNF).

Myrtle for bronchosis:

- Antibacterial: 1,8-cineole; acetic-acid; alpha-pinene; alpha-terpineol; bornyl-acetate; carvacrol; caryophyllene; delta-3-carene; delta-cadinene; dipentene; ellagic-acid; eugenol; gallic-acid; geranial; geraniol; limonene; linalool; methyl-eugenol; myrcene; myricetin; myricitrin; neral; nerol; nerolidol; p-cymene; terpinen-4-ol
- Antibronchitic: 1,8-cineole; borneol; gallic-acid
- Antihistaminic: linalool; myricetin
- Antiinflammatory: alpha-pinene; beta-pinene; borneol; carvacrol; caryophyllene; caryophyllene-oxide; delta-3-carene; ellagic-acid; eugenol; gallic-acid; myricetin; myricitrin
- Antioxidant: camphene; carvacrol; ellagic-acid; eugenol; gallic-acid; gamma-terpinene; linalyl-acetate; methyl-eugenol; myrcene; myricetin
- Antipharyngitic: 1,8-cineole
- Antipyretic: borneol; eugenol
- Antispasmodic: 1,8-cineole; borneol; bornyl-acetate; camphor; carvacrol; caryophyllene; eugenol; geraniol; limonene; linalool; linalyl-acetate; myrcene
- Antitussive: 1,8-cineole; carvacrol; terpinen-4-ol
- Antiviral: alpha-pinene; ar-curcumene; beta-bisabolene; bornyl-acetate; dipentene; ellagic-acid; gallic-acid; geranial; limonene; linalool; myricetin; neryl-acetate; p-cymene
- Bronchodilator: gallic-acid
- Bronchorelaxant: linalool
- COX-2-Inhibitor: eugenol
- Candidicide: 1,8-cineole; beta-pinene; carvacrol; caryophyllene; eugenol; geraniol; myricetin; myricitrin
- Candidistat: limonene; linalool
- Cyclooxygenase-Inhibitor: carvacrol; gallic-acid
- Decongestant: camphor
- Expectorant: 1,8-cineole; acetic-acid; alpha-pinene; bornyl-acetate; camphene; camphor; carvacrol; dipentene; geraniol; limonene; linalool
- Immunostimulant: gallic-acid

Myrtle for edema:

- Analgesic: borneol; camphor; eugenol; gallic-acid; myrcene; p-cymene
- Anesthetic: 1,8-cineole; camphor; carvacrol; eugenol; linalool; linalyl-acetate; methyl-eugenol; myrcene
- Antiedemic: caryophyllene; caryophyllene-oxide; eugenol
- Antiinflammatory: alpha-pinene; beta-pinene; borneol; carvacrol; caryophyllene; caryophyllene-oxide; delta-3-carene; ellagic-acid; eugenol; gallic-acid; myricetin; myricitrin
- COX-2-Inhibitor: eugenol
- Cyclooxygenase-Inhibitor: carvacrol; gallic-acid
- Diuretic: myricetin; myricitrin; terpinen-4-ol
- Lipoxygenase-Inhibitor: myricetin

Myrtle for infection:

- Analgesic: borneol; camphor; eugenol; gallic-acid; myrcene; p-cymene
- Anesthetic: 1,8-cineole; camphor; carvacrol; eugenol; linalool; linalyl-acetate; methyl-eugenol; myrcene
- Antibacterial: 1,8-cineole; acetic-acid; alpha-pinene; alpha-terpineol; bornyl-acetate; carvacrol; caryophyllene; delta-3-carene; delta-cadinene; dipentene; ellagic-acid; eugenol; gallic-acid; geranial; geraniol; limonene; linalool; methyl-eugenol; myrcene; myricetin; myricitrin; neral; nerol; nerolidol; p-cymene; terpinen-4-ol
- Antiedemic: caryophyllene; caryophyllene-oxide; eugenol
- Antiinflammatory: alpha-pinene; beta-pinene; borneol; carvacrol; caryophyllene; caryophyllene-oxide; delta-3-carene; ellagic-acid; eugenol; gallic-acid; myricetin; myricitrin
- Antiseptic: 1,8-cineole; alpha-terpineol; beta-pinene; camphor; carvacrol; carvone; ellagic-acid; eugenol; furfural; gallic-acid; geraniol; hexanal; limonene; linalool; methyl-eugenol; myricetin; nerol; terpinen-4-ol
- Antiviral: alpha-pinene; ar-curcumene; beta-bisabolene; bornyl-acetate; dipentene; ellagic-acid; gallic-acid; geranial; limonene; linalool; myricetin; neryl-acetate; p-cymene
- Astringent: ellagic-acid; gallic-acid
- Bacteristat: gallic-acid; malic-acid
- COX-2-Inhibitor: eugenol
- Cyclooxygenase-Inhibitor: carvacrol; gallic-acid
- Fungicide: 1,8-cineole; acetic-acid; alpha-phellandrene; camphor; carvacrol; caryophyllene; caryophyllene-oxide; eugenol; furfural; geraniol; linalool; methyl-eugenol; myrcene; p-cymene; terpinen-4-ol; terpinolene
- Fungistat: limonene; methyl-eugenol
- Immunostimulant: gallic-acid
- Lipoxygenase-Inhibitor: myricetin

Other Uses (Myrtle) — Grown since ancient times for the fragrant, aromatic flowers, leaves, and bark. *The Big Book of Herbs* (TAD) leads off with an interesting quote, "Sardinia's favorite flavoring is myrtle, a preference which may well go back to the Stone Age." In Sardinia, the wood is used for the fires to spit cook or pit cook whole animals, especially pig. "Gallina col mirto" is boiled chicken withdrawn from the pot, covered with myrtle leaves, consumed cold the following day. Yes, they use myrtle almost as a substitute for bay leaf (TAD). Myrtle has been known since ancient times for the fragrant, aromatic flowers, leaves, and bark. In Jerusalem and Damascus, the flowers, leaves, and fruits are sold for making perfume. In Italy, the leaves are used as a spice, and the flower buds are eaten. Leaves, made into tea, are considered an alternative to "buchu." The sprigs were formerly added to wine to increase its potency (FAC). The leaves are used for massage to work up a glowing skin. The fragrant oil obtained from the leaves is used in perfumery and as a condiment, especially when mixed with other spices. Oil of myrtle is used in all kinds of culinary compositions, especially table sauces. The oil is also used in toilet waters, especially eau de cologne and eau d'ange. Green and dried fruits sometimes are used as a condiment. Dried fruits, leaves, and flower buds are used for flavoring meats, poultry, sauces, liqueurs, and syrups. Widely used in Sardinia, Corsica, and Crete. In Corsica, the myrtle liqueur called "myrthe" flavors pâté de merles. Around Rabat, they mix the leaves with shampoo, believed to darken the hair. In India, the fixed oil from the berries is alleged to strengthen and promote the growth of hair. Seeds, ground and mixed with antimony, are used to color the eyelids (DEP). Turkish and Russian leather is tanned with the bark and roots, imparting a distinctive odor. The wood is very hard and of interesting texture and grain, growing in Mediterranean climate. In Biblical times, Jews collected myrtle to adorn their sheds and booths at the Feast of Tabernacles, chiefly as a symbol of divine generosity. It is considered emblematic of peace and joy in the Bible. To ancient Jews, it was symbolic not only of peace, but also of justice. Arabs say that myrtle is one of the three plants taken from the garden of Eden, because of its fragrance. Greeks consider it a symbol of love and immortality and used it for crowning their priests, heros, and outstanding men (BIB, DEP, FAC, PH2).

For more information on activities, dosages, and contraindications, see the *CRC Handbook of Medicinal Herbs*, ed. 2, Duke et al., 2002.

Cultivation (Myrtle) — Plants are often grown for ornament, as it makes a good hedge that is often everblooming in proper climates. It is said to be hardy to zone 9, doing best in full sun, moist but not wet soils, and well-drained garden loams, at pH 5.5–8.2 (average 6.8). Propagated mainly by cuttings of half-ripe or partly woody shoots about 10 cm long, taken in July with a slight heel of old wood. Cuttings inserted in sand bed in glass-frame or greenhouse. Roots form in a few weeks. Planting best done in spring. Young plants potted or planted out in sandy peaty soil. When grown as a potted plant in the North, it should be kept in a frost-proof greenhouse or other light but cool place during the winter and spring, and then grown outdoors in summer months. Overgrown plants should be pruned in early spring immediately before new growth appears.

Chemistry (Myrtle) — Here are a few of the more notable chemicals found in myrtle. For a complete listing of the phytochemicals and their activities, see the CRC phytochemical compendium, Duke and duCellier, 1993 (DAD) and the USDA database http://www.ars-grin.gov/duke/.

1,8-Cineole — See also *Elettaria cardamomum*.

Limonene — See also *Carum carvi*.

Linalool — ADI = 500 µg/kg; Acaricide; Anesthetic 0.01–1 µg/ml; Antiallergic; Antianaphylactic; Antibacterial MIC = 1600 µg/ml; Anticancer; Anticariogenic MIC = 1600 µg/ml; Anticonvulsant 200 mg/kg ipr mus; Antihistaminic; Antipyretic; Antiseptic 5 ×phenol; Antishock; Antispasmodic; Antiviral; Barbiturate-Synergist; Bronchorelaxant; Candidistat; Expectorant; Fungicide; Gabaergic; Hypnotic; Insectifuge; Insecticide; Irritant; Motor-Depressant; Nematicide MLC = 1 mg/ml; Prooxidant; Sedative ED = 1–32 mg/kg, 200 mg/kg ipr mus (1% as active as diazepam); Termitifuge; Trichomonicide LD100 = 600 µg/ml; Tumor-Promoter; LD50 = 2790 mg/kg orl rat; LD50 = 459 mg/kg ipr mus.

Methyl-Chavicol — Antipyretic; Calcium-Antagonist IC50 = 258 µM; DNA-Binder; Hepatocarcinogenic; Insecticide.

Verbenone — Allelochemic; Coleopterifuge; Insectifuge; LDlo = 250 ipr mus.

N

Nigella sativa L. (Ranunculaceae)
Black Caraway, Black Cumin, Fennel Flower, Nutmeg Flower, Roman Coriander

Medicinal Uses (Black Cumin) — This herb apparently may be even more important to the Muslims than to the Christians and Jews. According to an Arab proverb, "in the black seed is the medicine for every disease except death." Small wonder that the literature regards it as carminative, diaphoretic, digestive, diuretic, emmenagogue, excitant, lactagogue, laxative, resolvent, stimulant, stomachic, tonic, and vermifuge. Over a century ago, we find the following formula for eczema and head lice: 2 oz bruised *Nigella* seed, 2 oz bruised *Psoralea* seed, 2 oz bdellium, 2 oz Coscini root, 1 oz sulfur in two bottles of coconut oil (DEP). In Ayurvedic medicine, where used as a purgative adjunct, the herb is considered aperitif, aromatic, carminative, emmenagogue, and vermifuge. In Unani, it is further considered abortifacient and diuretic and used for ascites, coughs, eye-sores, hydrophobia, jaundice, paralysis, piles, and tertian fever. Algerians take the roasted seeds with butter for cough, with honey for colic. Lebanese took the seed extract for liver ailments. In Indonesia, the seeds are added to astringent medicines for abdominal disorders. In Malaya, the seeds are poulticed onto abscesses, headache, nasal ulcers, orchitis, and rheumatism. Ethiopians mix the seed with melted butter, wrap it in a cloth, and sniff it for headache. Arabian women use the seeds as a galactagogue. Mixing curry (which contains fenugreek) with black cumin for nurses, the Indians had a double whammy lactagogue. In large quantities, the seed are also used to induce abortion. The lipid portion of the ether extract of black cumin seeds has shown lactagogue activity in rats, verifying its folk usage as a lactagogue. And Nigellone in the oil protects guinea pigs against histamine induced bronchospasms, suggesting the rationale behind its use in asthma, bronchitis, and cough.

El Tahir et al. (1993b) reported that the volatile seed oil (ivn 4–32 µl/kg) dose-dependently stimulated respiratory rate and intratracheal pressure. Strangely, ivn thymoquinone (1.6–6.4 mg/kg) significantly increased intratracheal pressure but did not affect the respiratory rate. The oil-induced respiratory effects may be due to histamine release and indirect activation of muscarinic cholinergic mechanisms. They suggest that removing the useful thymoquinone would make the residual oil a useful respiratory stimulant (X8270170). Gilani et al. (2001) note that the "Kalonji" seed are used traditionally for asthma. The crude extract caused a dose-dependent (0.1–3.0 mg/ml) relaxation of spontaneous contractions in rabbit jejunum. The petroleum ether fraction was some 10 times more active than the crude extract. The seeds exhibit bronchodilator and antispasmodic activities mediated possibly through calcium channel blockade, proving the utility of the seed for diarrhea and asthma (X11381824).

De et al. (1999) screened 35 Indian spices for antimicrobial activity. Black cumin, bishop's weed, camboge, celery, chilli, clove, cinnamon, cumin, garlic, horseradish, nutmeg, onion, pomegranate, tamarind, and tejpat proved to have potent antimicrobial activities against the test organisms *Bacillus subtilis* (ATCC 6633), *Escherichia coli* (ATCC 10536), and *Saccharomyces cerevisiae* (ATCC 9763). Such results confirm traditional uses of spices as food preservatives, disinfectants, and antiseptics (X10548758).

Cardiovascular activity of the volatile oil has been examined. Intravenous administration of the oil at 4–32 µl/kg or thymoquinone at 0.2–1.6 mg/kg lowered arterial blood pressure and heart rate dose-dependently. The oil has potential as being an antihypertensive agent (X8270171). Oil administration significantly decreased serum total cholesterol, low density lipoprotein, and triglycerides and significantly elevated serum HDL (X11050701).

El-Dakhakhny et al. (2000) reported positive effects of the seed oil on gastric secretion and ethanol induced ulcers in rats. The oil led to significant increases in mucin and glutathione decreases in mucosal histamine. Ethanol administration produced 100% ulcers in all rats. They also note that black cumin oil protects against induced hepatotoxicity and improves serum lipid profile in rats. Daily oil administration (800 mg/kg orally for 4 weeks) did not adversely effect alkaline phosphatase, serum bilirubin or prothrombin activity, or serum transaminases (ALT and AST). Thymoquinone has been shown to be hepatoprotective, at least against hepatotoxicity of carbon tetrachloride in mice (Mansour, 2000).

Abuharfeil et al. (2001) found that aqueous extracts of nigella, (used in Jordanian traditional medicine for cancer) augmented natural killer cells and showed cytotoxic activity against tumor targets. The seed extracts augmented splenic NKCs (62.3% cf 52.6% for garlic extracts, and 30.6% for onion extracts) (PR15:109). Worthen et al. (1998) found that both thymoquinone and dithymoquinone are cytotoxic for several types of human tumor cells, and may not be MDR substrates, making them useful either alone or in conjunction with drugs to MDR has developed. Thymoquinone inhibited induced forestomach carcinogenesis in mice (Badary and Gamal, 2001). Thymoquinone also prevented fibrosarcoma induced by 20-methylcholanthrene in male Swiss albino mice (Badary and Gamal, 2001). The seeds are rich in sterols (5100 ppm), of which 63.1% is the antitumor sterol, beta-sitosterol. *Nigella* seems effective at killing certain cancer cells while leaving normal cells intact. It helps reduce or prevent toxicity of some toxic anticancer drugs, e.g., protecting bone marrow against chemotherapy. Seed extract enhance immune response, with increased activity of T-helper and NMKCs cells, and interferon-like responses. U.S. Patent No. 5,482,711 (January 1996) and No.

5,653,981 (August 1997) concern immunostimulant features of black cumin extracts. Cisplatin, a widely used chemotherapeutic drug, can be toxic to the kidney. When administered ip for 5 alternate days with 3 mg/kg cisplatin, cysteine (20 mg/kg) together with vitamin E (2 mg/rat), an extract of *Crocus sativus* stigmas (50 mg/kg), and *Nigella sativa* seed (50 mg/kg) significantly reduced blood urea nitrogen (BUN) and serum creatinine levels as well as cisplatin-induced serum total lipids increases. Given together with cisplatin, the protective agents led to an even greater decrease in blood glucose than seen with cisplatin alone. The serum activities of alkaline phosphatase, lactate dehydrogenase, malate dehydrogenase, aspartate aminotransferase, and alanine aminotransferase of cisplatin-treated rats were significantly decreased, whereas the activities of glutathione reductase and isocitrate dehydrogenase were significantly increased. The combination also partially reversed many of the kidney enzymes changes induced by cisplatin and tended to protect from cisplatin-induced diminutions in leucocyte counts, haemoglobin levels, and mean osmotic fragility of erythrocytes. They concluded that cysteine and vitamin E, *Crocus sativus* and *Nigella sativa* could be useful reducing cisplatin-toxic side effects including nephrotoxicity (El Daly, 1998).

The crude fixed oil and its pure thymoquinone both inhibited the cyclooxygenase and 5-lipoxygenase pathways of arachidonate metabolism (in rat peritoneal leukocytes stimulated with calcium ionophore A23187, as shown by dose-dependent inhibition of thromboxane B2 and leukotriene B4, respectively). The thymoquinone was quite potent, with approximate IC50 values against 5-lipoxygenase and cyclooxygenase of <1 µg/ml, and 3.5 µg/ml respectively. Thymoquinone was some tenfold more potent. Since the inhibition of eicosanoid generation and lipid peroxidation by the fixed oil is greater than would be expected from its content of thymoquinone (ca. 0.2% w/v), there must be synergy or other contributing phytochemicals, supporting the traditional use of the seed and derivatives for rheumatism and other related inflammatory diseases (Houghton et al., 1995). The seeds contain ca. 1.5% melanthine, a bad smelling fish poison.

Indications (Black Cumin) — Achylia (f; MAD); Allergy (f; HAD); Ameba (1; MPI); Amenorrhea (f; KAP); Anorexia (1; HAD); Arthrosis (1; HAD); Ascites (f; BIB); Asthma (1; HAD; HHB; MAD; SKJ; WOI); Bacillus (1; X10548758); Bacteria (1; HAD); Biliousness (f; KAP); Bite (f; HAD); Bronchosis (1; FNF; HAD; HHB; WOI); Bronchospasm (1; WOI); Cachexia (f; SKJ); Callus (f; BIB; JLH); Cancer (1; ABS; BIB; FNF; HAD); Cancer, abdomen (1; FNF; JLH); Cancer, colon (1; FNF; JLH); Cancer, eye (1; FNF; JLH); Cancer, liver (1; FNF; JLH); Cancer, nose (1; FNF; JLH); Cancer, uterus (1; FNF; JLH); Candida (1; ABS); Cardiopathy (1; X8270171); Catarrh (f; DEP; HHB); Childbirth (f; SUW); Cholera (1; MPI); Cold (f; DEP); Colic (f; BIB); Constipation (f; SKJ); Corn (f; BIB; JLH); Cough (1; SKJ; WOI); Cramp (1; FNF; HHB; MAD); Dermatosis (f; HAD; SUW; WOI); Diabetes (1; HAD); Diarrhea (1; MAD; X11381824); Dysentery (f; HHB; SKJ); Dysmenorrhea (f; DEP; KAP); Dyspepsia (f; BIB); Eczema (f; DEP); Emaciation (f; SKJ); Enterosis (f; BIB; MAD); Eruption (f; BIB); Escherichia (1; KAP; MPI); Fever (1; BIB; MAD; SUW; WOI); Fibrosarcoma (1; X11531013); Flu (f; BIB); Fungus (1; X10548758); Gas (f; MAD); Gout (1; HHB); Headache (f; BIB); Hemorrhoid (f; BIB); Hepatosis (1; BIB; JLH; MAD; X10883736); High Blood Pressure (1; MPI; X82701710); High Cholesterol (1; HAD); HIV (1; HAD); Hydrophobia (f; BIB); Hyperlipidemia (1; X10755708); Induration (f; JLH; MAD); Infection (1; HAD); Inflammation (1; X10552840); Jaundice (f; BIB; HHB; MAD); Leprosy (f; SKJ); Leukemia (1; X1270717); Leukorrhea (f; MAD); Lice (f; DEP); Malaria (f; KAP); Mycosis (1; X10548758); Myrmecia (f; BIB); Nephrosis (1; X10755708); Obesity (1; FNF); Ophthalmia (f; HAD); Orchosis (f; BIB); Pain (1; HAD); Paralysis (f; BIB); Parasite (1; HAD); PMS (1; HAD); Proctosis (f; SKJ); Prolapse (f; SKJ); Proteinuria (1; X10755708); Pityriasis (f; DEP); Puerperium (1; WOI); Pulmonosis (f; HAD; HHB; MAD); Rhinosis (f; BIB); Salmonella (1; HAD); Sclerosis (f; BIB); Smallpox (f; SKJ); Snakebite (f; BIB); Sniffle (f; MAD); Splenosis (f; MAD); Staphylococcus (1; HAD; MPI); Sting (f; HAD; SUW); Stomachache (f; BIB; MAD); Stomatosis (f; HAD); Swelling (f; BIB); Syphilis (f; SKJ); Toothache (f; MAD); Tumor (f; BIB; HAD); Vibrio (1; MPI); Virus (1; HAD); Worm (f; MAD); Wound (f; HAD) Yeast (1; X10548758).

Black Cumin for asthma:

- Antiasthmatic: nigellone; thymoquinone
- Antibronchitic: nigellone; thymol; thymoquinone
- Antihistaminic: nigellone; rutin; thymoquinone
- Antioxidant: rutin; thymol; thymoquinone
- Antispasmodic: nigellone; rutin; thymol; thymoquinone
- Cyclooxygenase-Inhibitor: thymol; thymoquinone
- Expectorant: astragalin; thymol
- Lipoxygenase-Inhibitor: rutin; thymoquinone

Black Cumin for cancer:

- Antiaggregant: thymol
- Anticancer: carvone; d-limonene; indole-3-acetic-acid; rutin
- Antihepatotoxic: rutin
- Antiinflammatory: alpha-spinasterol; cycloartenol; hederagenin; rutin; thymol
- Antileukemic: astragalin
- Antimelanomic: d-limonene; thymol
- Antimutagenic: rutin
- Antioxidant: rutin; thymol; thymoquinone
- Antitumor: d-limonene; rutin
- Antiviral: rutin
- Cyclooxygenase-Inhibitor: thymol; thymoquinone
- Immunostimulant: astragalin
- Lipoxygenase-Inhibitor: rutin; thymoquinone
- Sunscreen: rutin
- Topoisomerase-II-Inhibitor: rutin

Other Uses (Black Cumin) — Widely cultivated for its aromatic seeds, used whole or ground as a flavoring, especially in oriental cookery. The tiny seeds are very hot to the palate and are sprinkled on food like pepper. In Europe, they are sometimes mixed with real pepper. They were used like black pepper before the introduction of black pepper. French once used the black cumin in lieu of black pepper under the name "uatre epices" or "toute epices." Whole seeds (fitches) are used in Russian rye bread and for flavoring Turkish breads. Called "siyah daneh" or "onion seed," they are sprinkled on cakes, flatbreads, and rolls. Arabs mix the seed with honey as a confectionary. Seeds also used in Armenian string cheese, Tel banir, Haloumi, other cheeses, chutneys, curries, pickles, preserved lemons, and vegetables. Ethiopians add them to capsicum pepper sauces. Seeds may be used as a stabilizing agent for edible fats. A reddish-brown and semi-drying fatty oil is obtained from the seeds with benzene and subsequent steam distillation of the extract to remove about 31% of the volatile oil. Ethiopians may add *Nigella, Aframomum, Piper,* and *Zingiber* to local alcoholic beverages. Indians use them in a spice mixture called "panch phoran" (FAC). Sudanese use in such fermented foods as abreh, hulu-mur, and mish (FAC). Seeds are sprinkled among woolens as a moth repellant (BIB, WOI). In the Bible, Ezekiel recites a recipe I have seen nowhere else, "Take thou unto thee wheat, and barley, and beans, and lentiles, and millet, and fitches, and put them in one vessel, and make thee bread thereof... And thou shalt eat it as barley cakes...." Bean bread, good for the heart; the more you eat the less you infarct.

For more information on activities, dosages, and contraindications, see the *CRC Handbook of Medicinal Herbs*, ed. 2, Duke et al. 2002.

Cultivation (Black Cumin) — More often wild crafted than cultivated, this annual weed only partially fits the spice definition, but it is the seeds that are used. Seed can be sown in spring or fall, *in situ*, in well-drained sunny soil. A short-lived annual, it seeds within less than three months.

Chemistry (Black Cumin) — The EO alone contains four antioxidants that scavenge free radicals: antheole, carvacrol, 4-terpineol, and thymoquinone (Burits and Bucar, 2000). Here are a few of the more notable chemicals found in black cumin. For a complete listing of the phytochemicals and their activities, see the CRC phytochemical compendium, Duke and duCellier, 1993 (DAD) and the USDA database http://www.ars-grin.gov/duke/.

Anethole — See also *Osmorhiza* spp.

Arginine — Antidiabetic; Antiencephalopathic; Antihepatosis; Antiinfertility 4 g/day; Antioxidant; Aphrodisiac 3 g/day; Diuretic; Hypoammonemic; Pituitary-Stimulant; Spermigenic 4 g/day.

Carvacrol — See also *Cunila origanoides.*

Carvone — Antiacetylcholinesterase IC50 = 1.4–1.8 mM; Anticancer; Antiseptic (1.5 × phenol); Carminative; CNS-stimulant; Insecticide; Insectifuge; Motor-Depressant; Nematicide MLC = 1 mg/ml; Sedative; Trichomonicide LD100 = 300 µg/ml; Vermicide. LD50 = 1640 orl rat.

Methionine — Anticancer; Anticataract; Antidote (Acetominaphen; Paracetamol) 10 g/16 hr/man/orl; Antieczemic; Antihepatosis; Antioxidant; Antiparkinsonian 1–5 g day; Emetic; Essential; Glutathionigenic; Hepatoprotective; Lipotropic; Urine-Acidifier 200 mg/3× day/man/orl; Urine-Deodorant.

Nigellone — Antiasthmatic; Antibronchitic; Antihistaminic; Antispasmodic.

4-Terpineol — Antiasthmatic; Antioxidant; Antiradicular; Diuretic; Insectifuge.

Thymoquinone — Antibronchitic; Antifibrosarcomic 15 µM; Antihistaminic; Antinephrotic; Antioxidant; Antiproteinuric; Antiradicular; Antispasmodic; Antitumor IC50 = 78–393 µM; Cardiodepressant 0.2–1.6 mg/kg ivn rat; Chemopreventive 15 µM; Choleretic; Detoxicant 15 µM; Glutathiogenic 15 µM; Hepatoprotective; Hypolipidemic; Hypotensive 0.2–1.6 mg/kg ivn rat; MDR-Inhibitor IC50 = 78–393 µM; Quinone-Reductase-Inducer 15 µM; Toxic; Uricosuric.

N

O

Osmorhiza spp. (Apiaceae)
ANISEROOT, CLAYTON'S SWEETROOT, HAIRY SWEET CICELY, SMOOTH SWEET CICELY, WILD SWEET CICELY

Unable to distinguish the two species in Maryland, *O. claytonii* (Michx.) C. B. Clarke and *O. longistylis* (Torr.) DC., I have merged data from DEM and others for the two common Maryland species, doubting that others can distinguish them any better than I can.

Medicinal Uses (Sweetroot) — In his great book, Moerman (1998) maintains the two species as distinct, with several overlaps in usages. He notes that Chippewa poultice Clayton's Sweetroot on running sores and gargle the root decoction for sore throat. Chippewa women take the root infusion to bring on the period. Menominee use the root as a fattening agent and in an eyewash (collyrium) for sore eyes. Ojibwa use the root decoction in childbirth and take the infusion for sore throat. Tlingit use warm tea of the whole plant for cough. Moerman cites Cheyenne usage of longstyle sweetroot for bloated or disordered stomach. They use the roots for kidney ailment.

Meskwaki use the leaf infusion as a fattener and as an eye remedy. They grate the root with salt for equine distemper. Omaha and Winnebago apply the pounded root to boils. Pawnee take the root decoction for debility and general weakness. Potawatomi use as collyrium and use the root infusion as a stomachic. But Moerman cites no Cherokee usages. For Cherokee usage, we can thank my friend David Winston (2001), himself part Cherokee. He describes it, in western terms, as carminative, demulcent, expectorant, immunotonic, and nutritive. He adds that Cherokee have long considered it important for increasing disease resistance, strength, and weight. The root tea (1 tsp dry root to 8 oz water, cooked 3–4 hours or steeped 2 hours, 2–3 cups/day) is used for cold, dry cough, flu, gas, gastritis, and indigestion. "Sweet cicely strengthens what the Chinese call the 'wei qi,' making it useful for preventing colds and other external pernicious influences. The root can be used as a substitute for licorice or astragalus with many similar applications" (Winston, 2001). Alfs (2001) quotes a Canadian herbalist who recommends the herb for tonifying the mucous membranes. Many of the medicinal uses may be rationalized by the high anethole content. One of my herbal friends says that he, like me, does not distinguish the two species but doesn't use it if it doesn't smell like anise. That makes sense to reductionistic me, because I attribute much of the medicinal activity to the anethole, which makes many of the individuals smell of anise.

JAH2 data were referred to *O. claytoni*, DEM data to one or both species, Hussain et al.'s (1990) data to *O. longistylis*. All 1 ratings below were based on anethole, common to both species.

Indications (Sweetroot) — Alactea (1; CAN); Amenorrhea (f; DEM); Bacteria (1; FNF); Bloat (f; DEM); Boil (f; DEM; FAD); Childbirth (f; CEB; DEM; FAD); Cold (f; JAH2(2):45); Colic (f; ALF); Conjunctivosis (f; FAD); Cough (f; FAD; JAH2(2):45); Cramp (1; FNF); Debility (f; DEM; FAD; JAH2(2):45); Diarrhea (f; ALF); Distemper (f; DEM); Dyspepsia (1; DEM; FNF; JAH2(2):45); Fever (f; ALF); Flu (f; JAH2(2):45); Fungus (1; JNP49:156); Gas (f; CEB; JAH2(2):45); Gastrosis (f; FAD; JAH2(2):45; DEM); Hepatosis (1; FNF); Infection (1; FNF; JAH2(2):45); Mucososis (f; ALF); Mycosis (1; JNP49:156); Nausea (f; JAH2(2):45); Nephrosis (1; DEM; TOP139:177); Ophthalmia (f; CEB; DEM); Sore (f; DEM; FAD); Sore Throat (f; ALF; DEM; JAH2(2):45); Wound (f; DEM; FAD).

Other Uses (Sweetroot) — Tanaka (1976) notes that the roots and shoots are eaten and used to make tea. Facciola (1998) says roots and stems of *O. claytonii* are eaten as vegetable, and roots and unripe seeds used as anise-like flavorings. Re *O. longistylis*, he adds that leaves and green seeds may be added to salads (FAC). The dry seeds are used in cakes, candies, and liqueurs (FAC). In Wisconsin, the roots are stored whole, dry, and cleaned, and scraped before using, thereby retaining more of their anethole. The anise flavor is often so strong as to make it more spice than food. Where is that elusive dividing line? But for possible confusion by amateurs with poisonous members of the Apiaceae, this could clearly serve as a famine food. I have enjoyed it steeped in gin or vodka, making a poor man's All-American-Anisette, improved with a dash of lemon and sweetener, even a licorice swirl stick for those avoiding sugar. According to Foster and Duke (2000), the root of *O. longistylis* is eaten or soaked in brandy. Alfs (2001) enjoys nibbling on young pods and young stems as well. Roots are eaten raw or cooked. He notes that the Menominee relished the roots, especially for fattening slim people. But they recommended slow consumption—only a section at a time. Omaha and Ponca reportedly use the aromatic root to lure or attract and capture horses. Potawatomi add chopped roots to oats or the like to fatten their ponies. Chippewa wash a dog's nostril with the aromatic root decoction to improve the dogs power of scent (DEM). Root can be used as a substitute for astragalus or licorice, with many similar applications (Winston, 2001). Hussain et al. (1990) showed how this and several other anethole-containing herbs were useful as sweeteners.

For more information on activities, dosages, and contraindications, see the *CRC Handbook of Medicinal Herbs*, ed. 2, Duke et al. 2002.

Cultivation (Sweetroot) — On my property, there are aromatic and non-aromatic varieties that are not distinguishable morphologically. Some of both are hairy; others more frequent are glabrous. Some seem biennial, some seem perennial. The Longstyle Sweetroot, alias Hairy Sweet Cicely, *O. longistylis*, has rather similar aroma, morphology, and utility. Offhand, I doubt that these species are as distinct as the keys would make them, splitting hairs and measuring stylopodia. They seem easy enough to transplant even in December. I have a specimen, dug December 3, 2001, from which I removed enough root and leaf material for chromotographic analysis and then replanted it. As a crop, I suspect it could be treated as a biennial, with root harvest in late fall. Or perhaps those enamored of the local "anise" might dig, cut off lateral roots, and replant.

Chemistry (Sweetroot) — Hussain et al. (1990), in *Economic Botany*, reported analyses of seven anise-scented and/or anethole-containing herbs. Trans-anethole was dominant among five of the seven herbs. They found 5700 ppm EO in the roots dominated by anethole (5440 ppm), making this highest (on a ppm basis) among the seven plants they analyzed. There are higher herbs, like fennel seed, in my database. They attributed the sweet taste of fennel, star anise, sweet cicely, and our *Osmorhhiza* to anethole. They note that trans-anethole, of which some 70 tons a year were being used in the U.S., has GRAS status and is used in baked goods, beverages, candies, and chewing gums in levels up to 1500 ppm. Trans-anethole was not acutely toxic for mice and was non-mutagenic. It is 10–20 times sweeter than sucrose (Hussain et al., 1990). Here are a few of the more notable chemicals found in sweetroot. For a complete listing of the phytochemicals and their activities, see the CRC phytochemical compendium, Duke and duCellier, 1993 (DAD) and the USDA database http://www.ars-grin.gov/duke/.

Anethole — Antibacterial; Anticancer; Antihepatosis; Antiinflammatory; Antinephrotic; AntiNF-kB; Antiseptic; Antispasmodic; AntiTNF; Antitumor; Carminative; Dermatitogenic; Digestive; Estrogenic; Expectorant; Fungicide; Gastrostimulant; Hepatotoxic 695 mg/kg orl mus/days; Immunostimulant; Insecticide; Lactagogue; Leucocytogenic; Mutagenic; Secretagogue; Secretolytic; Sweetener (13 × sucrose); Sympathomimetic; LD50 = 3000 orl mus; LD50 = 2090 orl rat; LD50 = 900 ipr rat.

Carene — Antibacterial; Antiseptic; Fungicide; Irritant; LD50 = 4800 mg/kg orl rat.

Limonene — See also *Carum carvi*.

Trans-Anethole — Antigenotoxic 40–400 mg/kg orl mus; Antioxidant; Antiradicular; Sweetener (13 × sucrose); LD50 = 900 mg/kg ipr rat; LD50 = 1000 mg/kg ipr mus; LD50 = 1800–5000 mg/kg orl mus; LD50 = 2000–3000 mg/kg orl rat; LD50 = 2100 mg/kg orl gpg.

P

Persea americana Mill. (Lauraceae)
AVOCADO

Medicinal Uses (Avocado) — Considered abortifacient, antifertility, antiseptic, aperient, aphrodisiac, astringent, calmant, carminative, diuretic, emmenagogue, parasiticide, piscicide, poison, raticide, resolvent, rodenticide, rubefacient, stomachic, suppurative, and vermifuge. Fruit pulp is used as an aphrodisiac and emmenagogue. Pulp is used for tumors in Mexico. Avocado is used for labial tumors in Peru. Pulverized seed used as a rubefacient. Decoction of the seed used locally to relieve toothache. Powdered seed is used for dandruff. Seed oil used for skin eruptions. Rind used as vermifuge and liniment for intercostal neuralgia. Fruit skin considered antiseptic and is used for dysentery and worms. Leaves are chewed to alleviate pyorrhea. Leaf poulticed onto wounds; heated leaves applied to forehead for neuralgia. Leaf juice also considered antiseptic. Leaf decoction taken for diarrhea, sore throat, hemorrhage, and used to stimulate and regulate menstruation. Decoction of the shoots is used for cough. Boiled leaves or shoots of purple-skinned type are used as abortifacient (DAD).

And on April 10, 2001, America lost its greatest ethnobotanist, Richard Evans Schultes, who spent more than 12 years botanizing in the Amazon basin, where some of our spices occur naturally (avocado, cacao, capsicum, culantro, tonka, and vanilla), and others are so long introduced as to appear native (garlic, ginger, lemongrass, and turmeric). Like Dr. Schultes, I love the Amazon and the Native Americans there. And, like Schultes, I was always delighted when new science tended to back up some old folklore he had picked up in the Amazon. Schultes and Raffauf (1990) reported that Native Americans there used the avocado for liver problems. Japanese scientists in 2001 produced some solid science singling out the avocado as a food "farmaceutical" with hepatoprotective properties. As measured by changes in the levels of plasma alanine aminotransferase (ALT) and aspartate aminotransferase (AST), avocado showed potent hepatoprotective activity. Five active compounds were isolated. We still have much to learn about the Amazonian food farmacy (X11368579).

Kim et al. (2000a, 2000b) isolated persenone A from the fruits as an effective inhibitor of both nitric oxide (NO) and superoxide (O_2) generation, at least in cell culture systems. It suppressed lipopolysaccharide- and interferon-y-induced inducible nitric oxide synthase (iNOS), and cyclooxygenase (COX-2) in a mouse cell line. Persenone A (20 μM) almost completely suppressed both iNOS and COX-2 protein expression. In mouse skin, double treatments with persenone A (810 nmol) significantly suppressed double 12-O-tetradecanoylphorbol-13-acetate (TPA), 8.1 nmol application-induced hydrogen peroxide (H_2O_2) generation. Persenone A, as a COX-2-Inhibitor (JAD), could help prevent such inflammatory diseases as Alzheimer's, arthritis, and cancer, including colon cancer (X11193428).

Stucker et al. (2001) concluded that vitamin B(12) preparation containing avocado oil may be suitable for use in long-term therapy of psoriasis.

Little et al. (2001) evaluated the effectiveness of herbal therapies in osteoarthritis. Data were extracted independently by the same two reviewers. No serious side effects were reported.

Indications (Avocado) — Alopecia (f; DAV); Alzheimer's (1; X11193428); Ameba (f; DAV); Amenorrhea (f; JFM; TRA); Anemia (f; DAV; JFM); Arthrosis (1; X11069724) Atherosclerosis (1; FNF; JNU); Bacteria (1; FNF; WOI); Bleeding (f; DAD); Bruise (f; DAD); Calculus (f; DAV); Cancer (1; JLH; JNU; X11193428); Cancer, colon (1; X11193428); Cancer, labial (f; JLH); Cancer, skin (1; X11193428); Catarrh (f; DAD; JFM); Cold (1; AAB; FNF; JFM); Cough (f; AAB; DAD; JFM); Dandruff (f; DAD; DAV); Dermatosis (1; DAD; FNF; PH2); Diabetes (1; DAD; DAV; FNF); Diarrhea (f; AAB; JFM); Dysentery (f; DAV; JFM); Dysmenorrhea (f; AAB; DAD); Dyspepsia (1; AAB; FNF); Enterosis (f; AAB); Escherichia (1; WOI); Fertility (f; DAV); Fever (f; AAB; DAD; JFM); Frigidity (f; JFM); Fungus (1; FNF; X10872209); Gas (f; JFM); Gout (f; JFM); Headache (1; AAB; FNF; JFM); Hematoma (f; DAD); Hemorrhoid (f; JFM); Hepatosis (1; DAD; DAV; JFM; X11368579); High Blood Pressure (1; AAB; DAD; FNF; JFM); High Cholesterol (1; FNF; JNU); High Triglyceride (1; JNU); Ichthyosis (1; PHR; PH2); Impotence (f; JFM); Infection (1; FNF; WOI); Inflammation (1; FNF; X11193428); Malaria (f; DAD); Metrorrhagia (f; DAD); Mucososis (f; JFM); Mycosis (1; X10872209); Neuralgia (f; DAD); Poor Milk Supply (1; TRA); Psoriasis (1; X11586013); Pulmonosis (f; DAD); Pyorrhea (f; DAD); Rheumatism (f; AAB; DAD; JFM); Scabies (f; DAD); Snakebite (f; DAV); Sore Throat (f; DAD); Sprain (f; AAB; DAD); Stone (f; DAV); Toothache (f; DAD); Whitlow (f; JFM); Worm (f; JFM); Wound (f; DAD).

Avocado for atherosclerosis:

- Antiaggregant: caffeic-acid; estragole; pyridoxine; quercetin; salicylates; serotonin; tyramine
- Antianginal: carnitine
- Antiatherosclerotic: lutein; proanthocyanidins; pyridoxine; quercetin
- Antibacterial: 1,2,4-trihydroxyheptadeca-16-ene; anethole; caffeic-acid; chlorogenic-acid; cycloartenol; p-coumaric-acid; pinene; quercetin
- Anticoronary: carnitine; folic-acid
- Antidiabetic: pyridoxine; quercetin
- Antiedemic: caffeic-acid; proanthocyanidins
- Antihomocystinuric: pyridoxine
- Antiinflammatory: caffeic-acid; chlorogenic-acid; cycloartenol; quercetin; salicylates
- Antiischemic: carnitine; proanthocyanidins
- Antileukotriene: caffeic-acid; chlorogenic-acid; quercetin
- Antilipoperoxidant: quercetin
- Antioxidant: caffeic-acid; catechol; chlorogenic-acid; lutein; p-coumaric-acid; proanthocyanidins; quercetin
- Antiplaque: folic-acid; proanthocyanidins; quercetin
- Antistroke: proanthocyanidins
- Antithrombic: quercetin
- Bacteristat: quercetin
- Calcium-Antagonist: caffeic-acid; estragole; methyl-chavicol
- Cardioprotective: proanthocyanidins
- Cardiotonic: dopamine
- Cyclooxygenase-Inhibitor: quercetin
- Diuretic: caffeic-acid; chlorogenic-acid; dopamine
- Hypocholesterolemic: caffeic-acid; carnitine; chlorogenic-acid; cycloartenol
- Hypotensive: phylloquinone; quercetin; subaphyllin; valeric-acid
- Lipoxygenase-Inhibitor: caffeic-acid; chlorogenic-acid; p-coumaric-acid; quercetin
- Metal-Chelator: caffeic-acid; chlorogenic-acid
- Natriuretic: dopamine
- Phospholipase-Inhibitor: quercetin
- Prostaglandin-Synthesis-Inhibitor: p-coumaric-acid; quercetin

P

- Sedative: caffeic-acid; valeric-acid
- Vasodilator: dopamine; proanthocyanidins; quercetin
- cAMP-Phosphodiesterase-Inhibitor: quercetin

Avocado for dermatosis:

- Analgesic: caffeic-acid; chlorogenic-acid; pyridoxine; quercetin; serotonin
- Antibacterial: 1,2,4-trihydroxyheptadeca-16-ene; anethole; caffeic-acid; chlorogenic-acid; cycloartenol; p-coumaric-acid; pinene; quercetin
- Antidermatitic: biotin; pyridoxine; quercetin
- Antiinflammatory: caffeic-acid; chlorogenic-acid; cycloartenol; quercetin; salicylates
- Antiseptic: anethole; caffeic-acid; catechol; chlorogenic-acid; pinene; proanthocyanidins
- COX-2-Inhibitor: quercetin
- Cyclooxygenase-Inhibitor: quercetin
- Fungicide: anethole; caffeic-acid; chlorogenic-acid; p-coumaric-acid; phylloquinone; pinene; propionic-acid; quercetin

Avocado for hepatosis:

- AntiEBV: chlorogenic-acid
- Antiedemic: caffeic-acid; proanthocyanidins
- Antiendotoxic: serotonin
- Antihepatosis: anethole
- Antihepatotoxic: caffeic-acid; chlorogenic-acid; p-coumaric-acid; quercetin
- Antiherpetic: caffeic-acid; chlorogenic-acid; quercetin
- Antiinflammatory: caffeic-acid; chlorogenic-acid; cycloartenol; quercetin; salicylates
- Antileukotriene: caffeic-acid; chlorogenic-acid; quercetin
- Antilipoperoxidant: quercetin
- Antioxidant: caffeic-acid; catechol; chlorogenic-acid; lutein; p-coumaric-acid; proanthocyanidins; quercetin
- Antiperoxidant: caffeic-acid; chlorogenic-acid; p-coumaric-acid; quercetin
- Antiprostaglandin: caffeic-acid
- Antiradicular: caffeic-acid; chlorogenic-acid; lutein; quercetin
- Antiviral: caffeic-acid; catechol; chlorogenic-acid; nonacosane; proanthocyanidins; quercetin; subaphyllin
- COX-2-Inhibitor: quercetin
- Cholagogue: caffeic-acid; chlorogenic-acid
- Choleretic: caffeic-acid; chlorogenic-acid; p-coumaric-acid
- Cyclooxygenase-Inhibitor: quercetin
- Cytoprotective: caffeic-acid
- Hepatoprotective: caffeic-acid; chlorogenic-acid; proanthocyanidins; quercetin
- Hepatotropic: caffeic-acid
- Immunostimulant: anethole; caffeic-acid; chlorogenic-acid; folic-acid
- Interferonogenic: chlorogenic-acid
- Lipoxygenase-Inhibitor: caffeic-acid; chlorogenic-acid; p-coumaric-acid; quercetin
- Previtamin-A: alpha-carotene
- Vitamin-A-Activity: alpha-cryptoxanthin; cryptoxanthin

Other Uses (Avocado) — Fruit is highly nutritious and yields highest energy value of any fruit. It is eaten fresh, seasoned with salt, pepper, sugar, and lime juice; used in guacamole, ice cream, salads, sandwiches, soups, spreads, with tortillas, and made into wine (FAC). In Bali, the fruits are

P

added to sugar and water to form a popular iced drink known as "es apokat" and are mixed with other fruits, chipped ice, and brilliant red sugar syrup in the dessert called "es campur" (FAC). It also freezes well. Pulp yields 3–30% of oil, used in moisturizing gels in cosmetics. Oil has excellent keeping quality. Crude oil used as a dressing for the hair, without refining; used in making soap, face cream, hand lotion, and as a salad oil. The non-allergenic oil, similar to lanolin in its penetrating and skin-softening action, may filter out the tanning rays of the sun. Pulp residue may be used as stock feed after oil extraction. An indelible red-brown or blackish ink, provided by the milky fluid of the seed, was once used to write documents and to mark cotton and linen textiles. Leaf used as a tea (DAD). Leaves are used in broths, stews, and molés (sauces), especially good with fish, chicken, and beans. The soaked leaves are used as smoke flavorings in Mexican barbecues (FAC). Lightly toast avocado leaves slowly in a low oven or quickly on a grill. The leaves may then be used like a bay leaf; or they may be ground fine with the other seasonings in a mole, such as the famous thick, black, chocolate mole of Oaxaca. Larger leaves used as wrappers for foods to be steamed, such as chicken or fish. Hoja santa can substitute for avocado leaf and vice versa (AAR). The worse the fruit, the more flavorful the leaves. Leaves can be dried and carefully stored for up to a year. Flowers provide a dark, thick honey produced by honeybees. Wood used for construction, boards, turnery, and carving.

For more information on activities, dosages, and contraindications, see the *CRC Handbook of Medicinal Herbs*, ed. 2, Duke et al. 2002.

Cultivation (Avocado) — Seeds are viable 2–3 weeks after removal from tree; for longer periods, store in dry peat at 5.5°C. Do not allow to dry out. Avocados are usually propagated by veneer-grafting, shield-budding, and cleft grafts. Seeds are planted in boxes, pots, or in open ground with pointed end uppermost and barely covered with soil. Two to four months after planting, when stems are 0.6–1.3 cm in diameter, seedlings are ready for budding. Bud wood should be selected as recent growth but not so sappy that it will break when bent. Buds should be cut smoothly with a very sharp knife. The graft is inserted, wrapped in waxed tape or raffia, leaving the eye exposed. When bud unites with stock, wrap is loosened. When bud sprout is 15 cm or more long, the old stock should be cut off cleanly, close to the bud. Trees are transplanted to orchard at 9–30 months, spaced 6–12 m each way. Clean cultivate or intercrop with cover crops. Fertilize when young with nitrogen, phosphorus, potash, and magnesium. Avocados should be fertilized four or five times the first year after planting, giving them 450 g of fertilizer containing about 5% nitrogen, 6–7% phosphoric acid, and 3–4% potash, and increased to 900 g the second year. Full bearing, they should receive from 18–27 kg annually. It is difficult to overfeed an avocado. Vegetatively propagated trees begin to bear fruit in 3–4 years but tend to bear biannually. Yields are variable, 100–500 or more fruits per tree. Orchards yield 6750–13,500 kg/ha.

Chemistry (Avocado) — Bown (2001) notes that it is the fruit richest in protein; I note that it is the fruit richest in MUFAs (up to 69% oleic-acid). Here are a few of the more notable chemicals found in avocado. For a complete listing of the phytochemicals and their activities, see the CRC phytochemical compendium, Duke and duCellier, 1993 (DAD) and the USDA database http://www.ars-grin.gov/duke/.

Linoleic-Acid — See also *Ceratonia siliqua*.

Mufas — Anemiagenic; Anticancer; Antiinflammatory IC50 = 21 µM; Antileukotriene-D4 IC50 = 21 µM; Choleretic 5 ml/man; Dermatitigenic; Hypocholesterolemic; Insectifuge; Irritant; Percutaneostimulant; LD50 = 230 ivn mus; LDlo = 50 ivn cat.

Oleic-Acid — See also *Dipteryx odorata*.

Palmitic-Acid — 5-Alpha-reductase-inhibitor; Antifibrinolytic; Hemolytic; Hypercholesterolemic; Lubricant; Nematicide; Soap; LD50 = 57 ivn mus.

Persenone-A — Antiinflammatory 20 µ*M*; Antioxidant; Antiradicular; Antitumor 20 µ*M*; Antitumor-Promoter IC50 = 1.4 µ*M*; Chemopreventive; COX-2-Inhibitor 20 µ*M*; Nitric-Oxide-Inhibitor IC50 = 1.2 µ*M*; Superoxide-Inhibitor.

Pufas — Antiacne; Antieczemic; AntiMS; Antipolyneuritic.

Peumus boldus Molina (Monimiaceae)
BOLDO

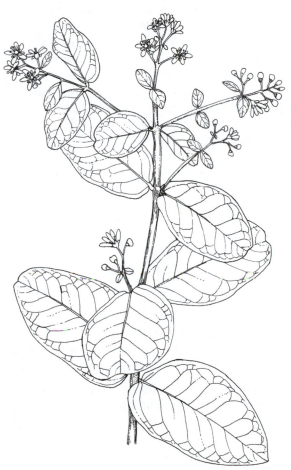

Synonyms — *Boldea fragrans* (Ruiz & Pav). Gay, *Peumus fragrans* Ruiz & Pav.

Medicinal Uses (Boldo) — A site was found in Monte Verde, Chile, 10,000 miles south of the Bering Land Bridge, where man's remains (I, and probably only I, call him Boldo Man) were co-mingled with slabs of mastodon meat, clay-lined fire-pits, boldo leaves (today used for GI disorders), chewed remains of seaweeds (today used for stomach distress), and 24 more medicinal plants, and 700 stone tools, including bolo stones (maybe I should call him Bolo Man) used with slings for hunting animals. I suppose a big mastodon kill and meat surfeit could lead to dyspepsia in those days. If Boldo Man was there 12,500 years ago, then his ancestors may have crossed the Bering Land Bridge considerably earlier than the usually accepted 12,000 years ago. According to Tim Friend, *USA Today,* February 11, 1997, "DNA studies of Native Americans also suggest migrations occurred 20,000 to 40,000 years ago. A site at Meadowcroft, Pennsylvania is carbon dated as 19,000 years old, but that date remains controversial."

Boldo, a Chilean tree used traditionally in folk medicine, mainly for the treatment of liver ailments, has recently been the subject of increasing attention. Boldo is listed in several pharmacopoeia, perhaps due to the alkaloid boldine. The EO is used medicinally (and possibly dangerously) as an anthelminthic, because it contains ascaridole, a proven anthelminthic. Certain as yet unnamed components of boldo relax smooth muscle and prolong intestinal transit (HAD). Boldo leaf and its preparations stimulate the production of bile, and its secretion from the gall bladder, and the secretion of gastric juice. Boiling water is poured over 1–2 g of the finely cut drug and strained after 10 min; a cupful is drunk two or three times a day as a choleretic. Losses of 80% or more of the aromatic boldine may occur. Standardized preparations are hence preferred. European tests of Hapatran, a trademarked hydroalcoholic extract of the herb, revealed its hepatoprotective activity, probably involving boldine, and a dose-related antiinflammatory activity, not necessarily involving the boldine (BIS). All this analgesic, antiinflammatory, and uricosuric activity suggests strongly to me that this might be considered as a mixture with celeryseed extract as a gout preventive and cure.

Boldine ([s]-2,9-dihydroxy-1, 10-dimethoxyaporphine) is the major alkaloid found both in the leaves and bark. It has pronounced antiinflammatory, antioxidant potential. Jang et al. (2000) look to it for antidiabetic potential as well. Increased oxidative stress may lead to pathogenesis and progression of diabetic tissue damage. Boldine attenuates development of hyperglycemia and weight loss STZ-induced in rats. Boldine treatment restores altered enzyme activities in liver and pancreas (but not kidney). It decomposes superoxide anions, hydrogen peroxides, and hydroxyl radicals dose-dependently. It attenuates production of superoxide anions, hydrogen peroxide, and nitric oxide caused by liver mitochondria (Jang et al., 2000). Jimenez et al. (2000) note that boldine prevents free radical-induced erythrocyte lysis, protecting intact red cells from hemolytic damage induced by free radical initiators (2, 2′-azobis-(2-amidinopropane) Boldine dose-dependently prevents leakage of hemoglobin into the extracellular medium. It is cytoprotective whether added 1 hr prior to, or simultaneously with, the inducer (X10925398). Dried hydro-alcoholic extracts show hepatoprotective, antiedemic, and antiinflammatory effects in mice and rats. Boldine, exhibits a dose-dependent antiinflammatory activity in the carrageenan-induced guinea pig paw edema test (ED50 = 34 mg/kg org gpg). Boldine also reduces bacterial hyperthermia in rabbits (ED51–98 = 60 mg/kg orl rbt). Boldine effectively inhibits prostaglandin biosynthesis IC53 = 75 μM. Boldine has mild diuretic, uric-acid excretory (perhaps useful in gout, JAD), and weak hypnotic effects. Injected boldine paralyzes both motor and sensory nerves and muscle fibers. Still, it is used veterinarially for jaundice.

Indications (Boldo) — Aging (1; APA); Anorexia (2; PHR); Atherosclerosis (1; APA; FNF); Autoimmune Disease (1; APA); Bacteria (1; FNF); Biliousness (2; APA; CAN; SHT); Cancer (1; APA; FNF); Cholecystosis (f; BGB; CAN; HHB); Cholelithiasis (1; CAN; HHB); Cold (f; CRC); Cough (f; CRC); Cramp (2; APA; BRU; FNF; KOM; SHT); Cystosis (1; CAN; PNC); Diabetes (1; X10987997); Dyspepsia (2; APA; BGB; BRU; FNF; KOM; PH2); Earache (f; CRC); Enterosis (2; APA; BOW; KOM); Gallstone (1; BOW; CAN; HHB; PNC); Gastrosis (2; CRC; KOM); Gonorrhea (1; CAN; GMH; HHB); Gout (f; APA; BGB; CRC); Head Cold (f; CRC); Heartburn (f; BGB; BRU); Hepatosis (2; APA; CAN; CRC; FNF; HHB; PHR); Hyperglycemia (1; X10987997); Hypertonia (2; KOM); Infection (1; CAN; CRC; EFS; FNF); Inflammation (1; APA; FNF); Insomnia (f; APA; CAN; EFS); Jaundice (f; CRC; GMH); Lethargy (f; EFS); Nephrosis (f; BGB); Nervousness (1; FNF); Obesity (f; BOW; PNC); Pain (1; BGB; FNF); Parasite (1; BOW); Prostatosis (f; BGB); Rheumatism (f; APA; BGB; CAN); Stomachache (1; APA); Stone (1; BRU); Syphilis (f; CRC; HHB); Urogenitosis (f; GMH); UTI (1; BOW); VD (f; CRC; HHB); Water Retention (1; APA; BGB; CAN; FNF); Worm (1; APA; CRC); Wound (f; CRC).

Boldo for atherosclerosis:

- Antiadrenergic: isocorydine
- Antiaggregant: coumarin; eugenol; thymol

- Antiatherosclerotic: carvacrol; thymol
- Antibacterial: 1,8-cineole; alpha-pinene; alpha-terpineol; alpha-thujone; benzaldehyde; beta-thujone; bornyl-acetate; carvacrol; cuminaldehyde; eugenol; geraniol; isorhamnetin; limonene; linalool; methyl-eugenol; nerol; p-cymene; reticuline; terpinen-4-ol; thymol
- Antiedemic: boldine; coumarin; eugenol
- Antiinflammatory: alpha-pinene; beta-pinene; boldine; carvacrol; coumarin; eugenol; guaiazulene; isorhamnetin; sinoacutine; sparteine; thymol
- Antioxidant: boldine; camphene; carvacrol; eugenol; gamma-terpinene; isorhamnetin; methyl-eugenol; sparteine; thymol
- Antioxidant (LDL): carvacrol; thymol
- Antiplaque: carvacrol; thymol
- Antithromboxane: eugenol
- Bacteristat: coumarin; terpinen-4-ol
- Beta-Adrenergic Receptor Blocker: boldine
- Calcium-Antagonist: eugenol
- Cardiotonic: coumarin
- Cyclooxygenase-Inhibitor: carvacrol; thymol
- Diuretic: boldine; sparteine; terpinen-4-ol
- Hypotensive: 1,8-cineole; benzyl-benzoate
- Prostaglandin-Synthesis-Inhibitor: boldine; eugenol
- Sedative: 1,8-cineole; alpha-pinene; alpha-terpineol; ascaridole; benzaldehyde; bornyl-acetate; coumarin; eugenol; farnesol; geraniol; isocorydine; limonene; linalool; methyl-eugenol; nerol; norisocorydine; p-cymene; thymol

Boldo for dyspepsia:

- Analgesic: ascaridole; camphor; coumarin; eugenol; p-cymene; reticuline; thymol
- Anesthetic: 1,8-cineole; benzaldehyde; camphor; carvacrol; eugenol; linalool; methyl-eugenol; pronuciferine; thymol
- Antiemetic: camphor
- Antiinflammatory: alpha-pinene; beta-pinene; boldine; carvacrol; coumarin; eugenol; guaiazulene; isorhamnetin; sinoacutine; sparteine; thymol
- Antioxidant: boldine; camphene; carvacrol; eugenol; gamma-terpinene; isorhamnetin; methyl-eugenol; sparteine; thymol
- Antipeptic: benzaldehyde; guaiazulene
- Antiulcer: eugenol; guaiazulene
- Carminative: ascaridole; camphor; carvacrol; eugenol; thymol; thymyl-acetate
- Gastrostimulant: boldine
- Secretagogue: 1,8-cineole
- Sedative: 1,8-cineole; alpha-pinene; alpha-terpineol; ascaridole; benzaldehyde; bornyl-acetate; coumarin; eugenol; farnesol; geraniol; isocorydine; limonene; linalool; methyl-eugenol; nerol; norisocorydine; p-cymene; thymol
- Tranquilizer: alpha-pinene

Boldo for hepatosis:

- Antiedemic: boldine; coumarin; eugenol
- Antihepatosis: boldine
- Antiherpetic: thymol
- Antiinflammatory: alpha-pinene; beta-pinene; boldine; carvacrol; coumarin; eugenol; guaiazulene; isorhamnetin; sinoacutine; sparteine; thymol

P

- Antioxidant: boldine; camphene; carvacrol; eugenol; gamma-terpinene; isorhamnetin; methyl-eugenol; sparteine; thymol
- Antiperoxidant: sparteine
- Antiprostaglandin: carvacrol; eugenol
- Antiradicular: boldine; carvacrol; eugenol; methyl-eugenol; thymol
- Antiviral: alpha-pinene; bornyl-acetate; limonene; linalool; p-cymene
- COX-2-Inhibitor: eugenol
- Choleretic: 1,8-cineole; boldine; eugenol
- Cyclooxygenase-Inhibitor: carvacrol; thymol
- Hepatoprotective: boldine; eugenol; isorhamnetin
- Hepatotonic: 1,8-cineole
- Immunostimulant: benzaldehyde; coumarin
- Phagocytotic: coumarin

Other Uses (Boldo) — The sweet, aromatic fruits are eaten. Leaves and the bark used as a spice in Chile (FAC). It looks a little iffy to me, though approved by Commission E. It is listed by the council of Europe as a natural source of food flavoring. Category N3 implies that boldo can be added to foodstuffs "in the traditionally accepted manner, although insufficient information is available for an adequate assessment of potential toxicity" (CAN). The EO, like that of ascaridole-containing wormseed (*Chenopodium ambrosioides*), is used in the food and liqueur industries as a flavoring (Miraldi et al., 1996). While I have consumed the tea, with modest pleasure, in Chile, I can't endorse it—a little nervous about the camphor and more nervous about the ascaridole and thujone. Boldo the herb, with up to 1.1% of the endoperoxide, ascaridole, has been approved for certain indications by both the British (CAN, PNC) and German (KOM, PHR) herbal gurus, while our wormseed, alias epasote, *Chenopodium ambrosioides*, was excluded. With up to 1.8% ascaridole, wormseed has largely fallen by the wayside. I'll wager a case of beano that boldo would do two-thirds as well as the wormseed, known to Latinos as epazote, at preventing the gas of bean dishes (JAD). Bark is used for tanning and dyeing fibers. The wood used for charcoal (GMH).

For more information on activities, dosages, and contraindications, see the *CRC Handbook of Medicinal Herbs*, ed. 2, Duke et al. 2002.

Cultivation (Boldo) — The Chilean tree (exports 1000 tons dry leaves per year) was introduced to Europe ca. 1870. It is thriving in California and temperate frost-free zones elsewhere. According to Bown (2001), the shrub or small tree does well, sown by seed in spring, or by semi-ripe cuttings in summer, in acid, sandy, sunny, well-drained soils, zone 9. Leaves are harvested during the growing season. Bark may be collected for alkaloid extraction.

Chemistry (Boldo) — Here are a few of the more notable chemicals found in boldo. For a complete listing of the phytochemicals and their activities, see the CRC phytochemical compendium, Duke and duCellier, 1993 (DAD) and the USDA database http://www.ars-grin.gov/duke/.

Ascaridole — Analgesic; Ancylostomicide; Carcinogenic; Carminative; Fungicide; Nematicide; Sedative; Vermifuge; LDlo = 250 orl rat.

Boldine — Abortifacient; Antiedemic ED50 = 34 mg/kg; Antihepatosis; Antiinflammatory ED50 = 34 mg/kg; Antilithic; Antioxidant IC50 = 17–33 µg/ml; Antiprostaglandin IC53 = 75 µ*M*; Antipyretic ED51–98 = 60 mg/kg orl rbt; Antiradicular; Beta-Blocker 0.1 µ*M*; Choleretic; Convulsant; Diuretic; Gastrostimulant; Hepatoprotective; Hypnotic; Immunomodulator; Myorelaxant; Neuroleptic; Teratogenic; Uricosuric.

Laurotetanine — Curaroid.

Pimenta dioica (L.) Merr. (Myrtaceae)
Allspice, Clove Pepper, Jamaica Pepper, Pimienta, Pimento

Synonyms — *Myrtus dioica* L., *M. pimenta* L., *Pimenta officinalis* Lindl., *P. pimenta* (L.) H. Karst., *P. vulgaris* Lindl.

Medicinal Uses (Allspice) — Reported to be analgesic, antibacterial, antioxidant, aromatic, carminative, fungicidal, stimulant, and stomachic. Jamaicans take the fruit decoction for colds, menorrhagia, and stomachache. Costa Ricans take the leaf infusion as a carminative and stomachic, useful for diabetes. Guatemalans apply it externally for bruises and rheumatic pain. Cubans drink the refreshing tea as depurative, stimulant, and tonic. I was pleased and surprised to find the late Prof. Varro Tyler rationalizing allspice for athlete's foot in *Prevention* (Apr. 2000). Dr. Tyler rightly noted that allspice oil may contain 70% of the fungicide eugenol. "Eugenol has both analgesic and antiseptic properties. It could possibly inhibit the growth of athlete's foot fungus," Dr. Tyler is quoted as saying. He suggested topical application, sprinkling ground allspice on the toes. Allspice is loaded with eugenol, and, perhaps more importantly, other antiseptic and fungicidal compounds, e.g., 1,8-cineole, alpha-terpineol, aromadendrene, ascorbic-acid, beta-pinene, chavicol, cinnamaldehyde, eugenol, hexanol, limonene, linalool, methyl-eugenol, myrcene, p-cymene, proanthocyanidins, selenium, terpinen-4-ol, and terpinolene with limonene classified as fungistatic.

Eugenol, the principal constituent of leaves and fruits, is toxic in large quantities and causes contact dermatitis. Allspice itself is irritant to the skin. Rinzler (1990) reports a study of 408 eczema patients, in which 19 reacted positively to allspice patch tests (RIN).

Indications (Allspice) — Arthrosis (1; RIN); Athlete's Foot (1; AAB; FNF); Bacteria (1; APA; FNF); Bruise (f; CRC); Cold (f; CRC); Colic (1; APA); Convulsion (1; APA; FNF); Corn (f; CRC; JLH); Cramp (1; AAB; APA; FNF); Diabetes (f; CRC; JFM); Diarrhea (f; APA); Dysmenorrhea (1; AAB; CRC; JFM); Dyspepsia (1; AAB; APA; CRC; FNF); Enterosis (f; APA); Exhaustion (1; AAB); Fever (f; JFM); Fungus (1; APA; FNF); Gas (1; AAB; APA; CRC); Gingivosis (1; APA); High Blood Pressure (1; ABS; FNF); Infection (1; APA; FNF); Mycosis (1; AAB; FNF); Myosis (1; APA); Nervousness (1; FNF); Neuralgia (f; CRC); Neurasthenia (f; BOW); Pain (1; AAB; APA; CRC; FNF); Parasite (1; APA); Pulmonosis (f; BOW); Rheumatism (1; AAB; CRC; FNF); Stomachache (1; APA; CRC); Stomatosis (1; APA); Toothache (1; APA); Vaginosis (1; APA); Virus (1; FNF); Vomiting (1; APA; FNF); Water Retention (1; FNF); Yeast (1; APA).

Allspice for high blood pressure:

- Antiaggregant: cinnamaldehyde; eugenol; isoeugenol; salicylates
- Antioxidant: eugenol; gamma-terpinene; isoeugenol; isoquercitrin; methyl-eugenol; myrcene; proanthocyanidins; quercitrin

- Antistroke: proanthocyanidins
- Antithromboxane: eugenol
- Cardioprotective: proanthocyanidins
- Cardiotonic: quercitrin
- Diuretic: isoquercitrin; quercitrin; terpinen-4-ol
- Hypotensive: 1,8-cineole; cinnamaldehyde; isoquercitrin; quercitrin
- Sedative: 1,8-cineole; alpha-pinene; alpha-terpineol; caryophyllene; cinnamaldehyde; eugenol; geranyl-acetate; isoeugenol; limonene; linalool; methyl-eugenol; p-cymene
- Vasodilator: cinnamaldehyde; proanthocyanidins

Allspice for infection:

- Analgesic: eugenol; myrcene; p-cymene
- Anesthetic: 1,8-cineole; cinnamaldehyde; eugenol; linalool; methyl-eugenol; myrcene
- Antibacterial: 1,8-cineole; alpha-pinene; alpha-terpineol; caryophyllene; cinnamaldehyde; delta-3-carene; delta-cadinene; eugenol; isoquercitrin; limonene; linalool; methyl-eugenol; myrcene; p-cymene; quercitrin; terpinen-4-ol
- Antiedemic: caryophyllene; caryophyllene-oxide; eugenol; proanthocyanidins; quercitrin
- Antiinflammatory: alpha-pinene; beta-pinene; caryophyllene; caryophyllene-oxide; cinnamaldehyde; delta-3-carene; eugenol; quercitrin; salicylates
- Antiseptic: 1,8-cineole; alpha-terpineol; aromadendrene; beta-pinene; eugenol; hexanol; limonene; linalool; methyl-benzoate; methyl-eugenol; proanthocyanidins; terpinen-4-ol
- Antiviral: alpha-pinene; ar-curcumene; cinnamaldehyde; limonene; linalool; p-cymene; proanthocyanidins; quercitrin
- Bacteristat: isoeugenol; malic-acid
- COX-2-Inhibitor: eugenol
- Circulostimulant: cinnamaldehyde
- Cyclooxygenase-Inhibitor: cinnamaldehyde
- Fungicide: 1,8-cineole; alpha-phellandrene; beta-phellandrene; caryophyllene; caryophyllene-oxide; chavicol; cinnamaldehyde; eugenol; linalool; methyl-eugenol; myrcene; p-cymene; terpinen-4-ol; terpinolene
- Fungistat: isoeugenol; limonene; methyl-eugenol
- Lipoxygenase-Inhibitor: cinnamaldehyde

Allspice for rheumatism:

- Analgesic: eugenol; myrcene; p-cymene
- Anesthetic: 1,8-cineole; cinnamaldehyde; eugenol; linalool; methyl-eugenol; myrcene
- Antiedemic: caryophyllene; caryophyllene-oxide; eugenol; proanthocyanidins; quercitrin
- Antiinflammatory: alpha-pinene; beta-pinene; caryophyllene; caryophyllene-oxide; cinnamaldehyde; delta-3-carene; eugenol; quercitrin; salicylates
- Antiprostaglandin: eugenol
- Antirheumatalgic: p-cymene
- Antispasmodic: 1,8-cineole; caryophyllene; cinnamaldehyde; eugenol; limonene; linalool; myrcene; quercitrin
- COX-2-Inhibitor: eugenol
- Collagenase-Inhibitor: proanthocyanidins
- Counterirritant: 1,8-cineole
- Cyclooxygenase-Inhibitor: cinnamaldehyde
- Elastase-Inhibitor: proanthocyanidins

- Lipoxygenase-Inhibitor: cinnamaldehyde
- Myorelaxant: 1,8-cineole; eugenol-methyl-ether; methyl-eugenol

Other Uses (Allspice) — Allspice of commerce is the dried unripe fruit, used as a condiment; in baked goods, chutney, ice cream, ketchup, mixed spices, pickles, sauces, soups; and in flavoring sausages and curing meats. Allspice powder consists of whole ground dried fruit. It's called "allspice" because it supposedly embraces the aromas of cinnamon, cloves, and nutmeg. To make an allspice substitute, merely combine one part nutmeg, two parts cinnamon, and two parts clove (RIN). Mexican Indians used allspice to flavor chocolate. I use it to flavor eggnog. Allspice is an essential ingredient of the rubs and marinades used in seasoning Jamaican jerked foods, which are also flavored by the smoke of pimento wood fires (FAC). In Jamaica, a local drink called "pimento dram" is made of ripe fruits and rum. It is regarded as a panacea. Allspice is used in such liqueurs as Benedictine and Chartreuse. A volatile oil, extracted from the spice and leaves, is used to flavor essences and perfumes and as a source of eugenol and vanillin. The oil is also used in flavoring beverages, candies, chewing gums, liqueurs, meats and sauces, and in Asian perfumery and shaving lotions. Bahamians make a pleasing tea from the leaves. Costa Ricans use the leaves as a spice. Many ethnic groups use the leaves in tea. Saplings are used as walking sticks and umbrella handles (DAD, FAD).

For more information on activities, dosages, and contraindications, see the *CRC Handbook of Medicinal Herbs*, ed. 2, Duke et al. 2002.

Cultivation (Allspice) — Fresh seeds germinate readily but lose viability quickly. Seeds extracted from fruits are planted in nursery beds with 50% shade. Seedlings transplanted to pots at the two- to four-leaf stage and outplanted when 30–40 cm high (ca. 10 months old). Allspice may also be propagated by bud propagation, which permits clonal female trees to predominate, with few male trees as pollinators. Allspice may be intercropped with banana or other crops for 5 years and then removed. Bown (2001) recommends rich, sandy, sunny, well-drained soils, zone 10, min. temp. 15–18°C (59–64°F). Nitrogenous manuring is particularly favorable for heavy leaf crop. A mixture of 10:10:10, ca. 100 g/tree, twice a year, is reasonable fertilization for young trees, with up to nearly 2.5 kg/tree for trees over 10 years old. For berry production, plants should be spaced 8–10 m apart each way; if grown for leaf oil, they may be planted much closer (2 m) and trimmed to bush size. It is not advisable to harvest berries and leaves from the same tree. Trees may be planted in pastures but must be protected and weeded. Trees should be trained low and spreading. Fruits ripen 3–4 months after flowering (DAD). Fruits should be harvested at full size, but not ripe (ca. 3–4 months after flowering). They lose aroma when mature. Trees can bear at age 5–8 (–12) years but are not fully producing until 20–25 years. They continue for several more years. Twigs with berries are broken off by hand and the berries stripped off, sun-dried for 7–10 days, and winnowed. An average yield is about 1 kg of dried berries per tree/year. Yields fluctuate, with a good crop every 3 years. Trees 10–15 years old may yield 0.9–25 kg/year. Mature trees at times yield 70 kg of green berries, but this is higher than one should expect (DAD).

Chemistry (Allspice) — Purseglove et al. (1981) give a long tabular comparison of the berry and leaf oils. Tucker et al. (2000) characterize the leaf oil as principally composed of eugenol (0.9–83.5%), methyl eugenol (0.1–84.9%), and myrcene (0.2–14.0%). Here are a few of the more notable chemicals found in allspice. For a complete listing of the phytochemicals and their activities, see the CRC phytochemical compendium, Duke and duCellier, 1993 (DAD) and the USDA database http://www.ars-grin.gov/duke/.

Chavicol — Fungicide; Nematicide.

Eugenol — ADI = 2.5 mg/kg; Analgesic; Anesthetic 200–400; Antiaggregant IC50 = 0.3 μM; Antiarachidonate; Antibacterial MBC = 400 µg/ml, 500 ppm; Anticancer; Anticonvulsant; Antiedemic; Antifeedant; Antigenotoxic 50–500 mg/kg orl mus; Antiherpetic IC50 = 16.2–25.6 µg/ml;

Antiinflammatory 11 µ*M*, IC~97 = 1000 µ*M*; Antikeratotic IC50 = 16.2–25.6 µg/ml; Antimitotic; Antimutagenic; Antinitrosating; Antioxidant 10 µ*M*, IC65 = 30 ppm; Antiprostaglandin 11 µ*M*, IC50 = 9.2 mM; Antipyretic 3 ml/man/day; Antiradicular EC50 = 2 µl/l; Antisalmonella MIC = 400 µg/ml; Antiseptic 400 µg/ml; Antispasmodic; Antithromboxane; Anti-TNF; Antitumor; Antiulcer; Antiviral IC50 = 16.2–25.6 µg/ml; Apifuge; Calcium-Antagonist IC50 = 224 µ*M*; Candidicide; Carcinogen; Carminative; Choleretic; CNS-Depressant; COX-1-Inhibitor IC97 = 1000 µ*M*; COX-2-Inhibitor IC > 97 = 1000 µ*M*; Cytochrome-p450-Inhibitor; Cytotoxic 25 µg/ml; Dermatitigenic; Enterorelaxant; Fungicide; Hepatoprotective 100 ppm; Herbicide; Insectifuge; Insecticide; Irritant; Juvabional; Larvicide; Motor-Depressant; Nematicide MLC = 2000 µg/ml; Neurotoxic; Sedative; Trichomonicide LD100 = 300 µg/ml; Trichomonistat IC50 = 10 µg/ml; Trypsin-Enhancer; Ulcerogenic; Vasodilator; Vermifuge; LD50 = 2680 orl rat; LD50 = 3000 orl mus; LDlo = 500 orl rat.

Methyleugenol — Anesthetic; Antibacterial; Anticancer; Anticonvulsant; Antidote (Strychine); Antifeedant; Antioxidant EC50 = 100 µl/l; Antiradicular EC50 = 100 µl/l; Antiseptic; Carcinogenic; Fungicide; Fungistatic 100 µg/ml; Insectifuge; Myorelaxant; Narcotic; Nematicide MLC = 1 mg/ml; Sedative.

Pimenta racemosa (Mill.) J. W. Moore (Myrtaceae)
BAYRUM TREE, WEST INDIAN BAY

Medicinal Uses (Bayrum Tree) — Reported to be analgesic, antiseptic, carminative, digestive, expectorant, and stomachic. Cubans decoct four seed in a cup of water as a stimulant. Bahamians drink the hot leaf tea as a stimulant, using the cooled tea as a skin bracer. Jamaicans take the tea for cold and fever; Grenadans for diarrhea; Trinidad natives for chest colds, flu, pneumonia, and stroke. Leaves are pasted with macerated leaves of *Caesalpinia coriaria* to relieve toothache.

According to Teissedre and Waterhouse (2000), the EO of *P. racemosa* inhibited LDL oxidation (IC71 = 2 µ*M*; IC94 = 5 µ*M*). That's close to tops in their survey of 23 EOs. They first assessed EOs as antibacterial and antifungal, second as flavorants and preservatives in foods, and third as cosmetics, for their aromatic and antioxidant properties. In their survey of 23 EOs, star anise was most potent (IC83 = 2 µ*M*) and Spanish red thyme (*Thymus zygis*) least potent (JAF4:3801).

Indications (Bayrum Tree) — Adenosis (f; CRC); Alopecia (f; BOW); Arthrosis (1; FNF; JFM); Bacteria (1; FNF); Bite (f; CRC); Bruise (f; CRC); Cancer (f; CRC); Cancer, breast (f; JLH); Cancer, uterus (f; JLH); Candida (1; FNF); Caries (1; FNF); Chest Cold (f; CRC; JFM); Chill (f; BOW); Cold (f; CRC); Cramp (1; FNF); Dandruff (f; BOW); Dermatosis (f; JFM); Diarrhea (f; CRC; JFM); Dyspepsia (f; CRC); Dysuria (f; CRC; JFM); Edema (f; CRC); Elephantiasis (f; CRC);

Fever (f; CRC; JFM); Flu (f; CRC; JFM); Fungus (1; FNF); Gas (f; JFM); Grippe (1; FNF; JFM); Headache (f; CRC); Incontinence (f; CRC); Induration (f; JLH); Infection (1; CRC; FNF); Inflammation (1; FNF); Lethargy (f; JFM); Mycosis (1; FNF); Myosis (1; FNF; JFM); Nausea (f; CRC); Neuralgia (f; BOW); Nicotinism (f; JFM); Pain (1; FNF; JFM); Pleurisy (f; CRC; JFM); Pneumonia (f; CRC; JFM); Rheumatism (1; FNF; JFM); Scirrhus (f; JLH); Smoking (f; CRC; JFM); Sore Throat (f; CRC); Spasm (1; CRC; FNF); Stroke (f; CRC; JFM); Toothache (1; CRC; FNF; JFM); Tumor (f; JLH); Uterosis (f; JLH); Varicosis (f; CRC); Vertigo (f; CRC); Virus (1; FNF); Water Retention (1; FNF).

Bayrum Tree for infection:

- Analgesic: camphor; eugenol; myrcene; p-cymene
- Anesthetic: 1,8-cineole; camphor; eugenol; linalool; methyl-eugenol; myrcene
- Antibacterial: 1,8-cineole; alpha-pinene; alpha-terpineol; caryophyllene; citral; delta-cadinene; eugenol; geranial; geraniol; limonene; linalool; methyl-eugenol; myrcene; neral; nerol; p-cymene; terpinen-4-ol; thujone
- Antiedemic: caryophyllene; caryophyllene-oxide; eugenol
- Antiinflammatory: alpha-pinene; beta-pinene; caryophyllene; caryophyllene-oxide; eugenol
- Antiseptic: 1,8-cineole; alpha-terpineol; beta-pinene; camphor; citral; eugenol; geraniol; hexanol; limonene; linalool; methyl-eugenol; nerol; terpinen-4-ol; thujone
- Antiviral: alpha-pinene; geranial; limonene; linalool; neryl-acetate; p-cymene
- Bacteristat: isoeugenol
- COX-2-Inhibitor: eugenol
- Fungicide: 1,8-cineole; alpha-phellandrene; beta-phellandrene; camphor; caryophyllene; caryophyllene-oxide; chavicol; citral; eugenol; geraniol; linalool; methyl-eugenol; myrcene; p-cymene; terpinen-4-ol
- Fungistat: isoeugenol; limonene; methyl-eugenol

Bayrum Tree for rheumatism:

- Analgesic: camphor; eugenol; myrcene; p-cymene
- Anesthetic: 1,8-cineole; camphor; eugenol; linalool; methyl-eugenol; myrcene
- Antiedemic: caryophyllene; caryophyllene-oxide; eugenol
- Antiinflammatory: alpha-pinene; beta-pinene; caryophyllene; caryophyllene-oxide; eugenol
- Antiprostaglandin: eugenol
- Antirheumatalgic: p-cymene
- Antispasmodic: 1,8-cineole; camphor; caryophyllene; eugenol; geraniol; limonene; linalool; myrcene; thujone
- COX-2-Inhibitor: eugenol
- Counterirritant: 1,8-cineole; camphor; thujone
- Myorelaxant: 1,8-cineole; eugenol-methyl-ether; methyl-eugenol

Bayrum Tree for toothache:

- Analgesic: camphor; eugenol; myrcene; p-cymene
- Anesthetic: 1,8-cineole; camphor; eugenol; linalool; methyl-eugenol; myrcene
- Antibacterial: 1,8-cineole; alpha-pinene; alpha-terpineol; caryophyllene; citral; delta-cadinene; eugenol; geranial; geraniol; limonene; linalool; methyl-eugenol; myrcene; neral; nerol; p-cymene; terpinen-4-ol; thujone
- Anticariogenic: alpha-terpineol; caryophyllene; delta-cadinene; geraniol; linalool

P

- Antiinflammatory: alpha-pinene; beta-pinene; caryophyllene; caryophyllene-oxide; eugenol
- Antioxidant: eugenol; gamma-terpinene; isoeugenol; methyl-eugenol; myrcene
- Antiseptic: 1,8-cineole; alpha-terpineol; beta-pinene; camphor; citral; eugenol; geraniol; hexanol; limonene; linalool; methyl-eugenol; nerol; terpinen-4-ol; thujone
- COX-2-Inhibitor: eugenol
- Candidicide: 1,8-cineole; beta-pinene; caryophyllene; eugenol; geraniol
- Candidistat: isoeugenol; limonene; linalool
- Fungicide: 1,8-cineole; alpha-phellandrene; beta-phellandrene; camphor; caryophyllene; caryophyllene-oxide; chavicol; citral; eugenol; geraniol; linalool; methyl-eugenol; myrcene; p-cymene; terpinen-4-ol
- Fungistat: isoeugenol; limonene; methyl-eugenol
- Vulnerary: terpinen-4-ol

Other Uses (Bayrum Tree) — Leaves distilled for the spicy bay oil, used in perfumery and in the preparation of bay rum. Formerly, leaves were distilled in rum and water, now oil is dissolved in alcohol or, in Dominica, only in water. Bay rum, with soothing and antiseptic qualities, is also used in toilet preparations and as a hair tonic. Bay rum is occasionally drunk. Oil also used on a limited scale for flavoring foods, chiefly in table sauces. Both bark and fruits used for flavoring in the Caribbean, e.g., in "blaff" (a fish broth). A leaf held in the mouth is said to discourage smoking. The dried green berries have a flavor. Wood is moderately hard and heavy, with a fine compact texture.

For more information on activities, dosages, and contraindications, see the *CRC Handbook of Medicinal Herbs*, ed. 2, Duke et al. 2002.

Cultivation (Bayrum Tree) — Much as in allspice. Bown (2001) recommends rich, sandy, sunny, well-drained soils, zone 10, min. temp. 15–18° C (59–64° F). Propagated by seeds sown when ripe, or semi-ripe cuttings in summer (BOW).

Chemistry (Bayrum Tree) — Here are a few of the more notable chemicals found in bayrum tree. For a complete listing of the phytochemicals and their activities, see the CRC phytochemical compendium, Duke and duCellier, 1993 (DAD) and the USDA database http://www.ars-grin.gov/duke/.

Chrysanthenone — Dentifrice (tobacco-stain remover).

Estragole — Antiaggregant IC50 = 320 µM; Anticancer; Antipyretic; Carcinogenic 4.4 µM/15 mos scu mus; DNA-Binder; Hepatocarcinogenic 0.23–0.46% diet/12mos; Insecticide; Insectifuge; Mutagenic; LD50 = 1250 orl mus; LD50 = 1820 orl rat.

Eugenol — See also *Pimenta dioica.*

Methyl-Eugenol — See also *Pimenta dioica.*

Thujone — Abortifacient; Antibacterial; Antiseptic; Antispasmodic ED50 = 0.127 mg/ml; Cerebrodepressant; Convulsant 40 mg/kg; Counterirritant; Emmenagogue; Epileptigenic; Hallucinogenic; Herbicide IC50 = 22 mM; Neurotoxic; Respirainhibitor; Toxic; Vermifuge; LD50 = 395 mg/kg orl gpg; LD50 = 192 mg/kg orl rat; LD50 = 230 mg/kg orl mus; LD50 = 140 mg/kg ipr rat; LDlo = 120 ipr rat; LD50 = 73–134 scu mus.

Piper auritum Kunth. (Piperaceae)
ANISILLO, HOJA SANTA

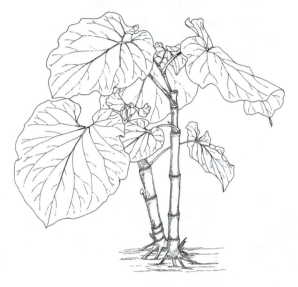

Medicinal Uses (Hoja Santa) — Half spice, half medicine, the aromatic leaves are widely used. The leaf decoction is used as a diaphoretic, diuretic, emollient, and stimulant. Extracts of the leaves are anesthetic, making the topical application rational for sundry aches and pains. In Costa Rica, our guides and ecotourists apply the leaf to headache, much as is done with *Pothomorphe peltata*. Both work. Warmed leaves are applied to inflammations. Leaves heated in almond oil are applied over the liver for colic (JFM). Leaf tea and/or pulverized seed is used to settle the stomach, e.g., the tea from half a leaf is taken after eating, almost like a digestive after-dinner mint. In Nicaragua, the leaf infusion and juice are administered orally or topically for aches and pains, childbirth and pregnancy, fever, digestive disorders, and burns.

The TRAMIL Commission (in my opinion, the Caribbean equivalent of Germany's Commission E) mentions that the tea is used for high blood pressure but warns of the toxicity of the aromatic ingredient safrole. Safrole and elemicin may exert hallucinogenic and/or psychotropic activities (TRA). At 0.1 ml/kg intravenously in dogs, the aqueous extract is hypotensive (TRA). The 95% ethanolic extract at 0.33 ml/l is spasmogenic and uterotonic *in vitro* (TRA). At 3 ml/l, the aqueous extracts have a vasodilator effect on lab rats (TRA). Myrcene has anesthetic activities. As a alpha-2-adrenergic agonist, myrcene affects arterial blood pressure.

Indications (Hoja Santa) — Angina (1; JFM; FNF); Bacteria (1; FNF); Colic (f; JFM); Cramp (1; FNF); Erysipelas (f; JFM); Fever (1; FNF; JFM); Fungus (1; FNF); Gonorrhea (f; JFM); Gout (f; JFM); Headache (1; FNF; JFM; TRA); High Blood Pressure (1; FNF; TRA); Infection (1; FNF); Inflammation (1; FNF; JFM); Nervousness (1; FNF); Pain (1; FNF; TRA); VD (f; JFM); Water Retention (1; FNF); Wound (1; FNF; JFM).

Hoja Santa for headache:

- Analgesic: borneol; camphor; eugenol; myrcene
- Anesthetic: 1,8-cineole; benzoic-acid; camphor; eugenol; linalool; myrcene; myristicin; safrole
- Antiinflammatory: alpha-pinene; beta-pinene; borneol; caryophyllene; caryophyllene-oxide; eugenol; myristicin
- Antineuralgic: camphor

- Antispasmodic: 1,8-cineole; borneol; camphor; caryophyllene; eugenol; limonene; linalool; myrcene; myristicin
- COX-2-Inhibitor: eugenol
- Counterirritant: 1,8-cineole; camphor
- Myorelaxant: 1,8-cineole; borneol
- Sedative: 1,8-cineole; alpha-pinene; borneol; caryophyllene; eugenol; limonene; linalool; myristicin
- Tranquilizer: alpha-pinene

Hoja Santa for wound:

- Analgesic: borneol; camphor; eugenol; myrcene
- Anesthetic: 1,8-cineole; benzoic-acid; camphor; eugenol; linalool; myrcene; myristicin; safrole
- Antibacterial: 1,8-cineole; alpha-pinene; benzoic-acid; caryophyllene; delta-cadinene; eugenol; limonene; linalool; myrcene; safrole
- Antiedemic: caryophyllene; caryophyllene-oxide; eugenol
- Antiinflammatory: alpha-pinene; beta-pinene; borneol; caryophyllene; caryophyllene-oxide; eugenol; myristicin
- Antipyretic: benzoic-acid; borneol; eugenol
- Antiseptic: 1,8-cineole; benzoic-acid; beta-pinene; camphor; eugenol; limonene; linalool; safrole
- Antiviral: alpha-pinene; beta-bisabolene; lignans; limonene; linalool
- COX-2-Inhibitor: eugenol
- Fungicide: 1,8-cineole; alpha-phellandrene; benzoic-acid; beta-phellandrene; camphor; caryophyllene; caryophyllene-oxide; eugenol; linalool; myrcene; myristicin; terpinolene
- Fungistat: limonene
- Vulnerary: benzoic-acid

Other Uses (Hoja Santa) — The leaves, with flavor and aroma (and carcinogen) of sassafras, are used fresh for seasoning soups, wild game such as armadillo, freshwater snails (called "jute" in Guatemala), and many other dishes (FAC, JAD, JFM). The leaves are not used dried (AAR). With pleasant anise flavor with minty overtones, leaves are also are used as wrappers for steamed or baked fish, shrimp, chicken, or cheese. Fresh leaves are wrapped around tamale dough before it is packaged in corn leaves and steamed (FAC). These leafy wrappers are eaten right along with the filling. Discard the tough central vein and use the two large pieces as wrappers (AAR). In Vera Cruz, leaves are ground with chiles, garlic, and roast tomatoes to make a feisty fish sauce. Leaves are also sauteed with shrimp and roasted peppers (TAD). Other Mexicans season a mole for pork and sometimes add them to posole verde. With a blended hoja santa sauce, you can improve fish or chicken. Puree one cup of leaves and one-third cup chicken broth; season with salt and pepper, garlic, onions, green chiles. Fry the sauce, as they do in Mexican cooking, with a small amount of oil bringing to a boil. In Honduras, young leaves are sometimes cooked and eaten as greens. The natives of Panama trap river fish using the leaves and/or fruits as bait. After feeding on it regularly, the flesh of the fish takes on the flavor of the leaf. Hence, the plant is recommended for aquaculture applications in parts of Central America (AAR, FAC).

For more information on activities, dosages, and contraindications, see the *CRC Handbook of Medicinal Herbs*, ed. 2, Duke et al. 2002.

Cultivation (Hoja Santa) — Tucker and Debaggio (2000) note that it is rarely hardy above zone 9 in the U.S. Bown (2001) suggests minimum temperatures of 10°C (50°F), seeds sown at 20–24°C (66–75°F). It seems to fare best on well-drained garden loam, moist but not constantly wet, tolerating some shade and full sun. They suggest propagating by summer cuttings. I too have had great luck (better than with kava-kava, which is not so pleasantly aromatic) with this in my garden, bringing it in to the greenhouse in winter. It makes cuttings readily. Pot bound plants make root sprouts that can be used for propagation. In the tropics, it can form pure stands attaining 20 ft in two years, much like kava-kava. Harvest as needed.

Chemistry (Hoja Santa) — Here are a few of the more notable chemicals found in hoja santa. For a complete listing of the phytochemicals and their activities, see the CRC phytochemical compendium, Duke and duCellier, 1993 (DAD) and the USDA database http://www.ars-grin.gov/duke/.

Borneol — Allelochemic; Analgesic; Antiacetytlcholine; Antibacterial MIC = 125–250 µg/ml; Antibronchitic; Antiescherichic MIC = 125 µg/ml; Antifeedant; Antiinflammatory; Antiotitic; Antipyretic; Antisalmonella MIC = 125 µg/ml; Antispasmodic ED50 = 0.008 mg/ml; Antistaphylococcic MIC = 250 µg/ml; Candidicide MIC = 250 µg/ml; Choleretic; CNS-Stimulant; CNS-Toxic; Fungicide; Hepatoprotectant; Herbicide IC50 = 470 mM; IC50 = 470 µM; Inhalant; Insectifuge; Irritant; Myorelaxant; Negative Chronotropic 29 µg/ml; Negative Inotropic 29 µg/ml; Nematicide MLC = 1 mg/ml; Sedative; Tranquilizer; LDlo = 2000 orl rbt.

Elemicine — Antiaggregant IC50 = 360 µM; Antidepressant ihl; Antifeedant; Antihistaminic; Antiserotonic; Antistress ihl; DNA-Binder; Fungicide MIC = 8 µg; Hallucinogenic; Hypotensive ihl; Insecticide 100 ppm; Insectifuge; Larvicide 100 ppm; Neurotoxic; Schistosomicide.

Eugenol — See also *Pimenta dioica.*

Myrcene — Analgesic; Anesthetic 10–20 mg/kg ipr mus, 20–40 mg/kg scu mus; Antibacterial; Anticonvulsant; Antimutagenic; Antinitrosaminic; Antioxidant; Antipyretic; Antispasmodic; Fungicide; Insectifuge; Irritant; p450(2B1)-Inhibitor IC50 = 0.14 µM.

Safrole — See also *Sassafras albidum.*

Piper nigrum L. (Piperaceae)
BLACK PEPPER

Medicinal Uses (Black Pepper) — Considered antipyretic, aromatic, carminative, rubefacient, and stimulant. Modern medicine in India utilizes black pepper as an aromatic stimulant in cholera, weakness following fevers, vertigo, coma, etc.; as a stomachic in gas and indigestion; as an alterative in arthritic diseases and paraplegia; and as an antimalarial. Rinzler (1990) rationalizes, "Because pepper irritates mucous membranes, highly spiced foods may be beneficial when you have hay fever or a head cold. The spice irritates tissues inside your nose and throat, causing them to weep a water secretion that makes it easier for you to cough up mucus or to blow your nose" (RIN). Pepper root, in the form of ghees, powders, enemas, and balms, is a folk remedy for abdominal tumors. Chinese poultice the leaves onto headaches and use them for urinary calculus as well. Powdered fruits are said to alleviate "superfluous flesh." An electuary prepared from the seed is said to help hard tumors, while a salve prepared from the seed is said to help eye indurations and internal tumors. The grain, with warm wine and egg, is said to help indurations of the stomach. A poultice, made from the pepper, salt, and vinegar, may help corns. Pepper is also poulticed, like mustard plasters, for colic, headache, parturition, puerperium, and rheumatism. A heavy dose of pepper with wild bamboo shoots is said to induce abortion. Although pepper contains the carcinogen safrole, it is at very low levels compared to sassafras.

Some unidentified compounds from pepper are even more potent that pepper phytosterols at lowering cholesterol, possibly some via HMG-CoA-reductase inhibition. Black pepper also speeds up transit through the gut, more than just increasing peristalsis. TCM practitioners boil 1 g black pepper with 15 g ginger 25 min in a liter of water to drink for idiopathic diarrhea. They use both black and red pepper for frostbite (Lin, 1994). Ayurvedics often prescribe black pepper in a synergistic triad called "trikatu" (three acrid spices, *Piper nigrum, P. longum,* and *Zingiber officinale*). Trikatu is of great interest, since it increases the bioavailability of other drugs. Trikatu may increase bioavailability either by (1) promoting absorption from the gastrointestinal tract, (2) protecting the drug from being metabolized/oxidized in the first passage through the liver after being absorbed, (3) a combination of these two mechanisms, or (4) causing increased production of bile. Under the influence of piperine (30 mg/kg), experimental blood levels of sparteine were increased by more than 100%. In human volunteers, 20 mg piperine increases bioavailability of curcumin twentyfold (MAB).

One pungent principle of pepper is piperine, present at levels of 2–6%. According to Rinzler (1990), chavicine, piperidine, and piperine are all diaphoretic (none of them were in my database at home as such) (RIN). Like other pungent principles, capsaicin from *Capsicum* and zingerone (from *Zingiber*), piperine is an adrenergic secretagogue, enhancing the secretion of catecholamines, especially epinephrine, from the adrenal gland, thereby leading to a warming sensation. Oral black pepper, or piperine, increases the flow of bile in rats. Desai and Kalro (1985) demonstrated under their experimental conditions that powdered black pepper does not damage the gastric mucosa, but that *Ferula* does, based on the rate of exfoliation of human gastric mucosa surface epithelial cells. Piperine may be useful as an analeptic in barbiturate poisoning. It has a central stimulant action in frogs, mice, rats, and dogs. Piperine, at 1 mg/ml, decreased the contraction of isolated guinea pig ileum. When injected i.v. into dogs at 1 mg/kg, it decreased blood pressure and respiration rate. When given orally to rats at 100 mg/100 g, it showed slight antipyretic activity. Piperine interacts with nitrite *in vitro* under slightly acidic conditions at 37° to form carcinogenic nitrosamines. Piperine is a stimulant. Piperine is also used in synthesizing heliotropine, which has its own medicinal indications. Piperine should not be combined with astringents, which may be rendered inert. Curcumin and piperine were also concluded to be radioprotective.

Isolated piperine has an inhibiting effect on *Lactobacillus plantarum, Micrococcus specialis,* and two fecal micro-organisms (*E. coli* and *Streptococcus faecalis*). It is mutagenic with *Leptospira*; with large doses, a bactericidal effect is produced. Piperine inhibits the ubiquitous, deadly bacterium *Clostridium botulinum.*

And piperine is more toxic to houseflies than pyrethrin. A mix of 0.05% piperine and 0.01 pyrethrins is more toxic than 0.1% pyrethrin (WOI). It improves insecticidal activity of other oils, like eucalyptus oil. Piperine is synergistically insecticidal to rice weevils and cowpea weevils. As insect repellent, spray with 0.5 tsp ground pepper in one qt warm water (RIN).

Lin (1994) has some novel observations. Merely holding 100 mg pepper in the mouth causes a transient increase of 13 mm Hg in systolic, 18 mm Hg in diastolic pressure. A parallel result has resulted from chewing 1 g ginger 11 for systolic, 14 for diastolic (MAB). Aqueous leaf extract raised blood pressure in dogs modestly (not stated whether oral or injected) (MPI).

The EO reportedly inhibits *Alternaria oryzae, A. tenuis, Aspergillus oryzae, Beauveria* sp., *Cryptococcus neoformans, Fusarium solani, Histoplasma capsulatum, Microsporum gypseum, Nocardia brasiliensis, Penicillium javanicum, P. striatum, Staphylococcus* "albus," *Trichoderma viride, Trichophyton mentagrophytes,* and *Vibrio cholera*. Alcoholic, aqueous, and ether extracts have taenicidal activity at 1:100 concentrations. Sharma et al. (2000) showed that, though bactericidal on their own, extracts of black pepper, capsicum, and turmeric partially protected *Bacillus megaterium, Bacillus pumilus,* and *Escherichia* from radiation, probably protecting their DNA. Chile was strongest. Karen Dean has been preparing her interesting column, Plant Patents, in *HerbalGram*, e.g., in *HerbalGram* #50, where she covers three new patents. One was for treating various mycoses, e.g., athlete's foot, candida, dermatomycoses, favus, jock itch, ringworm, all based on a "new class of general antiinfective agents" extracted from black-pepper (*Piper*), ginger (*Zingiber*), and other plant species containing vanillyl and piperidine ring structures typical of some of the pungent principals found in pepper and ginger. I suppose this does represent a better mousetrap, but the spices themselves have quite a history as fungicides. Staggs patent claims that his compositions constitute a less expensive, more effective, less toxic treatment for mycoses in humans and livestock (than widely used antifungal medications). I suspect we might do almost as well right there in the spice rack, grabbing the pungent compounds from our red and black peppers, and ginger and turmeric. The piperine from black pepper has been shown to synergize with a lot of natural and synthetic medications, including some fungicides. Stagg's compositions contain the pungent active ingredient in an oleoresin base, formulated into bath products, drops, douches, infusions, lotions, plasters, powders, shampoos, and tinctures. I'd like to compare Staggs composition with my Hot Foot [a mix of five powdered hot spices, black and red pepper, ginger and turmeric, and the stinking rose, the Biblical garlic, which adds the proven candidicidal and/or fungicidal compounds (ajoene (IC100 = 100 µg/ml)), allicin, caffeic-acid MIC = 400 g/ml, chlorogenic acid, citral, p-coumaric-acid, diallyl-disulfide, eruboside-B (MIC = 25 g/ml), ferulic acid, geraniol IC93 = 2 mM, p-hydroxybenzoic acid, linalol, phloroglucinol, and sinapic-acid].

Here I'd like to list a few of the comments from a hundred-year-old volume (*Dictionary of the Economic Products of India*, six vols. by George Watt. 1889–1892). It devotes seven fine print pages to black pepper and/or trikatu. Trikatu, used externally for dermatosis (and alopecia), might do just as well as Stagg's combo. Externally, it is valued as a rubefacient, for sore throat, piles, and some skin diseases. Moreover, Watt quotes a historical spice formula for poor eyesight. "He that hath a bad or thick sight, let him use pepper cornes, with annis, fennel seed, and cloves, for thereby the mystinesse of the eyes which darken the sight is cleared and driven away" (DEP). I like that quote, more than 100 years old. It might even help with glaucoma. But some eager beaver neonate might try to patent it. I hope not. It's an all spice formula, food farmacy, which belongs in the public domain. I'll not try to patent my "Peppery Phytopotentiator." It's a mix of grapefruit juice spiced up with cayenne or tabasco for its capsaicin, black pepper (for its piperine), licorice (for its glycyrrhizin, a saponin which saponifies making more readily available many medicines), cardamom (for its cineole which speeds up transdermal and probably transgut availability), and turmeric (for its curcumin). Odds are that if you are running low on a pharmaceutical or phytochemical medicine, this will stretch it, potentiating by one- to fivefold the medicine, Be careful with hard core pharmaceuticals! My friends Alan Tillotsen and his wife Naixim, MD, both also clinical herbalists, make the following suggestion to some of his indigent patients. Take a pound of turmeric (about $3), and add 5% ground black pepper (11 level teaspoons), which makes 460 g of medicine for less than $4. Dose is 2.5 grams twice a day (a huge dose = 250 mg black pepper), 92 days (3 months) for $4 or $1.35 per month. The piperine augments significantly the availability of curcumin, one of nature's Cox-2-Inhibitors, of great potential in Alzheimer's,

P

arthritis, and cancer, especially the colon cancer to which I am genetically targeted. Nor will I try to patent my High Colonic Tea or Gobo Gumbo, a food farmacy approach to Alzheimer's, arthritis, colon cancer, and lymphoma. Again, it emphasizes the black pepper to increase the availability of the curcumin. I don't believe in patenting nature or Father Nature's Farmacy. But my database can be helpful to those trying to break (or make) a patent. The updated database at home can lead you to the early references and recipes for a given indication for a given herb. The lawyers I have worked for are silent and, like me, sworn to secrecy, but I'll bet I helped them break two million-dollar patents recently.

And regarding toxicity, levels of 60,000–110,000 ppm piperine, 14,000 ppm piperetine, and 8000 ppm chavicine, levels higher than those, e.g., in the *Wealth of India,* have been assumed. It is hinted that chavicine, piperine, and piperetine may be like pepper's safroles and tannins, carcinogenic, accompanied by limonene, phellandrene, and pinene, which "have been shown to be either oncogens, cooncogens, or tumor promoters…. In addition, the high concentration of terpenes in black pepper oil suggests a high tumor promoting activity." Since piperine and other pepper alkaloids have chemical structures similar to that of the mutagenic urinary safrole metabolite, 3-piperidyl-1-(3′,4′-methylenedioxyphenyl)-1 propanone, pungent components of black pepper are suspected to be mutagenic and/or carcinogenic. Russians have suggested that the consumption of too much tea flavored with black pepper may have contributed to the unusually high incidence of esophageal cancer in the their Aktibinsk region. Like safrole, piperine stimulates hepatic regeneration in partially hepatectomized rats. Topical application of pepper extract to mice skin has increased the incidence of total malignant tumors. Reviewing the work on safrole, Buchanan (1978) concluded that it is the most thoroughly investigated methylenedioxybenzene derivative. The major flavoring constituent in sassafras root bark, safrole also occurs in basil, black pepper, cinnamon leaf oil, cocoa, mace, nutmeg, parsley, and star anise oil. When safrole was identified as a "low-grade hepatocarcinogen," it was banned in root beer, and the FDA in 1976 banned interstate marketing of sassafras for sassafras tea. The oral LD_{50} for safrole in rats is 1950 mg/kg body weight, with major symptoms including ataxia, depression, and diarrhea, death occurring in 4–5 days. Ingestion of relatively large amounts of sassafras oil produced psychoactive and hallucinogenic effects persisting several days in humans. With rats, dietary safrole at levels of 0.25%, 0.5%, and 1% produced growth retardation, stomach and testicular atrophy, liver necrosis, and biliary proliferation, and primary hepatomas. But at 10 mg/rat/day, safrole stimulates liver regeneration, like isosafrole and dihydrosafrole at 15 mg/rat/day, piperine at 25 mg/rat/day and estragole at 50 mg/rat/day. Is it conceivable that small doses help the liver while large doses hurt? Does this suggest hormetics or homeopathy? Also reviewing research on myristicin, which occurs in nutmeg, mace, carrot seed, celery seed, and parsley, as well as black pepper, Buchanan (1978) noted that the psychoactive and hallucinogenic properties of mace, nutmeg, and purified myristicin have been studied. It has been hypothesized that myristicin and elemicin can be readily modified in the body to amphetamines. The oral LD50 for nutmeg oil in rats, mice, and hamsters is 2600, 5620, and 6000 mg/kg, respectively. Based on LD50s, caffeine at 192 is much more dangerous. For more data on the phytochemicals and their activities, see the CRC phytochemical compendia, Duke and duCellier, 1993.

Indications (Black Pepper) — Adenitis (f; CRC; DAA); Alopecia (f; DEP); Amenorrhea (f; FEL); Anorexia (1; EFS; FNF); Arthrosis (1; CRC; DAD; DEP; FNF; PH2); Asthma (f; PH2; SKJ); Athlete's Foot (1; HG50); Atony (f; FEL); Bacteria (1; CRC; FNF; JBU); Bite (f; DEP; SKJ); Boil (f; DEP); Bronchosis (1; FNF; PHR); Calculus (1; CRC; DAD); Cancer (1; CRC; DAA; FNF); Cancer, abdomen (f; CRC; JLH); Cancer, anus (f; JLH); Cancer, breast (f; CRC; JLH); Cancer, colon (f; CRC; JLH); Cancer, eye (f; CRC; JLH); Cancer, face (f; CRC; JLH); Cancer, gum (f; CRC; JLH); Cancer, liver (f; CRC; JLH); Cancer, mouth (f; CRC; JLH); Cancer, nose (f; CRC; JLH); Cancer, parotid (f; CRC; JLH); Cancer, sinew (f; CRC; JLH); Cancer, spleen (f; CRC; JLH); Cancer, stomach (f; CRC; JLH); Cancer, throat (f; CRC; JLH); Cancer, uvula (f; CRC; JLH); Candida (1;

HG50); Catarrh (f; PH2); Chill (f; BOW); Cholera (1; CRC; DAD; FEL; SKJ); Cold (1; CRC); Colic (f; CRC; DEP); Coma (f; DEP); Condyloma (f; JLH); Constipation (1; CRC; DAD; FEL); Congestion (f; RIN); Convulsion (1; FNF; SKJ; SPI); Corn (f; JLH); Cough (1; CRC; PH2; SKJ); Cramp (1; FNF); Debility (f; DEP); Dermatosis (1; DEP; FNF; PH2; SKJ); Diarrhea (f; CRC; DEP; PH2; SPI); Dog Bite (f; SKJ); Dry Mouth (1; PHR); Dysentery (f; CRC; PH2); Dysmenorrhea (f; CRC; FEL); Dyspepsia (1; DAD; DEP; EFS; FEL; FNF; PHR; PH2); Dysuria (f; CRC); Epididymosis (1; SPI); Epilepsy (f; BOW); Escherichia (1; CRC); Favus (1; HG50:30); Fever (1; CRC; DAD; PH2); Frostbite (1; SPI); Fungus (1; FNF); Furunculosis (f; CRC); Galactorrhea (f; PH2); Gas (1; DAD; EFS; FEL; PH2); Gastrosis (f; FEL; PHR; PH2); Gingivosis (f; JLH); Gonorrhea (f; DEP); Gravel (f; CRC); Hay Fever (1; RIN); Headache (1; CRC; PHR); Head Cold (1; RIN); Hemorrhoid (f; DEP; HHB; PH2; SKJ); Hepatosis (f; JLH); Hiccup (f; PH2); High Blood Pressure (1; CRC; FNF); High Cholesterol (1; FNF; LIN); Induration (f; JLH); Infection (1; CRC; JBU); Inflammation (1; BOW; FNF); Itch (f; DEP); Leishmaniasis (1; PHR); Lethargy (1; DAD); Malaria (f; CRC; DEP); Mucososis (f; PH2; RIN); Mycosis (1; FNF; HG50); Nausea (f; CRC); Nervousness (1; FNF); Neuralgia (1; HHB; PHR; PH2); Onychiosis (1; HG50:31); Ophthalmia (f; JLH); Pain (1; FNF; JBU); Paralysis (f; CRC; DEP); Paraplegia (1; CRC; DAD; DEP; WOI); Parturition (f; CRC); Phymata (f; JLH); Prolapse (f; DEP); Respirosis (f; SPI); Rhinosis (f; SKJ); Ringworm (1; HG50); Scabies (1; PHR; PH2); Scarlatina (1; CRC; DAD); Scirrhus (f; JLH); Sinusosis (f; BOW); Snakebite (f; SKJ); Sore Throat (f; DEP; SKJ); Splenosis (f; JLH); Staphylococcus (1; MPI); Stomachache (f; DAA); Swelling (f; JLH); Tinea (1; HG50); Toothache (1; DEP; FNF); Tumor (1; CRC; FNF); Ulcer (1; FNF; JLH); Urethrosis (f; PH2); Urolithiasis (1; CRC); Vertigo (f; CRC); Vomiting (f; PH2); Wart (f; JLH); Water Retention (f; PNC); Wen (f; JLH); Yeast (1; HG50).

Black Pepper for dermatosis:

- Analgesic: borneol; caffeic-acid; camphor; eugenol; myrcene; p-cymene; piperine; quercetin
- Anesthetic: 1,8-cineole; benzoic-acid; camphor; carvacrol; cinnamic-acid; eugenol; linalool; linalyl-acetate; methyl-eugenol; myrcene; myristicin; piperine; safrole
- Antibacterial: 1,8-cineole; acetophenone; alpha-pinene; alpha-terpineol; benzoic-acid; caffeic-acid; carvacrol; caryophyllene; cinnamic-acid; citral; citronellal; citronellol; delta-3-carene; delta-cadinene; eugenol; hyperoside; isoquercitrin; kaempferol; limonene; linalool; methyl-eugenol; myrcene; nerolidol; p-coumaric-acid; p-cymene; perillaldehyde; piperine; quercetin; quercitrin; rhamnetin; rutin; safrole; terpinen-4-ol; terpinyl-acetate
- Antidermatitic: hyperoside; quercetin; rutin
- Antiinflammatory: alpha-linolenic-acid; alpha-pinene; beta-pinene; borneol; caffeic-acid; carvacrol; caryophyllene; caryophyllene-oxide; cinnamic-acid; cuparene; delta-3-carene; eugenol; hyperoside; kaempferol; myristicin; n-hentriacontane; piperine; quercetin; quercitrin; rutin; salicylates
- Antiitch: camphor
- Antiseptic: 1,8-cineole; alpha-terpineol; benzoic-acid; beta-pinene; caffeic-acid; camphor; carvacrol; carvone; citral; citronellal; citronellol; eugenol; kaempferol; limonene; linalool; methyl-eugenol; oxalic-acid; piperine; safrole; terpinen-4-ol
- Antistress: gaba; myristicin
- COX-2-Inhibitor: eugenol; kaempferol; quercetin
- Cyclooxygenase-Inhibitor: carvacrol; kaempferol; n-isobutyl-octadeca-trans-2-trans-4-dienamide; pellitorine; quercetin
- Fungicide: 1,8-cineole; acetophenone; alpha-phellandrene; benzoic-acid; beta-phellandrene; caffeic-acid; camphor; carvacrol; caryophyllene; caryophyllene-oxide; cinnamic-acid; citral; citronellol; eugenol; linalool; methyl-eugenol; myrcene; myristicin; p-coumaric-acid; p-cymene; perillaldehyde; quercetin; terpinen-4-ol; terpinolene

Black Pepper for high cholesterol:

- Antiaggregant: alpha-linolenic-acid; caffeic-acid; eugenol; kaempferol; myristicin; quercetin; safrole; salicylates
- Antiatherogenic: rutin
- Antiatherosclerotic: carvacrol; quercetin
- Antidiabetic: quercetin; rutin
- Antiischemic: ubiquinone
- Antilipoperoxidant: quercetin
- Antioxidant: caffeic-acid; camphene; carvacrol; eugenol; gamma-terpinene; hyperoside; isoquercitrin; kaempferol; linalyl-acetate; methyl-eugenol; myrcene; myristicin; p-coumaric-acid; quercetin; quercitrin; rhamnetin; rutin; ubiquinone
- Antioxidant (LDL): carvacrol
- Cholagogue: caffeic-acid
- Choleretic: 1,8-cineole; benzoic-acid; caffeic-acid; cinnamic-acid; eugenol; kaempferol; p-coumaric-acid; quercitrin
- Diuretic: caffeic-acid; gaba; hyperoside; isoquercitrin; kaempferol; myristicin; n-hentriacontane; quercitrin; terpinen-4-ol
- Hepatoprotective: borneol; caffeic-acid; eugenol; piperine; quercetin
- Hypocholesterolemic: caffeic-acid; d-limonene; rutin
- Hypotensive: 1,8-cineole; acetyl-choline; alpha-linolenic-acid; astragalin; gaba; hyperoside; isoquercitrin; kaempferol; myristicin; piperine; quercetin; quercitrin; rutin
- Thyrostimulant: piperine

Other Uses (Black Pepper) — Pepper has been used as a condiment and medicine since times of Hippocrates. Spices were so valued in the past that when Alarich, a Gothic leader, laid siege to Rome in A.D. 408, he demanded as ransom 3000 lb of pepper as well as various precious metals. Marco Polo, Magellan, and Columbus all took hazardous journeys seeking better routes to spice-growing countries. "Black pepper…is the world's most widely used spice even though *Piper nigum* is an exclusively tropical plant with several useful properties" (Sherman and Flaxman, 2001). In 4578 meat-based recipes, 93% call for at least one spice. Some lacked spices entirely, and others had up to 12 spices. In 10 countries (Ethiopia, Kenya, Greece, India, Indonesia, Iran, Malaysia, Morocco, Nigeria, and Thailand), every meat-based recipe…called for at least one spice. Black pepper and onion were called for most frequently, in 63% and 65% of all meat-based recipes, then garlic, 35%; chilis, 24%; lemon and lime juice, 23%; parsley, 22%; ginger, 16%; and bay leaf, 13% (Sherman and Billing, 1999).

Black pepper is the whole unripe dried fruit; white pepper is obtained by removal or blanching of the outer coating (pericarp). Both are available whole, cracked, coarsely ground, or finely ground. Both have numerous culinary uses, including seasoning and flavoring of soups, meats, poultry, fish, eggs, vegetables, salads, sauces, and gravies. Both are employed commercially in preparation of processed meats of all kinds, soups, sauces, pickles, salad dressings, mayonnaise, and other foods. Tellicherry and Muntok are famous grades of black and white pepper, respectively. Mignonette pepper or shot pepper is a mixture of white and black pepper widely used in France. Poivre gris is finely ground mignonette. Unripe, green peppercorns are pickled in vinegar and used as a relish (FAC). Pepato is a sheep's milk cheese from Sicily, flavored with whole black peppercorns (FAC). Seeds yield an EO used to flavor sausages, pickles, canned foods, and beverages (FAC). The pepper oil also serves in perfumery. Rinzler (1990) suggests a pinch of pepper where I would not have thought of pepper, i.e., apple pie, applesauce, baked apples or pears, eggnog, gingerbread, hot chocolate, and spiced wine punch. She wisely reminds us to grind peppercorns just before using to retain maximum flavor. She suggests using metal or plastic pepper mills, since wood absorbs the aromatic oils and is harder to keep clean and fresh (DAD, FAC, RIN).

And how often have you heard it said that spices were important before refrigeration to slow the spoilage of foods; spices were the original antioxidants? Lin (1994) laments that many expert committees, based only on the tocopherol content of pepper, concluded that it could not slow spoilage. Lin however, thinks of the whole antioxidant complex, noting that, in addition to 0.54% mixed tocopherols in the oleoresin (including 0.1% alpha-tocopherol), pepper contains five phenolic amides that are superior as antioxidants to alpha tocopherol *in vitro* (SPI). If you consult my database, you'll find many more antioxidants, suggesting the possibility of synergy in antioxidant effects. Interestingly, Lin (1994) suggests that the pungencies of black (piperine) and red pepper (capsaicin) are synergic, too.

For more information on activities, dosages, and contraindications, see the *CRC Handbook of Medicinal Herbs*, ed. 2, Duke et al. 2002.

Cultivation (Black Pepper) — Pepper is usually propagated by cuttings from straight shoots of vines less than two years old. However, cuttings from vines as old as 10 years have been used successfully. Terminal buds are clipped with leaves and branches of the third through seventh nodes. When the terminal bud regenerates, after about 10 days, the 60-cm cuttings are planted obliquely to the shaded side of the support, with three to four nodes below the surface, no deeper than 15 cm, and shaded. Stolons, marcots, and grafts are also used. Seedlings are rarely used for propagation. Scientists report ca. 80% germination with fresh seed after one month, adding that seedlings from seed obtained from cultivated plants are fairly uniform. Like vanilla, pepper needs support and may be grown upon some commercial tree crop, as betel palm, coffee, mango, *Strychnos*, or *Erythrina*. In Sarawak, *Eusideroxylon zwageri,* the Borneo ironwood, is used for stakes because of its resistance to termites. In Ponape, tree ferns 15–20 years old were recommended, but they are harder and harder to come by. Environmentalists would frown on destruction of tree ferns. Catch crops of capsicum, ginger, peanuts, soy, or tobacco may be grown if removed well before the vines reach the top of the post. One should avoid too dense a shade. Concrete posts, though expensive, are recommended in some areas. Elmo Davis (pers. comm., 1990) suggests that newcomers start out with several germplasm lots, growing on living fenceposts at first, later going into expanded production with the better germplasm and technology. There is some evidence to show that vines grown on posts establish themselves more quickly and yield more pepper berries than do plants grown on living trees. However, it does cost more to purchase and set posts than it does to grow plants on living trees. In some areas, when vines reach top of posts, they are detached from support and coiled around the base of support and buried, thus increasing the vigor of the plants. Fairly heavy manuring is given. Height of posts and spacings differ in different countries. Wood supports are commonly used. In Asia, wood supports may be 3.3–4.6 m high and spaced 2.3–2.6 m apart with same distance between rows. In Brazil, they are 2.6–3.3 m high and spaced 2.6–3.3 m apart each way. In Sarawak, higher yields have been obtained by growing pepper plants as hedges 2 m high in rows 2.6 m apart with plants 1.3 m apart in hedge with seven leading shoots trained up wires. For mechanized farms, 3.3 m avenues are minimal. On new plantings in infertile soils, preplant recommendations are 4000–4500 kg/ha crushed coral limestone, 1000–1200 kg/ha superphosphate, or 450–500 kg/ha treble superphosphate, 3000–3500 kg/ha muriate of potash, 10–12 kg/ha zinc as zinc sulfate, all disced in to a depth of 10–15 cm. On planting, each plant should receive ca. 100 g magnesium sulfate and 250 g nitrogen as ammonium sulfate. Each plant should receive ca. 600 g 10/10/10 fertilizer, divided in six applications, where labor is cheap and rainfall is high. Once full production is attained, individual fertilizations should be doubled. Old pepper plantings and half-dead vines may be rejuvenated by using a heavy mulch of compost and an application of fertilizer at frequent intervals. Guano with small amounts of sulfate of ammonia and superphosphate are added. Organic fertilizers are extensively used today, such as guano, fish meal, soybean cake, and more recently blood-bone meal. After fruiting begins, guano is applied up to 2 kg per vine, applied four times per year. Inorganic fertilizers produce as good yields as organic fertilizers, provided

P

they contain trace elements, as iron, copper, zinc, manganese, boron, and molybdenum. A mixture of NPK in proportion of 12:5:14 with additional magnesium is recommended as a basic application for all soil types. Sarawak recommendations are 11–13% N, 5–7% P_2O_5, 16–18 percent K_2O, and 4–5 percent MgO. Pepper is a nutrient-exhaustive crop. In Sarawak, mulching with cut grass seems to increase yields. In nineteenth century Malaysia, gambir (*Uncaria gambir*) residues were used to mulch the plants. Elsewhere, *Calopogonium mucunoides* is grown as a nitrogen-fixing cover crop between the plants. Plants should be pruned when 60 cm tall. Pruning encourages lateral fertile branches and dense foliage indicates a maximum number of fruiting branches. After vine has begun to bear, only two (or three) stems should be allowed to grow from a single root (DAD). Vines may start producing in the third year, bearing on until the fifteenth year. Flowering is preceded by flushing, usually at the beginning of the rainier season. First harvest is usually 3 years after planting, when the vines have reached the top of their posts. All fruiting spikes, ripe and unripe, are picked. That way, vines will fruit more evenly in the following season. If a whitish liquid oozes from a thumbnail injury to the fruit, the "corn" is too green. Whole fruiting spikes are picked after some fruits have reddened and others are green or yellow. Vines are picked once a week over the ripening period, or only in one big picking (e.g., in India). When partially dry, "corns" are separated from the stalks by rubbing the spikes between the palms of the hands or by trampling underfoot. For black pepper, the drupes are sun dried for 3–4 days. For white pepper, lightly-crushed riper spikes are sacked and soaked in running water for 7–10 days, then trampled underfoot, and the fruits separated by hand and sun dried for 3–4 days. White pepper can also be made by mechanical abrasion of dried black peppercorns. Small plantings of hedges, 2.5 m apart with plants 1.2 m apart in the hedge and 2 m high, have yielded 35,000 kg green pepper/ha. For each 100 kg of green pepper, 25–28 kg of white or 33–37 kg of black pepper can be produced. Well-cultivated vines yield to 1.8 kg green pepper the third year to 9 kg until the seventh year, declining to 2.3 kg in the eighth to fifteenth year, when plantings should be abandoned. In India, production may continue for 25 years, even 100 years in backyard gardens.

Chemistry (Black Pepper) — Patel and Srinivasan (1985) noted that dietary piperine significantly increased lipase, maltase, and sucrase activities. Here are a few of the more notable chemicals found in black pepper. For a complete listing of the phytochemicals and their activities, see the CRC phytochemical compendium, Duke and duCellier, 1993 (DAD) and the USDA database http://www.ars-grin.gov/duke/.

Piperidine — Antienzymatic; CNS-Depressant; Diaphoretic; Hepatotropic; Insectifuge 50 ppm; Spinoconvulsant; Urate-Solvent; LD50 = 400 orl rat; LD50 = 0.52 ml/kg.

Piperine — Abortifacient; Adrenergic; Analeptic; Analgesic; Anesthetic; Antiaflatoxin; Antibacterial; Anticancer; Anticlastogen; Anticonvulsant 50–400 mg/kg ipr; Antiedemic 10 mg/kg; Antifertility 12.5 mg/kg; Antiglucuronidase IC20 = 1 μM, IC20 = 25 μM, IC42 = 50 μM, IC57 = 100 μM; Antiimplantation; Antiinflammatory 10 mg/kg; Antileishmannic; Antimutagenic; Antinarcotic; Antipyretic; Antiseptic; Antispasmodic; AntiSubstance-P; Aryl-Hydrocarbon-Hydroxylase (AHH)-Inhibitor 125 mg/kg; ATPase-Stimulant; Carcinogenic; Cardiotonic; Carminative; Catecholaminogenic 650 nM/kg; Choleretic; CNS-stimulant; Diaphoretic; Digestive; Endorphinogenic; Epinephreninergic; Glutathione-Sparing; Hepatoprotective; Hepatoregenerative; Hypertensive; Hypotensive 1 mg/kg iv dog; Insecticide; Insectifuge; Lactase-Promoter; Maltase-Promoter; Mutagenic; Myocontractor; Myorelaxant; Positive Chronotropic; Positive Inotropic; p450-Inducer; p450-Inhibitor; Parasiticide; Peristaltic; Plasmodicide; Radioprotective; Respirastimulant 5 mg/kg ivn; Secretagogue; Sedative; Serotoninergic; Spasmogenic; Spermogenic; Stimulant; Sucrase-Promoter; Thermogenic 5 mg/man; Thyrostimulant; LD50 = 349 mg/kg ipr rat; LD50 = 514–800 mg/kg orl rat; LD50 = 330–1639 mg/kg orl mus.

Piperonal — Anticancer; Pediculicide; LD50 = 2700 orl rat.

Pistacia lentiscus L. (Anacardiaceae)
CHIOS MASTICTREE, MASTIC

Medicinal Uses (Mastic) — The best grades of mastic, yellowish-white translucent tears, are medicinally employed as an aromatic astringent. The resin is painted over wounds to protect them. Frequently cited in the cancer folklore, using the resin or juice from mastic. Used for diarrhea in children. Regarded as analgesic, antitussive, aperitif, aphrodisiac, astringent, carminative, diaphoretic, diuretic, expectorant, hemostatic, stimulant to the mucous membranes, and stomachic (JLH). Lebanese dissolve the resin in alcohol, adding lemon juice, for gall bladder and liver trouble. Algerians use the root decoction for cough. The oil obtained by hot extraction of the nuts is used for itch and rheumatism (BIB).

Indications (Mastic) — Adenosis (f; JLH); Aposteme (f; CRC; JLH); Bacteria (1; FNF; X8808717); Bleeding (f; CRC); Blennorrhea (f; CRC); Boil (f; BIB; CRC); Bronchosis (f; FEL); Cancer (1; CRC; FNF; JLH); Cancer, anus (f; CRC; JLH); Cancer, breast (f; CRC; JLH); Cancer, liver (f; CRC; JLH); Cancer, parotid (f; CRC; JLH); Cancer, spleen (f; CRC; JLH); Cancer, stomach (f; CRC; JLH); Cancer, testicle (f; CRC; JLH); Cancer, throat (f; CRC; JLH); Cancer, uterus (f; CRC; JLH); Candida (1; HH3; X8808717); Canker (1; BIB; CRC; FNF); Carbuncle (f; CRC); Caries (1; CRC; FEL; FNF); Catarrh (f; CRC; FEL; HH3); Cholecystosis (f; BIB; CRC); Cirrhosis (f; CRC; HH3); Condyloma (f; CRC; JLH); Cough (f; BIB); Cramp (1; FNF); Debility (f; CRC); Dermatosis (f; GHA); Diarrhea (f; CRC; HH3); Dysentery (f; CRC; HH3); Escherichia (1; HH3); Fever (1; GHA); Fungus (1; FNF; HH3; X8808717); Gastrosis (f; BIB; CRC); Gingivosis (1; FEL; FNF; PHR; PH2); Gonorrhea (f; CRC; HH3); Gout (f; HH3); Halitosis (1; BIB; CRC; DEP; FEL; FNF; PHR); Heart (f; CRC); Hepatosis (f; BIB; CRC; HH3); High Blood Pressure (1; FNF; HH3; X1409845); Induration (f; CRC; JLH); Infection (1; FNF; X8808717); Inflammation (1; FNF; JLH); Itch (f; BIB); Leukorrhea (f; CRC; HH3); Mastosis (f; CRC); Mucososis (f; CRC); Mycosis (1; FNF; HH3; X8808717); Myosis (f; BOW); Nephrosis (f; FEL); Nervousness (1; FNF); Pain (f; CRC; GHA); Phymata (f; CRC); Rheumatism (f; BIB; HH3); Ringworm (f; BOW); Sclerosis (f; CRC); Scirrhus (f; CRC; JLH); Sore (f; HH3); Staphylococcus (1; HH3); Toothache (1; CRC; FNF); Tumor (1; CRC; FNF); Ulcer (1; FNF; PH2; X3724207); VD (f; CRC; HH3); Virus (1; FNF); Water Retention (1; FNF); Wound (1; GHA); Yeast (1; HH3; X8808717).

Mastic for infection:

- Analgesic: myrcene; quercetin
- Anesthetic: linalool; myrcene
- Antibacterial: alpha-pinene; aucubin; cycloartenol; kaempferol; linalool; myrcene; myricetin; oleanolic-acid; quercetin
- Antiedemic: lupeol; oleanolic-acid
- Antiinflammatory: alpha-pinene; aucubin; beta-pinene; cycloartenol; kaempferol; lupeol; myricetin; oleanolic-acid; quercetin
- Antiseptic: beta-pinene; kaempferol; linalool; myricetin; oleanolic-acid
- Antiviral: alpha-pinene; kaempferol; linalool; lupeol; myricetin; oleanolic-acid; quercetin
- Bacteristat: quercetin
- COX-2-Inhibitor: kaempferol; oleanolic-acid; quercetin
- Cyclooxygenase-Inhibitor: kaempferol; oleanolic-acid; quercetin
- Fungicide: linalool; myrcene; quercetin
- Lipoxygenase-Inhibitor: kaempferol; myricetin; quercetin

Mastic for ulcer:

- Analgesic: myrcene; quercetin
- Anesthetic: linalool; myrcene

- Antibacterial: alpha-pinene; aucubin; cycloartenol; kaempferol; linalool; myrcene; myricetin; oleanolic-acid; quercetin
- Antiinflammatory: alpha-pinene; aucubin; beta-pinene; cycloartenol; kaempferol; lupeol; myricetin; oleanolic-acid; quercetin
- Antioxidant: aucubin; kaempferol; lupeol; myrcene; myricetin; oleanolic-acid; quercetin
- Antiseptic: beta-pinene; kaempferol; linalool; myricetin; oleanolic-acid
- Antispasmodic: kaempferol; linalool; myrcene; quercetin
- Antiulcer: kaempferol; oleanolic-acid
- Antiviral: alpha-pinene; kaempferol; linalool; lupeol; myricetin; oleanolic-acid; quercetin
- Bacteristat: quercetin
- COX-2-Inhibitor: kaempferol; oleanolic-acid; quercetin
- Cyclooxygenase-Inhibitor: kaempferol; oleanolic-acid; quercetin
- Fungicide: linalool; myrcene; quercetin
- Lipoxygenase-Inhibitor: kaempferol; myricetin; quercetin

Other Uses (Mastic) — Cultivated primarily for the resin, it is used as a licorice-flavored masticatory and a medicine. Women living in harems use the resin obtained from the bark by incision. They chew it to sweeten their breath and strengthen their gums. Also used to harden gums and alleviate toothache. Eastern children buy this for a chewing gum. Used for filling dental caries. Also used in cosmetics and depilatory creams and perfumes (DEP). Used in the manufacture of confectionary, liqueurs, and varnishes. Mastic also provides a flavor option for the jelly-like candy called "lokum," or Turkish delight, and to flavor masticha liqueur, rahat lokum, puddings, almond paste, cookies, Nabulsi cheese, cakes, and candies (FAC). All over the Middle East, mastic is used to flavor milk puddings. It enriches sweet apricot pudding made of dried apricot "leather." In the Middle East, mastic is as common for ice cream as vanilla in America (AAR). Many Saudi cooks drop a nugget of this spice into meaty stews, soup, or broth (AAR). The Romans used the fruits as an aromatic seasoning (FAC). In Arab countries, a little mastic is mixed with ground dates, butter, and some chopped walnuts for cookies. Mastikha, an alcoholic drink of Chios, similar to ouzo, is flavored with mastic as well as anise. Greeks made a liqueur, mastiche, flavoring the mastic with grape skins. Syrians also make a mastic beverage. A nonalcoholic drink is made by stirring a spoonful of sweet mastic jam into a glass of water. Poorer grades are used for varnish, used for coating metals and paintings, for lithography for retouching negatives, and for microscopy (mounts). Egyptians used mastic as an embalming agent. Sometimes used in incense. Oil of mastic used in cosmetics. The wood and leaves burn green when burned; both the fruit and wood give out a pleasant aroma. In Sardinia, wild boar and other meats are roasted with the wood, smoke of which contributes its own aroma and flavor. The oil expressed from the berries is used by the Arabs for both food and illumination. It is known as shina oil of Cyprus. Leaves, containing 10–12% tannic acid, gathered for dyeing and tanning (FEL). The twigs are used in basketry. (AAR, BIB, FAC, FEL).

For more information on activities, dosages, and contraindications, see the *CRC Handbook of Medicinal Herbs*, ed. 2, Duke et al. 2002.

Cultivation (Mastic) — Rarely cultivated, e.g., in the Canary Islands. It grows in southern Europe, northern Africa, Algeria, and the East. It grows wild along shores of the Adriatic Sea. In Algeria, it forms dense copses along the coast. Bown (2001) suggests cultivating in alkaline, dry, sandy or rocky, sunny soils, zone 9. Seeds are sown in spring at ca. 25°C (77°F), or green cuttings are taken in late spring or early summer, or semiripe cuttings later in summer. Trees are trimmed in spring to keep them small. One major source has been the island of Chios at elevations up to 500 m and annual rainfall 70–75 cm. Only male trees are propagated by cuttings, as the females have an inferior resin. Wild crafters may harvest the naturally exuded resin. Commercial, vertical incisions are made in the trunk, from which the resin exudes. It is collected after it hardens in about 3 weeks (WOI). According to Bown (2001), trees are tapped for 5–6 weeks, by making some 200–300 vertical incisions ca. 2 cm long in a tree. In Chios, harvesting is restricted by law to between July 15 and October 15.

Chemistry (Mastic) — Sixty-nine constituents were identified in the EOs; alpha-pinene, myrcene, trans-caryophyllene, and germacrene D were the major components. The *in vitro* antimicrobial activity of the EOs and resin (total, acid and neutral fraction) against six bacteria and three fungi were reported (PM65:749). Here are a few of the more notable chemicals found in mastic. For a complete listing of the phytochemicals and their activities, see the CRC phytochemical compendium, Duke and duCellier, 1993 (DAD) and the USDA database http://www.ars-grin.gov/duke/.

Alpha-Pinene — Allelochemic; Antibacterial; Anticancer; Antifedant; Antiflu; Antiinflammatory; Antispasmodic; Antiviral; Coleoptiphile; Expectorant; Herbicide IC50 = 30 μ*M*; Insectifuge 50 ppm; Insectiphile; Irritant; p450(2B1)-Inhibitor IC50 = 0.087 μ*M*; Sedative; Spasmogenic; Tranquilizer; Transdermal.

Shikimic-Acid — Analgesic; Anticancer; Antioxidant (7 × quercetin); Antiradicular (7 × quercetin); Antispasmodic; Antitumor; Bruchifuge; Carcinogenic; Ileorelaxant; Mutagenic; LD50 = 1000 ipr mus.

Prunus dulcis (Mill.) D. A. Webb (Rosaceae)
ALMOND, BITTER ALMOND, SWEET ALMOND

Synonyms — *Amygdalus communis* L., *A. dulcis* Mill., *Prunus amygdalus* Batsch, *P. communis* (L.) Arcang., *P. dulcis* var. *amara* (DC). Buchheim.

Medicinal Uses (Almond) — Regarded as alterative, astringent, carminative, cyanogenetic, demulcent, discutient, diuretic, emollient, laxative, lithontryptic, nervine, sedative, stimulant, and tonic. It is no surprise that the seeds and/or oil (containing amygdalin or benzaldehyde) are widely acclaimed as folk cancer remedies, for all sorts of cancers and tumors, calluses, condylomata, and corns (Duke, 1983; Hartwell, 1982). Lebanese use the oil for skin trouble, including white leukoderma-like patches on skin. Throughout the Middle East, the oil is used as an emollient and to alleviate itching. Raw oil from the bitter variety is used for acne. Almond and honey was given for cough. Thin almond paste was added to wheat porridge to pass gravel or stone. Lebanese believe that almonds and/or almond oil restore virility. Iranians make an ointment from bitter almonds for furuncles. Bitter almonds, when eaten in small quantity, sometimes produce nettle rash and, when taken in large quantity, they may cause poisoning. Ayurvedics consider the fruit, the seed, and its oil aphrodisiac, using the oil for biliousness, headache, the seed as a laxative. Unani use the seed for ascites, bronchitis, colic, cough, delirium, earache, gleet, hepatitis, headache, hydrophobia, inflammation, renitis, skin ailments, sore throat, and weak eyes. Burnt almond shells are used as a dentifrice. Unripe fruits are applied as an astringent to gums and mouth. Bitter almond in vinegar is plastered onto neuralgia. The Biblical pair, almonds and figs, are considered laxative and useful at allaying intestinal pain (DEP).

Like many other plants (not just green tea, grape, pinebark, and peanut skins), almonds, especially green almonds, are well endowed with antioxidant OPCs. De Pascual et al. (1998) note that green almond extracts contain two monomers, (+)-catechin and (–)-epicatechin, and 15 oligomeric procyanidins (6 dimers, 7 trimers, and 2 tetramers). And bitter almonds are not the only source of laetrile, or for that matter amygdalin. Over 100 years, George Watt, (DEP, 1892) compared the amygdalin content of bitter almond and peach pits (2.3–2.5%); cherry, laurel leaves 1.38%; apple, cherry, plum seeds, and *Rhamnus frangula*, less than 1%. Laetrile is a semisynthetic derivative of amygdalin. Claims for laetrile were based on three different theories. Theory (1) claimed that cancerous cells contained copious beta-glucosidases, which release HCN from laetrile via hydrolysis. Normal cells were reportedly unaffected, because they contained low concentrations of beta-glucosidases and high concentrations of rhodanese, which converts HCN to the less toxic thiocyanate. Later, however, it was shown that both cancerous and normal cells contain only trace amounts of beta-glucosidases and similar amounts of rhodanese. Also, it was thought that amygdalin was not absorbed intact from the gastrointestinal tract. Theory (2) proposed that, after ingestion, amygdalin was hydrolyzed to mandelonitrile, transported intact to the liver and converted to a beta-glucuronide complex, which was then carried to the cancerous cells, hydrolyzed by beta-glucuronidases to release mandelonitrile and then HCN. This was believed an untenable theory. Theory (3) called laetrile vitamin B-17, suggesting that cancer is a result of B-17 deficiency. It postulated that chronic administration of laetrile would prevent cancer. No evidence was adduced to substantiate this hypothesis. It was even claimed that patients taking laetrile reduced their life expectancy, both through of lack of proper medical care and chronic cyanide poisoning (which might be useful in sickle cell anemia patients). In order to reduce potential risks to the general public, amygdalin was made a prescription-only medicine in 1984 (CAN). The NCI-favored taxol sold for over $1 billion in 2000, but some cancer researchers still study amygdalin and/or laetrile "... laetrile stopped the spread of cancer, the metastases, about 80% of the time" [Kanematsu Suguira, one pioneer of chemotherapy who had worked at Sloane Kettering since 1917; as quoted in Mason (2001)]. Meanwhile, the official position of Sloane Kettering: "Its laetrile studies were negative" (Mason, 2001). I personally straddle the fence; I think laetrile is probably as useful as some of the other plant-derived drugs, like perhaps etoposide and taxol, yet not so toxic. If diagnosed with cancer, I'd be eating almonds (or apple seeds or wild cherries) for their amygdalin, and brazilnuts for their selenium, and maybe some hazelnuts, for homeopathic doses of taxol. Moss (2001), like me, recommends more fruits and vegetables than does the NIH, but he recommends fewer fruits and more veggies for cancer patients "an abundance of sugar—even natural sugars—might have an adverse effect."

Fatalities from cyanide poisoning have been reported for apple, apricot, and bitter almond seeds. And reportedly, feeding 2.5 g almond seed to rabbits resulted in hypoglycemic activity (JAC7:405).

Indications (Almond) — Acne (f; BIB); Adenosis (1; JLH); Ascites (f; BIB); Asthma (1; BIB; FNF); Bacteria (1; FNF); Biliousness (f; BIB); Bronchosis (1; BIB; FNF); Callus (f; BIB; JLH); Cancer (1; BIB; FNF; JLH); Cancer, bladder (1; APA); Cancer, breast (1; APA; JLH); Cancer, colon (1; ABS); Cancer, gland (1; FNF; JLH); Cancer, liver (1; FNF; JLH); Cancer, mouth (1; APA); Cancer, spleen (1; FNF; JLH); Cancer, stomach (1; FNF; JLH); Cancer, uterus (1; FNF; JLH); Cardiopathy (1; APA; FNF); Cold (f; BIB); Colic (f; BIB); Condyloma (f; BIB; JLH); Constipation (1; APA); Corn (f; BIB; JLH); Cough (f; BIB; DEP; PH2); Cramp (1; BIB; FNF); Cystosis (f; BIB; JLH); Delirium (f; BIB); Dermatosis (1; BIB; FNF; PH2; WOI); Diabetes (f; DAA); Dysmenorrhea (f; DEP); Dyspnea (f; BIB); Earache (f; BIB); Enterosis (f; DEP); Fungus (1; FNF); Furuncle (f; BIB); Gallstone (f; BOW); Gingivosis (1; BIB; FNF); Gleet (f; BIB); Gravel (f; BIB); Headache (f; BIB; DEP); Heartburn (f; BIB); Hepatosis (1; BIB; DEP; FNF; JLH); High Cholesterol (1; APA; FNF); Hydrophobia (f; BIB); Immunodepression (1; FNF); Impotence (f; BIB); Induration (f; BIB; JLH); Infection (1; FNF); Inflammation (1; BIB; FNF); Itch (f; BIB; WOI); Kidney Stone (f; BOW); Leukoderma (f; BIB); Nausea (f; PH2); Nephrosis (f; BIB); Nervousness (1; FNF); Neuralgia (f; DEP); Ophthalmia (f; DEP); Pain (f; DEP); Pulmonosis (f; BIB); Respirosis (f; EFS); Sclerosis (f; JLH) Sore (f; BIB; JLH); Sore Throat (f; BIB); Splenosis (f; BIB; DEP; JLH); Staphylococcus (1; MPI); Stomatosis (1; BIB; FNF); Stone (f; BOW); Streptococcus (1; MPI); Swelling (f; JLH); Tumor (1; FNF); Ulcer (1; BIB; FNF); VD (f; BIB); Virus (1; FNF); Vomiting (f; PH2); Water Retention (1; FNF).

Almond for cancer:

- 5-Alpha-Reductase-Inhibitor: alpha-linolenic-acid
- AntiHIV: caffeic-acid; quercetin
- Antiaggregant: alpha-linolenic-acid; caffeic-acid; eugenol; ferulic-acid; kaempferol; phytic-acid; quercetin; salicylates; triolein

- Antiarachidonate: eugenol
- Anticancer: alpha-linolenic-acid; amygdalin; benzaldehyde; caffeic-acid; eugenol; ferulic-acid; folic-acid; geraniol; kaempferol; mucilage; p-coumaric-acid; phytic-acid; quercetin; quercitrin
- Anticarcinogenic: caffeic-acid; ferulic-acid
- Anticervicaldysplasic: folic-acid
- Antiestrogenic: ferulic-acid; quercetin
- Antifibrosarcomic: quercetin
- Antihepatotoxic: caffeic-acid; ferulic-acid; p-coumaric-acid; quercetin; quercitrin
- Antiinflammatory: alpha-linolenic-acid; amygdalin; caffeic-acid; eugenol; ferulic-acid; kaempferol; quercetin; quercitrin; salicylates
- Antileukemic: daucosterol; kaempferol; quercetin
- Antileukotriene: caffeic-acid; quercetin
- Antilipoperoxidant: quercetin
- Antimelanomic: geraniol; quercetin
- Antimetaplastic: folic-acid
- Antimetastatic: alpha-linolenic-acid
- Antimutagenic: caffeic-acid; eugenol; ferulic-acid; kaempferol; quercetin; quercitrin
- Antineoplastic: ferulic-acid
- Antinitrosaminic: caffeic-acid; ferulic-acid; p-coumaric-acid; quercetin
- Antioxidant: caffeic-acid; cyanidin; eugenol; ferulic-acid; gamma-tocopherol; kaempferol; p-coumaric-acid; phytic-acid; quercetin; quercitrin
- Antiperoxidant: caffeic-acid; p-coumaric-acid; quercetin
- Antipolyp: folic-acid
- Antiproliferant: quercetin
- Antiprostaglandin: caffeic-acid; eugenol
- Antithromboxane: eugenol
- Antitumor: benzaldehyde; caffeic-acid; daucosterol; eugenol; ferulic-acid; geraniol; kaempferol; p-coumaric-acid; phytic-acid; quercetin; quercitrin
- Antiviral: caffeic-acid; ferulic-acid; kaempferol; quercetin; quercitrin
- Apoptotic: kaempferol; quercetin
- COX-2-Inhibitor: eugenol; kaempferol; quercetin
- Cyclooxygenase-Inhibitor: kaempferol; quercetin
- Cytoprotective: caffeic-acid
- Cytotoxic: caffeic-acid; eugenol; p-coumaric-acid; quercetin
- Hepatoprotective: caffeic-acid; eugenol; ferulic-acid; quercetin
- Hepatotonic: quercitrin
- Immunostimulant: alpha-linolenic-acid; benzaldehyde; caffeic-acid; ferulic-acid; folic-acid
- Lipoxygenase-Inhibitor: caffeic-acid; kaempferol; p-coumaric-acid; quercetin
- Lymphocytogenic: alpha-linolenic-acid
- Mast-Cell-Stabilizer: quercetin
- Ornithine-Decarboxylase-Inhibitor: caffeic-acid; ferulic-acid; quercetin
- p450-Inducer: quercetin
- PTK-Inhibitor: quercetin
- Prostaglandigenic: caffeic-acid; ferulic-acid; p-coumaric-acid
- Protein-Kinase-C-Inhibitor: quercetin
- Sunscreen: caffeic-acid; ferulic-acid
- Topoisomerase-II-Inhibitor: kaempferol; quercetin
- Tyrosine-Kinase-Inhibitor: quercetin

P

Almond for dermatosis:

- Analgesic: caffeic-acid; eugenol; ferulic-acid; quercetin
- Anesthetic: benzaldehyde; eugenol
- Antibacterial: benzaldehyde; caffeic-acid; eugenol; ferulic-acid; geraniol; kaempferol; p-coumaric-acid; quercetin; quercitrin
- Antidermatitic: quercetin
- Antiinflammatory: alpha-linolenic-acid; amygdalin; caffeic-acid; eugenol; ferulic-acid; kaempferol; quercetin; quercitrin; salicylates
- Antiseptic: benzaldehyde; caffeic-acid; cresol; eugenol; geraniol; kaempferol; oxalic-acid
- COX-2-Inhibitor: eugenol; kaempferol; quercetin
- Cyclooxygenase-Inhibitor: kaempferol; quercetin
- Demulcent: mucilage
- Fungicide: caffeic-acid; eugenol; ferulic-acid; geraniol; p-coumaric-acid; phytic-acid; quercetin

Other Uses (Almond) — The almond was very important, even in Biblical times. It has been inferred that almonds did not grow naturally in Egypt, since Jacob's sons took almonds to Joseph. Nowadays, the almond is widespread in the Holy Land. Almond branches were reportedly used as divining rods to locate hidden treasure. There is the legendary story of Charlemagne's troops' spears (almond) sprouting in the ground overnight and shading the tents the next day. Almonds are one of the earliest trees to flower in Tuscany. They are also one of the first to flower in the Palestinian spring. Because of their association with spring, the almond flower is also associated with life after death, or immortality. Modern English Jews carry branches of flowering almonds into the synagogue on spring festival days. Almond is a sophisticated flavoring, hence spice, sweet (var. *dulcis*) and bitter (var. *amara*). Raw, bitter almonds are not usually available commercially in the U.S. Almond extract provides the bitter flavor without the toxicity. A small amount of extract should always be added to the sweet almonds in a recipe. A hint of almond flavor will liven up whipped cream (one-fourth teaspoon extract per cup unwhipped cream) (AAR). Seeds are a favorite of mine, raw, salted, roasted, or sprouted. They are used in baked goods, cakes, confectionery, and pastry. Seed are ground and blended with water to make almond milk. In the Middle Ages, "almond milk" was a staple. Simmer blanched, chopped, or coarsely ground almonds in water (half a pound in two cups of water) for 10 min. This "milk" will give a recipe the desired almond flavor. Add a teaspoon of almond extract to supply the bitter almond taste (AAR). The milk can be made into almond butter. The tender kernels of young almonds, picked before they mature, are a traditional delicacy in the Middle East. Almond paste is used in amaretto, macaroons, marzipan, etc. Toast almonds by spreading whole nuts on a cookie sheet in a 350°F oven, no more than 8 min; sliced or slivered almonds, only 5 min. Do not let them brown (AAR). The dish "picada," a mix of almonds, garlic and olive oil, and chopped parsley, is important in Catalan cuisine. Edible almond oil, expressed from the seeds, is sweet and nutty, and serves well with salads and vegetables. It is used in flavoring, perfumery, and medicines. Benzaldehyde may be used for almond flavoring, usually being cheaper than almond oil. The fatty oil is used in cosmetic ointments and other natural cosmetics. Almond and olive oil are particularly recommended for people with dry skin; and dry skins are what we expect in Biblical country. It is used in cold creams, nourishing creams, and skin creams (WOI). For aging skin, Aubrey Hampton, author of one of the better books on natural cosmetics, recommends cleaning with olive oil castille soap and then moisturizing with almond oil. Both bitter and sweet almond oil contain 35–55% fixed oils, but the bitter oil may contain 3–40% amygdalin. Aubrey classifies the sweet oil as an excellent emollient for chapped hands and face lotions. Kernels of apricot, cherry, peach, and plum yield EOs almost identical to almond oil. Flowers are the source of a smooth, caramel-colored honey with an excellent nutty flavor. On La Palma, in the Canary Islands, where transatlantic flights often refuel, there's a special goat cheese that is smoked over a fire fueled with almond shells. Shells of almond (and other related *Prunus*) are used in soft grit blasting for machine parts and moulds in various industries. I remember chewing the gum that exudes from the related peach. The gum exuding from almond trees is used as a substitute for tragacanth (BIB, FAC, PH2, WOI).

For more information on activities, dosages, and contraindications, see the *CRC Handbook of Medicinal Herbs*, ed. 2, Duke et al. 2002.

Cultivation (Almond) — In India, trees are raised from seedlings, the seeds usually having a chilling requirement. Seeds are sown in nurseries, the seedlings transplanted after about 1 year. For special types, as in the U.S., scions are budded or grafted on to bitter or sweet almond, apricot, myrobalan, peach, or plum seedlings. Trees are planted 6–8 m apart and irrigated, in spite of their drought tolerance. Application of nitrogenous and/or organic fertilizers is said to improve yield. Trees should be pruned to a modified leader system. All types are self-sterile, so cvs or seedlings should be mixed (NUT, WOI).

Chemistry (Almond) — Here are a few of the more notable chemicals found in almond. For a complete listing of the phytochemicals and their activities, see the CRC phytochemical compendium, Duke and duCellier, 1993 (DAD) and the USDA database http://www.ars-grin.gov/duke/.

Amygdalin — Anticancer; Antiinflammatory; Antispasmodic; Antitussive; Bitter; Cyanogenic; Expectorant; Toxic.

Lysine — Antialkalotic; Antiherpetic 0.5–3 g/day; Essential; Hypoarginanemic 250 mg/kg/; LD50 = 181 ivn mus.

Mufas — Anemiagenic; Anticancer; Antiinflammatory IC50 = 21 μM; Antileukotriene-D4 IC50 = 21 μM; Choleretic 5 ml/man; Dermatitigenic; Hypocholesterolemic; Insectifuge; Irritant; Percutaneostimulant; LD50 = 230 ivn mus; LDlo = 50 ivn cat.

Prunasin — Cyanogenic.

Pufas — Antiacne; Antieczemic; AntiMS; Antipolyneuritic.

P

R

Rhus coriaria L. (Anacardiaceae)
SICILIAN SUMAC, SUMAC, TANNER'S SUMAC

Medicinal Uses (Sicilian Sumac) — Considered astringent and tonic (DEP).

Indications (Sicilian Sumac) — Bacteria (1; FNF); Biliousness (f; DEP); Candida (1; FNF); Cholera (f; DEP); Diarrhea (f; DEP; FNF); Dysentery (f; DEP; FNF); Dyspepsia (f; DEP; FNF); Gastrosis (f; EB49:406); Gingivosis (f; DEP); Hematemesis (f; DEP); Hemoptysis (f; DEP); Hepatosis (1; FNF); Infection (1; FNF); Inflammation (1; FNF); Leucorrhea (f; DEP); Ophthalmia (f; DEP); Sore (f; DEP); Stomachache (f; EB49:406); Ulcer (f; EB49:406; FNF); Virus (1; FNF); Vomiting (f; DEP).

Sicilian Sumac for dysentery:

- Antibacterial: ellagic-acid; gallic-acid; isoquercitrin; myricetin; myricitrin; quercetin; quercitrin; tannic-acid
- Antidiarrheic: tannic-acid
- Antidysenteric: tannic-acid
- Antihepatotoxic: gallic-acid; quercetin; quercitrin
- Antiseptic: avicularin; ellagic-acid; gallic-acid; myricetin; tannic-acid
- Antispasmodic: quercetin; quercitrin
- Antiviral: ellagic-acid; gallic-acid; myricetin; quercetin; quercitrin; tannic-acid
- Astringent: ellagic-acid; gallic-acid; tannic-acid
- Candidicide: myricetin; myricitrin; quercetin
- Hepatoprotective: ellagic-acid; quercetin
- Hepatotonic: quercitrin
- Protisticide: myricitrin

Sicilian Sumac for dyspepsia:

- Analgesic: gallic-acid; quercetin
- Antigastric: myricetin; quercetin
- Antiinflammatory: ellagic-acid; gallic-acid; myricetin; myricitrin; quercetin; quercitrin
- Antioxidant: ellagic-acid; gallic-acid; isoquercitrin; myricetin; quercetin; quercitrin; tannic-acid
- Antiulcer: quercitrin; tannic-acid

Sicilian Sumac for ulcer:

- Analgesic: gallic-acid; quercetin
- Antibacterial: ellagic-acid; gallic-acid; isoquercitrin; myricetin; myricitrin; quercetin; quercitrin; tannic-acid

R

- Antiinflammatory: ellagic-acid; gallic-acid; myricetin; myricitrin; quercetin; quercitrin
- Antioxidant: ellagic-acid; gallic-acid; isoquercitrin; myricetin; quercetin; quercitrin; tannic-acid
- Antiseptic: avicularin; ellagic-acid; gallic-acid; myricetin; tannic-acid
- Antispasmodic: quercetin; quercitrin
- Antiulcer: quercitrin; tannic-acid
- Antiviral: ellagic-acid; gallic-acid; myricetin; quercetin; quercitrin; tannic-acid
- Astringent: ellagic-acid; gallic-acid; tannic-acid
- Bacteristat: gallic-acid; quercetin
- COX-2-Inhibitor: quercetin
- Cyclooxygenase-Inhibitor: gallic-acid; quercetin
- Fungicide: quercetin
- Lipoxygenase-Inhibitor: myricetin; quercetin

Other Uses (Sicilian Sumac) — Fruits, crushed together with *Origanum syriacum*, are a main ingredient of a Middle Eastern spice mixture called "za'tar." In Syria, berries are soaked in milk. Immature fruits may be used as a caper substitute (FAC).

For more information on activities, dosages, and contraindications, see the *CRC Handbook of Medicinal Herbs*, ed. 2, Duke et al. 2002.

Cultivation (Sicilian Sumac) — In Sicily, it is grown as a cultivar with other drought-tolerant trees, like almond, carob, and pistachio. Sometimes, these trees are intercropped with *Opuntia ficus-indica*.

Chemistry (Sicilian Sumac) — Here are a few of the more notable chemicals found in sicilian sumac. For a complete listing of the phytochemicals and their activities, see the CRC phytochemical compendium, Duke and duCellier, 1993 (DAD) and the USDA database http://www.ars-grin.gov/duke/.

Astragalin — Aldose-Reductase-Inhibitor IC30 = 1 μM, IC62 = 10 μM; Antileukemic; Expectorant; Hypotensive; Immunostimulant.

Avicularin — Aldose-Reductase-Inhibitor IC32 = 1 μM, IC84 = 10 μM; Antiseptic; Cyclooxygenase-Activator 100 μM; Diuretic.

Myricetin — Allelochemic IC82 = 1 mM; Antiallergic; Antibacterial MIC = 20–500 μg/ml; Anticancer; Antifeedant IC52 = <1000 ppm diet; Antigastric; Antigingivitic MIC = 20 μg/ml; Antigonadotropic; Antihistaminic; AntiHIV; Antiinflammatory; Antimutagenic ID50 = 2–5 nM; Antioxidant IC99 = 200 ppm; Antioxidant 1.4 μM; Antiperiodontic MIC = 20 μg/ml; Antiplaque MIC = 20 μg/ml; Antiseptic MIC = 20 μg/ml; Antiviral; Apoptotic 60 μM; Candidicide MIC = 150 μg/ml; COMP-Inhibitor; Diuretic; Hypoglycemic; Larvistat IC50 = 2.6–3.5 mM/kg diet; Lipoxygenase-Inhibitor; Metalloproteinase-Inhibitor; Mutagenic; NEP-Inhibitor IC50 = >42 μM; Oxidase-Inhibitor; Quinone-Reductase-Inducer 36 μM; Topoisomerase-I-Inhibitor IC50 = 11.9 μg/ml; Topoisomerase-II-Inhibitor IC50 = 11.9 μg/ml; Tyrosine-Kinase-Inhibitor; Vasodilator.

Myricitrin — Antibacterial MIC = 500 μg/ml; Antiinflammatory; Antimutagenic ID50 = 10–40 nM; Antitumor; Candidicide MIC = 400 μg/ml; Choleretic; Diuretic; Paramecicide; Protisticide; Spermicide.

Quercetin — See also *Alpinia officinarum*.

S

Sassafras albidum (Nutt.) Nees (Lauraceae)
SASSAFRAS

Synonyms — *Laurus albida* Nutt., *Sassafras officinale* Nees & C. H. Eberm., *S. variifolium* (Salisb.) Kuntze.

Medicinal Uses (Sassafras) — Early on, sassafras was also called ague wood, cinnamon wood, and smelling tree. Erichsen-Brown (1989) says that, as early as 1575–1577, "The Spaniards did begin to cure themselves with the water of this tree and it did in them greate effectes, that it is almost incredible, for with the naughtie meates and drinkying of the rawe waters, and slepying in the dewes, the most parts of them came to fall into continual agues." I know of no better source of early American information on the eastern medicinal plants than Erichsen-Brown, in her excellent *Medicinal and Other Uses of North American Plants* (1989). According to David Winston, part Cherokee himself, Cherokee traditionally used sassafras as a carminative and eyewash, for gout, rheumatism, and skin problems (Winston, 2001). Tea, made from the oil or root bark was applied to cancer, corns, osteosarcomas, tumors, and wens. Sassafras tea was used in Appalachia as a diaphoretic and diuretic for bronchitis, gastritis, and indigestion, and to slow down the milk of nursing mothers. South Carolina blacks gave it to children to "bring out the measles." Pith of sassafras was once official in the U.S. as a mucilaginous demulcent, used for eye inflammation. The herb is alterative, analgesic, antiseptic, aromatic, carminative, depurative, diaphoretic, demulcent, diuretic, emmenagogue, and stimulant. Externally, sassafras has been used as a rubefacient on bruises, rheumatism, sprains, and swellings. Sassafras oil is applied externally as a pediculicide. Once used in dentistry to disinfect root canals, sassafras oil is also used to relieve stings and bites. Tucker et al. (1994) dredge up some interesting indications that sent me scurrying to Webster's. "It healeth opilations. It comforteth the liver and stomach and doth disopilate" (not there in Webster's, defined as Chagas Disease in my *Dorland's Medical Dictionary*; but the English Spelling with a double p, oppilation mere constipation). Sassafras extracts show very good activity against *Ancylostoma* and *Strongyloides*.

There is one interesting cancer "cure," that echos some of the cancer-preventative suggestions recently emanating from Congress and the NIH… "Let him drink Sassafras Tea every Morning, live temperately, upon light and innocent Food, and abstain entirely from strong liquor. The Way to prevent this Calamity, is, to be very sparing in eating Pork, to forbear all salt, and high season'd Meats, and live chiefly upon the Garden, The Orchard, and the Hen-House" (Cancer in Virginia, 1734, as quoted by Lewis and Elvin-Lewis, 1977). After quoting the "Spring Ode" by Donald Robert Perry Marquis, "Fill me with sassafras, nurse, and juniper juice! Let me see if I'm still any use!" Tyler (1994) waxed pessimistic on sassafras, "As a result of research conducted in the early 1960s, safrole was recognized as a carcinogenic agent in rats and mice." The major flavoring constituent in sassafras root bark, safrole, is identified as a "low grade hepatocarcinogen." No one really knows just how harmful it is to human beings, but it has been estimated that one cup of strong sassafras tea could contain as much as 200 mg of safrole, more than four times the minimal amount believed hazardous to man if consumed on a regular basis." Based on Bruce Ames', Herp Index (Ames et al., 1987), I calculated that a can of old fashioned root beer with its safrole, was about $^1/_{13}$th as carcinogenic as a can of brew, with its ethanol. It was banned in root beer, and the FDA, in 1976, banned interstate marketing of sassafras for sassafras tea. But remember, the oral LD50 for safrole in rats is 1950 mg/kg body weight, with major symptoms including ataxia, depression, and diarrhea, death occurring in 4–5 days. The oral LD50 for caffeine is less than 10% that of safrole. Still, I fear that safrole may be more dangerous than caffeine (Duke, 1985). Sassafras oil's LD50 is about the same as safrole, 1520–2370 mg/kg orally in rat, ten times less toxic than caffeine, at least under these circumstances.

My hundred-year-old grandmother didn't know that sassafras was carcinogenic, but she was moderate in all things, even sassafras tea, and could have founded the WCTU. I never knew her husband, my maternal grandfather, but my cousin said Grandpa Truss took two highballs before dinner and then sat down with the family around a stern family dinner. Ironically, sassafras is said to be antagonistic to the narcotic effects of alcohol.

> Living for a century,
> Gramma sipped her sassafras tea,
> Perhaps she wouldn't've grown so old,
> Had she been told 'bout old safrole,
> CarcinogeniciTEA!

Indications (Sassafras) — Acne (f; APA; CRC); Ague (f; CEB; DEM); Allergy (1; FNF); Anorexia (f; DEM); Arthrosis (f; FAD; SPI); Bacteria (1; FNF); Bite (f; BOW); Bronchosis (1; APA; CRC; FAD; FNF); Bruise (f; CRC; DEM; FEL); Burn (f; DEM); Cancer (f; CRC); Cancer, bone (f; JLH); Cataract (f; DEM); Catarrh (f; CRC; PHR); Childbirth (f; DEM; FEL); Cold (1; DEM; FAD; FNF); Congestion (1; FNF); Conjunctivosis (f; CRC); Constipation (f; DEM; SPI); Cough (1; DEM; FNF); Cramp (1; FNF); Cystosis (f; DEM; FEL); Dermatosis (f; APA; CRC; FAD; PH2); Diarrhea (f; DEM); Dropsy (f; CRC); Dysentery (f; CRC); Dysmenorrhea (f; CRC; FEL); Dyspepsia (f; DEM); Dysuria (f; DEM); Enterosis (f; FAD; FEL); Fever (1; CAN; DEM; FAD; FNF); Flu (1; APA; FNF); Fungus (1; FNF); Gallstone (f; DEM); Gangrene (f; FEL); Gastrosis (f; CRC; SPI); Gleet (f; CRC; FEL); Gonorrhea (f; CRC; FEL); Gout (f; APA; FAD; HH2); Gravel (f; SPI); Heart (f; DEM); Hepatosis (f; CRC; FAD; SPI); High Blood Pressure (f; APA; CRC; FAD); Impotence (f; DEM); Infection (1;

FNF); Infertility (f; CEB; SPI); Inflammation (1; CRC; FNF; PH2); Lice (1; BOW); Malaria (f; CEB); Mastosis (f; APA); Measles (f; APA; CRC; DEM); Mucososis (f; PH2); Mycosis (1; FNF); Nausea (f; DEM); Nephrosis (f; CRC; FAD; FEL); Obesity (f; DEM); Ophthalmia (f; CRC; DEM; FAD; FEL); Osteosarcoma (f; JLH); Pain (1; APA; CAN; DEM; FNF); Parotosis (f; CRC); Pneumonia (f; CRC); Poison Ivy (f; APA; FEL); Puerperium (f; APA); Pulmonosis (f; FAD); Rash (f; DEM); Respirosis (f; CRC; HH2); Rheumatism (1; APA; CAN; FAD; FEL; FNF; HH2; PH2); Scarlet Fever (f; DEM); Scrofula (f; FEL); Sore (f; DEM); Sore Throat (f; DEM); Sprain (f; CRC; FEL); Stomachache (f; DEM; FAD); Stone (f; SPI); Swelling (f; CRC; DEM; FEL); Syphilis (f; APA; CRC; FEL; PHR; PH2); Tapeworm (f; DEM); Typhus (f; CEB; CRC); UTI (f; PHR; PH2); VD (f; CRC; PH2); Virus (1; FNF); Water Retention (1; FNF); Worm (f; DEM); Wound (f; DEM).

Sassafras for cold/flu:

- Analgesic: camphor; eugenol; myrcene; p-cymene; reticuline
- Anesthetic: 1,8-cineole; camphor; eugenol; linalool; myrcene; myristicin; safrole
- Antiallergic: 1,8-cineole; citral; linalool; terpinen-4-ol
- Antibacterial: 1,8-cineole; alpha-pinene; anethole; caryophyllene; citral; delta-cadinene; eugenol; limonene; linalool; myrcene; p-cymene; reticuline; safrole; silver; terpinen-4-ol; thujone
- Antibronchitic: 1,8-cineole
- Antiflu: alpha-pinene; limonene; p-cymene
- Antihistaminic: citral; elemicin; linalool
- Antiinflammatory: alpha-pinene; beta-pinene; boldine; caryophyllene; eugenol; myristicin
- Antioxidant: boldine; camphene; eugenol; gamma-terpinene; myrcene; myristicin
- Antipharyngitic: 1,8-cineole
- Antipyretic: apiole; asarone; boldine; eugenol
- Antiseptic: 1,8-cineole; anethole; aromadendrene; beta-pinene; camphor; citral; eugenol; limonene; linalool; safrole; terpinen-4-ol; thujone
- Antistress: elemicin; myristicin
- Antitussive: 1,8-cineole; terpinen-4-ol
- Antiviral: alpha-pinene; limonene; linalool; p-cymene
- Bronchorelaxant: citral; linalool
- COX-2-Inhibitor: eugenol
- Decongestant: camphor
- Expectorant: 1,8-cineole; alpha-pinene; anethole; beta-phellandrene; camphene; camphor; citral; limonene; linalool
- Immunostimulant: anethole

Sassafras for rheumatism:

- Analgesic: camphor; eugenol; myrcene; p-cymene; reticuline
- Anesthetic: 1,8-cineole; camphor; eugenol; linalool; myrcene; myristicin; safrole
- Antiedemic: boldine; caryophyllene; eugenol
- Antiinflammatory: alpha-pinene; beta-pinene; boldine; caryophyllene; eugenol; myristicin
- Antiprostaglandin: eugenol
- Antirheumatalgic: p-cymene
- Antispasmodic: 1,8-cineole; anethole; apiole; asarone; camphor; caryophyllene; eugenol; limonene; linalool; myrcene; myristicin; reticuline; thujone
- COX-2-Inhibitor: eugenol
- Counterirritant: 1,8-cineole; camphor; thujone
- Myorelaxant: 1,8-cineole; asarone

Other Uses (Sassafras) — Colonists used sassafras sticks as henroosts to repel lice, for bedsteads to keep away bedbugs, long lasting dugouts, ox-yokes, barrels, and fence posts. Tucker et al. (1994) recall the story of some early Roanoke Virginians who, having traveled out of reach of their cache of food, lived first off a soup of sassafras and god, finally simply on sassafras soup. Following his visit to Roanoke in 1586, Sir Frances Drake took home a load of sassafras, the "wondrous root which kept the starving alive and in faire good spirit" (SPI). Once used to make "Godfrey's Cordial," a mixture of opium and sassafras. The oil is also used to flavor dentifrices, masticatories, mouthwashes, soaps, candies, root beers, and "sarsaparillas," as well as tobaccos. Twigs used for cleaning the teeth (LIL). A condiment is prepared by boiling the dried root bark with sugar and water until it forms a thick paste (FAC). Roots are added to maple sap, or sweetened with sugar, and brewed into a pleasant tea. Root bark tea, mixed with milk and sugar, was once called "saloop." I greatly enjoy a tea made from sassafras roots and sumac berries all winter long in Maryland. Strong tea can be made into jelly, especially if mucilaginous leaves are extracted as well, but that's for summer time. Two or three leaves in a glass of water yield a mucilaginous beverage. Young leaves are used in salads, or dried and powdered to form file powder, used in Creole cooking for thickening soups, stews, chowders, and gravies. My "Filé Jumbo" has sumac berries, basil leaves, wild ginger, with leaves of sassafras; my "Safrole No-No" has the roots of sassafras, with the safrole-containing bark intact. South Carolina blacks make a soup from young sassafras leaves with *Viola palmata* and *V. septemloba*. Young leaf buds are also eaten. Flowers were used in teas or brewed into beers. Root beer classically contained sassafras root bark (SPI). Safrole-free extracts GRAS at 10–290 ppm; leaves GRAS at 30,000 ppm. Sassafras wood is combined with hickory wood to smoke another of my favorite sins, southern dry-cured, country hams. The bark extract dyes wool orange (Duke, 1985, FAC, LIL).

For more information on activities, dosages, and contraindications, see the *CRC Handbook of Medicinal Herbs*, ed. 2, Duke et al. 2002.

Cultivation (Sassafras) — I suppose there are studies on how to cultivate sassafras, but where I come from the birds are much better at planting the seed of this perennial tree than I am, dropping the seeds, well fertilized, below their fence row perches. Once, on a cold winter day, I hastily uprooted a tree over 2 in thick at ground level, took it home, cut off several big pieces of root for my root booster tea, and then planted the remains. That little sapling came back the following year, better luck than I've had with the small yearlings I have replanted on occasion. We've not been real lucky with the related spicebush, *Lindera benzoin*. Hardy zone 4–8 (Bown, 2001). Tucker and DeBaggio's (2000) say sassafras is hardy to zone 5, doing best in full sun, in moist but not constantly wet, well-drained, rich organic soils. Bown (2001) suggests neutral to acid soil, in sun or shade. They recommend starting from seed (which may take 2 years to germinate), or transplanting seedlings before they develop their taproots (TAD). I have had good luck with root suckers and stump sprouts. As with sarsaparilla, I feel a good stand could easily yield 1–2 tons/a root.

Chemistry (Sassafras) — Here are a few of the more notable chemicals found in sassafras. For a complete listing of the phytochemicals and their activities, see the CRC phytochemical compendium, Duke and duCellier, 1993 (DAD) and the USDA database http://www.ars-grin.gov/duke/.

Anethole — See also *Osmorhiza* spp.

Asarone — Anticonvulsant; Antielleptic; Antipyretic; Antispasmodic; Cardiodepressant; CNS-Depressant; Emetic; Fungicide; Hypothalmic-Depressant; Mutagenic; Myorelaxant; Psychoactive; Sedative; Tranquilizer; LD50 = 275 ipr gpg; LD50 = 310 ipr mus; LD50 = 417 orl mus.

Safrole — Anesthetic; Antiaggregant IC50 = 110 µ*M*; Antibacterial; Anticancer; Anticonvulsant; Antipyretic; Antiseptic; Carcinogenic; Carminative; CNS-Depressant; CNS-Stimulant; Calcium-Antagonist IC50 = 58 µ*M*; Controlled; Cytochrome-P-450-Inducer; Cytochrome-P-488-Inducer; DNA-Binder; Hepatocarcinogen; Hepatoregenerative; Hepatotoxic; Mutagenic; Nematicide MLC

= 1 mg/ml; Neurotoxic; Pediculicide; Psychoactive; Tremorigenic; LD50 = 1950 orl rat; LD50 = 2350 orl mus; LD50 = 3400 orl mus.

Schinus terebinthifolius Raddi (Anacardiaceae)
BRAZILIAN PEPPER TREE, CHRISTMASBERRY TREE, FLORIDA HOLLY

Medicinal (Brazilian Pepper Tree) — Reported to be antibacterial, antiseptic, antiviral, aphrodisiac, astringent, stimulant, and tonic. The balsam, bark, fruits, and leaves are used in folk remedies for tumors, especially of the foot, in Brazil. Brazilians poultice the dried leaves on ulcers. The leaf and fruit, both said to possess antibiotic activity, have been used to bathe sores and wounds. The bark decoction is added to the bath water of rheumatic and sciatic patients. Macerated root juice is applied to contusions and/or ganglionic tumors (CRC).

Fruits "can wreak havoc on the human digestive system, with such after effects as vomiting, diarrhea, and hemorrhoids…extremely upset stomach…violent headaches, swollen eyelids, and shortness of breath. Birds are said to become intoxicated when they eat the 'peppers' and fish die in ponds bordered by the Brazilian pepper plant" (CRC). Horses may develop fatal colic upon ingesting the berries (CRC).

Indications (Brazilian Pepper Tree) — Adenosis (f; CRC); Arthrosis (f; CRC); Atony (f; CRC); Bacteria (1; CRC; FNF); Bronchosis (1; CRC; FNF); Bruise (f; CRC); Cancer (f; FNF); Chill (f; CRC); Dermatosis (f; CRC); Diarrhea (f; CRC); Enterosis (f; CRC); Frigidity (f; CRC); Ganglion (f; CRC); Gout (f; CRC); Hemoptysis (f; CRC); Impotence (f; CRC); Infection (1; CRC; FNF; WOI); Inflammation (1; FNF); Pain (f; CRC); Rheumatism (f; CRC; WOI); Sciatica (f; CRC); Sore (f; CRC; HH2); Swelling (f; CRC); Syphilis (f; CRC; WOI); Tendinitis (f; CRC); Tumor (f; CRC); Ulcer (1; CRC; FNF); Virus (f; CRC; FNF); Wound (f; CRC; HH2).

Brazilian Pepper Tree for cancer:

- AntiHIV: myricetin; quercetin
- Antiaggregant: kaempferol; quercetin
- Anticancer: kaempferol; myricetin; quercetin
- Antiestrogenic: quercetin
- Antifibrosarcomic: quercetin
- Antihepatotoxic: quercetin
- Antiinflammatory: kaempferol; myricetin; quercetin
- Antileukemic: kaempferol; quercetin
- Antileukotriene: quercetin
- Antilipoperoxidant: quercetin
- Antimelanomic: quercetin
- Antimutagenic: kaempferol; myricetin; quercetin
- Antinitrosaminic: quercetin
- Antioxidant: kaempferol; myricetin; quercetin
- Antiperoxidant: quercetin
- Antiproliferant: quercetin
- Antitumor: kaempferol; quercetin
- Antiviral: kaempferol; myricetin; quercetin
- Apoptotic: kaempferol; myricetin; quercetin
- COX-2-Inhibitor: kaempferol; quercetin
- Cyclooxygenase-Inhibitor: cardol; kaempferol; quercetin
- Cytotoxic: quercetin
- Hepatoprotective: quercetin

S

- Lipoxygenase-Inhibitor: kaempferol; myricetin; quercetin
- Mast-Cell-Stabilizer: quercetin
- Ornithine-Decarboxylase-Inhibitor: quercetin
- p450-Inducer: quercetin
- PTK-Inhibitor: quercetin
- Protein-Kinase-C-Inhibitor: quercetin
- Topoisomerase-II-Inhibitor: kaempferol; myricetin; quercetin
- Tyrosine-Kinase-Inhibitor: myricetin; quercetin

Brazilian Pepper Tree for infection:

- Analgesic: quercetin
- Antibacterial: kaempferol; myricetin; quercetin
- Antiinflammatory: kaempferol; myricetin; quercetin
- Antiseptic: kaempferol; myricetin
- Antiviral: kaempferol; myricetin; quercetin
- Bacteristat: quercetin
- COX-2-Inhibitor: kaempferol; quercetin
- Cyclooxygenase-Inhibitor: cardol; kaempferol; quercetin
- Fungicide: quercetin
- Lipoxygenase-Inhibitor: kaempferol; myricetin; quercetin

Other Uses (Brazilian Pepper Tree) — According to the *Wall Street Journal*, much recent material sold here at more than $125/kilo as "pink peppercorns" or "red peppercorns" is, in fact, nothing more than fruits of the Brazilian pepper plant. This tree was introduced to Florida as an ornamental, but now many Floridians regret the introduction. Said to be a good honey plant, making the plant more appealing to beekeepers than to those who are allergic to the plant. Reports on the marginal pulping potential are included in an interesting account by Morton (1977). Goats graze the foliage with impunity, but cattle may develop enteritis. Stem is source of a resin called "Balsamo de Missiones" (CRC).

Chemistry (Brazilian Pepper Tree) — Here are a few of the more notable chemicals found in Brazilian pepper tree. For a complete listing of the phytochemicals and their activities, see the CRC phytochemical compendium, Duke and duCellier, 1993 (DAD) and the USDA database http://www.ars-grin.gov/duke/.

Alpha-Pinene — See also *Amomum compactum*.

Amentoflavone — ACE-Inhibitor IC60–90 = 333 µg/ml; Aldose-Reductase-Inhibitor IC25 = 10 µ*M*; Anesthetic; Antibradykinic; Antiedemic >¾ indomethacin; AntiHIV IC97 = 200 µg/ml, IC50 = 119 µg/ml; Antiinflammatory >¾ indomethacin; Antileukemic IC50 = 10 µ*M*; Antioxidant; Antiperoxidant IC50 = 38 µ*M*; Antiulcer; Antiviral IC50 = 10 µ*M*; cAMP-Phosphodiesterase-Inhibitor (5–10 × papaverine); Fungicide; Phosphodiesterase-Inhibitor.

Beta-Pinene — See also *Amomum compactum*.

Cardol — Antifilarial; Cyclooxygenase-Inhibitor IC28–98 = 10 µ*M*; Molluscicide ED = 7–15 ppm; Toxic; Tyrosinase-Inhibitor.

Carene — Antibacterial; Antiseptic; Fungicide; Irritant; LD50 = 4800 mg/kg orl rat.

Delta-3-Carene — Antibacterial; Antiinflammatory ipr; Antiseptic; Dermatitigenic; Insectifuge; Irritant.

Terpinolene — Allelochemic; Antifeedant; Antinitrosaminic; Deodorant; Fungicide; LD50 = 4390 mg/kg orl rat.

Sesamum indicum L. (Pedaliaceae)
BENI, BENNESEED, SESAME

Synonyms — *S. mulayanum* N. C. Nair, *S. orientale* L.

Medicinal Uses (Sesame) — More spice than medicine in my book, still it has some strong medicinal folk lore and is generally viewed as a health food. Medicinally, seeds are emollient, diuretic, tonic, lactogenic, and useful in the treatment of piles. Seed decoction is used for emmenagogue and as a poultice applied to burns and ulcers. Combined with linseed, the seed decoction is used as an aphrodisiac. Seeds and oil are demulcent and used in dysentery and urinary complaints with other medicines. Leaf infusion used as demulcent in southern U.S. Leaves contain a gummy matter and, when submerged in water, form a mucilage-like substance also used in the treatment of diarrhea and dysentery. The leaf infusion is also said to promote hair growth. In India, the leaves are used for bladder, eye, kidney, and skin complaints. Oil is allowed as substitute for olive oil in pharmaceuticals but is purgative in large doses. The seed oil is a folk remedy for cacoethes, abdominal tumors, and indurated tumors. The oil of the whole plant beaten with boiled egg is said to help tumors of the eye. A cataplasm of the seed is said to help indurated tumors. The leaf, used in a poultice or formentation, is said to remedy painful tumors (CFR). Rinzler (1990) notes that oil of sesame seed, like many seed oils, is a good source of tocopherol, which blocks the formation of carcinogenic nitrosamines in lab animals, protects the lining of the lungs from some air pollutants, and may deter oxidative deterioration of cells. She denies any human studies demonstrating such activities in humans, or that vitamin E alleviates menopausal hot flashes, vaginal dryness, or infertility, or that it improves male sexual performance (RIN). More doctors take vitamin E than prescribe it. The health food people might prefer the vegetable oils with their tocopherols, maybe even the palm oils with their tocotrienols. Dietary sesame seeds elevate the concentrations of both tocopherols and tocotrienols in adipose tissue and skin, but not in plasma or other tissues (X11694614).

Weed (2002) suggests heavy consumption of certain spices when estrogen levels are down. Seeds like caraway, celery, coriander, cumin, poppy and sesame, mustard and anise, fennel and fenugreek all contain phytoestrogens, as do their oils, says Weed. She suggests using these seeds "lavishly" when cooking or making tea with any one of them, drinking 3–4 cups a day "for best results" (Weed, 2002).

Indications (Sesame) — Alopecia (f; DAA; JFM); Amenorrhea (f; FEL; KAP; WOI); Aneuria (f; DAA); Arthrosis (1; FNF; JFM; KAB); Asthma (f; KAB); Bacteria (1; FNF); Bleeding (f; KAB);

S

Boil (f; BOW); Bronchosis (1; FNF; JFM); Burn (1; FNF; KAB; WOI); Cachexia (f; DAA); Cacoethes (f; JLH); Cancer (1; FNF; JLH); Cancer, abdomen (f; JLH); Cancer, breast (f; JLH); Cancer, colon (f; JLH); Cancer, eye (f; JLH); Cancer, stomach (f; JLH); Caries (f; BOW); Catarrh (f; FEL; JFM; KAP); Cholera (f; KAP); Cold (1; DAA; FNF; JFM); Condylomata (f; DAA); Conjunctivosis (f; JFM); Constipation (f; DAA; KAB; PH2); Cough (f; KAB; WOI); Cramp (1; FNF); Cystosis (f; FEL; KAP); Dermatosis (1; FEL; FNF; JFM; PH2); Diabetes (f; BOW); Diarrhea (1; FEL; FNF; JFM); Dizziness (f; BOW); Dyschezia (f; PH2); Dysentery (f; FEL; KAB; SKJ); Dysmenorrhea (f; DAA; KAP; WOI); Dysuria (f; KAB; SKJ); Edema (1; FNF; JFM); Enterosis (f; JLH; KAP); Fungus (1; FNF); Gastrosis (f; JLH); Gout (f; KAB); Gray Hair (f; DAA; KAB); Headache (f; BOW); Hemorrhoid (f; DAA; KAB; SKJ; WOI); Hepatosis (1; BOW; FNF); HIV (1; FNF); High Blood Pressure (f; DAA); Immunodepression (1; FNF); Impotence (f; DAA); Induration (f; JLH); Infection (1; FNF); Inflammation (1; FNF; JFM; KAB); Laxative (f; JFM); Malaria (f; KAB); Mastosis (f; JLH); Menorrhagia (f; KAB); Migraine (f; KAB); Nephrosis (f; FEL); Neurosis (f; DAA); Neuroparalysis (f; DAA); Ophthalmia (f; FEL; JLH); Osteoporosis (f; BOW); Otorrhea (f; DAA); Pain (1; FNF); Proctorrhagia (f; WOI); Pulmonosis (f; KAB); Respirosis (f; KAB); Rheumatosis (1; FNF; JFM; PH2); Scab (f; PH2); Scabies (f; KAB); Smallpox (f; KAB); Snakebite (f; KAB); Sore (f; DAA; SKJ; WOI); Sore Throat (f; KAB); Splenosis (f; KAB); Strangury (f; KAB; KAP); Swelling (f; PH2); Syphilis (f; KAB); Tinnitus (f; BOW); Tumor (1; FNF); Urethrosis (f; FEL); Uterosis (f; DAA); Uterorrhagia (f; JFM); VD (f; KAB); Vertigo (f; DAA; KAB); Virus (1; FNF); Wart (1; DAA; JLH; JAC7:405).

Sesame for cancer:

- 5-Alpha-Reductase-Inhibitor: alpha-linolenic-acid
- AntiEBV: chlorogenic-acid
- AntiHIV: caffeic-acid; chlorogenic-acid; tannic-acid
- Antiaggregant: alpha-linolenic-acid; caffeic-acid; ferulic-acid; pyridoxine; salicylates
- Anticancer: alpha-linolenic-acid; caffeic-acid; chlorogenic-acid; ferulic-acid; folic-acid; p-coumaric-acid; phenol; squalene; trans-ferulic-acid; vanillic-acid
- Anticarcinogenic: caffeic-acid; chlorogenic-acid; ferulic-acid
- Anticervicaldysplasic: folic-acid
- Antiestrogenic: ferulic-acid
- Antihepatotoxic: caffeic-acid; chlorogenic-acid; ferulic-acid; p-coumaric-acid; protocatechuic-acid; verbascoside
- Antiinflammatory: alpha-amyrin; alpha-linolenic-acid; caffeic-acid; chlorogenic-acid; cycloartenol; ferulic-acid; gentisic-acid; protocatechuic-acid; salicylates; vanillic-acid; verbascoside
- Antileukemic: asarinin; sesamin; verbascoside
- Antileukotriene: caffeic-acid; chlorogenic-acid
- Antimetaplastic: folic-acid
- Antimetastatic: alpha-linolenic-acid
- Antimutagenic: caffeic-acid; chlorogenic-acid; ferulic-acid; protocatechuic-acid; saponins; sesaminol; tannic-acid
- Antineoplastic: ferulic-acid
- Antinitrosaminic: caffeic-acid; chlorogenic-acid; ferulic-acid; p-coumaric-acid; tannic-acid
- Antioxidant: caffeic-acid; chlorogenic-acid; ferulic-acid; gamma-tocopherol; p-coumaric-acid; phenol; pinoresinol; protocatechuic-acid; sesamin; sesaminol; sesamol; sesamolin; sesamolinol; squalene; tannic-acid; trans-ferulic-acid; vanillic-acid; verbascoside
- Antiperoxidant: caffeic-acid; chlorogenic-acid; p-coumaric-acid; protocatechuic-acid
- Antipolyp: folic-acid

S

- Antiprolactin: pyridoxine
- Antiprostaglandin: caffeic-acid
- Antiradiation: pyridoxine
- Antitumor: alpha-amyrin; caffeic-acid; chlorogenic-acid; ferulic-acid; nepetin; p-coumaric-acid; squalene; vanillic-acid; verbascoside
- Antiviral: caffeic-acid; chlorogenic-acid; ferulic-acid; gentisic-acid; phenol; protocatechuic-acid; tannic-acid
- Chemopreventive: squalene
- Cytoprotective: caffeic-acid
- Cytotoxic: alpha-amyrin; caffeic-acid; p-coumaric-acid; tannic-acid; verbascoside
- Hepatoprotective: caffeic-acid; chlorogenic-acid; ferulic-acid; sesamin
- Immunostimulant: alpha-linolenic-acid; caffeic-acid; chlorogenic-acid; ferulic-acid; folic-acid; protocatechuic-acid; squalene; tannic-acid
- Interferonogenic: chlorogenic-acid
- Lipoxygenase-Inhibitor: caffeic-acid; chlorogenic-acid; p-coumaric-acid; squalene; verbascoside
- Lymphocytogenic: alpha-linolenic-acid
- Ornithine-Decarboxylase-Inhibitor: caffeic-acid; chlorogenic-acid; ferulic-acid
- PKC-Inhibitor: verbascoside
- Prostaglandigenic: caffeic-acid; ferulic-acid; p-coumaric-acid; protocatechuic-acid
- Sunscreen: caffeic-acid; chlorogenic-acid; ferulic-acid; paba; squalene
- Tocopherol-Synergist: sesamin

Sesame for dermatosis:

- Analgesic: caffeic-acid; chlorogenic-acid; ferulic-acid; gentisic-acid; phenol; pyridoxine; verbascoside
- Anesthetic: guaiacol; phenol
- Antibacterial: asarinin; caffeic-acid; chlorogenic-acid; cycloartenol; cycloeucalenol; ferulic-acid; gentisic-acid; guaiacol; hexenal; o-coumaric-acid; p-coumaric-acid; paba; phenol; protocatechuic-acid; sesamin; squalene; tannic-acid; vanillic-acid; verbascoside
- Antidermatitic: biotin; guaiacol; pyridoxine
- Antiinflammatory: alpha-amyrin; alpha-linolenic-acid; caffeic-acid; chlorogenic-acid; cycloartenol; ferulic-acid; gentisic-acid; protocatechuic-acid; salicylates; vanillic-acid; verbascoside
- Antiseptic: caffeic-acid; chlorogenic-acid; guaiacol; hexanal; hexenal; oxalic-acid; phenol; tannic-acid; verbascoside
- Astringent: tannic-acid
- Fungicide: caffeic-acid; chlorogenic-acid; ferulic-acid; hexenal; o-coumaric-acid; p-coumaric-acid; phenol; protocatechuic-acid; verbascoside
- Immunomodulator: saponins

Other Uses (Sesame) — Sesame seed are eaten dry, toasted, sprinkled on breads, cakes, cookies, crackers, hamburger buns, fermented into tempeh, made into nut milks and seed yogurt, or used for halava and other confections. Crushed seeds, hulled or unhulled, constitute sesame paste, sesame butter, and tahini. They serve to a lesser extent in candy, confections, and in making a refreshing beverage, called "horchata." Benne Wafers, crisp cookies sprinkled with sesame seeds, have been popular in the Low Country of South Carolina since Colonial times. Roasted seeds employed as a flavoring in oriental cuisine. In West Africa and the Southern U.S., sesame seed meal is used for thickening soups and stews. Gomashio, or sesame salt, is a seasoning made of the roasted, ground seeds, mixed with salt. Sesame sprouts are eaten in salads. The leaves are also eaten raw in salads,

S

cooked as a green, or in beverages and soups. Sesame butter, or tahini, is rather reminiscent of peanut butter. Rinzler tells us how to make a half cup tahini: grind 2 tbsp in blender until smooth, adding 1/2 tsp sesame oil, 1/4 tsp salt, slowly pour in 1/4 cup tepid water. And Tucker and DeBaggio (2000) remind us that sesame oil was anciently expressed in Urartu (now Armenia) as early as 3900 B.P. (TAD). Seeds, which contain 45–55% oil and 24–29% protein, are sources of Bene, Teel, or Gingili Oil, which is edible, fixed, and semi-drying, used in cooking, shortening, margarine, as an illuminating oil, as a vehicle in pharmaceuticals, insecticides, soaps, paints, lubricants, cosmetics, and as a synergist for pyrethrum. Oil is sometimes mixed with olive oil. Press cake, containing about 20% oil, is an excellent feed for cattle and poultry. The presscake is sometimes made into confections or fermented into dageh and sigda. Cake and meal high in vitamin B, calcium, and phosphorus. In Sudan, mineral-rich filtered ashes from burnt stalks are added to stews. Sesamol has been used as an antioxidant in lard and vegetable oils (CFR). Though sesame is GRAS, some people may be anaphylactically allergic to sesame (CFR, FAC, RIN, TAD).

For more information on activities, dosages, and contraindications, see the *CRC Handbook of Medicinal Herbs*, ed. 2, Duke et al. 2002.

Cultivation (Sesame) — Sesame seed require no rest period before germination. Only certified pure seeds should be used, and then they should be treated against rot (according to inorganic agriculturists) with 0.05 kg "Orthocide 75" per 80 kg seed before planting. There are ca. 360 seed/g (10,200/oz). Sesame in the home garden seem to do best in circumneutral, well-drained garden loams, moist but not constantly wet (TAD). Bown (2001) recommends sowing in spring at 18–24°C (65–75°F). Minimum temps 15°C (59°F). Time for sowing seed varies with locality: Mexico, March–August; Venezuela, August–November; Arizona and Texas, when all frost is past; India, May–July; and Africa, at beginning of rains. Plant when soil temperature is about 24°C, in moist, weed-free, carefully worked seedbed. Sesame has poor ability to compete with weeds. Plants are killed by frost (TAD). Rate to sow seed in the U.S. is 1–3 kg/ha; in Kenya, 8–10 kg/ha. Seed may be sown broadcast or in rows 50–75 cm apart. When planted with regular row crop planting equipment, rows should be 45–75 cm apart, seeds 2.5–5 cm deep, and 75 cm apart in the rows. Germination requires 6–10 days. Growth is slow, so adequate shallow cultivation in early stages is necessary for weed control. Erect branched types require wider row spacing than dwarf, single-stem types. Sesame responds well to irrigation but tolerates drought well; one or two waterings may be desirable during the growing season. Irrigation raises yield in arid and desert areas; however, dry periods at germination and fruit formation may be injurious. Water containing high salt concentration is undesirable and may kill plants. Seed yield is low on poor soils, therefore application of 5:10:5 NPK fertilizer at planting time, at rate of 1.5–2 kg/ha, is advised. In addition, 36–50 kg/ha of nitrogen should be applied to crop when first flowers appear. Excess nitrogen must be avoided, as it promotes rank growth, delays maturity, and causes plants to become topheavy and to fall. Harvesting occurs 90–150 days after planting. Non-shattering sesame pods don't "Open Sesame" and are, of course, much better, especially in mechanized agriculture. Yields of ca. 1000–2300 lb/acre (1200–2600 kg/ha) are normal (CFR, TAD).

Chemistry (Sesame) — Here are a few of the more notable chemicals found in sesame. For a complete listing of the phytochemicals and their activities, see the CRC phytochemical compendium, Duke and duCellier, 1993 (DAD) and the USDA database http://www.ars-grin.gov/duke/.

Arginine — See also *Allium sativum*.

Sesamin — Antibacterial; Antileukemic IC50 = 2.90 µg/ml; Antioxidant IC50 = 58 µ*M*; Cytostatic ID50 = 10–100 µg/ml; Desaturase-Inhibitor; Hepatoprotective 100 mg/kg; Hypocholesterolemic; Immunosuppressant IC50 = 0.33 µg/ml (cf Prednisolone 0.06 µg/ml); Insecticide-Synergist; Juvabional; Piscicide; Tocopherol-Synergist.

Sesaminol — Antimutagenic; Antioxidant (= tocopherol).

Sesamol — Antioxidant IC50 = <58 µ*M* (> tocopherol), IC70 = 30 ppm; Carcinogenic; Tumorigenic; LD50 = 678 µg/kg ivn dog.

Sesamolin — Antioxidant; Insecticide-Synergist.

Sesamolinol — Antioxidant (>tocopherol).

Tocopherol — ADI = 2 mg/kg; Analgesic 100 IU 3 × day; Antiaging; Antiaggregant; Antialzheimeran 2000 IU; Antianginal 1067 mg/man/day; Antiatherosclerotic; Antibronchitic; Anticancer; Anticariogenic; Anticataractic; Antichorea; Anticoronary 100–200 IU/day; Antidecubitic; Antidermatitic; Antidiabetic 600–1200 mg/day; Antidysmenorrheic; Antiepitheleomic; Antifibrositic; Antiglycosation; Antiherpetic; Antiinflammatory; Antiischaemic; Antileukemic 100–250 µ*M*; Antileukotrienic; Antilithic 600 mg/day; Antilupus; Antimastalgic; AntiMD; AntiMS; Antimyoclonic; Antineuritic; Antinitrosaminic; Antiophthalmic; Antiosteoarthritic; Antioxidant IC50 = 30 µg/ml, IC95 = 650 µ*M*; Antiparkinsonian; AntiPMS 300 IU 2 × day; Antiproliferant IC50 = 150 µg/ml; Antiradicular; Antiretinopathic; Antisenility; Antisickling; Antispasmodic 300 mg/man/day; Antisterility; Antistroke; Antisunburn; Anti-Syndrome-X; Antithalassemic; Antithrombic 600 IU/day; Anti-Thromboxane-B2; Antitoxemic; Antitumor 7 µ*M* ckn; Antitumor (Breast) IC50 = 125 µg/ml, 100–250 µ*M*; Antitumor (Colorectal) 500–10,000 µ*M*; Antitumor (Prostate) 100–250 µ*M*; Antiulcerogenic 67 mg/man/3 x/day/orl; Apoptotic 100–250 µ*M*; Cerebroprotective; Circulostimulant; Hepatoprotective; 5-HETE-Inhibitor; Hypocholesterolemic 100–450 IU/man/day; Hypoglycemic 600 IU/man/day; Immunostimulant 60–800 IU; Insulin-Sparing 1000 IU; Lipoxygenase-Inhibitor; ODC-Inhibitor 400 mg/kg; p21-Inducer 500–10,000 µ*M*; Phospholipase-A2-Inhibitor; Protein-Kinase-C-Inhibitor 10–50 µ*M*, IC50 = 450 µ*M*; Vasodilator; RDA = 2–10 mg/day; PTD = 800 mg/day.

Syzygium aromaticum (L.) Merr. and L. M. Perry (Myrtaceae)
CLAVOS, CLOVE, CLOVETREE

Synonyms — *Caryophyllus aromaticus* L., *Eugenia aromatica* (L.) Baill., *E. caryophyllata* Thunb., *E. caryophyllus* (Spreng.) Bullock & Harrison.

Medicinal Uses (Clove) — Reported to be analgesic, anesthetic, antibacterial, antidotal, antioxidant, antiperspirant, antiseptic, carminative, deodorant, digestive, rubefacient, stimulant, stomachic, tonic, and vermifuge. Sold in oriental bazaars as a carminative and stimulant and to relieve the irritation of sore throat. Has been used as an expectorant in bronchitis and phthisis. As an aromatic, powdered cloves or an infusion thereof have been given for emesis, gas, and dyspepsia. In China,

S

crushed flower buds have been used for nasal polyps, in Malaya for callous ulcers, and in California for warts. When warts were rubbed three times a day with sesame flowers collected with the dawn dew still on them, the treatment was described as 97.2% effective in a study of more than 200 cases (JAC7:405). The USDA, of all places, reported that the equivalent 500 mg clove/day in humans could increase insulin efficiency, at least in models of human NIDDM. Clove oil is locally irritant and stimulates peristalsis. A powerful antiseptic, perhaps dangerously so, it has been applied as a local anesthetic for toothache. Mixed with catnip, ground cloves and sassafras are applied as a poultice to aching teeth. Eugenol is mixed with zinc oxide and used as the temporary filling to disinfect root canals from which the pulp has been removed prior to permanent restoration.

Clove and some of its active principles are cholagogue, and eugenol and acetyl eugenol possess antiaggregant activity (DAD). Srivastava and Malhotra (1990) report eugenol's effect on platelet aggregation and arachidonic acid metabolism to be antiaggregatory. Ultraconservatively, Rinzler (1990) lists no medicinal benefits for cloves and several nitpicking negatives, "Eugenol is closely related to safrole, a known carcinogen that causes liver cancer in laboratory models…. Because eugenol can be irritating to the intestinal tract, cloves are usually excluded from a bland diet. Contact with cloves may cause contact dermatitis (itching, burning, stinging, reddened or blistered skin)" (RIN). Clove bud oil is reported to have an oral LD_{50} of 2650 mg/kg body weight in rats (equaling that of the major ingredient, eugenol, which sensitizes some people, causing contact dermatitis). Unconservatively, I refer you to the USDA database to find dozens of potential positive activities of eugenol. Since your body co-evolved with eugenol, I expect that your body has homeostatic balances to deal with modest overdoses (or underdoses) of eugenol, which occurs in hundreds of culinary plants.

French enologists, Teissedre and Waterhouse (2000), studied antioxidant activities of EOs of spices and culinary herbs. They note that EOs are important as antibacterials and antifungals, as flavorants and preservatives when added to foods, and as cosmetics, for their aromatic and antioxidant properties. In their survey of 23 EOs, star anise was most potent (IC83 = 2 μM) and Spanish red thyme (*Thymus zygis*) least potent. Clove EO was relatively impotent (IC22 = 2 μM, IC38 = 5 μM) (JAF4:3801).

For inhibiting restriction endonucleases, hot water extracts of clove were 10 times more potent at a MIC of 2 μg/ml than peppermint or oregano (MIC = 20 μg/ml), which in turn were 10 times more potent than hot water extracts of bay, eucalypt, lemon balm, or nutmeg (MIC = 200 μg/ml) (Kato, 1990). And clove proved antiseptic to *Bacillus subtilis, Escherichia coli,* and *Saccharomyces cerevisiae,* confirming traditional uses of spices as food preservatives, disinfectants, and antiseptics (De et al., 1999). Takechi and Tanaka (1981) report the antiviral substance, eugenin, from the buds. And of eight oils studied by Zhu et al. (2001), clove was most toxic to subterranean termites, killing 100% of the termites in 2 days at 50 μg/c². Antitermite activity and volatility of the oils were inversely associated. Listed in decreasing order of volatility, the major constituents of the eight oils were eucalyptol, citronellal, citral, citronellol, cinnamaldehyde, eugenol, thujopsene, and both alpha- and beta-vetivone. Long lasting vetiver oil was the promising termiticide (Zhu et al., 2001).

Indications (Clove) — Alzheimer's (1; COX; FNF); Anorexia (f; PH2); Arthrosis (1; COX; FNF); Aspergillus (1; HH2); Athlete's Foot (15% tincture in 70% alcohol) (2; CAN; FNF); Bacillus (1; X10548758); Bacteria (1; FNF; TRA); Bite (f; BOW); Bronchosis (2; FNF; PHR); Bugbite (1; APA); Bunion (1; TGP); Callus (f; CRC); Cancer (1; COX; FNF); Candida (1; FNF; HH2); Caries (f; CRC); Childbirth (f; CRC); Chill (f; BOW); Cholera (f; CRC); Cold (2; PHR; PH2); Colic (1; CAN; PH2); Cough (2; PHR); Cramp (1; FNF); Cytomegalovirus (1; JAC7:405); Dermatosis (1; APA); Diarrhea (1; APA; CRC; HH2); Dyspepsia (f; CRC; HH2); Enterosis (f; CRC); Escherichia (1; HH2); Fever (2; PHR); Fungus (1; CRC; FNF; HH2; TRA); Gas (1; CRC; HH2; PH2); Gastrosis (f; CRC; PH2); Gingivosis (1; APA; FNF); Halitosis (1; LMP; PH2; TGP); Headache (1; HH2; PH2); Heart (f; CRC); Hernia (f; CRC); Herpes (1; TRA; JAC7:405); Hiccup (f; CRC); Impotence (f; BOW); Infection (2; APA; FNF; PHR; TRA); Infertility (f; CRC); Inflammation (2; COX; FNF;

KOM); Maculosis (1; TGP); Mucososis (1; APA); Mycosis (1; FNF; TRA); Myosis (f; HH2); Nausea (f; CRC); Nephrosis (1; BOW); Ophthalmia (f; PH2); Pain (2; APA; CAN; FNF; PIP); Parasite (f; BOW); Pharyngosis (2; APA; KOM; PH2; PIP); Phthisis (f; CRC); Polyp (f; CRC); Retinosis (1; TGP); Rhinosis (f; CRC); Sore (f; CRC); Sore Throat (f; PIP); Spasm (f; CRC); Staphylococcus (1; HH2); Stomatosis (2; APA; FNF; KOM; PH2; PIP); Teething (1; WAM); Toothache (2; APA; CAN; FNF; HH2; PH2; TRA); Trichomonas (1; HH2); Ulcer (f; PH2); Uterosis (f; CRC); Vaginosis (1; APA; HH2); Virus (1; CRC; FNF; JAC7:405); Vomiting (f; BOW); Wart (f; CRC); Water Retention (1; FNF); Worm (f; CRC); Wound (1; APA; CRC); Yeast (1; APA; HH2; TRA; X10548758).

Clove for bronchosis:

- Antibacterial: 3,4-dihydroxybenzoic-acid; alpha-terpineol; caryophyllene; delta-cadinene; ellagic-acid; eugenol; hyperoside; isoquercitrin; kaempferol; methyl-eugenol; myricetin; oleanolic-acid; procyanidin; quercetin; rhamnetin; terpinen-4-ol
- Antihistaminic: kaempferol; maslinic-acid; myricetin; quercetin
- Antiinflammatory: caryophyllene; caryophyllene-oxide; ellagic-acid; eugenol; hyperoside; kaempferide; kaempferol; maslinic-acid; myricetin; oleanolic-acid; quercetin
- Antioxidant: ellagic-acid; eugenol; hyperoside; isoeugenol; isoquercitrin; kaempferol; methyl-eugenol; myricetin; oleanolic-acid; pedunculagin; quercetin; rhamnetin
- Antipharyngitic: quercetin
- Antipyretic: eugenol
- Antiseptic: procyanidin; prodelphinidin
- Antispasmodic: caryophyllene; daucosterol; eugenol; kaempferol; quercetin
- Antitussive: terpinen-4-ol
- Antiviral: ellagic-acid; eugeniin; hyperoside; kaempferol; maslinic-acid; myricetin; oleanolic-acid; procyanidin; quercetin; rugosin-d; tellimagrandin-i
- COX-2-Inhibitor: eugenol; kaempferol; oleanolic-acid; quercetin
- Candidicide: caryophyllene; eugenol; myricetin; quercetin
- Candidistat: isoeugenol
- Cyclooxygenase-Inhibitor: kaempferol; oleanolic-acid; quercetin
- Expectorant: astragalin
- Immunostimulant: astragalin
- Phagocytotic: oleanolic-acid

Clove for toothache:

- Analgesic: eugenol; quercetin
- Anesthetic: benzyl-alcohol; eugenol; methyl-eugenol
- Antibacterial: 3,4-dihydroxybenzoic-acid; alpha-terpineol; caryophyllene; delta-cadinene; ellagic-acid; eugenol; hyperoside; isoquercitrin; kaempferol; methyl-eugenol; myricetin; oleanolic-acid; procyanidin; quercetin; rhamnetin; terpinen-4-ol
- Anticariogenic: alpha-terpineol; caryophyllene; delta-cadinene; ellagic-acid; oleanolic-acid; quercetin
- Antidermatitic: hyperoside; quercetin
- Antigingivitic: ellagic-acid; kaempferol; myricetin; oleanolic-acid
- Antiinflammatory: caryophyllene; caryophyllene-oxide; ellagic-acid; eugenol; hyperoside; kaempferide; kaempferol; maslinic-acid; myricetin; oleanolic-acid; quercetin
- Antioxidant: ellagic-acid; eugenol; hyperoside; isoeugenol; isoquercitrin; kaempferol; methyl-eugenol; myricetin; oleanolic-acid; pedunculagin; quercetin; rhamnetin
- Antiplaque: ellagic-acid; kaempferol; myricetin; oleanolic-acid; quercetin

S

- Antiseptic: alpha-terpineol; benzyl-alcohol; carvone; ellagic-acid; eugenol; furfural; kaempferol; methyl-benzoate; methyl-eugenol; myricetin; oleanolic-acid; terpinen-4-ol
- Astringent: ellagic-acid
- Bacteristat: isoeugenol; quercetin
- COX-2-Inhibitor: eugenol; kaempferol; oleanolic-acid; quercetin
- Candidicide: caryophyllene; eugenol; myricetin; quercetin
- Candidistat: isoeugenol
- Cyclooxygenase-Inhibitor: kaempferol; oleanolic-acid; quercetin
- Fungicide: caryophyllene; caryophyllene-oxide; eugenol; furfural; methyl-eugenol; quercetin; terpinen-4-ol
- Fungistat: isoeugenol; methyl-eugenol
- Immunostimulant: astragalin
- Vulnerary: terpinen-4-ol

Other Uses (Clove) — Cloves of commerce are the dried, unexpanded flower-buds, with a lower nail-shaped portion consisting of the calyx-tube enclosing the upper half of the ovary, the four calyx-teeth surrounding the unopened globular petals and stamens. It takes some 5000–7000 dried flower buds to make a pound (RIN). Cloves are used as a condiment or spice, in cordials, curries, hams, mincemeat, sausages, soups, sauces, tobaccos, masticatories, curries, pickles, preserves, desserts, cakes, and puddings. Ground cloves enter many spice mixtures, curry powders, pumpkin-pie spice, and sausage seasonings. In India, cloves may be chewed after meals. In Indonesia, cloves are used to make special cigarettes (kreteks) that crackle when burning. Cloves have been used in both alcoholic and nonalcoholic beverages, e.g., Benedictine and cola. Some sweet vermouths contain cloves. Whole or ground cloves are used in sachets, pomanders, and potpourris. Rinzler (1990) tells us how to make a scented pomander ball to aromatize our closets: stick clove nails (clavos) into a firm orange, all around and tightly packed. Roll the clove-studded orange in cinnamon powder, then wrap in tissue paper and place it on a shelf until the orange dries and shrinks. When it is completely dry, unwrap it, shake off any loose powder and hang the orange pomander in the closet. Clove oil, clove-stem oil, and clove-bud oil, obtained by steam distillation, are used in body lotions, insect repellents, mouthwashes, perfumes, soaps, toothpastes, and as an antiseptic. It is bactericidal. They contain eugenol, which is important in the manufacture of synthetic vanilla. Oil also used as a clearing agent in biomicroscopy (DAD). Fresh fruits are eaten, as are fruits of many other members of the myrtle family. Dried flowers are also consumed (DAD, FAC, RIN).

For more information on activities, dosages, and contraindications, see the *CRC Handbook of Medicinal Herbs*, ed. 2, Duke et al. 2002.

Cultivation (Clove) — Bown (2001) suggests fertile, sunny, well-drained soils, zone 10, minimum temp 15–18°C (59–64°F). Cloves are propagated by seeds which germinate in 12–14 days, with up to 90% germination, but some may take as long as 4–5 weeks. After that, germination quickly diminishes. When planting seeds, pulp of the mature fruit should be washed off and then planted with radicle downward and upper half exposed above the soil. Seeds should be planted ca. 20 cm apart each way in a shaded area. Bown (2001) suggests seed sowing at 27°C (81°F). Green cuttings in early summer, and nearly ripe cuttings in summer, are also useful. Watering and shade should be reduced when seedlings are about one year old to harden them. Plants should be outplanted in the field when 15–24 months old, spaced ca. 7 × 7 m. Interplanting with banana or cassava provides shade and some return in the years before the cloves bear. Sometimes, seeds are sown in nurseries and seedlings transplanted to a location in the shade of older clove trees when old trees are apparently about to die (DAD). From planting until bearing takes 4–7 years, with full bearing age at ca. 20 years. Tree may then bear for about 100 years. When buds turn reddish brown, they are ready for harvesting. They are picked carefully by hand, as branches are very brittle. If left unpicked, a fruit, called "mother of cloves," develops. Flower buds are dried four to five days on cement floors or drying mats. There may be two harvests per year, July–October and December–January. An average tree yields more than 3 kg dried cloves per year, but yields of 18 kg are not uncommon. Clove stems, stalks, leaves, fruit, and buds are used for distilling oil of cloves. It takes 11,000–15,000 cloves to make 1 kg spice. Cloves yield from 14–21% of volatile oil, high in eugenol.

Chemistry (Clove) — Here are a few of the more notable chemicals found in clove. For a complete listing of the phytochemicals and their activities, see the CRC phytochemical compendium, Duke and duCellier, 1993 (DAD) and the USDA database http://www.ars-grin.gov/duke/.

Eugeniin — Antiherpetic; Antilipolytic; Antiviral; Topoisomerase-II-Inhibitor IC100 = 0.5 μM.

Eugenol — See also *Pimenta dioica*.

Eugenyl-Acetate — Antiaggregant ED = 15 μM; Antiinflammatory; Antiprostaglandin ED50 = 3 μM; Antispasmodic; Irritant; Trypsin-Enhancer; LD50 = 1670 mg/kg orl rat.

Syringic-Acid — Antioxidant IC39 = 30 ppm; Ubiquict.

S

T

Tamarindus indica L. (Caesalpiniaceae)
INDIAN TAMARIND, KILYTREE, TAMARIND, TAMARINDO

Medicinal Uses (Tamarind) — For years, the pulp of the tamarind has been used, with good reason, as an antiscorbutic, laxative, and carminative. It is also used as a digestive and to treat bile disorders. Used in a gargle for sore throat, as a liniment for rheumatism when mixed with salt, applied on inflammations, administered to alleviate sunstroke and alcoholic intoxication, to aid the restoration of sensation in cases of paralysis, and as part of a vermifuge ointment. In Eritrea, the pulp is sold for dysentery and malaria; in Indonesia for hair ailments; in Madagascar for worms and stomach disorder; in Tanganyika for snakebite; in Sri Lanka for jaundice, eye diseases, and ulcers; in Cambodia for conjunctivitis; and in Brazil as a diaphoretic, emollient, laxative, and for hemorrhoids. The leaves and flowers are used as poultices for swollen joints, sprains and boils, and lotions or extracts of leaves and flowers for conjunctivitis, dysentery, jaundice, erysipelas, and hemorrhoids, and as antiseptic and vermifuge. Bahamians take the leaf tea for chills and fevers. In Curacao they take the leaf decoction for colds, coughs, diabetes, and sore throats. Jamaicans take it for fever, measles, and pain. The bark is an effective astringent, tonic, and antipyretic, used for asthma, caterpillar rashes, colic, eye inflammations, gingivitis, indigestion, and open sores. The astringent seed is used as a dysentery and chronic diarrhea remedy, and as a paste for drawing boils. Root infusion used for chest complaints and is an ingredient in leprosy prescriptions. Cubans take the root decoction for jaundice and hemorrhage. Many Latinos use the pulp as a gentle laxative, while Cubans use the powder of toasted seed to arrest diarrhea.

In their admirable effort to clear the safer and more efficacious of tropical American folk remedies, for so many people who can't afford modern prescription drugs, TRAMIL notes that ethanolic extracts are bactericidal, and that the diuretic pulp inhibits the gram(–) bacteria responsible for urinary infections. This suggests a food farmacy approach to cystitis combining the urinary antiseptic properties of cranberry and tamarind in a pleasant tart beverage. TRAMIL also reports antilipoperoxidative and hepatotropic properties, for the aqueous leaf extract, approved by TRAMIL for use in jaundice. They note antibacterial, antispasmodic, and vasodilator properties of the alcoholic extract (DAD, LEG, TRA, WOI).

Khan and Balick (2001) note human studies on tamarind for pain and worms. Tamarind increased bioavailability of other drugs and decreased pain (JAC7:405). Also, it can kill *Bacillus subtilis, Escherichia coli,* and *Saccharomyces cerevisiae* (De et al., 1999). Sambaiah and Srinivasan (1989) note that tamarind stimulates N-demethylase activity. It also stimulated liver microsomal cytochrome p450 dependent aryl hydroxylase.

Indications (Tamarind) — Abscess (f; WBB); Adenosis (f; JLH); Alcoholism (f; PH2); Amenorrhea (f; KAB; WBB); Anorexia (f; KAP; MAD); Apoplexy (f; DEP); Arthrosis (f; DAD); Asthma (f; DAD; KAB; WBB); Bacillus (1; X10548758); Bacteria (1; AAB; FNF; TRA); Biliousness (f; DEP; KAB; SUW; WOI); Bite (f; AAB); Bleeding (f; JFM; KAP; MAD; RYM); BO (f; KAB); Boil (f; AAB; DAD; DEP; IHB; WOI); Cancer (f; JLH; KAB); Cancer, abdomen (f; JLH); Cancer, colon (f; JLH); Cancer, gland (f; JLH); Cancer, spleen (f; JLH); Cancer, uterus (f; JLH); Cancer, vagina (f; JLH); Candida (1; APA; FNF); Chill (f; DAD; JFM); Cholecystosis (1; HH2; PHR; PH2); Cholera (1; AAB); Cold (f; JFM); Colic (f; AAB); Conjunctivosis (f; DAD; IHB; JFM; KAB); Constipation (1; APA; PH2); Cough (f; JFM; SKJ); Cramp (1; FNF); Dermatosis (f; AAB; IHB); Diabetes (f; JFM); Diarrhea (1; APA; FNF); Dizziness (f; HH2); Dysentery (f; DAD; DEP; JFM; WBB); Dysmenorrhea (f; MAD); Dyspepsia (f; KAB; SKJ); Dysuria (f; GMH; KAB); Earache (f; KAB); Eczema (f; MAD); Edema (f; WOI); Enterosis (f; WBB); Erysipelas (f; DAD); Escherichia (1; APA); Fever (1; APA; FNF; HH2; JFM; PHR; PH2; SUW); Fungus (1; APA; FNF); Furuncle (f; WBB); Gas (f; SKJ); Gastrosis (f; KAB); Gingivosis (f; DAD; WBB); Gonorrhea (f; WBB); Hangover (f; DEP; PH2, TGP); Headache (f; MAD); Heartburn (f; MAD); Hemorrhoid (f; DAD; DEP; PH2; WBB); Hepatosis (f; HH2; KAB; PHR; PH2); Hyperemesis gravidarum (f; BOW); Infection (1; AAB; DAD; FNF); Inflammation (1; DAD; DEP; FNF); Intoxication (f; DAD; DEP; KAB; PH2); Itch (f; MAD); Jaundice (1; DAD; JFM; MAD; TRA; WBB); Leprosy (f; DAD; WBB); Leukorrhea (f; MAD); Malaria (f; DAD; WBB); Measles (f; JFM); Morning Sickness (f; AAB; APA); Mucososis (f; IHB); Mycosis (1; AAB; FNF); Myosis (f; SKJ); Nausea (1; APA); Ophthalmia (f; DAD); Pain (1; DEP; FNF; JFM); Paralysis (f; DAD; KAB); Pharyngosis (f; PH2); Pulmonosis (f; DAD); Rash (f; AAB); Respirosis (f; DAD); Rheumatism (f; DAD; IHB; WBB); Ringworm (1; APA; KAB); Salmonella (1; AAB); Scabies (f; KAB); Schistosomiasis (1; AAB; APA); Smallpox (f; KAB); Snakebite (f; KAB; WBB); Sore (f; AAB; IHB); Sore Throat (f; AAB; DEP; JFM); Splenosis (f; JLH); Staphylococcus (1; AAB; APA); Sting (f; SKJ); Stomachache (f; PH2; SKJ); Stomatosis (f; IHB; KAB; PH2); Sunstroke (f; DEP; SKJ); Swelling (f; HH2; KAB; WOI); Syphilis (f; SKJ); Ulcer (f; DAD); UTI (f; DAD; TRA); Uvulosis (f; KAB); VD (f; WBB); Vertigo (f; HH2; KAB); Virus (1; FNF); Vomiting (f; PH2); Worm (1; APA; DAD); Wound (f; AAB; IHB); Yeast (1; APA; FNF; X10548758).

Tamarind for diarrhea:

- Antibacterial: (–)-epicatechin; acetic-acid; alpha-pinene; alpha-terpineol; beta-ionone; carvacrol; cinnamaldehyde; geranial; geraniol; limonene; linalool; myrcene; nerol; phenol; safrole; tamarindienal; terpinen-4-ol
- Antidiarrheic: hordenine
- Antispasmodic: carvacrol; cinnamaldehyde; geraniol; limonene; linalool; myrcene; pyridoxine

- Antiviral: (–)-epicatechin; alpha-pinene; cinnamaldehyde; geranial; limonene; linalool; phenol
- Candidicide: beta-pinene; carvacrol; cinnamaldehyde; geraniol; octanoic-acid; tamarindienal
- Candidistat: limonene; linalool
- Carminative: carvacrol; methyl-salicylate; safrole
- Demulcent: mucilage
- Hepatoprotective: hordenine
- Protisticide: acetic-acid
- Vermifuge: carvacrol

Tamarind for infection:

- Analgesic: methyl-salicylate; myrcene; phenol; pyridoxine
- Anesthetic: carvacrol; cinnamaldehyde; linalool; myrcene; phenol; safrole
- Antibacterial: (–)-epicatechin; acetic-acid; alpha-pinene; alpha-terpineol; beta-ionone; carvacrol; cinnamaldehyde; geranial; geraniol; limonene; linalool; myrcene; nerol; phenol; safrole; tamarindienal; terpinen-4-ol
- Antiinflammatory: (–)-epicatechin; alpha-pinene; beta-pinene; carvacrol; cinnamaldehyde; methyl-salicylate; orientin; vitexin
- Antiseptic: alpha-terpineol; aromadendrene; beta-pinene; carvacrol; furfural; geraniol; limonene; linalool; methyl-salicylate; nerol; oxalic-acid; p-cresol; phenol; safrole; terpinen-4-ol
- Antiviral: (–)-epicatechin; alpha-pinene; cinnamaldehyde; geranial; limonene; linalool; phenol
- Bacteristat: malic-acid
- Circulostimulant: cinnamaldehyde
- Cyclooxygenase-Inhibitor: carvacrol; cinnamaldehyde
- Fungicide: acetic-acid; beta-ionone; carvacrol; cinnamaldehyde; furfural; geraniol; linalool; myrcene; octanoic-acid; phenol; tamarindienal; terpinen-4-ol
- Fungistat: limonene
- Immunostimulant: (–)-epicatechin
- Lipoxygenase-Inhibitor: (–)-epicatechin; cinnamaldehyde

Other Uses (Tamarind) — Cultivated mainly for the pulp in the fruit, it is used to prepare beverages, to flavor confections, curries, and sauces (Worcestershire and Pickapeppa), and to make preserves and syrups. "Jugo" or "Fresco de Tamarindo" is a favorite beverage in many Latin American countries and is bottled commercially in some. Some Latins claim that this is the most important "secret" ingredient in sweet and sour sauces and in chutneys. Curacao natives use the pulp to make a soup with cinnamon and sugar (this could well be a food farmacy approach to alleviate mild diabetes). West Indian Dutch mix the pulp with ashes to make a food whose name translates to "cat feces." In Sri Lanka, the pulp is made into a brine for pickling fish. Javanese roll salted pulp into balls, then steam and sun-dry them. Some cook the whole unripe fruits in curries. Young, immature pods are eaten fresh mixed with spices or fish sauce, pickled like green mango, or added whole to soups, stews, and sauces, such as näm prik ma-kahm. In Africa, the pods are added to detoxify poisonous *Dioscorea* dishes. Elsewhere, it is used to deodorize fish dishes. In places, the flowers, seedlings, even the leaves are eaten as vegetables in curries. The bark is chewed as a delicacy, flowers are eaten raw in salads or cooked, and seedlings, when about a foot high, are used as vegetables (FAC). Leaves are eaten in curries, salads, and soups. Leaves are also eaten by cattle and goats. In India and West Africa, the leaves are used as fodder for silkworms. Leaves and flowers are useful as fixatives in dyeing. The leaves are used to bleach leaves of the buri palm (*Corypha elata* Roxb.), which are used for hat-making. The flowers are considered a good source of nectar for honeybees. Seeds, once boiled and peeled, are used as a starchy foodstuff.

T

Reportedly 70% kernel and 30% testa, testae are separated by roasting or soaking, and the kernels are boiled or fried before eaten like peanuts. Seeds are roasted and eaten, ground to flour, and used as a coffee substitute. Seed oil, though useful as a varnish, is said to be palatable and of culinary quality. Seeds are used for food in India and provide a source of carbohydrate for sizing cloth, paper, and jute products, and a vegetable gum used in the food processing industry. An infusion of the whole pods is added to the dye when coloring goat hides in West Africa. Pulp also used as a fixative with turmeric or annatto in dyeing and has served to coagulate rubber latex. Mixed with sea water, the pulp is used to clean silver, copper, and brass. Twigs used as chewsticks and the bark is chewed as a masticatory. The bark also used in tanning and dyeing, and burned to make an ink. Low-quality fiber from bark of young trees used to make twine or string. Wood of tamarind is very hard and durable, used locally for tool handles, rice pounders, oil and sugar mills, and furniture and turnery, and is said to produce the finest grade of gunpowder charcoal. Wood ashes used in tanning and in de-hairing goatskins. Frequently grown as a dooryard, roadside, or windbreak tree (DAD, FAC, LEG, WOI).

For more information on activities, dosages, and contraindications, see the *CRC Handbook of Medicinal Herbs*, ed. 2, Duke et al. 2002.

Cultivation (Tamarind) — Can be propagated by seed, sown at 21°C (70°F), in light, well-drained, sunny soils, zone 10, min. temp 15–18°C (59–64°F); or greenwood cuttings in spring or summer; or by air-layering or grafting (Bown, 2001). Seed for planting should be taken from trees yielding heavy crop of well formed rounded pods. Fresh seeds are not suitable for planting. Seeds gathered from April crop may be planted in September. Seeds are soaked 4–5 days in water, then planted about 4 cm deep in baskets, bamboo pots, or a nursery bed. When the seedlings are 60–70 cm tall, in about 9 or 10 months, they are transplanted during the rainy season to groves, where they are spaced 3–7 m apart each way. The richer the soil, the farther apart they should be planted. Often, the seedlings are planted in pits previously prepared with well decomposed manure and allowed to "weather." Once established, the plants require little attention, except watering during prolonged droughts, loosening of the soil, and an occasional weeding. Trees will grow like any wild tree in 4–5 years. Budding on young seedlings is also possible. Under favorable conditions, tamarinds start bearing at ca. 5 years, but four-year-olds may bear in Malagasy. Some Indian trees reach 14 years old before fruiting. Trees bear copiously up to age 50 or so and then may decline, though the tree may live to be 200 years old. Pods may be left on the tree for 6 months, as the moisture content dwindles to 20% or lower, but are generally gathered when ripe, and the hard pod shell is removed. Pods are picked or shaken from the trees, but it is best to clip the fruit stalks carefully. To preserve, the pods are shelled, the pulp placed in casks and covered with boiling syrup or packed carefully in stone jars with alternate layers of sugar, or the pulp layered with sugar or sprinkled with salt or molded into balls of pulp, to be stored in cool dry places. Mature trees can produce 150–225 kg fruit, half of which may be pulp. But average yields of 80–90 kg prepared pulp prevail per mature tree. Up to 170 kg of prepared pulp per tree has been reported in India and Ceylon. One hundred trees/ha each yielding 200 kg fruit yielding 100 kg pulp translates to 10 MT pulp/ha possible (DAD).

Chemistry (Tamarind) — Here are a few of the more notable chemicals found in tamarind. For a complete listing of the phytochemicals and their activities, see the CRC phytochemical compendium, Duke and duCellier, 1993 (DAD) and the USDA database http://www.ars-grin.gov/duke/.

Linoleic-Acid — See also *Ceratonia siliqua.*

Lysine — See also *Prunus dulcis.*

Mufa — See also *Persea americana.*

Tamarindienal — Antibacterial; Bitter; Candidicide; Fungicide.

Tartaric-Acid — ADI = 30 mg/kg; Acidifier; Additive; Antioxidant Synergist; Irritant; Sequestrant; LDlo = 5000 orl dog; LD50 = 4360 mg/kg orl rat.

Theobroma cacao L. (Sterculiaceae)
Cacao, Chocolate, Cocoa

Medicinal Uses (Cacao) — Reported to be antiseptic, diuretic, ecbolic, emmenagogue, and parasiticide. Mexican Amerindians apply the seed oil to wounds. Venezuelans use the oil or "butter" on burns, cracked lips, eruptions, mastalgia, sore genitals, inserting them into the rectum or vagina for proctosis or vaginosis. Kuna Indians use the leaves on wounds, the flowers for eye infestations, and the pulp of the fruits for parturitions. Colombians use the leaf tea as a diuretic cardiotonic. Decoctions of the bark and/or seeds are used in Amazonia for scalp and skin conditions. Cocoa butter is massaged onto to wrinkles in the hope of correcting them. And creams containing aminophylline are massaged onto cellulite, possibly reducing the unattractive cottage-cheese ripples of cellulite in middle aged matrons. Chocolate contains two or more xanthines rather related to aminophylline, also used in some topical cellulite creams. Like coffee and tea, chocolate is getting a lot of press as an antioxidant beverage.

Chocolate, as we take it in the U.S., might be more lipogenic than lipolytic. Chocolate can boost serotonin levels and boost endorphin levels with a powerful result; all brain chemicals are positioned at optimal levels for positive moods and renewed energy. Chocolate also contains phenylethylamine and is a good source of magnesium, important in serotonin manufacture and in stabilizing mood. If the pharmaceutical firms can promote their pharmaceutical serotoninergic drugs for obesity, saying that serotonin sends a satiety signal to the body, then I suppose our herbal serotoninergics could help as well. Ancient Mayans drank chocolate containing thermogenic caffeine and theobromine, mixed with hot peppers, containing thermogenic capsaicin, a potentially synergic mix of three thermogenics from two of our spices. Recently, Dullo et al. (1999), though working with green tea, showed that the epigallocatechin, which also occurs in chocolate, was potentially synergetic as a thermogen with caffeine and perhaps the other xanthines. They measured the 24-hour energy expenditure of 10 healthy men receiving 3 daily doses of caffeine (50 mg), or the extract (containing 50 mg caffeine and 90 mg epigallocatechin), or a placebo. Compared with placebo, the extract (in their case, green tea, not chocolate) induced significant increases (~4%) in energy expenditure. The extract contains a high amount of catechin polyphenols, which may work with other phytochemicals to speed up fat oxidation and thermogenesis. I did not treat coca, the source of cocaine, in this book on spices, but in a sense the coca leaf is a flavoring, being used for a century to flavor Coca-Cola. And the coca leaf (*Erythroxylum*), like the cocoa seed (*Theobroma*), contains several anorectic alkaloids that might help to curb hunger, and that might be synergistic with the pinca-pinca (an Andean *Ephedra*), used in teas there, and containing two more anorectic alkaloids, ephedrine and psuedoephedrine. Bring in the cathine and cathinone from the khat in Ethiopia, and the synephrine from orange peel, sweeten with *Stevia*, and you have my herbal AntiOBESitea, a thermogenic pot pourri that would surely be declared illegal, but would surely be thermogenic and anorectic. And probably as safe as some of the legal weight-loss pharmaceutical combinations. Coca-Cola, as originally formulated, would also be illegal. My dad used to call a Coca-Cola a "dope."

And more and more, science seems to back up cocoa reputation as an aphrodisiac, inducing a few addicted chocaholics. Even Aztec Indians considered it aphrodisiac. Somehow, chocolate may be more effective in women than men. The xanthine alkaloids, caffeine, theobromine, and theophylline have cGMP-phosphodiesterase inhibition activity, which some hint might be Viagra-like activity. That puts a different slant on the "Food of the Gods." There's even a marijuana-like antidepressant called anandamide. Technically, it is called an anesthetic. And it also helps headaches. In *JAMA* (Vol. 276{8}:584e, 1996) we read of topical treatment of erectile dysfunction with a cream containing 3% aminophylline, 0.25% isosorbide dinitrate, and 0.05% codergocrine mesylate. Thirty-six erectile dysfunction patients were given either active cream or placebo for alternating weeks; 21 reported full erection and satisfactory intercourse with the active cream. The active cream was more effective in psychogenic than organic impotence (eight of nine with psychogenic impotence achieved full erection vs. four of eight with neurogenic impotence and two of seven with arterial insufficiency). In the lab, the cream increased penile arterial flow and induced tumescence in 24 patients. Their conclusion might lead to more optimism for other more herbal topicals: "Topical treatment with a cream containing three different vasodilators might be considered before intracavernous injections of vasoactive agents, particularly in psychogenic impotence." One can look at my database and find several other promising vasodilators. I'd not be afraid to try celery's apigenin, chocolate's caffeine, theobromine, theophylline, onion's quercetin, poppies' papaverine, and/or ginger's zingerone, but I'd be leery of the rubefacient capsaicin.

According to an article in the *Chicago Sun Times,* people who suffer extreme depression as victims of unrequited love have an irregular production of phenylethylamine. Such individuals often go on a chocolate binge during periods of depression. Chocolate is particularly high in phenylethylamine, perhaps serving as medication. Theophylline is a potent CNS and cardiovascular stimulant with diuretic and bronchial smooth muscle relaxant properties. Recently, this drug was proven effective in preventing and treating apnea in premature infancy.

And Sutton (1981) reports the collapse and death of a three-year old bitch that had eaten a 250 g package of cocoa. Postmortem examination revealed congestion of lungs, liver, kidney, and pancreas, and petechial and ecchymotic hemorrhage of the thymus, all compatible with acute circulatory failure. The stomach contained high concentrations of theobromine and/or caffeine. Though used cosmetically, cocoa butter has been reported to have allergenic and comedogenic properties in animals. Cocoa extracts are GRAS (§182.20). Tyler (1994) produces a chart comparing various caffeine sources to which we have added other rounded figures:

Cup (6 oz.) espresso coffee	310 mg
Cup (6 oz.) boiled coffee	100 mg
Cup (6 oz.) instant coffee	65 mg
Cup (6 oz.) tea	10–50 mg
Cup (6 oz.) cocoa	13 mg
Can (6 oz.) cola	25 mg
Can (6 oz.) Coca-Cola	20 mg
Cup (6 oz.) mate	25–50 mg
Can (6 oz.) Pepsi Cola	10 mg
Tablet caffeine	100–200 mg
Tablet (800 mg) Zoom (*Paullinia cupana*)	60 mg

In humans, caffeine, 1,3,7-trimethylxanthine, is demethylated into three primary metabolites: theophylline, theobromine, and paraxanthine. Since the early part of the 20th century, theophylline has been used in therapeutics for bronchodilation, for acute ventricular failure, and for long-term control of bronchial asthma. At 100 mg/kg, theophylline is fetotoxic to rats, but no teratogenic abnormalities were noted. In therapeutics, theobromine has been used as a diuretic, as a cardiac stimulant, and for dilation of arteries. But at 100 mg, theobromine is fetotoxic and teratogen. Leung (1995) reports a fatal dose in man at 10,000 mg, with 1000 mg or more capable of inducing headache, nausea, insomnia, restlessness, excitement, mild delirium, muscle tremor, tachycardia, and extrasystoles. Leung also adds "caffeine has been reported to have many other activities, including mutagenic, teratogenic, and carcinogenic activities; …to cause temporary increase in intraocular pressure, to have calming effects on hyperkinetic children…to cause chronic recurring headache" (Leung, 1995).

Indications (Cacao) — ADD (1; DAD); Adenosis (f; HH2); Allergy (1; FNF); Alopecia (f; CRC); Angina (1; BOW); Asthma (1; APA; DAV; FNF); Bacteria (1; FNF); Bite (f; DAD); Bleeding (f; IED); Bronchosis (1; APA; FNF); Burn (f; APA; IED; JFM); Cellulite (1; BRU; FNF; HAD); Chafing (f; APA; FEL); Childbirth (f; CRC; DAD; JFM); Cold (1; APA; FNF); Congestion (1; APA); Cough (f; APA; CRC; DAD); Cramp (1; FNF); Cystosis (f; KOM; PHR; PH2); Debility (f; TRA); Dermatosis (f; IED); Diabetes (f; KOM; PHR; PH2); Diarrhea (1; APA; FNF; KOM; PHR; PH2); Eczema (f; DAV); Enterosis (1; APA; PHR; PH2); Eruption (f; JFM); Fever (f; APA; CRC); Flu (1; APA); Hemorrhoid (1; CRC; FNF); Hepatosis (1; FNF; PHR; PH2); High Blood Pressure (1; BOW; GMH); Hyperkinesis (1; DAD); Infection (1; APA; FNF; PHR); Inflammation (1; FNF); Malaria (f; CRC); Mastosis (f; APA; CRC; JFM); Nephrosis (f; CRC; PHR; PH2); Nipple (f; FEL); Obesity (f; BRU; FNF; HAD); Ophthalmia (f; CRC; DAD); Parturition (f; APA); Pregnancy (f; APA); Proctosis (f; JFM); Rheumatism (f; CRC); Scabies (f; DAV); Screw Worm (f; JFM); Snakebite (f; CRC); Thyrosis (f; HH2); Tumor (1; CRC); Vaginosis (f; JFM); Virus (1; FNF); Water Retention (1; FNF); Worm (f; CRC); Wound (f; DAD; JFM); Wrinkle (f; APA; CRC; DAD).

Cacao for asthma:

- Antiallergic: ferulic-acid; kaempferol; linalool; quercetin
- Antiasthmatic: caffeine; protocatechuic-acid; pyridoxine; quercetin; theobromine; theophylline
- Antibronchitic: theophylline
- Antihistaminic: caffeic-acid; chlorogenic-acid; kaempferol; linalool; luteolin; proanthocyanidins; quercetin; rutin; vitexin
- Antileukotriene: caffeic-acid; chlorogenic-acid; quercetin
- Antioxidant: (–)-epicatechin; caffeic-acid; caffeine; catalase; catechol; chlorogenic-acid; cyanidin; epigallocatechin; ferulic-acid; isovitexin; kaempferol; luteolin; p-coumaric-acid; p-hydroxy-benzoic-acid; polyphenols; proanthocyanidins; protocatechuic-acid; quercetin; quercitrin; rutin; vanillic-acid; vitexin
- Antipharyngitic: quercetin
- Antiprostaglandin: caffeic-acid
- Antispasmodic: caffeic-acid; ferulic-acid; kaempferol; linalool; luteolin; p-coumaric-acid; protocatechuic-acid; pyridoxine; quercetin; quercitrin; rutin; theophylline; valerianic-acid; valeric-acid
- Bronchodilator: theobromine; theophylline
- Bronchorelaxant: linalool
- Cyclooxygenase-Inhibitor: kaempferol; polyphenols; quercetin
- Expectorant: acetic-acid; linalool
- Lipoxygenase-Inhibitor: (–)-epicatechin; caffeic-acid; chlorogenic-acid; epigallocatechin; esculetin; kaempferol; luteolin; p-coumaric-acid; polyphenols; quercetin; rutin
- Mast-Cell-Stabilizer: quercetin

Cacao for cellulite:

- Anorectic: theobromine
- Anticellulitic: theophylline
- Antiedemic: caffeic-acid; coumarin; proanthocyanidins; quercitrin; rutin
- Antiobesity: caffeine
- Catabolic: caffeine; quercetin; rutin
- Collagen-Sparing: caffeic-acid; chlorogenic-acid
- Collagenic: proanthocyanidins
- Diuretic: caffeic-acid; caffeine; chlorogenic-acid; dopamine; glycolic-acid; kaempferol; luteolin; quercitrin; theobromine; theophylline
- Lipolytic: nicotinic-acid
- Phosphodiesterase-Inhibitor: caffeine

Cacao for diarrhea:

- Antibacterial: (–)-epicatechin; acetic-acid; caffeic-acid; chlorogenic-acid; esculetin; ferulic-acid; gentisic-acid; kaempferol; linalool; luteolin; p-coumaric-acid; p-hydroxy-benzoic-acid; polyphenols; protocatechuic-acid; quercetin; quercitrin; rutin; vanillic-acid
- Antidiarrheic: polyphenols
- Antispasmodic: caffeic-acid; ferulic-acid; kaempferol; linalool; luteolin; p-coumaric-acid; protocatechuic-acid; pyridoxine; quercetin; quercitrin; rutin; theophylline; valerianic-acid; valeric-acid
- Antiviral: (–)-epicatechin; caffeic-acid; caffeine; catechol; chlorogenic-acid; ergosterol; ferulic-acid; gentisic-acid; kaempferol; linalool; luteolin; polyphenols; proanthocyanidins; protocatechuic-acid; quercetin; quercitrin; rutin; theophylline
- Astringent: formic-acid
- Candidicide: ferulic-acid; quercetin
- Candidistat: linalool
- Hepatoprotective: caffeic-acid; chlorogenic-acid; ferulic-acid; luteolin; polyphenols; proanthocyanidins; quercetin
- Protisticide: acetic-acid; kaempferol

Other Uses (Cacao) — Fermented seeds, called cacao beans, are used for commercial cocoa, chocolate, and cocoa butter, which is widely used in baked goods, beverages, cakes, candy, confectionary, desserts, ice creams, pastries, and pudding. Fermented seeds are roasted, cracked, and ground to give a powdery mass from which fat is expressed. This is the cocoa from which a popular beverage is prepared. In the preparation of chocolate, this mass is mixed with sugar, flavoring, and extra cocoa fat. Mexican chocolate and the Spanish xocolata a la pedra are flavored with cinnamon. Traditionally, Native Americans mixed cacao with corn meal to make nutritious beverages and gruels. Toasted cacao beans are mixed with mamey seeds in making texate, a refreshing beverage. In Tabasco, ground cacao beans are mixed in dough to prepare pozol, a fermented maize dough. The resulting product is called "chorote." Pulp of the fruit is sucked as a sweet snack or preserved, crystallized, or made into alcoholic beverages and vinegar. The husk is toasted until black and made into atole negro. It also contains a pigment that is said to be useful as a food colorant (FAC). Milk chocolate incorporates milk as well. Cocoa extracts occur in my two very favorite liqueurs, Bailey's Irish Cream and Creme de Cacao. Cocoa powder is used to coat California's Dry Monterey Jack cheese. Cocoa butter is used in confections and in manufacture of tobacco, soap, and cosmetics. Cocoa butter has been described as the world's most expensive fat, used rather extensively in the emollient "bullets" used for hemorrhoids. "Sweatings," which flow from the pulp surrounding the beans during fermentation, are made into a jelly in Brazil. Juice, alcoholic beverages, vinegar, sugar, and pectin could also be produced. Every 1000 lb of dried cocoa beans produces 930 lb of dried pod husk, which can be used as a source of potash or ground and added to livestock feed, in limited amounts (DAD, FAC).

For more information on activities, dosages, and contraindications, see the *CRC Handbook of Medicinal Herbs*, ed. 2, Duke et al. 2002.

Cultivation (Cacao) — Propagation may be by cuttings, buddings, or graftings, but seeding is cheaper. Seeds germinate at maturity and are viable only a short time. They may be stored 10–13 weeks if moisture content is kept at 50%. Soon after picking, pulp is removed from seed, which are planted in shaded nursery beds or baskets. Transplant in a few months (when ca. 0.6 m tall) into shaded fields at 2.4 × 2.4 m or 3.6 × 3.6 m. Spacing is closer if soils are poor and elevations above 300 m. Fields should remain shaded for 3 years. Remove floral buds until trees are 5 years old. Cacao is often intercropped with other trees of economic value such as bananas, rubber, oil palm, or coconut. Weeding is by hand or herbicides. Irrigation may be practiced, but drain ditches should always be provided to prevent excess water. Responds to fertilizers, mostly in the absence of shade. Recommended is 5 cwt urea, 2.5 cwt triple superphosphate, 10 cwt potassium sulfate/ha. Windbreaks are usually provided. Capable of bearing pods by the end of the second year, maximum yields are usually obtained by 8–10 years, some producing pods for over 100 years. Although fruits mature throughout the year, usually only two harvests are made. In West Africa, the main harvest begins in September, extends to February, with a second smaller harvest in May–June. From fertilization to harvesting the fruit requires 5–6 months. Harvest season lasts about 5 months. Pods are cut from trees and allowed to mellow on the ground. Then, pods are cracked and the beans removed, the husks are burned. Beans are fermented in leaf-lined kegs 2–8 days before drying in sun, at which time they change from purple to brown. Beans are then bagged and shipped. Further processing includes roasting, crushing, and separating out the kernel, grinding the nibs, and extraction of about half of the fat. The world low production yield is 29 kg/ha in American Samoa, the international production yield 346 kg/ha, and world high production 2000 kg/ha in Haiti. Yields of 3375 kg/ha of dry beans are possible on good plantations. The oil content (35–50%) suggests potential oil yields of more than 1750 kg/ha. Average yields run 0.5–10 kg/tree; 0.5–2.5 MT beans/ha. Over 3375 kg/ha of dry cacao beans have been produced on plantations well manured, well shaded, and with excellent control of weeds, pests, and diseases. In 1980, the U.S. is estimated to have consumed more than 75,000 MT of cocoa butter, in a business amounting to nearly $600 million. Chocolate manufacturers consumed nearly half. One ton went into suppositories, 10–20% of which are made with a cocoa butter base. Malaysia and Brazil are the main sources for cocoa butter, providing over half of U.S. imports. Cocoa butter imports were 19,000 MT worth $279 million in 1990.

Chemistry (Cacao) — Caffeic acid occurs in the unfermented beans. Cocoa butter contains mainly triglycerides of fatty acids that consist primarily of oleic, stearic, and palmitic acids. Over 73% of the glycerides are present as monounsaturated forms (oleopalmitostearin and oleodistearin), the remaining being mostly diunsaturated glycerides (palmitodiolein and stearodiolein), with lesser amounts of fully saturated and triunsaturated (triolein glycerides).

Here are a few of the more notable chemicals found in cacao. For a complete listing of the phytochemicals and their activities, see the CRC phytochemical compendium, Duke and duCellier, 1993 (DAD) and the USDA database http://www.ars-grin.gov/duke/.

Caffeine — Adenosine Antagonist; Analeptic 200 scu mus; Analgesic-Synergist; Antiapneic; Antiapoptotic; Antiasthmatic 5–10 mg/kg orl/man; Anticancer; Anticarcinogenic; Anticariogenic; Antidermatitic; Antiemetic; Antifeedant; Antiflu; Antiherpetic; Antihypotensive 250 mg/day/orl/man; Antinarcotic Antiobesity; Antioxidant; Antirhinitic 140 mg/day/orl/man; Antiserotoninergic 40 ipr rat, 40 scu rat; Antitumor; Antitumor (Lung); Antivaccinia; Antiviral; Apoptotic; Arrhythmogenic

1500 mg/man; cAMP-Phosphodiesterase-Inhibitor; cGMP-Phosphodiesterase-Inhibitor; Cardiotonic 10–25 orl dog, 65–500 orl cat; Catabolic; Choleretic; Coronary-Dilator; CNS-Stimulant 30 orl rat, 100 orl mus; Diuretic; Energizer 20–200 mg/man; Ergotamine-Enhancer; Herbicide; Hypertensive; Hypoglycemic; Insecticide; Lipolytic; Myorelaxant; Neurotoxic; Phosphodiesterase-Inhibitor; Positive Inotropic; Respirastimulant; Spasmogenic 1500 mg/man; Stimulant; Tachycardic 1500 mg/man; Teratogenic 14 orl rat; Topoisomerase-I-Inhibitor 0.1 nM, 75 µM; Topoisomerase-II-Inhibitor 99 mM; Vasodilator; LDlo = 192 orl hmn; LD50 = 192 orl rat; LD50 = 127–1200 orl mus; LD50 = 200 orl rat; LD50 = 247–355 orl rat; LD50 = 224–246 orl rbt.

Linoleic-Acid — See also *Ceratonia siliqua.*

Theobromine — Anorectic; Antiasthmatic 10 mg/kg/orl/man; Anticellulitic; Arteriodilator; Bronchodilator; Cardiotonic; cAMP-Phosphodiesterase-Inhibitor; cGMP-Phosphodiesterase-Inhibitor; CNS-Stimulant; Diuretic 300–600 mg/man/day; Emetic; Fetotoxic 100 mg/kg; Herbicide IC41 = 100 ppm; Myocardiotonic; Myorelaxant; Myostimulant; Stimulant; Teratogenic 100 mg/kg; Vasodilator; LD50 = 200 orl cat; LD50 = 789 mg/kg ipr mus; LD50 = 1356 mg/kg orl mus; LD50 = 950 mg/kg orl rat.

Theophylline — ADI = 60–200 orl/man/day; Allergic; Antidote (Charcoal & Propanolol); Antiapneic; Antiasthmatic 5 mg/kg/orl/man; Antibradyarrhythmic; Antibronchitic; Anticellulitic; Antiemphysemic; Antineuralgic; Antispasmodic 100 µM; Antirhinitic; Antiviral; Arteriodilator; Bronchodilator; cAMP-Inhibitor IC50 = 0.06 mg/ml; Cardiovascular 30 mg/kg orl rat; cAMP-Phosphodiesterase-Inhibitor; cGMP-Inhibitor; Choleretic; CNS-Stimulant; Diuretic 15 mg/kg ivn rbt; Fetotoxic 100 mg/kg; Herbicide IC58 = 100 ppm; Hypertensive; Hyperuricemic; Hypoglycemic; Myocardiotonic; Myorelaxant 100 µM; Potitive Inotropic; Prostaglandin-Secretor 10 mg/kg orl rat; Stimulant; Tachycardic; Teratogenic; Vasodilator; LDlo = 100 orl cat; LD50 = 600 orl mus; LDlo = 350 orl rbt; LDlo = 115 ivn rbt; LD50 = 200 scu mus.

Trigonella foenum-graecum L. (Fabaceae)
FENUGREEK, GREEK CLOVER, GREEK HAY

Medicinal Uses (Fenugreek) — Said to be used for aphrodisiac, astringent, demulcent, diuretic, emmenagogue, emollient, expectorant, lactagogue, restorative, and tonic. Ecbolic and spermicidal activities have also been reported. Mucilaginous seeds are believed to be carminative, emollient, tonic, and vermifugal, and are used for chapped lips, diarrhea, oral ulcers, rheumatic conditions, and stomach irritation. Indian women believe the seeds promote lactation. Chinese use the seed for abdominal pain, chilblains, cholecystosis, fever, hernia, impotence, hypogastrosis, nephrosis, and rheumatism. Malayans poultice the seeds onto burns and use them for chronic coughs, dropsy, hepatomegaly, and splenomegaly. Crushed leaves are taken internally for dyspepsia (CRC, KOM, LIL, PH2, RIN, TAD).

Weed (2002) suggests heavy consumption of certain spices when estrogen levels are down. Seeds like caraway, celery, coriander, cumin, poppy and sesame, mustard and anise, fennel and fenugreek, all contain phytoestrogens, as do their oils, says Weed. She suggests using these seeds "lavishly" when cooking or making tea with any one of them, drinking 3–4 cups a day "for best results" (Weed, 2002). Taken as a tea, or used to season food, fenugreek seed are a cheap and pleasant way to ease menopausal symptoms, restore blood sugar balance, and increase libido. Brew a tablespoonful (15 mg) of seeds for no more than 15 min in a cup of hot water. "Make phytosterol-rich fenugreek seed tea your wake-up and good-night brew; you'll have easier hot flashes and flushes and, when you do flash, your sweat will smell like sweet maple syrup" (Weed, 2002). But she cautions that "Fenugreek promotes fertility." Fenugreek is one of three estrogenic plants in the original Lydia Pinkham's formula. Could Lydia Pinkham's in large doses have had mastogenic properties to go along with its antidysmenorrheic and anticlimacteric properties? Following my comments in *The Green Pharmacy* (Duke, 1997), about the potential that fenugreek might have for augmenting the size of micromastic breasts, I have received many letters and phone calls suggesting that yes, it worked. There now are at least three commercial breast-enhancing products, usually priced rather highly, containing fenugreek. I can't guarantee that it enlarges the breast, but I'll bet on it. Fenugreek is often recommended to enhance the flow of milk in women who are deficient, e.g., mothers of twins who have only enough milk for one. I'm convinced it helps. Many lactation consultants recommend it. In the *Integrative Herb Guide for Nurses*, Libster (2002) suggests fenugreek for hormone balance, noting that several mothers in her practice used fenugreek successfully to increase milk production and flow.

Fenugreek has been documented to lower the cholesterol in a few clinical trials. And containing at least five hypoglycemic compounds, it clearly has some promise in type 2 diabetes and Syndrome X. Three and one-half ounces of fenugreek flour in breads consumed daily was good at lowering fasting blood sugar. One-half ounce (15 g) powdered fenugreek in water lowers postprandial blood glucose (TAD). In Turkey, diabetics take 2 tsp ground seed with a glass of water, A.M. and P.M. (SPI). I think these activities more important than those approved by Germany's Commission E, dermatitis and loss of appetite. Commission E also acknowledges that it is antiseptic, rubefacient, and secretolytic. Treating rats with seed for four weeks significantly decreased the quantity of calcium oxalate deposited in the kidneys, supporting Saudi folklore (JE26:249).

Fenugreek is not often on the lips of cancer patients, but of the spices considered in this book, it is the best source of selenium, except for specially grown garlic and Brazil nut. According to Boik (2001), at least 35 *in vitro* studies suggest selenium has cytotoxic effects on different cancer lines, usually at levels of 3–9 μM. Twenty more show antitumor activity in animal studies. Epidemiology suggests that 200 µg selenium/day will prevent colon, lung, and prostate cancer. Selenium has been used in some studies with brain cancer patients, reducing the side effects of chemotherapy. All in all, studies suggest selenium supplementation may help prevent and treat cancer. Selenium, in small amounts, is needed for antioxidant glutathione peroxidase. Effects on cancer may also accrue to redox activity. Boik (2001) identifies selenium as a PTK-Inhibitor that may secondarily help affect CAMs, induce apotosis, inhibit angiogenesis, inhibit cell migration and invasion, inhibit eicosanoid effects, inhibit histaminic effects, inhibit NF-kB activity, inhibit platelet aggregation, and inhibit TNF. So we have one trace mineral that could reduce the incidence of cancer. Boik calculates tentative human dosages as 2.3–5.8 mg/day (as scaled from animal antitumor studies), suggesting a target dose of 3.7 mg/day selenium, way above the 200 µg normally prescribed to humans for noncancerous conditions (Boik, 2001).

And Trigonelline has been shown to significantly inhibit liver carcinoma in mice. It is reportedly used as a pessary for cervical cancer in China. In my database, fenugreek is the best source of diosgenin (to 1.9%), which is one starter material for natural progesterone, which, topically applied, better than oral progesterone, can help or help regulate breast cancer, cervical dysplasia, dysmenorrhea, endometrial cancer, endometriosis, hot flashes, menopause, metrorrhagia, osteoporosis, ovarian cysts, PMS, and uterine fibroids.

U.S. Patent No. 5,900,329 describes the oral administration of a fenugreek extract to strengthen skin and horny substances, particularly claws, hooves, and nails of animals, and to stimulate or revitalize growth of epidermal structures, like hairs, based on trigonelline, trigonellic acids and biotin. The patent claims it takes 3–4 months for fenugreek alone to accomplish this, while with ginseng and/or horsechestnut added, the stimulus is evident within 1 month. Thus, the patent covers propecic claims. Caveat emptor (Dean, K., 2000, Plant Patents, *Herbalgram* No. 50., p. 31).

According to Sambaiah and Srinivasan (1989) fenugreek stimulated liver microsomal cytochrome p450 dependent aryl hydroxylase. Patel and Srinivasan (1985) suggest that fenugreek decreased levels of phosphatase and sucrase activities.

Indications (Fenugreek) — Abscess (f; WOI); Adenosis (f; CRC; HHB); Aging (f; BOW); Alactea (1; PH2; WOI); Allergy (f; PED); Alopecia (1; APA; KAP; MAD); Anemia (1; GMH; SPI); Anorexia (2; APA; CAN; KOM; PH2; JAC7:405); Aposteme (f; JLH); Arthrosis (1; KOM); Atherosclerosis (1; BGB; FNF; SKY); Backache (f; BOW); Bacteria (1; FNF; WOI); Boil (f; BGB; GMH; KAP); Bronchosis (f; APA; PH2); Burn (f; CRC); Calculus (1; APA); Cancer (1; APA); Cancer, abdomen (1; FNF; JLH); Cancer, bladder (1; FNF; JLH); Cancer, breast (1; FNF; JLH); Cancer, cervix (1; BOW); Cancer, colon (1; FNF; JLH); Cancer, eye (1; FNF; JLH); Cancer, gland (1; FNF; JLH); Cancer, groin (1; FNF; JLH); Cancer, intestine (1; FNF; JLH); Cancer, kidney (1; FNF; JLH); Cancer, liver (1; FNF; JLH); Cancer, parotid (1; FNF; JLH); Cancer, rectum (1; FNF; JLH; MAD); Cancer, spleen (1; FNF; JLH); Cancer, stomach (1; FNF; JLH); Cancer, testes (1; FNF; JLH); Cancer, throat (1; FNF; JLH); Cancer, uterus (1; FNF; JLH); Cancer, uvula (1; FNF; JLH); Carbuncle (f; GMH; KAP); Catarrh (f; PH2); Cellulosis (f; BOW); Cervicosis (f; BOW); Childbirth (1; APA; BGB; BOW); Chilblain (f; CRC); Cholecystosis (f; CRC); Colic (f; KAP); Constipation (1; SKY; SPI); Colitis (f; PH2); Cough (f; APA; PED; PH2); Cystosis (f; JLH; SKJ); Debility (f; MAD); Dermatosis (2; APA; CRC; FNF; PHR; PH2); Diabetes (NIDDM) (2; APA; BRU; CAN; CRC; FNF); Diarrhea (1; APA; CRC; KAP); Dysentery (f; HHB; KAB); Dysgeuzia (f; KAB); Dysmenorrhea (f; BGB); Dyspepsia (2; APA; CAN; FNF; PHR); Eczema (1; BGB; PHR; PH2); Edema (1; BGB; BOW; FNF; SKJ); Enterosis (f; APA; BGB; CRC; PH2; WOI); Exhaustion (f; MAD); Fever (f; APA; CRC; PH2); Fistula (f; CRC); Fungus (1; FNF); Furunculosis (f; BGB; HHB; PHR); Gas (1; APA); Gastrosis (f; APA; BGB; CAN; GMH); Gout (f; BGB; CAN; CRC; GMH); Hay Fever (f; PED); Hemorrhoid (f; MAD); Hepatosis (1; CRC; FNF; JLH; KAP); Hernia (f; APA; BGB; CRC; PH2); High Blood Pressure (1; CAN; FNF); High Cholesterol (2; APA; BRU; CAN; FNF; SKY); High Triglycerides (1; BGB; FNF; SKY); Hyperglycemia (1; FNF); Hyperlipidemia (1; BGB; FNF); Impotence (f; APA; CRC; PH2); Infection (1; APA; FNF; WOI); Inflammation (2; APA; BRU; FNF; KOM; PH2); Kidney Stone (1; JE26:249); Leukorrhea (f; KAP); Lymphadenosis (f; BGB; CAN); Mastosis (f; JLH); Myosis (f; BGB; CAN); Nephrosis (f; APA; CRC; JLH); Neuralgia (f; APA; CRC); Neurasthenia (f; BOW; GMH); Ophthalmia (f; JLH); Orchosis (f; JLH); Osteomyelosis (f; HHB; MAD); Oxaluria (1; APA); Pain (1; PH2; TAD); Parotosis (f; JLH); PMS (f; BGB); Proctosis (f; JLH; MAD); Rachosis (f; MAD); Respirosis (f; APA; PH2); Rheumatism (f; APA; CRC); Sciatica (f; CRC); Scrofula (f; GMH; HHB); Smallpox (f; KAB; KAP); Sore (f; APA; BGB; MAD; PH2); Sore Throat (1; APA; CRC; MAD); Spermatorrhea (f; BOW); Splenosis (f; HHB; KAP); Splenomegaly (f; CRC; KAB); Stomatosis (f; APA); Stone (1; JE26:249); Stress (1; FNF); Swelling (f; HHB; KAB; PHR); Syndrome-X (1; SYN); Syphilis (f; SKJ); Tuberculosis (f; APA; CRC; HHB; MAD; SPI); Tumor (f; CRC); Ulcer (1; APA; PNC); Ulcus cruris (f; HHB); Uterosis (f; JLH); Vaginosis (f; BGB); VD (f; SKJ); Vomiting (f; PH2); Water Retention (1; FNF); Wound (f; BGB; HHB).

Fenugreek for dermatosis:

- Analgesic: coumarin; gentianine; pyridoxine; quercetin
- Antibacterial: apigenin; carpaine; gentianine; isoquercitrin; kaempferol; lignin; luteolin; p-coumaric-acid; quercetin; quercitrin; rutin

- Antidermatitic: apigenin; biotin; pyridoxine; quercetin; rutin; vitexin
- Antiinflammatory: apigenin; coumarin; diosgenin; fenugreekine; genistein; gentianine; gitogenin; kaempferol; luteolin; orientin; quercetin; quercitrin; rutin; salicylates; super-oxide-dismutase; vicenin-2; vitexin
- Antiseptic: kaempferol; oxalic-acid; trigonelline
- Antistress: diosgenin; gamma-aminobutyric-acid
- COX-2-Inhibitor: apigenin; kaempferol; quercetin
- Cyclooxygenase-Inhibitor: apigenin; kaempferol; quercetin
- Demulcent: mucilage
- Fungicide: coumarin; formononetin; genistein; p-coumaric-acid; phytic-acid; quercetin
- Immunomodulator: saponins
- MDR-Inhibitor: genistein

Fenugreek for diabetes:

- Aldose-Reductase-Inhibitor: coumarin; genistein; isoquercitrin; kaempferol; luteolin; p-coumaric-acid; quercetin; quercitrin; rutin; vitexin
- Antiaggregant: apigenin; coumarin; genistein; kaempferol; phytic-acid; pyridoxine; quercetin; salicylates
- Anticapillary-Fragility: quercetin; rutin
- Antidiabetic: pyridoxine; quercetin; rutin
- Antioxidant: apigenin; genistein; isoorientin; isoquercitrin; isovitexin; kaempferol; lignin; luteolin; orientin; p-coumaric-acid; phytic-acid; quercetin; quercitrin; rutin; super-oxide-dismutase; vitexin
- Antiperoxidant: isoorientin; p-coumaric-acid; quercetin
- Antiradicular: isoquercitrin; kaempferol; quercetin; rutin
- Antithrombic: quercetin
- Hypocholesterolemic: diosgenin; formononetin; genistein; lignin; nicotinic-acid; phytic-acid; rutin; trigonelline
- Hypoglycemic: coumarin; fenugreekine; gentianine; nicotinic-acid; quercetin; quercitrin; salicylates; trigonelline
- Insulinase-Inhibitor: nicotinic-acid
- Insulinogenic: quercetin
- Insulinotonic: nicotinic-acid

Fenugreek for high cholesterol:

- Antiaggregant: apigenin; coumarin; genistein; kaempferol; phytic-acid; pyridoxine; quercetin; salicylates
- Antiatherogenic: rutin
- Antiatherosclerotic: genistein; pyridoxine; quercetin
- Antidiabetic: pyridoxine; quercetin; rutin
- Antihomocystinuric: pyridoxine
- Antiischemic: genistein
- Antilipoperoxidant: quercetin
- Antioxidant: apigenin; genistein; isoorientin; isoquercitrin; isovitexin; kaempferol; lignin; luteolin; orientin; p-coumaric-acid; phytic-acid; quercetin; quercitrin; rutin; super-oxide-dismutase; vitexin
- Choleretic: apigenin; kaempferol; luteolin; nicotinic-acid; p-coumaric-acid; quercitrin
- Diuretic: apigenin; carpaine; fenugreekine; gamma-aminobutyric-acid; isoquercitrin; kaempferol; luteolin; quercitrin

- Hepatoprotective: diosgenin; luteolin; quercetin
- Hypocholesterolemic: diosgenin; formononetin; genistein; lignin; nicotinic-acid; phytic-acid; rutin; trigonelline
- Hypolipidemic: formononetin; phytic-acid
- Hypotensive: apigenin; carpaine; fenugreekine; gamma-aminobutyric-acid; gentianine; isoquercitrin; kaempferol; quercetin; quercitrin; rutin; vitexin

Other Uses (Fenugreek) — Widely cultivated as a condiment crop, fenugreek seeds, containing coumarin, are used in curries, salads, and soups (LIL). Seeds are also used in teas, blending nicely with mint flavored combinations (RIN). Roasted seeds are used as a coffee substitute. In Lebanon, a milkshake-like hypotensive beverage is made by grinding the green seed after soaking. In Greece, raw or boiled seeds are eaten with honey. In North Africa, fenugreek is mixed with breadstuffs. I made my own artificial maple syrup with fenugreek seed boiled in water, sweetened with stevia. It had that great maple scent, better on ice cream than on pancakes. The predominant scent in many prepared curry powders, it can override all other aromas in an Indian spice shop. In south India, it is found in chutneys, lentil dishes, pickles, and vegetables, and with potato, eggplant, and cauliflower (AAR). Pastirma (or basderma), a spiced dried beef in Armenia and Turkey, is cured with fenugreek powder, garlic, allspice, cumin, black pepper, cinnamon, nutmeg, and cloves, and colored red with paprika and ground red chiles. This spice mixture is called "chaiman" (or chemen), as is the fenugreek itself (AAR). Turks report that a boned piece of beef, roughly $40 \times 20 \times 10$ cm, is rubbed with salt and allowed to dry. The spice mix, 34% powdered garlic, 20% fenugreek, 6.5% cayenne, 1.5% cumin and mustard, and 38% water, is applied to the meat, drying to a red crust. The meat is then heated to 60-65°C for around 4 hours, then left to cure up to 12 days at ambient temperature (SPI). Jewish hilbeh is made by pouring boiling water over about 2 tbsp of ground fenugreek and letting it sit undisturbed for several hours overnight. The resulting jelly is beaten at high speeds, adding a pinch of salt, a little lemon juice, some cayenne pepper (or garlic, ginger, caraway or cardamom seeds, chopped tomatoes, finely chopped chiles, and coriander). Refrigerated hilbeh will keep for a week (AAR). Harem women are said to eat roasted fenugreek seed to attain buxomness. Mixed with cottonseed, the seed increases the flow of milk in cows but imparts the fenugreek aroma to milk. Plant serves as a potherb, much favored in India. Sprouts and seedlings are said to make good salad dressed with oil and vinegar. Sprouted seeds can be braised in olive oil with parboiled cardoon stalks. For sprouts, soak 1 or 2 tbsp seeds in warm water for a few hours, then lay them on wet paper towel in a sealed, opaque glass container. One-fourth inch is ample growth for the sprouts (AAR). In the Near East, sprouting seeds are added to a lamb stew traditionally flavored with honey. In India, green leaves are eaten as methi. Fresh methi, cooked as a green, is a favorite served with fish. Highly seasoned with turmeric, cumin, ajowan, fresh ginger, and chiles, the leaves are mixed with cornmeal and fried as fritters. In Iran, fresh or dried, leaves are used to add strong flavor to stews, soups and ashes, thick, nourishing, amin-course soups. Leaves are usually first sautéed in oil with other green vegetables and herbs, e.g., celery, leeks, parsley, scallions, or spinach. Fenugreek is in ghormeh sabzi, a rich, dark green mix of herbs, beans, dried limes, and some meat (AAR). With its bitter taste, somewhat like lovage and celery, it might be useful in vegetarian bouillon (LIL). Seeds and leaves may be brewed into a pleasant tea. For northerners, attempting to make liqueurs that call for tonka or vanilla, fenugreek seed may be used as a substitute (LIL). I often add the seed to homemade herbal liqueurs and teas. The liqueur I call "Lydia's Downer" has hypoglycemic fenugreek and several reportedly hypotensive herbs, fennel, elecampane, parsley, tarragon, and a dash of rue (caution: photosensitizer), steeped in the cheapest gin (LIL). I now suggest adding methi leaves to Lydia's Downer as an anesthetic. I also suggest a "Lydia's Lady's Liqueur" with angelica, anise, cohosh (not containing formononetin, as per studies in 2001), fenugreek, fennel, and red clover flowers as a poor feminists alternative to Remifemin. The celery-scented oil is used in butterscotch, cheese, licorice, pickle, rum, syrup, and vanilla flavors. Used also in cosmetics, hair preparations, and perfumery. Indians grow the plant as a forage. Considered a good soil renovator. Fenugreek is a source of diosgenin, used in the synthesis of hormones. As with other chemurgic crops, a large percentage must be thrown into the pot to extract a small percentage (1–2% on a dry weight basis) of pharmaceutical (diosgenin); proteins, fixed oils, oleoresins, e.g., coumarin, mucilages, and/or gums might also be extracted. Organic residues might be used for biomass fuels or manures, inorganic residues for "inorganic" chemical fertilizers. The husk of the seed might be removed for its mucilage, with the remainder partitioned into oil, sapogenin, and protein-rich fractions. Seed mucilage (ca. 45%) could be prepared from the marc left after extraction of the fixed oil (used as a lactagogue). Its relatively high viscosity makes it a good emulsifying agent to be used in pharmaceutical and food industries. Due to its neutral ionic properties, it is compatible with other drugs or compounds sensitive to acids. Plant is used to make horse hair shiny. Powdered seeds are used locally for a yellow dye (AAR, CRC, FAC, LEG).

For more information on activities, dosages, and contraindications, see the *CRC Handbook of Medicinal Herbs*, ed. 2, Duke et al. 2002.

Cultivation (Fenugreek) — For temperate zone instructions for North Americans, I readily recommend *The Big Book of Herbs* (TAD). The following comments are derived largely from other sources, oriented more to the third world. Seeds long retain their viability. Seed may be sown in the fall in mild climate, or in the spring farther north, and germinate in 4–5 days. Seed should be sown in close drills ca. 7.5 cm apart, the rows ca. 45 cm apart, or broadcast at ca. 22.5 kg/ha. Deep plowing and thorough harrowing are essential. Clean cultivation, either mechanical or manual, is necessary. In southern India, fenugreek is intercropped with coriander, gingelly, or Bengal gram. For seed production, potash and phosphoric acid have been recommended; for forage, nitrogenous manures. When grown as an irrigated crop, seeds are broadcast rather thickly onto beds at the rate of 25–30 kg/ha and then stirred into the soil. Irrigation should start immediately after sowing and continue when necessary. Seeds ripen some 3–5 months after planting. Pods retain their seed shell (don't readily shatter). Manually, the plants are uprooted and dried a few days before threshing. Seeds are then further dried. Seed yields of 500–3500 kg/ha may be expected; fodder/forage yields 20–10 MT/ha (LEG, LIL).

Chemistry (Fenugreek) — Here are a few of the more notable chemicals found in fenugreek. For a complete listing of the phytochemicals and their activities, see the CRC phytochemical compendium, Duke and duCellier, 1993 (DAD) and the USDA database http://www.ars-grin.gov/duke/.

Diosgenin — See also *Costus speciosus*.

Fenugreekine — Antiinflammatory; Antivaccinia; Antiviral; Cardiotonic; Diuretic; Hypoglycemic; Hypotensive; Viristat.

Selenium — Analgesic 200 µg/day; Anorectic; Antiacne 200 µg/day; Antiaggregant; Antiangiogenic 2 µM, 230 µg/kg orl rat; Anticancer; Anticirrhotic; Anticoronary 200 µg/day; Antidandruff; Antidote (Mercury); Antikeshan; Antileukemic 1.6 mg/kg ipr mus; Antileukotrienic; Antimelanomic 480 µg/kg; Antimetastatic 480 µg/kg; Antimyalgic 200 µg/day; Antiosteoarthritic; Antioxidant 100–200 (–400) µg/man/day; Antiproliferant 2 µM; Antiradicular 100–200 (–400) µg/man/day; Anti-Syndrome-X 100–200 (–400) µg/man/day; Antitumor 100–200 (–400) µg/man/day; Antitumor (Brain) 38–150 µg/kg; Antitumor (Breast) 0.8 mg/kg scu mus, 150 µg/kg diet rat, 230 µg/kg orl rat; Antitumor (Lung) 240 µg/kg diet; Antiulcerogenic; AP-1-Inhibitor 2–50 µM; Apoptotic; Depressant; Fungicide; Immunostimulant 100–200 (–400) µg/man/day; NF-kB-Inhibitor; ODC-Inhibitor; Polyamine-Synthesis-Inhibitor; Protein-Kinase-C-Inhibitor IC50 = 2–50 µM; Prostaglandin-Sparer; VEGF-Inhibitor; RDA = 10–75 µg/day; PTD = 1 mg/day.

Trigonelline — Antimigraine; Antiseptic; Antitumor (Cervix); Antitumor (Liver); Epidermal-Stimulant; Hypocholesterolemic; Hypoglycemic 500–3000 mg/man/day, 50 mg/kg orl rat; Mutagenic; Osmoregulator; Propecic; LDlo = 5000 scu rat; LD50 = 5000 orl rat.

V

Vanilla planifolia Jacks (Orchidaceae)
VANILLA

Synonyms — *Myrobroma fragrans* Salisb., *Vanilla fragrans* auct.

Medicinal Uses (Vanilla) — This is one of the few spices in this book that clearly has more culinary than medicinal interest. Still it was used as a galenical and a carrier for other less tasteful ingredients. Reported to be aphrodisiac, carminative, stimulant, and vulnerary. Venezuelans use the pods against fevers and spasms. Yucatanese use vanilla extract (pod steeped in alcohol) as aphrodisiac and stimulant. Argentineans use it as an antispasmodic, aphrodisiac, or emmenagogue. In Palau, vanilla is used for dysmenorrhea, fever, and hysteria. Vanilla is said to inhibit caries, probably because of the catechin.

Adesanya and Sofowora (1995) investigated several aromatic aldehydes known to form imine adducts with amino groups, thereby inhibiting the gelation of the hemoglobin. O-vanillin proved most active >2,3-dihydroxybenzaldehyde > salicylaldehyde >2,4-dihydroxybenzaldehyde > p-vanillin >m-anisaldehyde > p-hydroxybenzaldehyde > guaiacol. The activity of the o-vanillin is dose dependent (Adesanya and Sofowora, 1995). At least five of these compounds are found in vanilla. Regrettably, Adesanya and Sofowora (1995) did not report if a mix of these aldehydes was additive or synergistic. The very pleasant vanilla is my richest source of vanillin, the unpleasant asafetida my second richest source. Vanillic acid, also reportedly useful in sickle cell, is most abundant in Pircorrhiza but is generously represented in the spices, coriander at almost 1000 ppm (ZMB), onion at ca. 250, and horseradish at 45 ppm. Shaughnessy et al. (2001) elucidated the antimutagenic effects of vanillin, best represented in vanilla (ca. 3%, ZMB), and cinnamaldehyde, best represented in cinnamon (ca. 3%, ZMB). Vanillin is also choleretic (ABS, DAD, JFM).

Vanilla is not much better as an antiseptic than as a medicine. A study by Nielsen and Rios (2000) showed the volatiles of vanilla to be completely ineffective at inhibiting growth of various bread molds: *Penicillium commune, P. roqueforti, Aspergillus flavus,* and *Endomyces fibuliger,* using volatile EOs and oleoresins (OL) from spices and herbs. Mustard EO was most efficacious, with cinnamon, clove, and garlic highly active, oregano slightly active, and vanilla inactive (X11016611). But the vanillin seems to be synergistic with more repellant volatile oils, like turmeric, citronella, and citrus. Tawatsin et al. (2001) showed that volatile oil from turmeric repelled *Aedes aegypti, Anopheles dirus,* and *Culex quinquefasciatus,* especially with the addition of 5% vanillin, under cage conditions for up to eight hours. Various volatile oils can be formulated with vanillin as mosquito repellents to replace deet.

French enologists Teissedre and Waterhouse (2000) studied antioxidant activities of EOs of spices and culinary herbs. They note that EOs are important as antibacterials and antifungals, as flavorants and preservatives when added to foods, and as cosmetics, for their aromatic and antioxidant properties. In their survey of 23 EOs, star anise was most potent (IC83 = 2 μM); EO of vanilla was not so strong (IC40 = 2 μM, IC66 = 5 μM) (JAF4:3801).

The calcium oxalate crystals in the plant may cause dermatitis. Workers with vanilla may exhibit dermatitis, headache, and insomnia, all symptoms of vanillism. Several toxic compounds are present in minor quantity. GRAS § 182.10, 182.20, and 169.3.

Indications (Vanilla) — Bacteria (1; FNF); Cancer (1; FNF); Caries (1; CRC; DAD); Cramp (1; FNF); Dysmenorrhea (f; CRC; DAD; FNF; HHB); Fever (f; CRC; DAD); Fungus (1; FNF); Hepatosis (1; FNF); Hysteria (f; CRC; DAD; HHB); Immunodepression (1; FNF); Infection (1; FNF); Inflammation (1; FNF); Nervousness (1; FNF); Pain (1; FNF); Polyp (f; JLH); Rhinosis (f; JLH); Sickle-Cell Anemia (1; FNF); Tumor (1; FNF); Virus (1; FNF); Water Retention (1; FNF).

Vanilla for cancer:

- Antiaggregant: anisyl-alcohol; catechin; cinnamaldehyde; coumarin; eugenol; ferulic-acid; salicylates
- Antiandrogenic: coumarin
- Antiarachidonate: eugenol
- Anticancer: benzaldehyde; catechin; catechol; cinnamaldehyde; cinnamic-acid-ethyl-ester; coumarin; eugenol; ferulic-acid; limonene; p-cresol; phenol; piperonal; salicylic-acid; syringaldehyde; umbelliferone; vanillic-acid; vanillin
- Anticarcinogenic: ferulic-acid
- Antiestrogenic: ferulic-acid
- Antihepatotoxic: catechin; ferulic-acid; protocatechuic-acid
- Antiinflammatory: catechin; cinnamaldehyde; coumarin; eugenol; ferulic-acid; n-hentriacontane; protocatechuic-acid; salicylates; salicylic-acid; syringaldehyde; umbelliferone; vanillic-acid; zingerone
- Antileukemic: catechin; cinnamaldehyde
- Antilipoperoxidant: catechin
- Antimelanomic: coumarin
- Antimetastatic: coumarin
- Antimutagenic: anisaldehyde; cinnamaldehyde; cinnamyl-alcohol; coumarin; eugenol; ferulic-acid; limonene; n-nonacosane; p-cresol; protocatechuic-acid; umbelliferone; vanillin
- Antineoplastic: ferulic-acid
- Antinitrosaminic: ferulic-acid

- Antioxidant: catechin; catechol; eugenol; ferulic-acid; phenol; protocatechuic-acid; salicylic-acid; syringaldehyde; vanillic-acid; vanillin; zingerone
- Antiperoxidant: protocatechuic-acid
- Antiprostaglandin: catechin; eugenol; umbelliferone
- Antithromboxane: eugenol
- Antitumor: benzaldehyde; coumarin; eugenol; ferulic-acid; limonene; salicylic-acid; vanillic-acid; vanillin
- Antiviral: catechin; catechol; cinnamaldehyde; ferulic-acid; limonene; phenol; protocatechuic-acid; vanillin
- COX-2-Inhibitor: eugenol; salicylic-acid
- Chemopreventive: coumarin; limonene
- Cyclooxygenase-Inhibitor: catechin; cinnamaldehyde; salicylic-acid; zingerone
- Cytotoxic: cinnamaldehyde; eugenol
- Hepatoprotective: catechin; eugenol; ferulic-acid
- Immunostimulant: benzaldehyde; catechin; coumarin; ferulic-acid; protocatechuic-acid
- Lipoxygenase-Inhibitor: catechin; cinnamaldehyde; umbelliferone
- Lymphocytogenic: coumarin
- Lymphokinetic: coumarin
- Ornithine-Decarboxylase-Inhibitor: ferulic-acid; limonene
- Prostaglandigenic: ferulic-acid; protocatechuic-acid
- Sunscreen: ferulic-acid; umbelliferone

Vanilla for cramp:

- Analgesic: coumarin; eugenol; ferulic-acid; phenol; salicylic-acid
- Anesthetic: benzaldehyde; benzoic-acid; benzyl-alcohol; cinnamaldehyde; eugenol; guaiacol; phenol
- Anticonvulsant: eugenol
- Antiinflammatory: catechin; cinnamaldehyde; coumarin; eugenol; ferulic-acid; n-hentriacontane; protocatechuic-acid; salicylates; salicylic-acid; syringaldehyde; umbelliferone; vanillic-acid; zingerone
- Antispasmodic: benzaldehyde; cinnamaldehyde; eugenol; ferulic-acid; limonene; protocatechuic-acid; umbelliferone; vanillyl-alcohol
- COX-2-Inhibitor: eugenol; salicylic-acid
- Carminative: eugenol
- Cyclooxygenase-Inhibitor: catechin; cinnamaldehyde; salicylic-acid; zingerone
- Lipoxygenase-Inhibitor: catechin; cinnamaldehyde; umbelliferone
- Sedative: benzaldehyde; benzyl-alcohol; cinnamaldehyde; coumarin; eugenol; limonene
- Tranquilizer: cinnamaldehyde

Vanilla for infection:

- Analgesic: coumarin; eugenol; ferulic-acid; phenol; salicylic-acid
- Anesthetic: benzaldehyde; benzoic-acid; benzyl-alcohol; cinnamaldehyde; eugenol; guaiacol; phenol
- Antibacterial: acetic-acid; acetophenone; benzaldehyde; benzoic-acid; catechin; cinnamaldehyde; eugenol; ferulic-acid; guaiacol; limonene; o-coumaric-acid; phenol; protocatechuic-acid; salicylic-acid; umbelliferone; vanillic-acid
- Antiedemic: catechin; coumarin; eugenol; syringaldehyde

- Antiinflammatory: catechin; cinnamaldehyde; coumarin; eugenol; ferulic-acid; n-hentri-acontane; protocatechuic-acid; salicylates; salicylic-acid; syringaldehyde; umbelliferone; vanillic-acid; zingerone
- Antilymphedemic: coumarin
- Antiseptic: anisic-acid; benzaldehyde; benzoic-acid; benzyl-alcohol; catechol; cresol; ethyl-vanillate; eugenol; formic-acid; furfural; guaiacol; limonene; oxalic-acid; p-cresol; phenol; salicylic-acid; umbelliferone
- Antiviral: catechin; catechol; cinnamaldehyde; ferulic-acid; limonene; phenol; protocat-echuic-acid; vanillin
- Astringent: catechin; formic-acid
- Bacteristat: coumarin; malic-acid; n-hexacosane
- COX-2-Inhibitor: eugenol; salicylic-acid
- Circulostimulant: cinnamaldehyde
- Cyclooxygenase-Inhibitor: catechin; cinnamaldehyde; salicylic-acid; zingerone
- Fungicide: acetic-acid; acetophenone; anisaldehyde; benzoic-acid; catechin; cinnamal-dehyde; coumarin; ethyl-vanillate; eugenol; ferulic-acid; furfural; o-coumaric-acid; octanoic-acid; phenol; protocatechuic-acid; salicylic-acid; umbelliferone; vanillin
- Fungistat: formic-acid; limonene
- Immunostimulant: benzaldehyde; catechin; coumarin; ferulic-acid; protocatechuic-acid
- Lipoxygenase-Inhibitor: catechin; cinnamaldehyde; umbelliferone

Other Uses (Vanilla) — When the conquistadors entered Mexico, under Cortez, in the 16th century, they learned of the black pod "tlilxochitl," Aztec word for the vanilla pod. Legend has it that Montezuma gave Cortez honey-sweetened chocolate, flavored with vanilla, in golden goblets. Subjects of Montezuma offered him vanilla beans as tribute. Full-grown, unripe pods of this orchid, when properly dried or cured, contain vanillin used in flavoring all kinds of confectionery, usually in the form of vanilla essence, extract, or tincture. True vanilla beans possess a pure, delicate spicy flavor and a peculiar bouquet not duplicated exactly by the synthetic product. Vanilla extract is used to flavor beverages, cakes, chocolates, confections, custards, ice creams, liqueurs, puddings, soft drinks, syrups, and yogurts. It is also used in perfumes, sachet powders, and soap. Highest maximum use level is nearly 10,000 ppm for vanilla in baked goods. Vanilla is the main flavoring of Galliano, an Italian liqueur, containing several (up to 40) other botanicals. Galliano is an ingredient of the Harvey Walbanger cocktail (FAC). Vanilla sugar, made by storing vanilla beans in granulated or confectioner's sugar for ca. a week, is used as an aromatic, flavorful decoration for baked goods, fruit, and other desserts (FAC). Vanilla is the most popular flavor in the U.S., accounting for ca. $\frac{1}{3}$ of all ice cream sales. Ice cream is the largest user of natural vanilla, making up half the vanilla market. Vanilla is said to alleviate the sweet tooth of obese people fearing dental caries; the vanilla extract cutting back on the sugar requirement for fresh fruit salads in reducing diets. The catechin is said to curb caries. Most vanilla flavorings used in baking, confectionary, and many frozen desserts, contain some vanillin, ethyl vanillin, vanitrope, or a combination of these. Vanillin accounts for >90% of the U.S. market for vanilla flavorings. Most vanillin of commerce is prepared synthetically from lignin. It has served as a poison bait for fruit flies, grasshoppers, and melon beetles. A vanilla-scented cotton ball can serve as a perfuming sachet in a dresser drawer (RIN). "The scent of vanilla is a delicious perfume. Pour a few drops of vanilla extract on a small piece of absorbent cotton and tuck it in your bra" (Do not put the vanilla extract on your skin; it may be irritating)(DAD, RIN).

For more information on activities, dosages, and contraindications, see the *CRC Handbook of Medicinal Herbs*, ed. 2, Duke et al. 2002.

Cultivation (Vanilla) — Bown (2001) suggests moist, rich, well-drained soils in humid zones, with day temperatures 26–30°C (79–86°F), and minimum temperatures 16°C (61°F). Though propagated by seed, vanilla is more usually propagated by long cuttings, usually with 12–24 internodes, since these can bear fruit in 1–2 years if planted at onset of rainy season. Short cuttings do not fruit for 3 or 4 years. Place supports ca. 2.7 m apart, with one vine per prop; some 1600 vines/ha are possible. Plants may be spaced 1.2 m apart in rows 2.5 m apart, giving 4000 vines/ha. Many types of support trees are used, these providing about 50% shade. However, spacing of support plants and vanilla vines varies greatly in different areas. Trellises are also used in some areas. Weeds like Guatemala grass, or a fast-growing legume, may be used for mulch. Do not use animal manure. Potash and lime are necessary. Vines must be pruned to keep the fruits within reach of hand-pickers. Diseased plants must be removed. The tip of vine should be clipped about 9 months before flowering season. Flowering plants need daily attention. Flowers must be pollinated and the remaining buds removed from the inflorescence. After fruiting, old stems should be cut away. In Mexico, a small crop is gathered 3 years after planting, and crops increase until the ninth or tenth year, when decline begins. Plants have been known to bear for 30–40 years (DAD). Vanilla beans are hand cut or broken off as they ripen to a yellow color and develop a hard black tip. Picked beans are placed on racks for 24 hr before they are wrapped in mats or wool blankets for sweating and fermentation. Beans are repeatedly sweated between the blankets in the sun, or in ovens during the day, and packed in wool covered boxes at night for about two months, during which time the pods lose 70–80% of their original weight and take on the characteristic odor and color of the commercial vanilla. After beans are fermented, they are dried for 8-12 days. Yields average about 115 g cured bean per plant, more than 450 g of green bean. One hectare of vines (about 1500) yields 110 kg of cured pods in Mexico, 150 kg in Hawaii, and 100 kg in Seychelles. It takes 3.3 kilos of green vanilla fruit to yield 1 kilo of dried prepared pods (DAD).

Chemistry (Vanilla) — Sun et al. (2001) state that more than 180 volatile compounds have been identified from cured vanilla beans, more than one-third aromatic. Noting the relative bug-resistance of the vanilla leaves in the greenhouse, they studied alcoholic extracts of leaves and stems. The ethyl acetate fraction was more toxic to mosquito larvae, with 4-ethoxymethylphenol, 4-butoxymethylphenol, vanillin, 4-hydroxy-2-methoxycinnamaldehyde, and 3,4-dihydroxyphenylacetic acid. 4-ethoxymethylphenol was the predominant compound, but 4-butoxymethylphenol showed greater toxicity to mosquito larvae. Four of the five killed the mosquito larvae within 24 hr. Green vanilla beans contain two glycosides, glucovanillin (avenein) and glucovanillic alcohol. Vanillin is principal flavoring constituent, about 1.5–2% for Mexican and 2.6% for Bourbon. Purseglove et al. (1981) dedicate more than four pages of tabulations to the chemistry of vanilla. Here are a few of the more notable chemicals found in vanilla. For a complete listing of the phytochemicals and their activities, see the CRC phytochemical compendium, Duke and duCellier, 1993 (DAD) and the USDA database http://www.ars-grin.gov/duke/.

Catechin — Allelochemic IC86 = 1 mM; Anesthetic; Antiaggregant IC68 = 200 µg/ml; Antialcoholic 2000 mg/man/day; Antiarthritic; Antiatherosclerotic; Antibacterial MIC = >1000 µg/ml; Anticancer; Anticariogenic; Antiedemic; Antiendotoxic; Antifeedant; Antiflu; Antihepatosis 1 g/3×/day/man; Antihepatotoxic; Antiherpetic; Antihistaminic 1000 mg 5×/day/man; Antiinflammatory IC50 = 80 µM (cf indomethacin IC50 = 1 µM); Antileukemic IC50 = >10 µg/ml; Antilipoperoxidant; Antiosteotic 500 mg/3×/day; Antioxidant IC50 = 0.19 µM, 2.4 × Vit. E, $^2/_3$ quercetin; Antiperiodontal; Antiplaque; Antiprostaglandin IC50 = 80 µM (cf indomethacin IC50 = 1 µM); Antiradicular IC50 = 8 µM; Antisclerodermic; Antistress; Antiulcer 1 g/5×/day/man/orl; Antiviral; Astringent; Carcinogenic; COMP-Inhibitor; COX-1-Inhibitor IC50 = 80 µM (cf indomethacin IC50 = 1 µM); Fungicide ED50 = 2.9–4.6 µg/ml; Hemostat; Hepatoprotective; Hypocholesterolemic; Hypolipedemic; Immunostimulant; Lipoxygenase-Inhibitor IC96 = 5 mM; Neuroprotective 1–10 µM; Propecic; Xanthine-Oxidase-Inhibitor.

V

Vanillic-Acid — Antibacterial 1.5–15 mg/ml; Anticancer; Antifatigue; Antiinflammatory; Antioxidant IC21 = 30 ppm; Antiradicular (7 × quercetin); Antisickling; Antitumor; Ascaricide; Choleretic; Immunosuppressive; Laxative; Ubiquict; Vermifuge.

Vanillin — ADI = 10 mg/kg; Allelochemic IC50 = 4.26 mM; Anticancer; Antimutagenic; Antioxidant (= ascorbic acid); Antipolio; Antiradicular (7 × quercetin); Antiviral; Choleretic; Fungicide; Immunosuppressive; Insectifuge; Irritant; LD50 = 1580 orl rat.

W

Wasabia japonica (Miq.) Matsum. (Brassicaceae)
JAPANESE HORSERADISH, WASABI

Synonyms — *Alliaria wasabi* Prantl, *Cochlearia wasabi* Sieb., *Eutrema japonica* (Miq.) Koidz., *E. wasabi* Maxim., *Lunaria japonica* Miq., *Wasabia pungens* Matsum., *W. wasabi* (Maxim.) Makino.

Medicinal Uses (Wasabi) — The health benefits of other cruciferous veggies also accrue to the wasabi, making it one of those that are especially useful for preventing cancer. These are all generously endowed with health-giving, sulfur containing compounds, like isothiocyanates and sulforaphane, to name some receiving a lot of press lately. I think that most of the activities and indications given for horseradish could accrue as easily to wasabi were it as well studied among occidentals. Dr. Hideki Masuda, Ph.D., director of the Material Research and Development Laboratories at Ogawa & Co., Ltd., in Japan, reports (pers. comm.) that a phytochemical in wasabi prevents tooth decay in laboratory tests. Isothiocyanates, which are reported from wasabi, are known to possess antiseptic and bactericidal properties. Dr. Masuda hypothesized and demonstrated that isothiocyanates would inhibit the growth of *Streptococcus mutans*, the bacteria that causes dental caries, in test-tube studies. One of 12 isothiocyanates he isolated effectively inhibited the enzyme glucosyltransferase (GTF), an enzyme that catalyzes the formation of glucan from sucrose by which *Streptococcus mutans* form plaque on teeth. *Streptococcus mutans* produces lactic acid on plaque and initiates tooth decay. He also reports other health benefits for wasabi: prevention of prostate cancer, prevention of harmful blood clot formation, antiasthmatic properties, and anti-parasitic properties (Masuda, pers. comm. 2001). Andy Weil (*Self Healing* March 2001) suggests horseradish (and wasabi) for battling stuffy sinus: "You can thin mucus and congestion with fresh horseradish. Clean and peel a press root, grate or grind after chunking, adding enough vinegar to moisten." I'll add that, if making the sauce doesn't open your sinuses and make your nose run, then its time to eat it (Weil, A. 2001. And it can be used as a sinus-opening seafood sauce, of green wasabi and red ketchup, lavishly lashed with lemon juice, and served with boiled shrimp, if not sushi.

One report in New York, and another in California, recount serious adverse effects due to ingestion of a generous portion of wasabi: white face, confusion, profuse sweating, even collapse (TAD). This response could be serious in patients with weakened blood vessels in the brain or heart (RIN). One person reportedly suffered "vasomotor near collapse" after ingesting a small amount (all in one bite, since he did not know wasabi). Diners in Japanese restaurants should know that green wasabi paste is meant to be mixed with tamari (soy sauce) and used in very small amounts (Libster, 2002).

Indications (Wasabi) — Asthma (1; ABS); Bacteria (1; ABS); Cancer (1; FNF; TAD); Cancer, lung (1; X10822125); Cancer, prostate (1; ABS); Caries (1; ABS); Congestion (f; ABS); Fungus (1; X10571166); High Blood Pressure (1; FNF); Infection (1; X10571166); Mycosis (1; X10571166); Parasite (1; ABS); Sinusosis (1; ABS); Streptococcus (1; ABS); Thrombosis (1; TAD); Trypanosomiasis (1; FNF); Virus (1; FNF).

Wasabi for cancer:

- Anticancer: 6-methyl-sulfinyl-hexyl-isothiocyanate; allyl-isothiocyanate; sinigrin
- Anticarcinomic: 6-methyl-sulfinyl-hexyl-isothiocyanate
- Antimutagenic: 4-pentenyl-isothiocyanate; allyl-isothiocyanate

Wasabi for cold/flu:

- Antiseptic: allyl-isothiocyanate
- Decongestant: allyl-isothiocyanate
- Phagocytotic: sinigrin

Other Uses (Wasabi) — Occurs in Japan and eastern Siberia as a cultivar, the roots, twigs, and petioles used as a spice, especially with fish dishes. It's the hot green horseradish-like spice so often served with Japanese cuisine, e.g., sashimi (slices of raw fish) and sushi (a piece of raw seafood on a bed of rice). The fleshy rhizomes are grated into the attractive green paste. Japanese consider it distinct from, and pungently superior to, common horseradish. In the U.S., it is sold in Japanese-American stores as a can of dry powder, to which one merely adds ¾ tsp water to 1 tsp powder. Mixed with soy sauce, the powder becomes a piquant and tasty dip. It is used to decorate carrots and cucumbers and with such dishes as "nigri zushi" (small kneaded ball of sour rice with sliced fish), "norimaki zushi" (Japanese sour rice in sheets of nori, a marine red algae, with cucumber and/or shitake), and "soba" (cold buckwheat noodles). Leaves, flowers, leafstalks, and freshly sliced rhizomes are soaked in salt water and then mixed with sake (Japanese rice wine) lees to make a popular pickle called "wasabi-zuke" (FAC, TAD).

For more information on activities, dosages, and contraindications, see the *CRC Handbook of Medicinal Herbs*, ed. 2, Duke et al. 2002.

Cultivation (Wasabi) — This herbaceous perennial, to almost a foot and a half tall, is hardy to zone 8, according to Tucker and deBaggio (2000). Bown (2001) suggest zones 6–9 with temperatures of 10–15°C (50–59°F) in the growing season. Stream temperatures must be kept at 10–13°C (50–57°F) because at high temperatures the plants fail and diseases succeed. It tolerates 50–80% shade, slightly alkaline, organic soils, but is best grown rather like watercress, of the same family, in cool running water. Cool, spring-fed streams in limestone forest are best. Japanese artificially widen stream beds with rock walls and elaborate terraces especially for wasabi production. In Japan, propagated from two-year-old rhizome offsets, transplanted in spring, these will soon form large clumps with clusters of heart shaped leaves. Offsets are spaced about 10 × 18 in (25 × 45 cm). Nitrogen-rich organic fertilizers are applied in November and March. Plants are usually harvested in June in Japan, digging the plants, separating offshoots for next year's plantings, and washing the rhizomes and removing the leaves. Yields may attain 2 tons/a to 4.5 MT/ha (TAD).

Chemistry (Wasabi) — Morimitsu et al. (2002) found that wasabi is the richest source of 6-methylsulfinylhexyl isothiocyanate (6-HITC), an analogue of sulforaphane (4-methylsulfinylbutyl isothiocyanate) isolated from broccoli, as the major GST-Inducer in wasabi. 6-HITC is a potential activator of novel detoxification pathways (11706044). Here are a few of the more notable chemicals found in wasabi. For a complete listing of the phytochemicals and their activities, see the CRC phytochemical compendium, Duke and duCellier, 1993 (DAD) and the USDA database http://www.ars-grin.gov/duke/.

Allyl-Isothiocyanate — Antiasthmic; Anticancer; Antifeedant; Antimutagenic; Antiseptic; Counterirritant; Decongestant; Embryotoxic; Fungicide MIC = 1.8–3.5 µg/ml; Herbicide IC100 = 0.4 mM; Insectiphile; Mutagenic; Nematiovistat 50 µg/ml; Spice FEMA 1–80 ppm; LD50 = 339 orl rat.

6-Methylsulfinylhexyl-Isothiocyanate — Anticarcinomic; Antitumor.

4-Pentenyl-Isothiocyante — Antimutagenic.

X

Xylopia aethiopica (Dunal) A. Rich (Annonaceae)
ETHIOPIAN PEPPER, GUINEA PEPPER, NEGRO PEPPER, SPICE TREE

Medicinal Uses (Ethiopian Pepper) — Important in African folk medicine. In Nigeria, it is used as a cancer remedy (JLH). Fruit extracts and bark decoction are used for bronchitis, biliousness, dysentery, and painful febrile conditions. Congolese steep bark in palm wine, suggesting one to two glasses a day for asthma, rheumatism, and stomachache. Concentrated root decoction used as mouthwash in toothache. Fruits used to counter pain and as a carminative and laxative. Used also as anthelminthic, they are especially recommended in puerperium. Its analgesic property is useful for chest pain, dermatosis, headache, lumbago, neuralgia, parturition, rib ache, and sideache. The fruits were suspected to enhance fertility and aid delivery. It was even believed abortifacient. Fruits are particularly high in zinc, perhaps a solid rationale for recommending consumption during pregnancy and lactation (X8616672).

Xylopic acid and two other diterpene isolates were found to have antimicrobial properties when tested against five microorganisms, *Bacillus subtilis, Candida albicans, Escherichia coli, Pseudomonas aeruginosa,* and *Staphylococcus aureus* (X600023). Oxophoebine and liriodenine showed selective toxicity against DNA repair and recombination deficient mutants of the yeast *Saccharomyces cerevisae,* while oxoglaucine, O-methylmoschatoline, and lysicamine were inactive. Bioactive oxoaporphine alkaloids may act as DNA topoisomerase inhibitors (X8158166).

And extracts showed cardiovascular and diuretic activity and displayed low toxicity (LC50 = 0.5–5.0 ng/ml). The diterpene kaurenoids showed significant systemic hypotensive and coronary vasodilatory effect accompanied with bradycardia, perhaps due to calcium antagonistic mechanism. The diuretic and natriuretic effects found were comparable to chlorothiazide (X11535360).

Indications (Ethiopian Pepper) — Amenorrhea (f; UPW); Asthma (f; UPW); Bacteria (1; FNF); Biliousness (f; UPW); Boil (f; UPW); Bronchosis (f; FNF; UPW); Cancer (f; FNF; JLH; UPW); Cardiopathy (1; X11535360); Childbirth (f; UPW); Constipation (f; UPW); Convulsion (f; UPW); Cough (f; UPW); Dermatosis (f; FNF; UPW); Dysentery (f; FNF; UPW); Epilepsy (f; UPW); Headache (f; FNF; UPW); High Blood Pressure (1; X11535360); Infertility (f; UPW); Lumbago (f; UPW); Neuralgia (f; UPW); Pain (f; UPW); Pneumonia (f; FNF; UPW); Pyorrhea (f; UPW); Respirosis (f; UPW); Rheumatism (f; FNF; UPW); Rib Ache (f; UPW); Roundworm (f; UPW); Side Ache (f; UPW); Stomachache (f; UPW); Toothache (f; FNF; UPW); Water Retention (1; X11535360); Worm (f; UPW).

Ethiopian Pepper for cardiopathy:

- Antiatherogenic: rutin
- Antiedemic: piperine; rutin
- Antihemorrhoidal: rutin
- Antioxidant: rutin
- Cardiotonic: piperine
- Hypocholesterolemic: rutin
- Hypotensive: piperine; rutin

X

- Sedative: liriodenine
- Vasodilator: liriodenine

Ethiopian Pepper for dysentery:

- Antibacterial: liriodenine; piperine; rutin
- Antihepatotoxic: rutin
- Antiseptic: piperine
- Antispasmodic: piperine; rutin
- Antiviral: rutin
- Candidicide: liriodenine
- Carminative: piperine
- Hepatoprotective: piperine

Ethiopian Pepper for edema:

- Analgesic: liriodenine; piperine
- Anesthetic: piperine
- Anticapillary-Fragility: rutin
- Antiedemic: piperine; rutin
- Antiinflammatory: piperine; rutin
- Lipoxygenase-Inhibitor: rutin

Other Uses (Ethiopian Pepper) — Fruits probably most important as spice. Seeds and fruits once exported from Africa to Europe in the Middle Ages as a peppery spice. Now mostly of local use in Africa to season coffee, palm wine, and local dishes (FAC). Added to purify dirty water. Fruits even smoked like a pungent tobacco. Pulverized fruits added to snuffs to increase the pungency. Fruits were ground with red pepper (*Capsicum*) and *Cola* to repel the Kola weevil. Seeds also used as a spice, distinct from the fruits. Mixed with other spices, they are rubbed on the body as a cosmetic and scent. Leaves macerated in palm wine to make an intoxicant. Bark is used to make cordage, to wrap around torches, and in making fragile doors and partitions. Wood used in boats, bows, crossbows, joists, masts, oars, paddles, posts, and spars (UPW). Termite resistant wood is also used in home building. Burning with a hot flame, it is burned to fuel steamboats. Root wood can be used as corkage.

For more information on activities, dosages, and contraindications, see the *CRC Handbook of Medicinal Herbs*, ed. 2, Duke et al. 2002.

Cultivation (Ethiopian Pepper) — Sometimes left to stand in the villages.

Chemistry (Ethiopian Pepper) — Some 28 odor-active compounds in the flavor are linalol (floral), followed by (E)-beta-ocimene (flowery), alpha-farnesene (sweet, flowery), beta-pinene (terpeny), alpha-pinene (pine needle-like), myrtenol (flowery), and beta-phellandrene (terpeny). Vanillin (vanilla-like) and 3-ethylphenol (smoky, phenolic) were detected for the first time in the dried fruit (Tairu et al., 1999). Here are a few of the more notable chemicals found in Ethiopian pepper. For a complete listing of the phytochemicals and their activities, see the CRC phytochemical compendium, Duke and duCellier, 1993 (DAD) and the USDA database http://www.ars-grin.gov/duke/.

Liriodenine — Antibacterial; Antidermatophytic MIC 3.12 µg/mL; Antitumor; Candidicide; Channel-Blocker 0.1 mmol/l; Cytotoxic; Fungicide MIC 100 µg/mL; Topoisomerase-II-Inhibitor IC50 = 0.11 µ*M*; Vasodilator 0.1 mmol/l.

Piperine — See also *Piper nigrum.*

Xylopic-Acid — Antiseptic; Calcium-Antagonist; Diuretic; Hypotensive; Vasodilator.

Z

Zingiber officinale Roscoe (Zingiberaceae)
GINGER

Synonym — *Amomum zingiber* L.

Medicinal Uses (Ginger) — Almost as important medicinally as it is culinarily. Ginger, as our maps indicate, originated on the Indian subcontinent. Newmark and Schulick (2000) quote an Indian proverb that every good quality is contained in ginger. Confucius always had ginger with his meals. The surgeon general for Claudius and Nero used ginger for stomach problems. Mills and Bone (1999) cite *in vivo, in vitro,* and some clinical evidences re ginger's antiallergic, antiemetic, anti-hepatotoxic, antiinflammatory, antinauseant, antioxidant, antiparasitic, antiplatelet, antipyretic, antiseptic, antitussive, cardiovascular, digestive, and hypoglycemic activities (MAB). Reportedly carminative, aromatic, stimulant, stomachic, and tonic. Chinese have dozens of uses for the ginger. The bruised leaves are used as a digestive stimulant and for bruise. Sprouts are used for diarrhea, dysentery, marasmus, and worms. Chinese use the root for alopecia, bleeding, cancer, cholera, colds, congestion, diarrhea, dropsy, dysmenorrhea, nausea, rheumatism, snakebite, stomachache, and toothache. Chinese consider the root sialagogue when chewed, sternutatory when inhaled. It is also considered antidotal to aroid and mushroom poisonings. Cotton balls soaked in ginger juice are used in China for first- and second-degree burns; reported clinically successful in alleviating pain, blisters, and inflammation. In TCM, fresh ginger is "sheng jiang," an herb for wind chills; dry ginger is "gan jiang" more to warm the interior (Libster, 2002). Ginger and garlic are mixed with honey in one Indian cough and asthma remedy. Juice administered in Malay Peninsula against colic. Externally, the rhizome is an efficient rubefacient and counterirritant. The bark poultice, like the leaf cataplasm, is used for felons and inflamed tumors. The rhizome also shows up in folk remedies for cancer. Fresh root chewed and sucked to relieve thirst. Leaves pounded and poulticed warm onto bruises. Weed (2002) recommends ginger tea with honey as a warming drink for upset stomach. Fresh root, grated and steeped in boiling water, or a tablespoon of powdered ginger in a cup of hot water can be a pleasure (Weed, 2002). Ginger tea can even calm the heart. It may, however, increase sweating and flooding. Ginger will warm and help relieve constipation (which may contribute to urge incontinence). Ginger baths, soaks, and compresses can soothe sore and aching joints; hot ginger compresses for fibromyalgia (Weed, 2002). Weed (2002) says, "Ginger root tea warms and nourishes the entire pelvis. Try a cup/250 ml a day, sweetened with honey, for several weeks. Regular menses may be re-established, or the spotting may temporarily increase, then stop." At the same time, she warns that hot flashes can be triggered by black pepper, cayenne, and ginger.

Ginger is sometimes chewed after kola nut to enhance the latter, as aphrodisiac. Or in Latin America, roots macerated in aguadiente, as a male aphrodisiac, and, like some steroids, used for arthritis and rheumatism. Qureshi et al. (1989) concluded that ginger extracts significantly increase sperm motility and quantity. Ginger shares several phytochemicals with cardamom, regarded by some Arabs as aphrodisiac. Thus, there may be something behind the Root Booster formula, ginseng, ginger, sassafras, and sarsaparilla (Duke, 1985). Slave-master Portuguese colonialists apparently introduced ginger into West Africa, hoping to increase the slaves' fertility and fecundity. In Tibet, ginger is believed to stimulate the vital energies of the debilitated and weak. Chinese believe it balances hormonal flow (Schulick, 1994). Ginger certainly can be made pleasant to the taste. Schulick (1994) even recalls a fourteenth century sex manual, entitled *The Perfumed Garden,* which stated that a man who prepared himself for love with ginger and honey would give such pleasure to the woman that she would wish the act to continue forever. Adding a little cardamom, chile, and ginger to my vegetarian soups makes them warmer, if not more aphrodisiac. The ginger is said to piquantly pique the genitals of those who consume it.

In the U.S. ginger is best known for sea sickness and morning sickness. Since there is no pharmaceutical approved for morning sickness, my friend, Dr. Don Brown, recommends 250 mg of ginger root four times per day, saying it appears to be effective for reducing nausea and the incidence of vomiting. Results could take up to 48 hr. No adverse effects were noted on pregnancy outcome (Brown, 2001). EV.EXT 33, a patented ginger extract, was examined in Wistar SPF rats by oral gavage at 100, 333, and 1000 mg/kg, to pregnant female rats from days 6–15 of gestation. EV.EXT 33 was well tolerated. No deaths or treatment-related adverse effects were observed. Weight gain and food consumption were similar in all groups during gestation. Reproductive performance was not affected by treatment with EV.EXT 33. No embryotoxic or teratogenic effects were observed. EV.EXT 33, when administered to pregnant rats during the period of organogenesis, caused neither maternal nor developmental toxicity at daily doses of up to 1000 mg/kg body weight (X11137381). Vutyavanich et al. (2001), in a randomized, double-masked, placebo-controlled trial, of 32 women (taking 250 mg capsules, 4×/day), concluded that ginger is effective at relieving severity of nausea and vomiting in pregnancy. Dr. Diamond, M.D. (2001) does not contraindicate ginger in pregnancy, and neither do I.

And recently, there is evidence to suggest that powdered ginger is more effective than Dramamine for motion sickness. Double-blind crossover studies showed that 1 g/day/4 days powdered ginger diminished or eliminated symptoms of hyperemesis gravidarum. I concluded that a liter of ginger ale, if it contained the 0.525% ginger permitted in cookies, would contain five times the dose required to prevent motion sickness better than Dramamine. Two hundred grams of ginger snaps containing that 5250 ppm ginger would presumably be equally effective. Again, illustrating the controversy in the food farmacy literature. Yamahara et al. (1992) concluded that ginger extracts (75 mg/kg), 6-shogoal (2.5 mg/kg), and 6-, 8- or 10-gingerol (5 mg/kg) enhanced gastrointestinal motility of a charcoal meal. The effects were similar to or slightly weaker than those of metoclopramide and donperidone. Synergies could be working here, too. Ginger root significantly reduced the incidence of postoperative emetic sequelae compared to placebo, having the "same effect as metoclopramide." Still, there is controversy over whether any antiemetic should be administered prophylactically, even though the overall incidence of the nausea and vomiting is ca. 30%. Additionally, ginger has anticathartic activity. [6]-shogoal, [6]-dehydrogingerdione, [8]- and [10]-gingerol were found to have an anticathartic action. [6]-Shogaol was more potent than [6]-dehydrogingerdione, [8]- and [10]-gingerol (X2074539).

OB-200G, a polyherbal preparation containing aqueous extracts of *Garcinia cambogia*, *Gymnema sylvestre*, *Piper longum*, *Zingiber officinale*, and resin from *Commiphora mukul*, like each component herb, has thermogenic properties. OB-200G exerts antiobesity activities in animal models of obesity. New studies suggest the role of serotonin in mediation of satiety by OB-200G; hence its antiobesity effect (Kaur and Kulkarni, 2001). Eldershaw et al. (1992) showed that some gingerols and shogaols are thermogenic.

According to Sambaiah and Srinivasan (1989), ginger stimulated liver microsomal cytochrome p450 dependent aryl hydroxylase. Investigating antioxidant activity, Ahmed et al. (2000) found that ginger significantly lowered lipid peroxidation by maintaining the activities of the antioxidant enzymes superoxide dismutase, catalase, and glutathione peroxidase in rats. The authors concluded that ginger is as effective as ascorbic acid as an antioxidant.

Newmark and Schulick (2000) comment that ginger is the opposite of Celebrex. Celebrex has one molecule, designed to do one thing. Ginger has close to 500 identified constituents, and many more than that still unidentified (JAD). Ginger has several constituents that inhibit COX-2 and that inhibit the 5-lipoxygenase metabolism of arachidonic acid, thus depriving prostate cancer cells of their fuel for growth. Ginger has at least four prostaglandin-inhibitors that are stronger than indomethacin, the latter being structurally to melatonin. Seventeen pungent oleoresin principals of ginger exhibited a concentration and structure dependent inhibition of COX-2 (IC50 = 1–25 μM). [8]-paradol and [8]-shogaol and two synthetic analogues, 3-hydroxy-1-(4-hydroxy-3-methoxyphenyl)decane and 5-hydroxy-1-(4-hydroxy-3-methoxyphenyl)dodecane, showed strongest inhibitory effects on COX-2. Ginger metabolites, [6]- and [8]-series of gingerol, shogaol, and paradol strongly inhibit COX-2 in disrupted rat basophilic leukemia-1 cells. [6]-gingerol reported to reduce phorbol ester induced inflammation in mice when applied topically. Such potent COX-2 inhibition supports the use of ginger as an inflammatory, perhaps competitive with the synthetic COX-2-Inhibitors (Tjendraputra et al., 2001).

And ginger is a noteworthy source of natural melatonin, a potent antioxidant — more potent than glutathione in scavenging the hydroxyl radical and more potent than vitamin E in scavenging the peroxyl radical. It stimulates the main antioxidant enzyme of the brain, glutathione peroxidase. Melatonin is readily diffused into all tissues of the body, including intracellular membranes, due to its lipophilic structure. It protects DNA against free radical damage (Boik, 1995). Melatonin also inhibited human M-6 melanoma cells *in vitro*. The mechanism by which melatonin affects melanoma cells is uncertain; it may be due to an antiestrogen mechanism. Studies have shown that extremely low-frequency electromagnetic fields, like those in household electric equipment, block the ability of melatonin to inhibit the proliferation of breast cancer cells *in vitro* (Boik, 1995). Melatonin may inhibit cancer by augmenting interleukin-2's anticancer effects. In 90 patients with advanced solid neoplasms, low dose IL-2 (3 million IU) and 40 mg/day melatonin significantly decreased the proliferation of neoplastic cells as compared to IL-2 alone (Boik, 1995). More recently, Boik (2001) notes that melatonin inhibits proliferation of breast cancer and other cell lines *in vitro* at low concentrations (0.1–1 nM; higher and lower levels were inefficacious) (41 *in vitro* studies and 13 animal studies). Boik suggests a tentative anticancer dose of 3–20 mg/day. Because the target dose is achievable (but not with food farmacy, JAD), synergistic interactions may not be required for melatonin to produce an anticancer effect in humans. Nevertheless, melatonin may greatly benefit from synergism, and it makes sense to continue testing it in combination with other compounds (Boik, 2001).

Though ginger may be hot to the taste, it exerts great antiulcer potential — methinks even better in combination with licorice. Schulick (1994) notes that ginger can relieve inflammation while simultaneously protecting the digestive system from ulcers. When lab animals are exposed to severe stress, ginger extract, even at high doses, can inhibit ulcers by as much as 97.5%. For the 25 million American males and 12 million females suffering duodenal and/or gastric ulcers, daily ginger is an attractive alternative to cimetidine, famotidine, and ranitidine, accounting for sales of $2.8 billion a year. *U.S. News & World Report* (as early as 21 Feb. 1994) noted that the H2-receptor antagonists are among those with the most side effects and the highest prices. Upwards of 90% of users suffer high recurrence rates. Cost of consumption over 15 years may exceed $10,000 (Schulick, 1994). Furanogermenone also prevents gastric ulcers in rats (oral doses of 500 mg/kg). Ginger (water decoction), long pepper (water decoction), and *Ferula* species (colloidal solution) showed antiulcer in rats at oral doses of 50 mg/kg, 60 min prior to experiment, significant protection against gastric ulcers induced by 2-hr cold restraint stress, aspirin (200 mg/kg, 4 hr) and 4-hr pylorus ligation. The antiulcerogenic effect seemed due to augmentation of mucin secretion and decreased cell shedding rather than offensive acid and pepsin secretion which, however, were also found to be increased. Yamahara et al. (1992) note that beta-sesquiphellandrene, beta-bisabolene, ar-curcumene, and 6-shogaol also have antiulcer activities. In fact, ginger can contain more than a dozen each of antiinflammatory, antiulcer, and sedative compounds (USDA database), making the ginger, like the licorice, very promising with ulcers.

Hikino et al. (1985) reported antihepatotoxic actions of gingerols, shogaols, and diarylheptanoids. The [7]- and [8]-gingerols and shogaols exhibited greater antihepatotoxic activities. Turmeric has more notoriety as an antihepatotoxic, but I can see reason for mixing and matching our gingers and turmerics and other members of this tropical spice family.

Goto et al. (1990) showed that 6-shogoal and 6-gingerol synergistically killed larvae, perhaps again vindicating the folk usage of ginger as an antidote for seafood poisoning. Adewunmi et al. demonstrated that gingerol and shogoal have potent molluscicidal activity against *Biophalaria glabrata*. At 5 ppm, gingerol completely nullified the infectivity of *Schistosoma mansoni* miracidia and cercariae in *B. glabrata* and mice respectively, suggesting that the molluscicide is capable of interrupting schistosome transmission at concentrations lower than the molluscicidal concentration. According to Murakami (1994), 6-gingesulfonic acid has more potent antibiotic activity than 6-gingerol and 6-shogaol. Schulick (1994) notes that ginger is active against gram(–) and gram(+) bacteria, naming *Escherichia coli, Proteus vulharis, Salmonella typhimurium, Staphylococcus aureus*, and *Streptococcus viridans* (but no studies on helicobacter), while still serving as a probiotic, stimulating the growth of potentially useful *Lactobacillus*. In this latter regard, ginger was better than cinnamon, garlic, and pepper. Proteolytic enzymes like zingibain can enhance synthetic antibiotic bactericides as much as 50% (Schulick, 1994). When the CDC and FDA constantly tell us that herbs can't help against germ warfare, they forget their own words, that people who have compromised immune systems are more likely to be infected by germs. Ginger can kill gram(+) bacteria, if not anthrax, and boost the immune system at the same time. "Ginger not only maintains immune system functioning, but studies in Montreal and in Tokyo in 1955 and 1979 concluded that the spice may actually enhance immunity. Ginger apparently induced neutrophils which can stimulate host resistance to various diseases" (Schulick, 1994).

Some ginger compounds exhibited moderate growth regulatory and antifeedant activity against *Spilosoma obliqua* and significant antifungal activity against *Rhizoctonia solani*. [6]-dehydroshogaol exhibited maximum growth regulatory activity (EC50 = 3.55 mg/ml); dehydrozingerone imparted maximum antifungal activity (EC50 = 86.5 mg/l) (Agarwal et al. 2001). The antifungal diarylheptenones, gingerenones A, B, C and isogingerenone B, might possibly be synergistic, in ginger and man. Gingerenone A showed strong antifungal activity to *Pyricularia oryzae*, and moderate anticoccidium activity against *Eimeria tenella*.

In rats, oral doses of 50–100 mg/kg ethanolic extract had antiinflammatory activity comparable to aspirin, but the analgesic activity was only $^1/_{10}$th that of aspirin. Shogoal and gingerol are analgesic components. In one clinical trial, extract of ginger and greater galangal extract significantly (but modestly) reduced symptoms of oestoarthritis of the knee. There was a good safety profile, with some mild GI adverse events in the herbal group compared to controls (Altman and Marcussen, 2001). The extract was active against Gram(–) and Gram(+) bacteria.

Reportedly, 500–600 mg powdered ginger root can prevent migraine, based on a single case history. And merely chewing without swallowing, 1 g ginger can raise systolic pressure 11 mm Hg, diastolic, 14 (MAB). Paralleling similar reports from pepper; holding 100 mg pepper in the mouth transient increases systolic pressure 13 mm Hg c, diastolic 18 (Lin, 1994).

Under the able direction of Lionel Robineau, M.D., TRAMIL suggests that shogoal is intensively antitussive, compared to dihydrocodeine. All but "prescribing" ginger for colds, cough, flu, stomachache, and vomiting, TRAMIL classifies ginger in their "REC" category. Here is how TRAMIL explains their REC category:

> "We recommend certain usage of plants (or parts) very frequently used in cases of well defined ailments by the population...and for which the same indications for use are given in...the Caribbean, or in other tropical regions, and which have been the object of validative phytochemical pharmacological and/or toxicological work.

> Category C also includes plants that are well known as innocuous, whose biological activity for the cited indication is still to be proven, but that can be recommended as a placebo.

> In this last category, there are also new indications of 'TRAMIL' plants, whose use the participants in the TRAMIL 3 workshop had recommended and encouraged in view of the available scientific information for those species."

I suppose science has marched on since the days of Watt's "trikatu" for dermatoses, just before A.D. 1900. Now there's a patented new trick or two. U.S. Patent No. 6,063,381 describes antiinfective agents from black pepper, ginger, and other spices containing the vanillyl and piperidine ring structure. These compounds are useful with mycoses, tissue injuries, and abnormal proliferation of keratin. Various topical applications provide "outstanding results" in athlete's foot, candida, jock itch, ringworm, and favus. Caveat emptor (Dean, 2000).

Indications (Ginger) — Adenosis (f; KAB); Aging (f; WHO); Alcoholism (1; MAB); Allergy (1; FAY; FNF; MAB); Alopecia (f; DAA; DAD; FAY; WHO); Alzheimer's (1; COX; FNF); Anemia (f; DAA); Anorexia (2; JFM; KAB; PHR; WHO); Anxiety (1; MAB); Arthrosis (1; COX; MAB; SKY); Ascites (f; KAB); Asthma (f; FAY; JFM; MAD); Atherosclerosis (f; SKY); Backache (1; WHO); Bacteria (1; APA; FNF; MAB; TRA); Bite (f; DAA; KAB); Bleeding (f; DAA); Blister (1; DAD; DAA; FAY); Boil (f; KAB); Borborygmus (f; BGB); Bronchosis (1; AAB; BGB; FAY; FNF); Bruise (f; DAA; DAD); Burn (1; APA; DAD; FAY; MAB); Cancer (1; MAB); Candida (1; TRA); Cardiopathy (1; APA; FAY); Cataract (f; WHO); Catarrh (2; DAD; TRA); Chemotherapy (1; MAB; SKY); Chest Cold (1; AAB); Childbirth (f; AAB); Cholera (f; DAA; DAD); Cold (2; AKT; APA; BGB; FNF; MAD; TRA; WHO); Colic (1; PNC; BGB; SUW; WHO); Congestion (1; DAA; DAD; FNF; RIN); Convulsion (1; PNC); Corneosis (f; DAA); Cough (1; APA; BGB; FAY; FNF; PNC); Cramp (1; APA; BGB; KOM; MAB; PIP; PNC; TRA; WAM); Dandruff (f; APA); Depression (1; APA; DAA; MAB; WOI); Diabetes (1; DAA); Diarrhea (2; AAB; BGB; DAA; TRA; WHO); Dizziness (2; JAD); Dropsy (f; DAA; DAD); Dysmenorrhea (1; AAB; APA; DAA; JFM; MAB); Dyspepsia (2; FAY; FNF; KOM; PIP; MAD; SUW; TRA; WAM); Dyspnea (f; BGB; PH2); Earache (f; APA); Edema (1; MAB); Elephantiasis (f; KAB); Enterosis (1; APA; FAY; MAD;

Z

PNC); Epigastrosis (f; BGB; MAD); Epistaxis (f; FAY); Escherichia (1; HH3); Fever (2; APA; CAN; FAY; FNF; MAB; MAD; TRA); Flu (2; APA; BGB; FNF; TRA; VVG; WHO); Fungus (1; DAD; MAB; TRA); Gas (1; AAB; APA; MAB; MAD; PED; PHR; PH2; PNC; SUW; VVG); Gastrosis (2; APA; FAY; MAD; PHR; TRA); Headache (1; APA; FAY; KAP; MAB; WAM); Head Cold (f; JFM; RIN); Hemorrhoid (f; KAB; MAD; WHO); Hepatosis (1; APA; MAD); High Blood Pressure (1; APA; PNC); High Cholesterol (1; MAB; PED; PNC); Hoarseness (f; JFM); Hyperemesis (2; AKT); Immunodepression (1; FNF; PH2); Impotence (1; APA; MAB); Infection (1; DAD; FNF; MAB; TRA); Infertility (f; MAD); Inflammation (2; FAY; FNF; MAB; TRA; SKY; WAM; WHO); Insomnia (f; WHO); Kawasaki Disease (1; MAB); Low Blood Pressure (1; MAB); Lumbago (1; PNC); Malaria (f; JFM; MAD); Marasmus (f; DAA; DAD); Migraine (1; APA; FAY; MAB; PH2; SKY; WHO); Morning Sickness (2; FNF; KOM; MAB; PIP; WHO); Motion Sickness (2; FNF; KOM; MAB; PIP; WHO); Mycosis (1; DAD; HH3; MAB; TRA); Myosis (1; AAB; AKT; WAM; WHO); Nausea (2; BGB; DAA; FAY; FNF; TRA; WAM; WHO); Nephrosis (f; APA; DAA); Nervousness (1; FNF); Neuralgia (1; COX; FNF); Neurasthenia (f; MAD); Obesity (1; PH2); Opacity (f; JFM); Ophthalmia (f; JFM); Osteoarthrosis (1; AKT; COX); Pain (1; AKT; FAY; FNF; JBU; PED; PNC; TRA; WAM; WHO); Palpitation (f; FAY); Parasite (1; MAB; TRA); Pharyngosis (1; JFM; PH2; TRA); Postoperative Nausea (2; WHO); Pyrexia (f; PNC); Raynaud's (f; BGB); Rheumatism (1; FAY; MAB; MAD; PNC; SKY; WHO); Salmonella (1; HH3; TRA); Schistosomiasis (1; DAD; HH3; TRA); Seasickness (2; FNF; WHO); Snakebite (f; DAA; DAD); Sore Throat (1; APA); Splenosis (f; FAY); Staphylococcus (1; HH3; TRA); Stomachache (1; AAB; AKT; DAA; DAD; FNF); Stomatosis (f; MAD); Streptococcus (1; HH3); Stroke (1; APA); Swelling (1; FAY; HH3; MAB; WHO); Thirst (f; DAD); Thrombocytosis (1; MAB); Toothache (f; DAD; MAD; KAP; WHO); Trichomoniasis (1; DAA); Ulcer (1; APA; FAY; FNF; MAB; VVG); Vaginosis (1; DAA); Vertigo (1; MAB); Virus (1; APA; FNF; MAB; TRA; WAM); Vitiligo (f; FAY); Vomiting (3; KOM; PIP; WHO); Worm (f; DAA; DAD); Yeast (1; TRA).

Ginger for cold/flu:

- Analgesic: 6-gingerol; 6-shogaol; borneol; caffeic-acid; camphor; capsaicin; chlorogenic-acid; eugenol; ferulic-acid; gingerol; myrcene; p-cymene; quercetin; shogaol
- Anesthetic: 1,8-cineole; benzaldehyde; camphor; capsaicin; eugenol; linalool; myrcene
- Antiallergic: 1,8-cineole; 6-gingerol; 6-shogaol; citral; ferulic-acid; gingerol; kaempferol; linalool; quercetin; shogaol; terpinen-4-ol
- Antibacterial: 1,8-cineole; acetic-acid; alpha-pinene; alpha-terpineol; benzaldehyde; beta-ionone; beta-thujone; bornyl-acetate; caffeic-acid; caryophyllene; chlorogenic-acid; citral; citronellal; citronellol; curcumin; delta-cadinene; eugenol; ferulic-acid; geranial; geraniol; kaempferol; limonene; linalool; myrcene; myricetin; neral; nerol; nerolidol; p-coumaric-acid; p-cymene; p-hydroxy-benzoic-acid; patchouli-alcohol; perillaldehyde; quercetin; terpinen-4-ol; vanillic-acid
- Antibronchitic: 1,8-cineole; borneol; curcumin
- Antiflu: alpha-pinene; caffeic-acid; limonene; p-cymene; quercetin
- Antihistaminic: 6-shogaol; 8-gingerol; 8-shogaol; caffeic-acid; chlorogenic-acid; citral; gingerol; kaempferol; linalool; myricetin; quercetin; shogaol
- Antiinflammatory: 10-dehydrogingerdione; 10-gingerdione; 6-dehydrogingerdione; 6-gingerdione; alpha-curcumene; alpha-linolenic-acid; alpha-pinene; beta-pinene; borneol; caffeic-acid; capsaicin; caryophyllene; chlorogenic-acid; curcumin; eugenol; ferulic-acid; gingerol; kaempferol; myricetin; quercetin; salicylates; shogaol; vanillic-acid; zingerone
- Antioxidant: 6-gingerdiol; 6-gingerol; 6-shogaol; caffeic-acid; camphene; capsaicin; chlorogenic-acid; curcumin; delphinidin; eugenol; ferulic-acid; gamma-terpinene; gingerol; isoeugenol; kaempferol; melatonin; myrcene; myricetin; p-coumaric-acid; p-hydroxy-benzoic-acid; quercetin; vanillic-acid; vanillin; zingerone

- Antipharyngitic: 1,8-cineole; quercetin
- Antipyretic: 6-gingerol; 6-shogaol; borneol; eugenol; gingerol; salicylates; shogaol
- Antirhinoviral: ar-curcumene; beta-bisabolene; beta-sesquiphellandrene; zingiberene
- Antiseptic: 1,8-cineole; alpha-terpineol; aromadendrene; benzaldehyde; beta-pinene; caffeic-acid; camphor; capsaicin; chlorogenic-acid; citral; citronellal; citronellol; eugenol; furfural; geraniol; gingerol; hexanol; kaempferol; limonene; linalool; myricetin; nerol; oxalic-acid; paradol; shogaol; terpinen-4-ol
- Antistress: gamma-aminobutyric-acid
- Antitussive: 1,8-cineole; 6-gingerol; 6-shogaol; terpinen-4-ol
- Antiviral: alpha-pinene; ar-curcumene; beta-bisabolene; bornyl-acetate; caffeic-acid; chlorogenic-acid; curcumin; cyanin; ferulic-acid; geranial; kaempferol; limonene; linalool; myricetin; p-cymene; quercetin; vanillin
- Bronchorelaxant: citral; linalool
- COX-2-Inhibitor: curcumin; eugenol; kaempferol; melatonin; quercetin; 10-gingerol; 8-paradol; 6-shogaol; xanthorizol
- Cyclooxygenase-Inhibitor: 6-gingerol; capsaicin; curcumin; gingerol; kaempferol; melatonin; quercetin; shogaol; zingerone
- Decongestant: camphor
- Expectorant: 1,8-cineole; acetic-acid; alpha-pinene; beta-phellandrene; beta-sesquiphellandrene; bornyl-acetate; camphene; camphor; citral; geraniol; limonene; linalool
- Immunostimulant: alpha-linolenic-acid; benzaldehyde; caffeic-acid; chlorogenic-acid; curcumin; ferulic-acid; melatonin
- Interferonogenic: chlorogenic-acid
- Phagocytotic: ferulic-acid

Ginger for dyspepsia:

- Analgesic: 6-gingerol; 6-shogaol; borneol; caffeic-acid; camphor; capsaicin; chlorogenic-acid; eugenol; ferulic-acid; gingerol; myrcene; p-cymene; quercetin; shogaol
- Anesthetic: 1,8-cineole; benzaldehyde; camphor; capsaicin; eugenol; linalool; myrcene
- Antiemetic: 6-gingerol; camphor; gingerol; shogaol
- Antigastric: myricetin; quercetin
- Antiinflammatory: 10-dehydrogingerdione; 10-gingerdione; 6-dehydrogingerdione; 6-gingerdione; alpha-curcumene; alpha-linolenic-acid; alpha-pinene; beta-pinene; borneol; caffeic-acid; capsaicin; caryophyllene; chlorogenic-acid; curcumin; eugenol; ferulic-acid; gingerol; kaempferol; myricetin; quercetin; salicylates; shogaol; vanillic-acid; zingerone
- Antioxidant: 6-gingerdiol; 6-gingerol; 6-shogaol; caffeic-acid; camphene; capsaicin; chlorogenic-acid; curcumin; delphinidin; eugenol; ferulic-acid; gamma-terpinene; gingerol; isoeugenol; kaempferol; melatonin; myrcene; myricetin; p-coumaric-acid; p-hydroxy-benzoic-acid; quercetin; vanillic-acid; vanillin; zingerone
- Antipeptic: benzaldehyde; beta-eudesmol
- Antistress: gamma-aminobutyric-acid
- Antiulcer: 6-gingerol; 6-gingesulfonic-acid; 6-shogaol; alpha-zingiberene; ar-curcumene; beta-bisabolene; beta-eudesmol; beta-sesquiphellandrene; capsaicin; chlorogenic-acid; curcumin; eugenol; kaempferol; zingiberene; zingiberone
- Antiulcerogenic: caffeic-acid
- Anxiolytic: gamma-aminobutyric-acid
- Carminative: camphor; ethyl-acetate; eugenol; zingiberene
- Digestive: capsaicin
- Gastrostimulant: 6-shogaol; galanolactone; gingerol; shogaol
- Proteolytic: zingibain

Z

- Secretagogue: 1,8-cineole; p-hydroxy-benzoic-acid; zingerone
- Sedative: 1,8-cineole; 6-gingerol; 6-shogaol; alpha-pinene; alpha-terpineol; benzalde-hyde; borneol; bornyl-acetate; caffeic-acid; caryophyllene; citral; citronellal; citronellol; eugenol; farnesol; gamma-aminobutyric-acid; geraniol; geranyl-acetate; gingerol; isoborneol; isoeugenol; limonene; linalool; nerol; p-cymene; perillaldehyde; shogaol
- Sialagogue: capsaicin
- Tranquilizer: alpha-pinene; gamma-aminobutyric-acid

Other Uses (Ginger) — With its agreeable aroma and pungent taste, it is prepared from whole or partially peeled rhizomes, called "hands." Ginger is extensively used as condiment, in baked goods, beverages, cakes, candies, chutneys, curries, ginger ale, ginger beer, mincemeat, pastries, pickles, and preserves. Ginger is marketed whole, cracked, ground, powdered, and as a flavoring. It is said to be used as a vegetable substitute for rennet. Young rhizomes, called green ginger, stem ginger, or young ginger, are peeled and eaten raw in salads, pickled, or cooked in syrup and made into sweetmeats. Like garlic, ginger gets milder if cooked, bitter if burned. To make "pink ginger," the Japanese garnish, take very young ginger roots, scrape off the skin, saturate with lemon juice (which turns it pick), and season with salt (RIN). Pickled ginger, known as "amazu-shôga" or "gari," is frequently consumed with sushi, etc. Pickled and dyed red, they are known as "hajikami-shôga." Candied ginger, preserved in honey, sugar, or syrup, is a real treat. In Australia, the young rhizomes are preferred for making crystallized ginger. The juice of the rhizomes is nice in ginger ale, ginger beer, wine, brandy, and herbal teas. Young, spicy shoots are eaten as a potherb or puréed and used in sauces and dips. Young inflorescences are eaten raw in khaao yam. The leaves are used to wrap food for grilling (FAC). Ginger contains a proteolytic enzyme which, like ficin, bromelain, and papain, can be used for tenderizing meats. The proteolytic enzyme is present at levels of 2.26% of the fresh rhizome, such that 50 kg ginger can yield 1 kg of the enzyme; by contrast, it takes 8000 kg papaya (but remember, it's mostly water) to produce 1 kg papain, papaya's digestive enzyme. All these proteolytic enzymes, like the hydroxy fruit acids they often accompany, have cosmetic applications as well. The EO, called "oil of ginger," is used in food flavoring, beverages, and perfumes, especially men's toilet lotions (DAD, RIN, WOI). And Bown (2001) notes that shogaols, breakdown products of gingerol, produced as ginger dries, are almost twice as hot as gingerols. Hence the dried ginger, with half the water, may have more than twice the pungency. That may well be why the Chinese use the dry ginger for different purposes than the fresh.

Thanks to Fulder (1996) for dredging up some Confucian ginger recipe, something akin to beef jerky: They would beat the beef, removing the skinny parts. Then they laid it in a frame of reeds and sprinkled it with cinnamon, ginger, and salt. It is then eaten once dried. They also prepared deer, elk, and mutton similarly (Fulder, 1996). In *What Color is Your Diet*, Dr. David Heber (2001), professor of medicine and director of the UCLA Center for Human Nutrition says "Ginger and garlic are contrasting and can be used with cut up broccoli or Brussels sprouts." Heber notes that steamed vegetables with subtle mild tastes "or mildly unpleasant tastes" may need dressing up, and spices can be used to contrast two flavors. "A low fat diet based on fruits, vegetables, and whole grains offers the best possible diet for achieving optimum health. And the results show that people around the world who eat in this manner, exercise regularly, and avoid tobacco have the least amount of cancer, heart disease, hypertension, diabetes, and osteoporosis" (Heber, 2001).

For more information on activities, dosages, and contraindications, see the *CRC Handbook of Medicinal Herbs, ed. 2*, Duke et al. 2002.

Cultivation (Ginger) — In southern India, planting begins in May, when rains commence; later in other areas. Beds 2 m long and about 1 m wide are prepared. Small holes 12 cm deep and 7.5 cm apart are dug, and a rhizome planted in each. Holes are filled with dry cow-dung powder. When planting area is completed, beds are covered with green-leafed tree branches. As leaves wither and fall off, they act as a green fertilizer; bare branches are removed and beds weeded.

When shoots are 10–12 cm tall, fresh undried cow-dung and leaf mold are placed on beds. About a month after planting, edges of beds are raised to allow drainage from monsoon rains. Except for an occasional weeding, no further cultivation is required until harvesting in December or January. Plants require 9 months to mature in southern India. In West Indies, ginger is often grown on ridges about 1 m apart, plants are spaced 30–45 cm apart in the row. When planted in beds 45 cm wide, a row of ginger is planted along each edge, with sides of ridge nearly perpendicular. Planting done in March and April. Rhizomes are broken into pieces 2.5–5 cm long, each cutting having at least one bud. Before planting, holes or furrows are partly filled with manure, then cuttings put in, spaced 30–45 cm apart, and covered with 7.5–10 cm of soil. After planting, beds are shaded with branches, then leaf mold is added later. Plantings last several years but begin to decline to uneconomical levels after fifth year. After 3 years, ginger is usually rotated with corn, peas, yams, or lentils, to enrich soil, so that, at end of 5 years of rotation, the area may be planted to ginger again. Although fertilizers do increase yields, they make drying rhizome more difficult. Often cultivated in southern Queensland as an annual. In favorable locations, ratoon crops may be grown, but ratooning is practicable only with ginger grown for late harvesting. If crop is ratooned after early harvest, it shoots before winter, and new growth is destroyed by frost or dies off in near-frost temperatures. Ginger and turmeric are well suited for mixed cultivation. Purchased ginger roots can be planted in your window herb box and harvested much later when the sprouted leaves have gone dormant (DAD, RIN). Crop is ready for harvest when stem turns white, before rhizomes get fibrous and tough. Roots are dug like potatoes. After being dug, well-ripened large rhizomes are spread on mats 2.5–3.5 m above a fireplace and smoked to dry and preserve them. For rhizomes destined for market, fibrous roots are removed, and rhizomes are washed, decorticated, dried, and sometimes bleached. Rhizomes destined for replanting are heaped and covered with leaf mold or manure to keep them from drying out until planting. Rhizomes must be frequently turned and protected from night dew. In Hawaii, ginger is harvested from January to April, when prices are low due to large imports of ginger from other regions. One kg of planted rhizome yields 25–61.5 kg fresh rhizome, the quality depending on culture, soil, and weather conditions. As each rhizome weighs about 450 gm, and each plant produces about 9 such rhizomes, yields of 5 tons or more of green ginger and 0.5 tons or more of dried rhizome/ha are easily possible. By careful cultivation and manuring, yields can be increased (DAD). For more details see Duke and duCellier (1993) and Purseglove et al. (1981).

Chemistry (Ginger) — Patel and Srinivasan (1985) noted that dietary ginger significantly increased lipase, maltase, and sucrase activities. The protein is rich in threonine and proline but contains little or no tryptophan. Asparagine and pipecolic acid have been isolated from aqueous extracts of the rhizomes. Here are a few of the more notable chemicals found in ginger. For a complete listing of the phytochemicals and their activities, see the CRC phytochemical compendium, Duke and duCellier, 1993 (DAD) and the USDA database http://www.ars-grin.gov/duke/.

10-Gingerdione — Antiinflammatory (>indomethacin); Antiprostaglandin IC50 = 1.0 µM.

6-Gingerdione — Antiinflammatory (>indomethacin); Antiprostaglandin IC50 = 1.6 µM.

Gingerenone-A — Anticoccidioid 10 ppm; Fungicide 10 ppm.

Gingerenone-B — Fungicide.

Gingerenone-C — Fungicide.

Gingerol — Analgesic; Antiaggregant 0.5–20 µM, 10–100 µM; Antiallergic; Anticancer; Antiemetic; Antihepatotoxic; Antihistaminic; Antiinflammatory; Antioxidant; Antiprostaglandin 0.5–20 µM; Antipyretic; Antischistosomic 5 ppm; Antiseptic; Antithromboxane 0.5–20 µM; Cardiotonic 1–30 µM; Cholagogue; Cyclooxygenase-Inhibitor; Fungicide; Gastrostimulant; Hepatoprotective;

Hypotensive; Inotropic 1–30 μM; Molluscicide 5 ppm, LD20 = 12.5 ppm; Mutagenic; Nematicide; Positive Inotropic 1–30 $\mu g/ml$; Schistosomicide EC100 = 10 ppm; Sedative; Thermogenic.

6-Gingerol — See also *Aframomum melegueta.*

8-Gingerol — See also *Aframomum melegueta.*

Zingiberene — See also *Curcuma zedoaria.*

Zingiberol — See also *Curcuma xanthorrhiza.*

Zingiberone — Antimutagenic; Antiulcer.

Well, you have toured the spices with me, from *Aframomum* to *Zingiber*. Recently I appeared with Nina Simonds on a fundraiser for Maryland public TV. They showed Nina's video, *A Spoonfull of Ginger*, between pledge pitches. I told Nina she should get rights to use that famous trumpeter's version of "A Taste of Honey" and come out with a recipe for Honey Candied Ginger. I'd like to close with some of Nina's comments recounting how she, like me, had been converted to natural medicine. Her conversion was a yin-yang conversion in China, my conversion, more earthy, was in the rain forests of Panama. Now we both share Hippocrates dictum: let food be your farmacy [Sic! Stet!]

Nina's nice book combines the best of both food and food farmacy. Her charming introductory anecdote discussed her problem back in 1972, stomachache, from having consumed too many yin foods proportionate to the yang foods she consumed after arriving in Tapei for a 3.5-year sojourn of China. The cure, eat some yang foods. Among those yang spices, the ginger that cured her stomachache so well that she wrote the book, her *Spoonful of Ginger*, with a lot of recipes for curry, a hot yang combination of spices, including ginger's first cousin, turmeric.

She prefaces her book with the two quotes which close my book:

"Let food be your medicine and medicine be your food."—Hippocrates

"To take medicine only when you are sick is like digging a well when you are thirsty. Is it not already too late?"—The Yellow Emperor's Classic of Internal Medicine (~4500 B.C.).

Z

Reference Abbreviations

Frequently we will cite the three-letter abbraviation followed by the volume number and the first page of the citation as a shorthand uniquely identifying in the database, CRC Ed. 2, as well as this book.

(Book and journal abbreviations)

60P Desmarchelier and Witting Schaus (2000)
AAB Arvigo and Balick (1993)
AAR Arndt (1999)
ABS Siquera et al. (1998)
ACN *American Journal of Clinical Nutrition*
ACT *Alternative and Complementary Therapies*
AEH DeSmet et al. (1993, 1997)
AEL Leung's (Chinese) Herb News, followed by number and page (e.g., AEL31:3)
AHP McGuffin et al. (1997)
AJC *American Journal of Chinese Medicine*
AKT Tillotson et al. (2001)
AMA *Alternative Medicine Alert*
AOT Tucker et al. (1994)
APA Peirce (1999)
APP *Acta Physiologica et Pharmacologica Bulgica*
ARC Aloe Research Council
ATM *ATOMS Journal* publication
AYL Leung (1980), or Leung's newsletter when followed by numbers
BIB Duke (1983)
BGB Blumenthal et al. (2000)
BIS Bisset (1994)
BOI Boik (1995)
BJH *British Journal of Haematology* (e.g., BJH111:359; 2000, vol. 111, p. 359)
BO2 Boik (2001)
BOW Brown (2001)
BPCC *Biosyn. Prod. Cancer Chemotherapy*
BRI Brinker (1998)
BRU Bruneton (1999)
CAN Newall, Anderson, and Phillipson (1996)
CEB Erichsen-Brown (1989)
CFR Reed (1976)
COX Newmark and Schulick (2000)
CRC Duke (1985)
CTD Castner, Timme, and Duke (1998)
DAA Duke and Ayensu (1985)
DAD Duke and duCellier (1993)
DAV Duke and Vasquez Martinez (1994)
DAW Duke and Wain (1991)
DAZ De Lucca and Zalles (1992)

DEM Moerman (1998)

DEP Watt (1889–1892, reprint 1972)

DON Brown (2000)

EB *Economic Botany*

Econ. Bot. *Economic Botany*

EFS Steinmetz (1957)

EJH Hoffman (1999)

EMP *Economic & Medicinal Plant Research*

FAC Facciola (1998)

FAD Foster and Duke (1990)

FA2 Foster and Duke (2000)

FAS Foreign Agriculture Service (of USDA)

FAY Foster and Yue (1992)

FEL Felter and Lloyd (1898)

FFJ *Flavour and Fragrance Journal*

FIT *Fitoterapia*

FLP *Flora of Pakistan* (USDA PL480 Project)

FNF *Father's Nature's Farmacy* (online database available at, http://www.ars-grin.gov/duke/)

GEO Guenther (1948–1952)

GHA Ghazanfar (1994)

GMH Grieve (1931)

GMJ Grenand, Moretti, and Jacquemin (1987)

HAD An Herb a Day; by Jim Duke, published in *Business of Herbs*, and/or *Wild Food Forum*

HDN Neuwinger (1996)

HDR *Herbal Desk Reference*; online version under my Medical Botany Syllabus (MBS)

HD1 Duke (2000)

HEG Hegnauer (*Chemotaxonomie der Pflanzen*), Birkhauser Verlag, multiple volumes

HFH *Herbs for Health* followed by JA July Aug. '99 (1999)

HG *HerbalGram* (followed by number without space; e.g., HG17 *HerbalGram* no. 17)

HHB List and Hohammer (1969–1979)

HH2 *Hager's Handbuch*, 2nd ed., Hansel et al.

HH3 *Hager's Handbuch,* 3rd ed., Blaschek et al.

HIL Hildegard, translated by Throop, P. (1998)

HOB Hobbs (1996)

HOC Tyler (1994)

HOP Duke and Atchley (1986)

HOW Duke (1992)

HOX Ausubel (2000)

IED Duke (1986)

IHB Burkill (1966)

IJA *Indian Journal of Animal Sciences* (59, 1989)

IJC *International Journal of Cancer Research*

IJE *Indian Journal of Experimental Biology* (29, 1991)

IJI *International Journal of Integrative Medicine*

IJP *Indian Journal of Pharmaceutical Sciences* (52, 1990)

JAC *Journal of Alternative and Complementary Medicine*

JAD James A. Duke, personal commentary

JAF *Journal of Agricultural and Food Chemistry*

JAH *Journal of American Herbalist Guild*

JAH2 Winston (2001)

JAM *Journal of the American Medical Association*

JAMA *Journal of the American Medical Association*
JAR *International Journal of Aromatherapy*
JBH Harborne and Baxter (1983)
JBU Aromatherapy: A Place in Herbal Medicine, *HerbalGram*, Buckle, J. (Ed.)
JE *Journal of Ethnopharmacology*
JFM Morton (1977, 1981)
JLH Hartwell (1982)
JLR *Journal of Longevity Research*
JN *Journal of Nutrition*
JNE *Journal of Nutritional and Environmental Medicine*
JNP *Journal of Natural Products*
JNS *Journal of Nutrition Science,* Vitaminol
JNU Joseph, Nadeau, and Underwood (2001)
JPP *Journal of Pharmacy and Pharmacology*
KAB Kirtikar and Basu (reprint, 1975)
KAP Kapoor (1990)
KCH Huang (1993)
KEB Bone (1996)
KOM Blumenthal et al. (1998) (Commission E.)
LAB Lawson and Bauer (1998)
LAF Leung and Foster (1995)
LAW Koch and Lawson (1996)
LEG Duke (1981)
LEL Lewis and Elvin-Lewis (1977)
LIB Libster (2002)
LIL *Living Liqueurs,* Duke (1987)
LMP Perry (1980)
LRN *Lawrence Review of Natural Products*, looseleaf; periodically updated
LRNP *Lawrence Review of Natural Products*, looseleaf; periodically updated
M&I *Microbiology & Immunology*
M11 *Merck Manual,* 11th ed.
M12 *Merck Manual,* 12th ed.
M28 Martindales, 28th ed.
M29 Martindales, 29th ed.
M30 Martindales, 30th ed.
MAB Mills and Bone (1999)
MAD Madaus (1976)
MAM Miller and Murray (1998)
MAP Murray and Pizzorno (1991)
MB Blumenthal, personal communicatrion
MBB Bajracharya (see AKT; Tillotson's Nepalese Mentor)
MBC Martinez, Bernal, and Caceres (2000)
MIC Micmac Module of Duke's online Medical Botany Syllabus
MLM McCaleb, Leigh, and Morien (2000)
MPB Mors, Rizzini, and Pereira (2000)
MPG Gupta (1995)
MPI Indian Council of Medical Research (ICMR) (1976, 1987)
NAD Nadkarni (1976)
NH Nature's Herbs, personal communication, Grant or Rich
NR *Nutrition Reviews*
NUT Duke (1989)

OMM *Oriental Materia Medica* (three volumes by Dr. Hsu and collaborators)
PAM Pizzorno and Murray (1985)
PC *Phytochemistry* (29, 1990)
PCF Huang, Ho, and Lee (1992)
PDR *Physicians' Desk Reference* (Ed. 45, 1991)
PEA Purseglove et al. (1981)
PED Pedersen (1998)
PEP *Pakistan Encyclopedia Planta Medica* (1986)
PFH *Bland Food Hum. Nutr.*,
PH2 Gruenwald et al. (2000)
PHM *Phytomedicine*
PHR *PDR for Herbal Medicine,* 1st ed., Fleming, et al. (1998)
PIP Schilcher (1997)
PJB *Protocol Journal of Botanical Medicine,* (vol. 1, no. 1, 1995; since terminated)
PM *Planta Medica* (56, 1990)
PNC Williamson and Evans (1989)
PR *Phytotherapy Research*
PS *Plant Science* (Ireland)
QRNM *Quarterly Review of Natural Medicine*
RAR Rutter (1990)
RAT Robbers and Tyler (1999)
RFW Weiss (1988)
RIN Rinzler (1990)
RYM Roig y Mesa (1928)
SAB Sabinsa (1998)
SAR Schultes and Raffauf (1990)
SAS Tainter and Grenis (1993)
SHB Buhner (2000)
SF Foster (1996)
SF2 Foster (1998)
SHT Schulz, Hansel, and Tyler (1998)
SKJ Jain (1991)
SKY Lininger et al. (1998)
SN *Science News* (followed by number and page)
SPI Charalambous (1994)
SUW Suwal (1976)
SYN Challem, Berkson, and Smith (2000)
TAD Tucker and Debaggio (2000)
TAN Tanaka (1976)
TGP Duke (1997)
TIB Kletter and Kriechbaum (2001)
TMA Time-Life (1996)
TOM Crellin and Philpott (1990)
TOX Keeler and Tu (1991)
TRA Germosén-Robineau (1997)
UPW Burkill (1985–2000)
USA USDA's Ag Handbook 8 and sequelae
USD USDA's nomenclature database
VAG Van Wyk and Gericke (2000)
VVG Van Wyk, Van Oudtshoorn, and Gerike (1997)
WAF White et al. (2000)

WAG Wright and Gaby (1999)
WAM White and Mavor (1998)
WBB Watt and Breyer-Brandwijk (1962)
WER Werbach (1993)
WHO World Health Organization (1999)
WIC Wichtl (1984)
WOI Council of Scientific and Industrial Research (CSIR) (1948–1976)
WO2 *The Wealth of India,* revised; three new volumes published (1985–1992)
WO3 *The Wealth of India,* first supplement series (2000)
X As a prefix followed by a big number, a PubMed reference citation (e.g., X123456)
YAN Yanovsky (1936)
ZIM Gelfand et al. (1985)
ZUL Hutchings et al. (1996)

References

Abaul, J., Bourgeois, P., and Bessiere, J.M., Chemical composition of the essential oil of the chemotypes of *Pimenta racemosa* var. *racemosa* (P. Miller), J. W. Moore, Boids d'Inde of Guadeloupe (F.W.I.), *Flav. & Fragr. J.,* 10:319–21, 1995.

Abreu, P.M. and Noronha, R.G., Volatile constituents of the rhizomes of *Aframomum alboviolaceum Flav. & Fragr. J.,* 12:79–83, 1997.

Abuharfeil, N.M., Salim, M., and Von Kleist, S., Augmentation of natural killer cell activity *in vivo* against tumour cells by some wild plants from Jordan, *Phytother. Res.,* 15(2):109–113, 2001.

Adebajo, A.C. and Reisch, J., Minor furocoumarins of *Murraya koenigii, Fitoterapia,* 71(3):334–7, 2000.

Adegunloye, B.J. et al., Mechanisms of the blood pressure lowering effect of the calyx extract of *Hibiscus sabdariffa* in rats. *Afr. J. Med. Med. Sci.,* 25(3):235–8, 1996. (X10457797)

Adesanya, S.A. and Sofowora, A., Phytochemical investigation of candidate plants for the management of sickle cell anemia, in *Phytochemistry of Plants Used in Traditional Medicine,* Hostettmann, K. et al., Eds., (*Proc. Phytochem. Soc. Eur. 37*), Oxford Science Publications, New York, 1995.

Adewunmi et al., Molluscicidal and antischistosomal activities of Zingiber officinale, *Plants Med.,* 56(4):374, 1990.

Agarwal, M. et al., Insect growth inhibition, antifeedant and antifungal activity of compounds isolated/derived from *Zingiber officinale* Roscoe (ginger) rhizomes, *Pest Manag. Sci.,* 57(3):289–300), 2001. (X11455660)

Ahmed, R.S., Seth, V., and Banerjee, B.D., Influence of dietary ginger (*Zingiber officinalis* Rosc.) on antioxidant defense system in rat: comparison with ascorbic acid, *Indian J. Exp. Biol.,* 38(6):604–6, 2000. (X11116533)

Al-Bekairi, A.M. et al., Toxicity studies on *Allium cepa,* its effect on estradiol-treated mice and on epididymal spermatozoa, *Fitoterapia,* 60(11):301–6, 1991.

Alfs, M., *Edible & Medicinal Wild Plants of Minnesota & Wisconsin,* Old Theology Bookhouse, New Brighton, MN, 2001.

Altman, R.D. and Marcussen, K.C., Effects of a ginger extract on knee pain in patients with osteoarthritis, *Arthritis Rheum.,* 44(11):2531–8, 2001. (X11710709)

Al-Yahya, M.A. et al., Gastric antisecretory, antiulcer, and cytoprotective properties of ethanolic extract of *A. galanga* Willd. in rats, *Phytother. Res.,* 4(3):112–4, 1990. (PR)

Amagase, H. et al., Intake of garlic and its bioactive components, *J. Nutr.,* 131(3s):955S-62S, 2001. (X11238796)

Ames, B.N., Magaw, R., and Gold, L.S., Ranking possible carcinogenic hazards, *Science,* 236: 271–280, 1987.

Andrews, J., *Pepper — The Domesticated Capsicums,* University of Texas Press, Austin, 1995.

Anon., Coumarin containing plants in uilization abstracts, *Econ. Bot.,* 2:333–338, 1948.

Anon., Essential oils of peppermint, orange or lemongrass kill most strains of fungal and bacterial infections, *Posit. Health News,* 17:26–7, 1998. (X11366554)

Arndt, A., *Seasoning Savvy — How to Cook with Herbs, Spices, and Other Flavorings,* Haworth Herbal Press, New York, 1999.

Aruna, K. and Sivaramakrishnan, V.M., Plant products as protective agents against cancer, *Indian J. Exp. Biol.,* 28(11):1008–1011, 1990.

Arvigo, R. and Balick, M., *Rain Forest Remedies — One Hundred Healing Herbs of Belize,* Lotus Press, Twin Lakes, WI, 1993. (AAB)

Asai, A. and Miyazawa, T., Dietary curcuminoids prevent high-fat diet–induced lipid accumulation in rat liver and epididymal adipose tissue, *J. Nutr.,* 131:2932–2935, 2001.

Badary, O.A. and Gamal El-Din, A.M., Inhibitory effects of thymoquinone against 20-methylcholanthrene-induced fibrosarcoma tumorigenesis, *Cancer Detect. Prev.,* 25(4):362–8, 2001.

Badary, O.A. et al., Inhibition of benzo(a)pyrene-induced forestomach carcinogenesis in mice by thymoquinone, *Eur. J. Cancer Prev.,* 8(5):435–40, 1999.

Beckstrom-Sternberg, S.M. and Duke, J.A., Potential for Synergistic Action of Phytochemicals in Spices, in *Spices, Herbs and Edible Fungi*, Charalambous, G., Ed., Elsevier Science B.V., 1994.

Bergner, P., *The Healing Power of Garlic*, Prima Publishing, Rocklin, CA, 1996.

Blumenthal, M. et al., Eds., *The Complete German Commission E Monographs. Therapeutic Guide to Herbal Medicines*, American Botanical Council, Austin, TX and Integrative Medicine Communications, Boston, MA, 1998. (KOM)

Boakye-Yiadom, K., Fiagbe, N.I., and Ayim, J.S., Antimicrobial properties of some West African medicinal plants iv. Antimicrobial activity of xylopic acid and other constituents of the fruits of *Xylopia aethiopica* (Annonaceae), *Lloydia*, 40(6):543–5, 1977. (X600023)

Boik, J., *Cancer and Natural Medicine*, Oregon Medical Press, Princeton, MN, 1995. (BOI)

Boik, J., *Natural Compounds in Cancer Therapy — Promising Nontoxic Antitumor Agents from Plants and Other Natural Sources*, Oregon Medical Press, Princeton, MN., 2001. (BO2)

Bordia, A. et al., Effects of the essential oils of garlic and onion on alimentary hyperlipidemia, *Atherosclerosis*, 21(1):15–9, 1975.

Borek, C., Antioxidant health effects of aged garlic extract, *J. Nutr.*, 131(3s):1010S, 2001.

Bown, D., *New Encyclopedia of Herbs & Their Uses*, revised American ed., D.K. Publishing, Inc., New York, NY, 2001. (BOW)

Briggs, W.H. et al., Administration of raw onion inhibits platelet-mediated thrombosis in dogs, *J. Nutr.*, 131(10):2619–22, 2001.

Brown, D., Ginger alleviates nausea and vomiting of pregnancy, *HerbalGram*, 53:21–2, 2001.

Bruneton, J., *Pharmacognosy, Phytochemistry, Medicinal Plants,* 2nd ed., Lavoisier Publishing, Paris, 1999. (BRU)

Buchanan, R.L., Toxicity of spices containing methylenedioxybenzene derivatives: a review, *J. Food Safety*, 1:275, 1978.

Burits, M. and Bucar, F., Antioxidant activity of *Nigella sativa* essential oil. *Phytother. Res.*, 14(5):323–8, 2000.

Burkill, H.M., *The Useful Plants of West Tropical Africa*, Royal Botanical Gardens, Kew, 2nd ed., Vol. 1, A-D, 1985; Vol. 2, E-I, 1994; Vol. 3, J-L, 1995; Vol. 4, M-R, 1997; Vol. 5, S-Z, 2000. (UPW)

Caceres, A. et al., Pharmacologic properties of *M. oleifera* 2 — screening for antispasmodic, antiinflammatory and diuretic activity, *J. Ethnopharmacol.*, 36(3):233–7, 1992. (JE36:233)

Campos, M.G. et al., Xanthorrhizol induces endothelium-independent relaxation of rat thoracic aorta, *Life Sci.*, 67(3):327–33, 2000. (X10983876)

Charalambous, G., Ed., *Spices, Herbs and Edible Fungi*, Elsevier Science B.V., 1994. (SPI)

Chasan, R., Ed., Editorial comments on Sherman, P.W. and Billing, J., Darwinian gastronomy — why we use spices. Spices taste good because they are good for us, *BioScience*, 49(6):453–463, 1999.

Chen, H.C., Chang, M.D., and Chang, T.J., Antibacterial properties of some spice plants before and after heat treatment, *Zhonghua Min Guo Wei Sheng Wu Ji Mian Yi Xue Za Zhi*, 18(3):190–5, 1985. (X4064797)

Cheng, A.L. et al., Phase I clinical trial of curcumin, a chemopreventive agent, in patients with high-risk or pre-malignant lesions, *Anticancer Res.*, 21(4B):2895–900, 2001. (X11712783)

Chewonarin, T. et al., Effects of roselle (*Hibiscus sabdariffa* Linn.), a Thai medicinal plant, on the mutagenicity of various known mutagens in *Salmonella typhimurium* and on formation of aberrant crypt foci induced by the colon carcinogens azoxymethane and 2-amino-1-methyl-6-phenylimidazo[4,5-b]pyridine in F344 rats, *Food Chem. Toxicol.*, 37(6):591–601, 1999. (X10478827)

Chu, D.M. et al., Amebicidal activity of plant extracts from Southeast Asia on *Acanthamoeba* spp., *Parasitol Res.*, 84(9):746–52, 1998. (X9766904)

Chung, W.Y. et al., Antioxidative and antitumor promoting effects of [6]-paradol and its homologs, *Mutat. Res.*, 496(1–2):199–206, 2001.

CSIR (Council of Scientific and Industrial Research), *The Wealth of India — A Dictionary of Indian Raw Materials & Industrial Products*, 11 Vols., CSIR, New Delhi, 2nd ed. in process and three new Vol. published, 1948–1976. (WOI)

CSIR (Council of Scientific and Industrial Research), *The Wealth of India — Revised Edition, Raw Materials*, Vol. I–A, CSIR, New Delhi, 1985.

Davies, M.J. et al., Effects of soy or rye supplementation of high-fat diets on colon tumour development in azoxymethane-treated rats, *Carcinogenesis*, 20(6):927–31, 1999.

De, M., Krishna De, A., and Banerjee, A.B., Antimicrobial screening of some Indian spices, *Phytother. Res.*, 13(7):616–8, 1999.

Dean, K., Plant patents, *HerbalGram*, 50:3, 2000.

Deodhar, S.D., Sethi, R., and Shrimal, R.C., Preliminary studies on anti-rheumaticactivity of curcumin, *Ind. J. Med. Res.*, 71:632–634, 1980.

De Pascual-Teresa, S. et al., Characterization of monomeric and oligomeric flavan-3-ols from unripe almond fruits, *Phytochem. Anal.*, 9(1):21–27, 1998.

De Roos, N., Schouten, E., and Katan, M., Consumption of a solid fat rich in lauric acid results in a more favorable serum lipid profile in healthy men and women than consumption of a solid fat rich in trans-fatty acids, *J. Nutr.*, 131(2):242–5, 2001.

Desai, H.G. and Kalro, R.H., Effect of black pepper and asafoetida on the DNA content of gastric aspirates, *Ind. J. Med. Res.*, 82(1):325–9, 1985.

Diamond, W.J., *The Clinical Practice of Complementary, Alternative, and Western Medicine*, CRC Press, Boca Raton, FL, 2001.

Dominic, C.J. and Pandey, S.D., Failure of deodorized males to induce oestrus in the wild mouse, *Ann. Endocrinol.*, (Paris) 40(3):229–34, 1979. (X573093)

Duke, J.A., Ecosystematic Data on Economic Plants, *Quart. J. Crude Drug Res.*, 17(3/4):91–110, 1979. (LEG)

Duke, J.A., *Handbook of Legumes of World Economic Importance*, Plenum Press, New York, 1981.

Duke, J.A., *Medicinal Plants of the Bible*, Conch Publications, New York, 1983. (BIB)

Duke, J.A., *Culinary Herbs — A Potpourri*, Trado-Medic Books, Buffalo, NY, 1985, out of print.

Duke, J.A., *Herbalbum — An Anthology of Varicose Verse*, published by the author, 1985b.

Duke, J.A., *Living Liqueurs*, Quarterman Press, Lincoln, MA, 1987, out of print. (LIL)

Duke, J.A., The Spice Trade Before and After Columbus, AAAS Meeting, February 1991, *AAAS Pub 91–02S*, AAAS, Washington, D.C., 1991.

Duke, J.A., *Spice Rack/Medicine Chest — Five Hundred Years after Columbus*, lecture, widely repeated in the U.S. (e.g., at the Smithsonian and with Old Ways in Spain), 1992. (HOW)

Duke, J.A., *Herbs of the Bible — 2000 Years of Plant Medicine*, Interweave Press, Loveland, CO, 1999.

Duke, J.A., *The Green Pharmacy Herbal Handbook*, Rodale/Reach, 1997. (HD1)

Duke, J.A and Atchley, A.A., *Handbook of Proximate Analysis Tables of Higher Plants*, CRC Press, Boca Raton, FL, 1986. (HOP)

Duke, J.A. and duCellier, J.L., *Handbook of Alternative Cash Crops*, CRC Press, Boca Raton, FL, 1993. (DAD)

Duke, J.A. and Wain, K., *Medicinal Plants of the World*, USDA Computer Listing of Medicines of the World, now slightly updated and online at USDA (http://www.ars-grin.gov/duke/), 1981. (DAW)

Duke, J.A. et al., *CRC Handbook of Medicinal Plants*, 2nd ed., CRC Press, Boca Raton, 2002.

Dull, R.E.T., *U. S. Spice Trade*, USDA For. Ag. Service, FTEA 1–90, not cited specifically, but integral to my Columbian Exchange Lecture, 1990.

Dullo, A.G. et al., Efficacy of a green tea extract rich in catechin polyphenols and caffeine in increasing 24-h energy expenditure and fat oxidation in humans, *Am. J. Clinical Nutrition*, 70:1040–5, 1999.

El Daly, E.S., Protective effect of cysteine and vitamin E, *Crocus sativus* and *Nigella sativa* extracts on cisplatin-induced toxicity in rats, *J. Pharm. Belg.*, 53(2):87–93, discussion 93–5, 1998.

El-Dakhakhny, M. et al., Effects of *Nigella sativa* oil on gastric secretion and ethanol induced ulcer in rats, *J. Ethnopharmacol.*, 72(1–2):299–304, 2000.

El-Dakhakhny, M., Mady, N.I. and Halim, M.A., *Nigella sativa* L. oil protects against induced hepatotoxicity and improves serum lipid profile in rats. *Arzneimittelforschung*, 50(9):832–6, 2000.

Eldershaw, T.P. et al., Pungent principles of ginger (*Zingiber officinale*) are thermogenic in the perfused rat hindlimb, *Int. J. Obes. Relat. Metab. Disord.*, 16(10):755–63, 1992.

Elgayyar, M. et al., Antimicrobial activity of essential oils from plants against selected pathogenic and saprophytic microorganisms, *J. Food Prot.*, 64(7):1019–24, 2001. (X11456186)

el Tahir, K.E., Ashour, M.M., and al-Harbi, M.M., The cardiovascular actions of the volatile oil of the black seed (*Nigella sativa*) in rats: elucidation of the mechanism of action, *Gen. Pharmacol.*, 24(5):1123–31, 1993.

el Tahir, K.E., Ashour, M.M., and al-Harbi, M.M., The respiratory effects of the volatile oil of the black seed (*Nigella sativa*) in guinea-pigs: elucidation of the mechanism(s) of action, *Gen. Pharmacol.*, 24(5):1115–22, 1993b (X8270170)

Erichsen-Brown, C., *Medicinal and Other Uses of North American Plants — A Historical Survey with Special Reference to the Eastern Indian Tribes*, Dover Publications, Inc., New York, NY, 1989.

Facciola, S., *Cornucopia — A Source Book of Edible Plants*, Kampong Publications, Vista, CA, 1998. (FAC)

F.A.S. (Foreign Agriculture Service), USDA, Spice Import Statistics from their web site, 2002.

Fiorelli, G. et al., Estrogen synthesis in human colon cancer epithelial cells, *J. Steroid Biochem. Mol. Biol.*, 71(5–6):223–30, 1999.

Fleischauer, A.T. and Arab, L., Garlic and cancer: a critical review of the epidemiologic literature, *J. Nutr.*, 131(3s):1032S-40S, 2001.

Foster, S. and Duke, J.A., *A Field Guide to Medicinal Plants. Eastern and Central North America* (Peterson Field Guides), illustr. Houghton Miflin, Boston, MA, 1990. (FAD)

Foster, S. and Duke, J.A., *A Field Guide to Medicinal Plants and Herbs of Eastern and Central North America* (Peterson Field Guides), 2nd ed., illustr. Houghton Miflin, Boston, MA, 2000. (FA2)

Fukushima, S. et al., Suppression of chemical carcinogenesis by water-soluble organosulfur compounds, *J. Nutr.*, 131(3s):1049S-53S, 2001.

Fulder, S., *The Ginger Book — The Ultimate Home Remedy*, Avery Publishing Group, Garden City Park, NY, 1996.

Fulder, S., *The Garlic Book. Nature's Powerful Healer*, Avery Publishing Group, Garden City Park, NY, 1997.

Garcia, M.D. et al., Topical antiinflammatory activity of phytosterols isolated from *Eryngium foetidum* on chronic and acute inflammation models, *Phytother. Res.*, 13(1):78–80, 1999.

Germosén-Robineau, Ed., *Farmacopea Vegetal Caribeña*, Enda-caribe, Tramil, Ediciones Emile Desormeaux, Martinique, F.W.I., 1997. (TRA)

Ghazanfar, S.A., *Handbook of Arabian Medicinal Plants*, CRC Press, Boca Raton, FL, 1994. (GHA)

Gilani, A.H. et al., Bronchodilator, spasmolytic and calcium antagonist activities of *Nigella sativa* seeds (Kalonji) — a traditional herbal product with multiple medicinal uses, *J. Pak. Med. Assoc.*, 51(3):115–20, 2001.

Goto, C. et al., Lethal efficacy of extract from *Zingiber officinale* (traditional Chinese medicine) or [6]-shogaol and [6]-gingerol in Anisakis larvae *in vitro*, *Parasitol. Res.*, 76(8):653–6, 1990.

Grieve, M., *A Modern Herbal*, Hafner Press, New York, 1931; Dover Reprint, 2 Vols, first reprint 1971, Dover, New York. (GMH)

Gruenwald, J. et al., *PDR for Herbal Medicine*, 1st ed., Medical Economics Co., Montvale, NJ, 1998. (PHR)

Gruenwald, J. et al., *PDR for Herbal Medicines*, 2nd ed., Medical Economics Company, Inc., Montvale, NJ, 2000. (PH2)

Guenther, E., *The Essential Oils*, 6 volumes, D. van Nostrand, New York, 1948–1952. (GEO)

Habsah, M. et al., Screening of Zingiberaceae extracts for antimicrobial and antioxidant activities, *J. Ethnopharmacol.*, 72(3):403–10, 2000.

Haji Faraji, M. and Haji Tarkhani, A., The effect of sour tea (*Hibiscus sabdariffa*) on essential hypertension, *J. Ethnopharmacol.*, 65(3):231–6, 1999. (X10404421)

Haila, K.M., Lievonen, S.M., and Hienonen, M.I., Effects of lutein, lycopene, annatto, and gamma-tocopherol on autoxidation of triglycerides, *J. Agric. Food Chem.*, 44(8):2096–2100, 1996.

Hammer, K.A., Carson, C.F., and Riley, T.V., Antimicrobial activity of essential oils and other plant extracts, *J. Appl. Microbiol.*, 86(6):985–90, 1999. (X10438227)

Hansawasdi, C., Kawabata, J., and Kasai, T., Alpha-amylase inhibitors from roselle (*Hibiscus sabdariffa* Linn.) tea, *Biosci. Biotechnol. Biochem.*, 64(5):1041–3, 2000. (X10879476)

Hansawasdi, C., Kawabata, J., and Kasai, T., Hibiscus acid as an inhibitor of starch digestion in the Caco-2 cell model system, *Biosci. Biotechnol. Biochem.*, 65(9):2087–9, 2001. (X11676026)

Hansel, R. et al., Eds., *Hager's Handbuch der Pharmazeutischen Praxis*, Drogen (A–D0, Springer-Verlag, Berlin, 1992. (HH2).

Hansel, R. et al., Eds., *Hager's Handbuch der Pharmazeutischen Praxis*, Drogen (E–O), Springer-Verlag, Berlin, 1992.

Harrigan, G.G. et al., Isolation of bioactive and other oxoaporphine alkaloids from two annonaceous plants, *Xylopia aethiopica* and *Miliusa* cf. *Banacea*, *J. Nat. Prod.*, 57(1):68–73, 1994. (X8158166)

Hartwell, J.L., *Plants Used Against Cancer — A Survey*, Duke, J.A., Ed., Quarterman Publications, Inc., Lawrence, MA, 1982, reprinted from eleven different issues of *Lloydia* 1967–1971. (JLH)

Heber, D., *What Color is Your Diet?*, Regan Books—Harper Collins, New York, 2001.

Heisey, R.M. and Gorham, B.K., Antimicrobial effects of plant extracts on *Streptococcus mutans, Candida albicans, Trichophytum rubrum* and other micro-organisms, *Letts. Appl. Microbiol.*, 14:136–9, 1992.

Hikino, H. et al., Antihepatotoxic actions of gingerols and diarylheptanoids, *J. Ethnopharmacol.*, 14(1):31–9, 1985.

Horie, T. et al., Alleviation by garlic of antitumor drug-induced damage to the intestine, *J. Nutr.,* 131(3s):1071S, 2001.

Hostettmann, K. et al., Eds., *Phytochemistry of Plants Used in Traditional Medicine (Proc. Phytochem. Soc. Eur. 37),* Oxford Science Publications, New York, 1995.

Houghton, P.J. et al., Fixed oil of *Nigella sativa* and derived thymoquinone inhibit eicosanoid generation in leukocytes and membrane lipid peroxidation, *Planta Med.,* 61(1):33–6, 1995.

Hoult, J.R. and Paya, M., Pharmacological and biochemical actions of simple coumarins: natural products with therapeutic potential, *Gen. Pharmacol.,* 27(4):713–22, 1996.

Huang, M.-T., Ferraro, T., and Ho, C.T., Cancer Chemoprevention by Phytochemicals in Fruits and Vegetables — an overview in: *Food Phytochemicals for Cancer Prevention I, Fruits and Vegetables,* ACS Symposium Series 546, Huang, M.-T. et al., Eds., The American Chemical Society, Washington, D.C., 1994.

Hussain, R.A. et al., Sweetening agents of plant origin — phenylpropanoid constituents of seven sweet-tasting agents, *Econ. Bot.,* 44(2):174–182, 1990.

Hwang, J.K., Shim, J.S., and Pyun, Y.R., Antibacterial activity of xanthorrhizol from *Curcuma xanthorrhiza* against oral pathogens, *Fitoterapia,* 71(3):321–3, 2000.

Hwang, J.K. et al., Xanthorrhizol — a potential antibacterial agent from *Curcuma xanthorrhiza* against *Streptococcus mutans, Planta Med.,* 66(2):196–7, 2000.

ICMR (Indian Council of Medical Research), *Medicinal Plants of India,* Vol.1, Indian Council of Medical Research, Cambridge Printing Works, New Delhi, 1976; Vol.2, Indian Council of Medical Research, Cambridge Printing Works, New Delhi, 1987. (MPI)

Inouye, S., Uchida, K., and Yamaguchi, H., *In-vitro* and *in-vivo* anti-Trichophyton activity of essential oils by vapour contact, *Mycoses,* 44(3–4):99–107, 2001. (X11413931)

Jafri, M.A. et al., Evaluation of the gastric antiulcerogenic effect of large cardamom (fruits of *Amomum subulatum* Roxb), *J. Ethnopharmacol.,* 75(2–3):89–94, 2001. (X11297839)

Jain, M.K. and Apitz-Castro, R., Garlic — a matter for heart, in: *Spices, Herbs and Edible Fungi,* Charalambous, G., Ed., Elsevier Science B.V., Amsterdam, 1994.

Jain, S.K. and deFillips, R., *Medicinal Plants of India,* 2 Vols., Reference Publications, Algonac, MI, 1991. (SKJ)

Jang, M.K., Sohn, D.H., and Ryu, J.H., A curcuminoid and sesquiterpenes as inhibitors of macrophage TNF-alpha release from *Curcuma zedoaria, Planta Med.,* 67(6):550–2, 2001.

Jang, Y.Y. et al., Protective effect of boldine on oxidative mitochondrial damage in streptozotocin-induced diabetic rats, *Pharmacol. Res.,* 42(4):361–71, 2000.

Jeong, H.J. et al., Inhibition of aromatase activity by flavonoids, *Arch. Pharm. Res.,* 22(3):309–12, 1999. (X10403137)

Jimenez, I. et al., Protective effects of boldine against free radical-induced erythrocyte lysis, *Phytother. Res.,* 14(5):339–43, 2000.

Jin, Y.H. et al., Aloesin and arbutin inhibit tyrosinase activity in a synergistic manner via a different action mechanism, *Arch. Pharm. Res.,* 22(3):232–6, 1999.

Joseph, J., Nadeau, D., and Underwood, A., *The Color Code,* Hyperion, New York, 2002. (JNU)

Kajiya, K. et al., Role of lipophilicity and hydrogen peroxide formation in the cytotoxicity of flavonols, *Biosci. Biotechnol. Biochem.,* 65(5):1227–9, 2001.

Kasture, V.S., Chopde, C.T., and Deshmukh, V.K., Anticonvulsive activity of *Albizzia lebbeck, Hibiscus rosa sinesis* and *Butea monosperma* in experimental animals, *J. Ethnopharmacol.,* 71(1–2):65–75, 2000. (X10904147)

Kasuga, S. et al., Pharmacologic activities of aged garlic extract in comparison with other garlic preparations, *J. Nutr.,* 131(3s):1080S-4S, 2001.

Kato, J. et al., Inhibition of restriction endonucleases by hot water extract of spices, *Bull. Coll. Agr. & Vet. Med. Nihon. U.,* 47:84, 1990.

Kaufman, P.B. et al., *Natural Products from Plants,* CRC Press, Boca Raton, FL, 1998.

Kaur, G. and Kulkarni, S.K., Investigations on possible serotonergic involvement in effects of OB-200G (polyherbal preparation) on food intake in female mice, *Eur. J. Nutr.,* 40(3):127–33, 2001.

Kawai, T. et al., Anti-emetic principles of *Magnolia obovata* bark and *Zingiber officinale* rhizome, *Planta Med.,* 60(1):17–20, 1994.

Kazuho, A. and Hiroshi, S., Effects of saffron extract and its constituent crocin on learning behaviour and long-term potentiation, *Phytother. Res.,* 14(3):149–152, 2000.

Keville, K., The herb report — black cumin (*Nigella sativa*), The American Herb Association's *Quarterly Newsletter,* 16(2):3, 2000.

Khan, A. et al., Insulin potentiating factor and chromium content of selected foods and spices, *Biol. Trace Elem. Res.,* 24(3):183–8, 1990.

Khan, S. and Balick, M.J., Therapeutic plants of ayurveda — a review of selected clinical and other studies for 166 species, *J. Alt. Comp. Med.,* with 27 illustrations by Francesa Anderson, 7(5):405–515, 2001. (JAC)

Kim, K.I. et al., Antitumor, genotoxicity and anticlastogenic activities of polysaccharide from *Curcuma zedoaria*, *Mol. Cells,* 10(4):392–8, 2000.

Kim, O.K. et al., An avocado constituent, persenone A, suppresses expression of inducible forms of nitric oxide synthase and cyclooxygenase in macrophages, and hydrogen peroxide generation in mouse skin, *Biosci. Biotechnol. Biochem.,* 64(11):2504–7, 2000a..

Kim, O.K. et al., Inhibition by (-)-persenone A-related compounds of nitric oxide and superoxide generation from inflammatory leukocytes, *Biosci. Biotechnol. Biochem.,* 64(11):2500–3, 2000b. (X11193427)

Kloss, J., *Back to Eden*, 5th ed., Woodbridge Press Publishing Co., Santa Barbara, CA, 1975.

Koch, H.P. and Lawson, L.D., Eds., (org. copyright 1939.) *Garlic — The Science and Therapeutic Application of Allium sativum L. and Related Species*, Williams & Wilkins, Baltimore, MD, 1996. (LAW)

Kubo, I. and Kinst-Hori, I., Flavonols from saffron flower — tyrosinase inhibitory activity and inhibition mechanism, *J. Agric. Food Chem.,* 47(10):4121–4125, 1999.

Kubo, I. et al., Flavonols from *Heterotheca inuloides*: tyrosinase inhibitory activity and structural criteria, *Bioorg. Med. Chem.,* 8(7):1749–55, 2001.

Kulkarni, R.R. et al., Treatment of osteoarthritis with a herbomineral formula: a double-blind, placebo-controlled, cross-over study, *J. Ethnopharmacol.,* 33:91–95, 1991.

Lamm, D.L. and Riggs, D.R., Enhanced immunocompetence by garlic: role in bladder cancer and other malignancies, *J. Nutr.,* 131(3s):1067S-70S, 2001.

Lawson, L.D., Garlic — a review of its medicinal effects and indicated active compounds, in: *Phytomedicines of Europe—Chemicals and Biological Activity,* Lawson, L. and Bauer, R., Eds., ACS Symposium Series #691, American Chemical Society, Washington, D.C., 1998. (LAB)

Lee, E. and Surh, Y.J., Induction of apoptosis in HL-60 cells by pungent vanilloids, [6]-gingerol and [6]-paradol, *Cancer Lett.,* 134(2):163–8, 1998.

Leung, A.Y. and Foster, S., *Encyclopedia of Common Natural Ingredients*, 2nd ed., John Wiley & Sons, New York, 1995. (LAF)

Lewis, W. and Elvin-Lewis, M., *Medical Botany*, John Wiley & Sons, New York, 1977. (LEL)

Leung, A.Y., *Encyclopedia of Common Natural Ingredients used in Food, Drugs, and Cosmetics*, John Wiley & Sons, New York, 1980. (AYL)

Libster, M., *Delmar's Integrative Herb Guide for Nurses*, Delmar–Thomson Learnings, Albany, NY, 2002.

Lim, G.P. et al., The curry spice curcumin reduces oxidative damage and amyloid pathology in an Alzheimer transgenic mouse, *J. Neurosci.,* 21(21):8370–7, 2001.

Lin, R.I., Pharmacological properties and medicinal use of pepper (*Piper nigrum* L.), in: *Spices, Herbs and Edible Fungi*, Charalambous, G., Ed., Elsevier Science B.V. Amsterdam, 1994.

Lin, S.C. et al., Protective and therapeutic effects of *Curcuma xanthorrhiza* on hepatotoxin-induced liver damage, *Am. J. Chin. Med.,* 23(3–4):243–54, 1995.

List, P.H. and Hohammer, L., *Hager's Handbuch der Pharmazeutischen Praxis*, Vols. 2–6, Springer-Verlag, Berlin, 1969–1979. (HHB)

Little, C.V. and Parsons, T., Herbal therapy for treating osteoarthritis (Cochrane Review), *Cochrane Database Syst. Rev.,* 1:CD002947, 2001. (X11279783)

Liu, L.X., Durham, D.G., and Richards, R.M., Vancomycin resistance reversal in enterococci by flavonoids, *J. Pharm. Pharmacol.,* 53(1):129–32, 2001. (X11206187)

Ma, J. et al., TI Apoptosis induced by isoliquiritigenin in human gastric cancer MGC-803 cells, *Planta Med.,* 67(8):754–757, 2001.

Majeed, M. et al., *Curcuminoids — Antioxidant Phytonutrients*, NutriScience Publishing, Piscataway, NJ, 1995.

Mansour, M.A., Protective effects of thymoquinone and desferrioxamine against hepatotoxicity of carbon tetrachloride in mice, *Life Sci.,* 66(26):2583–91, 2000.

Marles, R.J., Compadre, C.M. and Farnsworth, N.R., Coumarin in vanilla extracts — its detection and significance, *Econ. Bot.,* 41(1):41–47, 1987.

Martinez-Tome, M. et al., Antioxidant properties of Mediterranean spices compared with common food additives, *J. Food Prot.,* 64(9):1412–9, 2001. (X11563520)

Mason, R., Questioning conventional oncology — an interview with cancer activist Ralph W. Moss, Ph.D., *Alt. & Com. Therapie,* 7(1):21–26, 2001.

Mata, R. et al., Biological and mechanistic activities of xanthorrizol and 4-(1′,5′-dimethylhex-4′-enyl)-2-methylphenol isolated from *Iostephane heterophylla, J. Nat. Prod.,* 64(7):911–4, 2001.

McCaleb, R., JAMA bashes herbs for surgical patients, *Herb Research News* 5(2):2, 2001.

McCormick & Co., Inc., *Spices of the World Cookbook,* Penguin Books, 1981.

Mercadante, A.Z., Steck, A., and Pfander, H., Isolation and identification of new apocarotenoids from annatto (*Bixa orellana*) seeds, *J. Agric. Food Chem.,* 45(4):1050–1054, 1997.

Micklefield, G.H., Greving, I., and May, B., Effects of peppermint oil and caraway oil on gastroduodenal motility, *Phytother. Res.,* 14(1):20–23, 2000.

Mills, S. and Bone, K., *Principles and Practice of Phytotherapy,* Churchill Livingstone, Edinburgh, 1999. (MAB)

Milner, J.A. and Rivlin, R.S., Eds., Recent advances on the nutritional effects associated with the use of garlic as a supplement, *J. Nutr.,* 131(3s):951s-1123s, 2001.

Miraldi, E. et al., *Peumus boldus* essential oil — new constituents and comparison of oils from leaves of different origin, *Fitoterapia,* 67(3):227–230, 1996.

Miyamura, M. et al., Seven aromatic compounds from bark of *Cinnamomum cassia, Phytochemistry,* 22(1):215, 1983.

Moerman, D.E., *Native American Ethnobotany,* Timber Press, Portland, OR, 1998. (DEM)

Moldenke, H.N., The economic plants of the Bible, *Econ. Bot.,* 8:152–163, 1954.

Moldenke, H.N. and Moldenke, A.L., *Plants of the Bible,* Chronica Botanica Co., Waltham, MA, 1952.

Morimitsu, Y. et al., A sulforaphane analogue that potently activates the Nrf2-dependent detoxification pathway, *J. Biol. Chem.,* 277(5):3456–63, 2002.

Morton, J.F., *Major Medicinal Plants,* C.C. Thomas, Publisher, Springfield, IL, 1977. (JFM)

Moss, R.W, as quoted in Mason, R., Questioning conventional oncology — an interview with cancer activist Ralph W. Moss, *Alt. & Com. Therapie,* 7(1):21–26, 2001.

Mowrey, D.B., *Guaranteed Potency Herbs. Next Generation Herbal Medicine,* Cormorant Books, Lehi, UT, 1988.

Murakami, N., Stomachic principles in ginger. III. An anti-ulcer principle, 6-gingesulfonic acid, and three monoacyldigalactosylglycerols, gingerglycolipids A, B, and C, from Zingiberis Rhizoma originating in Taiwan, *Chem. Pharm. Bull. (Tokyo),* 42(6):1226–30, 1994. (X8069973)

Murray, M.T and Pizzorno, J.E., *Encyclopedia of Natural Medicine,* Prima Publishing, Rocklin, CA, 1991. (MAP)

Mustalish, R.W. and Baxter, R.L., Mina Jao: A village green pharmacy in Amazonia, *HerbalGram,* 51:56–64, 2001.

Nakatani, N., Antioxidative and Antimicrobial Constituents of Herbs and Spices in: *Spices, Herbs and Edible Fungi,* Charalambous, G., Ed., Elsevier Science B.V., Amsterdam, 1994.

Negi, P.S. et al., Antibacterial activity of turmeric oil: a byproduct from curcumin manufacture, *J. Agric. Food Chem.,* 47(10):297–300, 1999.

Newmark, T.M. and Schulick, P., *Beyond Aspirin — Nature's Answer to Arthritis, Cancer, and Alzheimer's Disease,* Hohm Press, Prescott, AZ, 2000. (COX)

Nielsen, P.V. and Rios, R., Inhibition of fungal growth on bread by volatile components from spices and herbs, and the possible application in active packaging, with special emphasis on mustard essential oil, *Int. J. Food Microbiol.,* 60(2–3):219–29, 2000. (X11016611)

Nishiyama, N. et al., Ameliorative effect of S-allylcysteine, a major thioallyl constituent in aged garlic extract, on learning deficits in senescence-accelerated mice, *J. Nutr.,* 131(3s):1093S-5S, 2001. (X11238823)

Nohara, T. et al., Two novel diterpenes from bark of *Cinnamomum cassia, Phytochemistry,* 21(8):2130, 1982.

Ochse, J.J., *Vegetables of the Dutch East Indies.,* A. Asher & Co., B.V., Amsterdam, 1931 (reprint 1980).

Ohnishi, S.T. and Ohnishi, T., *In vitro* effects of aged garlic extract and other nutritional supplements on sickle erythrocytes, *J. Nutr.,* 131(3s):1085S-92S, 2001.

Onogi, T. et al., Capsaicin-like effect of (6)-shogaol on substance P-containing primary afferents of rats: a possible mechanism of its analgesic action, *Neuropharmacology*, 31(11):1165–9, 1992.

Parry, J.W., *Spices Volume 1*, Chemical Publishing Co., Inc., New York, 1969.

Patel, K. and Srinivasan, K., Influence of dietary spices or their active principles on digestive enzymes of small intestinal mucosa in rats, *Int. J. Food Sci. Nutrition*, 47(1):55–9, 1985.

Pattnaik, S., Subramanyam, V.R., and Kole, C., Antibacterial and antifungal activity of ten essential oils *in vitro*, *Microbios*, 86(349):237–46, 1996.

Peirce, A. *The APhA Practical Guide to Natural Medicines*, Stonesong Press Book, Wm. Morrow & Co., Inc., New York, 1999. (APA)

Pitasawat, B. et al., Screening for larvicidal activity of ten carminative plants, *Southeast Asian J. Trop. Med. Public Health*, 29(3):660–2, 1998.

Premkumar, K. et al., Inhibition of genotoxicity by saffron (*Crocus sativus* L.) in mice, *Drug Chem. Toxicol.*, 24(4):421–8, 2001.

Purseglove, J.W. et al., *Spices*, 2 Vols., Longman, London, 1981. (PEA)

Qureshi, S. et al., Studies on herbal aphrodisiacs used in Arab system of medicine, *Am. J. Chin. Med.*, 17(1&2):57–63, 1989.

Qureshi, S. et al., Toxicity studies on *Alpinia galanga* and *Curcuma longa*, *Planta Med.*, 58(2):124–7, 1992. (PM)

Rahman, K., Historical perspective on garlic and cardiovascular disease, *J. Nutr.*, 131(3s):977S, 2001.

Ram, A. et al., Hypolipidaemic effect of *Myristica fragrans* fruit extract in rabbits, *J. Ethnopharmacol.*, 55(1):49–53, 1996. (JE55:49)

Ramsewak, R.S. et al., Biologically active carbazole alkaloids from *Murraya koenigii*, *J. Agric. Food Chem.*, 47(2):444–7, 1999.

Reed, C.F., with Duke, J.A., Typescript reports on 1000 economic plants submitted to the Economic Botany Laboratory USDA, 1976. (CFR)

Rinzler, C.A., *The Complete Book of Herbs, Spices and Condiments*, Facts on File, New York, 1990. (RIN)

Rivlin, R.S., Historical perspective on the use of garlic, *J. Nutr.*, 131(3s):951S-4S, 2001.

Rosen, R.T. et al., Determination of allicin, S-allylcysteine and volatile metabolites of garlic in breath, plasma or simulated gastric fluids, *J. Nutr.*, 131(3s):968S-71S, 2001.

Rosengarten, F. Jr., *The Book of Spices*, Livingston Publishing Co., Pyramid Communications, Inc., New York, 1969, republished in paperback, 1973.

Rutter, R.A., *Catalogo de Plantas Utiles de la Amazonia Peruana*, Instituto Linguistico de Verano. Yarinacocha, Peru, 1990.

Saarinen, N. et al., No evidence for the *in vivo* activity of aromatase-inhibiting flavonoids, *J. Steroid Biochem. Mol. Biol.*, 78(3):231–9, 2001.

Sachdewa, A., Nigam, R., and Khemani, L.D., Hypoglycemic effect of *Hibiscus rosa sinensis* L. leaf extract in glucose and streptozotocin induced hyperglycemic rats, *Indian J. Exp. Biol.*, 39(3):284–6, 2001.

Saleem, M., Alam, A., and Sultana, S., Asafoetida inhibits early events of carcinogenesis: a chemopreventive study, *Life Sci.*, 68(16):1913–21, 2001. (X11292069)

Sambaiah, K. and Srinivasan, K., Influence of spices and spice principles on hepatic mixed function oxygenase system in rats, *Ind. J. Biochem. Biophys.*, 26(4):254–8, 1989.

Schulick, P., *Ginger — Common Spice & Wonder Drug*, 2nd ed., Herbal Free Press, Ltd., Brattleboro, VT, 1994.

Schultes, R.E. and Raffauf, R.F., *The Healing Forest*, Dioscorides Press, Portland, OR, 1990.

Sharaf, M. et al., Quercetin triglycoside from *Capparis spinosa*, *Fitoterapia*, 71:46–49, 2000.

Sharma, A., Gautam, S., and Jadhav, S.S., Spice extracts as dose-modifying factors in radiation inactivation of bacteria, *J. Agric. Food Chem.*, 48(4):1340–4, 2000. (X10775394)

Sharma, N. et al., Inhibition of benzo[a]pyrene- and cyclophoshamide-induced mutagenicity by *Cinnamomum cassia*, *Mutat. Res.*, 480–481:179–88, 2001.

Shaughnessy, D.T., Setzer, R.W., and DeMarini, D.M., The antimutagenic effect of vanillin and cinnamaldehyde on spontaneous mutation in Salmonella TA104 is due to a reduction in mutations at GC but not AT sites, *Mutat. Res.*, 1; 480–481:55–69, 2001.

Sherman, P.W. and Billing, J., Darwinian gastronomy — Why we use spices. Spices taste good because they are good for us, *BioScience*, 49(6):453–463, 1999.

Sherman, P.W. and Flaxman, S.M., Protecting ourselves from food, *American Scientist*, 89:142–151, 2001.

Sherman, P.W. and Hash, G.A., Why vegetable recipes are not very spicy, *Evolution and Human Behavior,* 22(3):147–163, 2001.

Shirwaikar, A. et al., Chemical investigation and antihepatotoxic activity of the root bark of *Capparis spinosa, Fitoterapia,* 67(3):200–4, 1996.

Shwaireb, M.H. et al., Inhibition of mammary gland tumorigenesis in the rat by caraway seeds and dried leaves of watercress, *Oncology Reports,* 2(4):689–92, 1995.

Simonds, N., *A Spoonful of Ginger — Irresistible, Health-Giving Recipes from Asian Kitchens,* Alfred A. Knopf, New York, 1999.

Sivam, G.P., Protection against *Helicobacter pylori* and other bacterial infections by garlic, *J. Nutr.,* 131(3s):1106S-8S, 2001. (X11238826)

Smith G.C. et al., Mineral values of selected plant foods common to southern Burkina Faso and to Niamey, Niger, west Africa, *Int. J. Food Sci. Nutr.,* 47(1):41–53, 1996. (X8616672)

Somova, L.I. et al., Cardiovascular and diuretic activity of kaurene derivatives of *Xylopia aethiopica* and *Alepidea amatymbica, J. Ethnopharmacol.,* 77(2–3):165–74, 2001. (X11535360)

Srivastava, K.C. and Malhotra, N., Characterization and effects of a component isolated from a common spice clove (*Caryophylli flos*) on platelet aggregation and eicosanoid production, *Thromb. Haemorrh Disorders,* 1/2:59, 1990.

Standish, L.J., Calabrese, C., and Galantino, M.L., Eds., *AIDS and Complementary & Alternative Medicine — Current Science and Practice,* Churchill Livingstone, Philadelphia, 2002.

Strandberg, T.E. et al., Birth outcome in relation to licorice consumption during pregnancy, *Am. J. Epidemiol.,* 153(11):1085–8, 2001. (X11390327)

Stucker, M. et al., Vitamin B(12) cream containing avocado oil in the therapy of plaque psoriasis, *Dermatology.,* 203(2):141–7, 2001. (X11586013)

Suekawa, M. et al., Pharmacological studies on ginger. IV. Effect of (6)-shogaol on the arachidonic cascade, *Nippon Yakurigaku Zasshi,* 88(4):263–9, 1986. (X1330955)

Sun, R.C. et al., Bioactive aromatic compounds from leaves and stems of *Vanilla fragrans, J. Agric. Food Chem.,* 49(11):5161–5164, 2001. (JAF)

Surh, Y., Molecular mechanisms of chemopreventive effects of selected dietary and medicinal phenolic substances, *Mutat. Res.,* 428(1–2):305–27, 1999.

Sutton, R.H., Cocoa poisoning in a dog, *Vet. Rec.,* 109(25–26):563–564, 1981.

Tachibana, Y. et al., Antioxidative activity of carbazoles from *Murraya koenigii* leaves, *J. Agric. Food Chem.,* 49(11):5589–5594, 2001.

Tainter, D.R. and Grenis, A.T., *Spices and Seasonings — a Food Technology Handbook,* VCH Publishers, Inc., NY, 1993. (SAS)

Tairu, A.O., Hofmann, T., and Schieberle, P., Characterization of the key aroma compounds in dried fruits of the West African peppertree *Xylopia aethiopica* (Dunal) A. Rich (Annonaceae) using aroma extract dilution analysis, *J. Agric. Food Chem.,* 47(8):3285–7, 1999.

Takechi, M. and Tanaka, Y., Purification and characterization of antiviral substance eugeniin from the bud of *Syzygium aromatica aromaticum, Planta Med.,* 42(1):69, 1981.

Tanaka, T., *Tanaka's Cyclopedia of Edible Plants of the World,* Keigaku Publishing Co., Tokyo, 1976. (TAN)

Tawatsin, A. et al., Repellency of volatile oils from plants against three mosquito vectors, *J. Vector Ecol.,* 26(1):76–82, 2001. (X11469188)

Teissedre, P.L. and Waterhouse, A.L., Inhibition of oxidation of human low-density lipoproteins by phenolic substances in different essential oils varieties, *J. Agric. Food Chem.,* 48(9):3801–5, 2000.

Thieret, J.W., Frankincense and Myrrh, *Lloydia,* 1(4):6–9, 1996.

Tisserand, R., New perspectives on essential oil safety in *Aroma'95 — One body — one mind,* Aromatherapy Publications, Susex, England, 1995.

Tjendraputra, E. et al., Effect of ginger constituents and synthetic analogues on cyclooxygenase-2 enzyme in intact cells, *Bioorganic Chemistry,* 29:156–163, 2001.

Travis, J., A spice takes on Alzheimer's disease, *Sci. News,* 160:362, 2001.

Tseng, T.H. et al., Induction of apoptosis by hibiscus protocatechuic acid in human leukemia cells via reduction of retinoblastoma (RB) phosphorylation and Bcl-2 expression, *Biochem. Pharmacol.,* 60(3):307–15, 2000.

Tucker, A.O. and Debaggio, T., *The Big Book of Herbs,* Interweave Press, Inc., Loveland, CO, 2000. (TAD)

Tucker, A.O., Maciarello, M.J., and Broderick, C.E., Filé and the Essential Oils of the Leaves, Twigs, and Commercial Root Teas of *Sassafras albidum* (Nutt). Nees, pp. 595–604 in *Spices, Herbs and Edible Fungi*, Charalambous, G., Ed., Elsevier Science, Amsterdam, 1994. (AOT)

Tyler, V.E., *Herbs of Choice — The Therapeutic Use of Phytomedicinals*, Pharmaceutical Products Press, New York, 1994. (HOC)

Unnikrishnan, M.C. and Kuttan, R., Tumour reducing and anticarcinogenic activity of selected spices, *Cancer Lett.*, 51(1):85–9, 1990. (X2110862)

Vernin, G. and Parkanyi, C., Studies of Plants in the Umbelliferae Family: II GC/MS Analysis of Celery Seed Essential Oils, pp. 329–345 in *Spices, Herbs and Edible Fungi*, Charalambous, G., Ed., Elsevier Science, Amsterdam, 1994.

Vimala, S., Norhanom, A.W., and Yadav, M., Anti-tumour promoter activity in Malaysian ginger rhizobia used in traditional medicine, *Br. J. Cancer*, 80(1–2):110–6, 1999. (X10389986)

Vutyavanich, T., Kraisarin, T., and Ruangsri, R., Ginger for nausea and vomiting in pregnancy — randomized, double-masked, placebo-controlled trial, *Obstet. Gynecol.*, 97(4):577–82, 2001.

Wang, C.J. et al., Protective effect of Hibiscus anthocyanins against tert-butyl hydroperoxide-induced hepatic toxicity in rats, *Food Chem. Toxicol.*, 38(5):411–6, 2000. (X10762726)

Watt, G., *Dictionary of the Economic Products of India*, 6 Vols., Nav Bharat Offset Process, Delhi, India, 1889–1892, reprint 1972. (DEP)

Weed, S.S., *New Menopausal Years — The Wise Woman Way*, Ash Tree Publishing, Woodstock, NY, 2002.

Weidner, M.S. and Sigwart, K., Investigation of the teratogenic potential of a *Zingiber officinale* extract in the rat, *Reprod. Toxicol.*, 15(1):75–80, 2001. (X11137381)

Wetherilt, H. and Pala, M., Herbs and spices indigenous to Turkey, pp. 285–307 in *Spices, Herbs and Edible Fungi*, Charalambous, G., Ed., Elsevier Science, Amsterdam, 1994.

White, L.B. et al., *The Herbal Drugstore*, Rodale Press, 2000.

White, L. and Mavor, S., *Kids, Herbs, Health — A Parent's Guide to Natural Remedies*, Interweave Press, 1998.

Winston, D., Nwoti — Cherokee Medicine and Ethnobotany, *J. Amer. Herbalists Guild*, 2(2): 45–9, 2001. (JAH2)

Wood, J.N., Ed., *Capsaicin in the Study of Pain*, Academic Press, NY, 1993.

Worthen, D.R., Ghosheh, O.A., and Crooks, P.A., The *in vitro* anti-tumor activity of some crude and purified components of blackseed, *Nigella sativa* L., *Anticancer Res.*, 18(3A):1527–32, 1998.

Wrobel, K., Wrobel, K., and Urbina, E.M., Determination of total aluminum, chromium, copper, iron, manganese, and nickel and their fractions leached to the infusions of black tea, green tea, *Hibiscus sabdariffa*, and *Ilex paraguariensis* (maté) by ETA-AAS, *Biol. Trace Elem. Res.*, 78(1–3):271–80, 2000. (X11314985)

Yamahara, J. et al., Stomachic principles in ginger. II. Pungent and anti-ulcer effects of low polar constituents isolated from ginger, the dried rhizoma of Zingiber officinale Roscoe cultivated in Taiwan — the absolute stereostructure of a new diarylheptanoid, *Yakugaku Zasshi*, 112(9):645–55, 1992.

Yarnell, E. and Abascal, K., Immunomodulators and HIV infection — an update, *Alt. & Comp. Therapie*, 321–324, 2000.

Yasni, S. et al., Identification of an active principle in essential oils and hexane-soluble fractions of *Curcuma xanthorrhiza* Roxb. showing triglyceride-lowering action in rats, *Food Chem. Toxicol.*, 32(3):273–8, 1994.

Yokoyama, K.M. et al., *U. S. Import Statistics for Agricultural Commodities 1981–1986*, Transaction Books, New Brunswick NJ, 1988. (Not cited but used in Columbian Exchange lectures.)

Yoshikawa, M. et al., Alcohol absorption inhibitors from bay leaf (*Laurus nobilis*) — structure-requirements of sesquiterpenes for the activity, *Bioorg. Med. Chem.*, 8(8):2071–7, 2000. (X11003152)

Yoshikawa, M. et al., Stomachic principles in ginger. III. An anti-ulcer principle, 6-gingesulfonic acid, and three monoacyldigalactosylglycerols, gingerglycolipids A, B, and C, from Zingiberis Rhizoma originating in Taiwan, *Chem. Pharm. Bull. (Tokyo)*, 42(6):1226–30, 1994. (X8069973)

Yoshioka, T. et al., Antiinflammatory potency of dehydrocurdione, a zedoary-derived sesquiterpene, *Inflamm. Res.*, 47(12):476–81, 1998.

Yoshitani, S.I. et al., Chemoprevention of azoxymethane-induced rat colon carcinogenesis by dietary capsaicin and rotenone, *Int. J. Oncol.*, 19(5):929–39, 2001. (X11604990)

Zhu, B.C. et al., Evaluation of vetiver oil and seven insect-active essential oils against the Formosan subterranean termite, *J. Chem. Ecol.*, 27(8):1617–25, 2001.

Zhu, N. et al., Furanosesquiterpenoids of *Commiphora myrrha*, *J. Nat. Prod.*, 64(11):1460–1462, 2001.

Zohary, M., *Plants of the Bible*, Cambridge University Press, New York, 1982.

Zohary, M., *Flora Palaestina*, Israel Academy of Sciences and Humanities, 8 Vols., 4 of plates to date, 1966 to present.

Scientific Name Index

A

Aframomum melegueta K. Schum., 33. *See also Amomum melegueta* Roscoe

Aframomum melegueta, 36, 322

Aframomum spectrum (Oliv. and D. Hanb.) K. Schum., 36

Alliaria wasabi Prantl, 309

Allium cepa, 23, 37, 54

Allium sativum, 20, 45, 80, 280

Allium, 10

Alpinia galanga (L.) Sw., 55

Alpinia galanga, 61, 154, 195, 197

Alpinia officinarum Hance, 58

Alpinia officinarum, 44, 88, 133, 197, 270

Amomum aromaticum Roxb., 62

Amomum cardamomum L., 159

Amomum compactum Soland. ex Maton, 63

Amomum compactum, 195, 276

Amomum kepulaga Sprague & Burk., 63

Amomum melegueta Roscoe, 33. *See also Aframomum melegueta* K. Schum.

Amomum subulatum Roxb., 62

Amomum zedoaria Christm., 148

Amomum zingiber L., 313

Amygdalus communis L., 263

Amygdalus dulcis Mill., 263

Andropogon citratus DC., 150

Anethum graveolens, 1, 19

Apium graveolens L., 66

Armoracia lapathifolia Gilib. ex Usteri, 71

Armoracia rusticana P. Gaertn. et al., 71

Artemisia dracunculus, 26

B

Balsamodendrum myrrha Nees, 123

Banksea speciosa J. König, 126

Benzoin aestivale (L.) Nees, 205

Bergera koenigii L., 213

Bixa orellana L., 77

Boldea fragrans (Ruiz & Pav). Gay, 241

Boswellia carteri Birdw., 80

Boswellia sacra Flueck., 80

Brassica sp., 1, 22

C

Capparis rupestris Sm., 85

Capparis spinosa, 15, 85

Capsicum fastigiatum Bl., 89

Capsicum minimum Blanco, 89

Capsicum, 3, 6, 89

Carum carvi, 16, 63, 95, 225, 235

Carum velenovskyi Rohlena, 95

Caryophyllus aromaticus L., 281

Ceratonia siliqua, 102, 240, 290, 296

Chalcas koenigii (L.) Kurz, 213

Cinchona calisaya, 105

Cinchona officinalis, 105

Cinchona pubescens, 105

Cinchona sp., 105

Cinnamomum aromaticum Nees, 109

Cinnamomum cassia auct., 109

Cinnamomum cassia, 17

Cinnamomum verum J. Presl, 114

Cinnamomum verum, 17, 114, 126

Cinnamomum zeylanicum Blume, 114

Cochlearia armoracia L., 71

Cochlearia wasabi Sieb., 309

Cocos nucifera L., 119

Commiphora molmol (Engl.) Engl., 123

Commiphora myrrha (Nees) Engl., 123

Commiphora myrrha var. *molmol* Engl., 123

Coriandrum sativum, 1, 19

Costus speciosus (J. König) Sm., 126

Costus speciosus, 144, 147, 301

Coumarouna odorata Aubl., 155

Coumarouna punctata Blake, 155

Crocus sativus L., 129

Cuminum cyminum, 1, 19

Cunila mariana L., 134

Cunila origanoides (L.) Britton, 134

Cunila origanoides, 101, 231

Cunila, 6

Curcuma domestica Valeton, 137

Curcuma longa, 27, 137, 147

Curcuma xanthorrhiza Roxb., 145

Curcuma xanthorrhiza, 322

Curcuma zedoaria (Christm.) Roscoe, 148

Curcuma zedoaria, 144, 322

Cymbopogon citratus (DC.) Staph, 150

D

Dipteryx odorata (Aubl.) Willd., 155

Dipteryx odorata, 88, 205, 240

E

Elettaria cardamomum (L.) Maton, 159

Elettaria cardamomum, 16, 58, 63, 65, 189, 205, 225

Eryngium foetidum L., 163

Eugenia aromatica (L.) Baill., 281

Eugenia caryophyllata Thunb., 281

Eugenia caryophyllus (Spreng.) Bullock & Harrison, 281

Common Name Index

A

Achiote. *See* Annatto
African myrrh. *See* Myrrh
Ajo. *See* Garlic
Alligator pepper. *See* Grains of Paradise
Allspice; Clove pepper, Jamaica pepper; Pimienta;
Pimento; *Pimenta dioica* (L.) Merr.
(Myrtaceae) (Synonyms: *Myrtus dioica* L., *M.*
pimenta L., *Pimenta officinalis* Lindl., *P.*
pimenta (L.) H. Karst., *P. vulgaris* Lindl.),
245
Almond; Bitter almond; Sweet almond; *Prunus dulcis*
(Mill.) D. A. Webb (Rosaceae) (Synonyms:
Amygdalus communis L., *A. dulcis* Mill.,
Prunus amygdalus Batsch, *P communis* (L.)
Arcang., *P. dulcis* var. *amara* (DC).
Buchheim.), 263
American dittany. *See* Frost mint
Aniseroot. *See* Sweetroot
Anisillo. *See* Hoja santa
Annatto; Achiote; Annoto; Arnato; Bija: Lipstick pad;
Lipstick tree; *Bixa orellana* L. (Bixaceae),
77
Annoto. *See* Annatto
Arnato. *See* Annatto
Asafetida; *Ferula assa-foetida* L. (Apiaceae), 167
Avocado; *Persea americana* Mill. (Lauraceae), 237

B

Bayleaf laurel. *See* Bayleaf
Bayleaf; Bayleaf laurel; Grecian laurel; Laurel; Sweet bay;
Laurus nobilis L. (Lauraceae), 199
Bayrum tree; West Indian Bay; *Pimenta racemosa* (Mill.)
J. W. Moore (Myrtaceae), 248
Bengal cardamom. *See* Nepalese cardamom
Beni. *See* Sesame
Benjamin bush. *See* Spicebush
Benseed. *See* Sesame
Benzolive Tree. *See* Horseradish tree
Bermuda onion. *See* Onion
Bija. *See* Annatto
Bird Chilli. *See* Hot Pepper
Bird pepper. *See* Hot Pepper
Bitter almond. *See* Almond
Black amomum. *See* Guinea grains
Black caraway. *See* Black cumin
Black cardamom. *See* Nepalese cardamom
Black cumin; Black caraway; Fennel flower; Nutmeg
flower; Roman coriander; *Nigella sativa* L.
(Ranunculaceae), 227
Black pepper; *Piper nigrum* L. (Piperaceae), 253

Boldo; *Peumus boldus* Molina (Monimiaceae) (Synonyms:
Boldea fragrans (Ruiz & Pav). Gay, *Peumus*
fragrans Ruiz & Pav.), 241
Brazilian Pepper Tree: Christmasberry Tree; Florida Holly;
Schinus terebinthifolius Raddi
(Anacardiaceae), 275

C

Cacao: Chocolate; Cocoa; *Theobroma cacao* L.
(Sterculiaceae), 291
Cane reed; Crepe ginger; Wild ginger; *Costus speciosus* (J.
König) Sm. (Costaceae. Also placed in
Zingiberaceae), 126
Caper; Caperbush; *Capparis spinosa* L. (Capparaceae)
(Synonym: *Capparis rupestris* Sm.), 85
Caperbush. *See* Caper
Caraway; *Carum carvi* L. (Apiaceae) (Synonym: *Carum*
velenovskyi Rohlena), 95
Cardamon; Malabar or Mysore Cardamon; *Elettaria*
cardamomum (L.) Maton (Zingiberaceae)
(Synonym: *Amomum cardamomum* L.), 159
Carob; Locust Bean; St John's-Bread; *Ceratonia siliqua* L.
(Fabaceae), 102
Cassia bark. *See* Cassia
Cassia lignea. *See* Cassia
Cassia; Cassia bark; Cassia lignea; China junk cassia;
Chinese cassia; Chinese cinnamon; Saigon
cinnamon; *Cinnamomum aromaticum* Nees
(Lauraceae) (Synonym: *Cinnamomum cassia*
auct.), 109
Celery; *Apium graveolens* L. (Apiaceae), 66
Ceylon cinnamon. *See* Cinnamon
Chili pepper. *See* Hot Pepper
China junk cassia. *See* Cassia
Chinese cassia. *See* Cassia
Chinese cinnamon. *See* Cassia
Chinese ginger. *See* Lesser Galangal
Chios Mastictree. *See* Mastic
Chocolate. *See* Cacao
Christmasberry tree. *See* Brazilian Pepper Tree
Cilantro. *See* Culantro
Cinchona; Redbark quinine; Red cinchona; Yellowbark
quinine; Yellow cinchona; *Cinchona* sp.
(Rubiaceae); *C. pubescens* (Vahl); *C. calisaya*
(Wedd.); formerly *C. officinalis* (Auct.), 105
Cinnamon: Ceylon cinnamon; *Cinnamomum verum* J. Presl
(Lauraceae) (Synonyms: *Cinnamomum*
zeylanicum Blume, *Laurus cinnamomum* L.),
114
Clavos. *See* Clove
Clayton's Sweetroot. *See* Sweetroot

Clove pepper. *See* Allspice

Clove; Clavos; Clovetree; *Syzygium aromaticum* (L.) Merr. and L. M. Perry (Myrtaceae) (Synonyms: *Caryophyllus aromaticus* L., *Eugenia aromatica* (L.) Baill., *E. caryophyllata* Thunb., *E. caryophyllus* (Spreng.) Bullock & Harrison.), 281

Clovetree. *See* Clove

Cluster cardamom. *See* Round cardamom

Cocoa. *See* Cacao

Coconut palm. *See* Coconut

Coconut; Coconut palm; Copra; Nariyal; *Cocos nucifera* L. (Arecaceae), 119

Copra. *See* Coconut

Crepe ginger. *See* Cane reed

Culantro; Cilantro; False coriander; Shadow beni; Stinkweed; *Eryngium foetidum* L. (Apiaceae), 163

Cumaru. *See* Tonka bean

Curry leaf; *Murraya koenigii* (L.) Spreng. (Rutaceae) (Synonyms: *Bergera koenigii* L., *Chalcas koenigii* (L.) Kurz., 213

D

Dittany. *See* Frost mint

Drumstick tree. *See* Horseradish tree

Dutch tonka bean. *See* Tonka bean

E

Ethiopian Pepper; Guinea Pepper; Negro Pepper; Spice Tree; *Xylopia aethiopica* (Dunal) A. Rich (Annonaceae), 311

F

False coriander. *See* Culantro

Fennel flower. *See* Black cumin

Fenugreek; Greek clover; Greek Hay; *Trigonella foenum-graecum* L. (Fabaceae), 296

Florida Holly. *See* Brazilian Pepper Tree

Frankincense; Olibanum tree; *Boswellia sacra* Flueck. (Burseraceae) (Synonym: *Boswellia carteri* Birdw.), 80

Frost flower. *See* Frost mint

Frost mint; American dittany; Dittany; Frost flower; Maryland dittany; Mountain dittany; Stone mint; *Cunila origanoides* (L.) Britton (Lamiaceae) (Synonym: *Cunila mariana* L., *Satureja origanoides* L.), 134

G

Galanga; *Kaempferia galanga* L. (Zingiberaceae), 197

Galangal. *See* Greater Galangal

Garlic; ajo; rocambole; serpent garlic; *Allium sativum* L. (Alliaceae), 45

Ginger; *Zingiber officinale* Roscoe (Zingiberaceae) (Synonym: *Amomum zingiber* L.), 313

Grains of Paradise; alligator pepper; guinea grains; melegueta pepper; *Aframomum melegueta* K. Schum. (Zingiberaceae) (Synonyms: *Amomum melegueta* Roscoe), 33

Greater cardamom. *See* Nepalese cardamom

Greater Galangal; galangal; languas; Siamese ginger; *Alpinia galanga* (L.) Sw. (Zingiberaceae) (Synonyms: *Languas galanga* (L.) Stuntz, *Maranta galanga* L.), 55

Grecian laurel. *See* Bayleaf

Greek Clover. *See* Fenugreek

Greek Hay. *See* Fenugreek

Green onion. *See* Onion

Guinea grains; black amomum; *Aframomum sceptrum* (Oliv. and D. Hanb.) K. Schum. (Zingiberaceae), 36. *See also* Grains of Paradise

Guinea Pepper. *See* Ethiopian Pepper

H

Hairy Sweet Cicely. *See* Sweetroot

Herbol myrrh. *See* Myrrh

Hoja santa; Anisillo; *Piper auritum* Kunth. (Piperaceae), 251

Horseradish tree; Benzolive tree; Drumstick tree; West Indian ben; *Moringa oleifera* Lam. (Moringaceae) (Synonyms: *Guilandina moringa* L., *Moringa moringa* (L.) Small, *M. pterygosperma* Gaertn.), 209

Horseradish; *Armoracia rusticana* P. Gaertn. et al. (Brassicaceae)(synonyms: *A. lapathifolia* Gilib. ex Usteri, *Cochlearia armoracia* L., *Nasturtium armoracia* (l.) Fr., *Radicula armoracia* (L.) B. L. Rob., *Rorippa armoracia* (L.) Hitchc., 71

Hot Pepper; Bird chili; Bird pepper; Chili pepper; Red Chili; Spur pepper; Tabasco pepper; *Capsicum* spp. L. (Solanaceae) (Synonyms: *Capsicum minimum* Blanco, *Capsicum fastigiatum* Bl.), 89

I

Indian cardamom. *See* Nepalese cardamom

Indian saffron. *See* Turmeric

Indian Sorrel. *See* Roselle

Indian Tamarind. *See* Tamarind

J

Jalpaiguri cardamom. *See* Nepalese cardamom

Jamaica pepper. *See* Allspice

Jamaica Sorrel. *See* Roselle

Japanese Horseradish. *See* Wasabi

Java cardamom. *See* Round cardamom

Juniper; *Juniperus communis* L. (Cupressaceae), 191

V

W

Z